HASHING IN COMPUTER SCIENCE

HASHING IN COMPUTER SCIENCE

FIFTY YEARS OF SLICING AND DICING

Alan G. Konheim

WILEY

JOHN WILEY & SONS, INC., PUBLICATION

Published by John Wiley & Sons, Inc., Hoboken, New Jersey
Published simultaneously in Canada

Library of Congress Cataloging-in-Publication Data

Konheim, Alan G., 1934–
Hashing in computer science : fifty years of slicing and dicing / Alan G. Konheim.
p. cm.
ISBN 978-0-470-34473-6
1. Hashing (Computer science) 2. Cryptography. 3. Data encryption (Computer science) 4. Computer security. I. Title.
QA76.9.H36K65 2010
005.8′2–dc22

2009052123

10 9 8 7 6 5 4 3 2 1

To my grandchildren …
Madelyn, David, Joshua, and Karlina

CONTENTS

PREFACE

0.1 WHAT IS HASHING

The *Merriam-Webster Dictionary* (1974) provides several definitions for *hash*[1]:

- *noun:* an American food made by combining and compressing *chopped-up* leftovers—meat, eggs, and potatoes.
- *verb:* to chop into small pieces; make into hash; mince.
- *noun:* colloquial for *hashish* (or *hashish*), the resin collected from the flowers of the cannabis plant. The primary active substance is THC (tetrahydrocannabinol), although several other cannabinoids are known to occur. Hash is usually smoked in pipes, water pipes, joints, and hookahs, sometimes mixed with cannabis flowers or tobacco. It can also be eaten.

A search for *hashing* on *Google* yields 1,720,000 hits; of course, it doesn't come close to the 670,000,000 hits for *sex* or even the 129,000,000 hits for *money*, but then ⋯

Although hashing may be exhilarating and even intoxicating, this book deals with its applications in computer science relating to the processing of information that is both *old* (1953+) and *new* (1990+).

0.2 MY HASHING ROOTS

Nat Rochester was the software project manager for the IBM 701 machine; together with Gene M. Amdahl, Elaine M. McGraw (née Boehme), and Arthur L. Samuel, they considered a key-value-to-address machine for the 701 assembler. Several sources agree that Amdahl invented linear open addressing, a form of hashing (*aka* scatter storage), when confronted with the problem of collisions.

I was a Research Staff member in the Department of Mathematical Sciences at the IBM Thomas J. Watson Research Center in Yorktown Heights (New York) from

[1] Denis Khotimsky, a former student, directed me to www.onin.com/hhh/hhhexpl.htm, which gives an entirely different meaning to hashing, as follows:

> Hashing ⋯ it's a mixture of athleticism and sociability, hedonism and hard work, a refreshing escape from the nine-to-five dweebs you're stuck with five days a week. Hashing is an exhilaratingly fun combination of running, orienteering, and partying, where bands of harriers and harriettes chase hares on eight-to-ten kilometer-long trails through town, country, and desert, all in search of exercise, camaraderie, and good times.

This type of hashing began in Kuala Lumpur, Malaysia, in 1938.

1960 to 1982. Sometime in 1964, I was asked by my manager to meet with Wes Peterson (1924–2009) to learn about his work on hashing. Wes was a former employee of and now consultant to IBM; in 1964, he was a faculty member at one of the University of Florida campuses. Wes used simulation to evaluate the performance of *linear probing*, an algorithm developed during the design of the IBM 701 machine. As a result of our conversation, I started to work with Benjamin Weiss, then a post-doc at IBM, now a Professor Emeritus at the Hebrew University in Jerusalem, to develop a combinatorial analysis of hashing with linear probing.

In a footnote on page 529 of *The Art of Computer Programming: Volume 3/ Sorting and Searching* (Addison-Wesley Publishing Company, 1973), Professor Donald Knuth wrote that his running-time analysis of Algorithm L "was the first non-trivial algorithm I had ever analyzed satisfactorily. ···" He also noted "that more than ten years would go by before the derivation got into print!" Algorithm L is central to the performance of *linear probing*, which is essentially the hashing algorithm proposed by Gene Amdahl. Ben Weiss and I found the formula $N_j \equiv (j + 1)^{(j-1)}$ (Theorem 13.2 is Knuth's Algorithm L), which counts the number of hash sequences inserting j keys in a hash table of size $j + 1$ leaving the last hash table address unoccupied. First, we computed by hand the values N_j for $j = 2, 3, 4$ until we recognized the form of the algebraic expression $(j + 1)^{(j-1)}$, a technique that I have suggested to students in other contexts.

I contacted Wes again in 1997, where he was and remains today a Professor of Computer Science at the University of Hawaii. I inquired about a sabbatical leave there in 1999. My wife and I visited Wes and his wife Hiromi at their home in December 1997 while celebrating our 40th anniversary. Wes had just been awarded the Japan Prize, and both Petersons were both contemplating sabbatical leaves in Japan starting in January 1999. Timing is everything! I was hired by Wes and Hiromi to watch over their dog Koko; I was certainly the highest paid dog sitter in the Hawaiian Islands. Koko was a constant source of innocent merriment during my 5-month sabbatical; he curiously developed a significant taste for *Hebrew National* salami, and our acquiescing to his new found tastes was criticized by both Wes and Koko's veterinarian. However, before leaving for Japan, Wes warned my wife that *Koko* might die before their return. The miraculous and curative powers of *Hebrew National* salami—We *Answer* To *a Higher Authority*—added 2 years to Koko's life. We saw Wes again when we returned to Hawaii to celebrate our 50th anniversary in July 2007.

0.3 THE ORIGINAL APPLICATION OF HASHING (DATA STORAGE)

When data are stored and manipulated in a computer system, a mechanism must be provided to access the data easily. We view information in storage as composed of *records* of variable sizes located in the system storage media in a variety of regions, especially in dynamic file systems. A unique *key* is associated with and stored in each record. This key could be a name or a suitable number, but in general, any string can serve as a unique identifier. The key is used by the programs processing the data to locate—to `SEARCH` for the address of—the desired record. When a telephone directory is used to determine a telephone number, `SEARCH` is somewhat easier than the general instance of file management because the records composed of the triple `(Name Address TelephoneNumber)` are stored *sorted* on the `Name` *field*.

The concept of hashing was first mentioned in a January 1953 internal IBM document by Hans Peter Luhn and several years later in an article by Arnold I. Dumey; the word *hashing* was first used in an article by Robert Morris in 1968.

There is considerable literature for the *old* hashing starting at least in 1963 and continuing until 2005. We describe various hashing methods in Chapters 7–15.

0.4 NEW APPLICATIONS FOR HASHING

Because technology evolves quickly, certain problems arise in many diverse new disciplines. Chapter 16 describes the work of Karp and Rabin. In a 1987 paper, they observed that hashing could be applied to the search of data for a particular string, for example, the word *bomb* in monitored e-mail messages.

The tracking of music performed on radio and television is part of agreements between the Music Performance Trust Fund and the recording industry. As part of this process, it is necessary to listen to the performance of a song on radio and to identify a song; is the performance by the Glenn Miller band or U2? Techniques for *audio fingerprinting*—the automated recognition of music—appeared in 2002 and will be described in Chapter 17.

The IBM Corporation introduced the Data Encryption Standard (DES) in 1973. It generated considerable criticism but resulted in important changes in the relationship among the government, the business world and the academic community. The Web offered services to about 300,000 hosts (processing systems) in 1990. Today, millions of users are connected by the Internet. The October 2006 Forrester Research report predicted that "Non-travel online retail revenues ··· **e-commerce** ··· on the Web will top the quarter-trillion-dollar mark by 2011." This has produced, a new criminal activity called identity theft. Cryptography offered a possible solution to these problems.

Transactions on the Web require the following:

- *Secrecy* to hide credit card information transmitted from a buyer to a seller
- *User authentication* to verify to each party in a transaction the identity of the other party
- *Message authentication* to detect unauthorized changes in transmitted data

Cryptography has been ported to the world of e-commerce, and hashing is one of the constituents of the secrecy/authentication protocols for e-commerce used to protect and defend Web transactions. We describe the exciting current work on hashing in e-Commerce in Chapter 18.

Not only are the mom-and-pop stores on main street disappearing, but it seems that only the large-box chain stores may survive. The villain or savior is the Web, which provides a convenient and often cost-effective way of purchasing some items and/or services. Among the products easily distributed in this medium are music, video, software, books, and data.

Because information in its native form can easily be copied, distributed, or altered without the owners or the purchasers of the information being aware, their unrestricted distribution poses serious issues. We show how hashing is used in Chapter 19 to inhibit these forms of misuse.

0.5 ABOUT THIS BOOK

This book will describe the basic applications of hashing, analyze hashing protocols, and illustrate the tools needed by a student to understand basic and fundamental problems in computer systems.

Why should students study hashing, and why this book? The purpose of higher education in computer science and mathematics is to prepare students for careers in industry, government, and the university. Computer science—like music and cooking—is not a spectator sport; it requires participation. We all learn by practice, attempting to solve new problems by adapting the solutions of old problems. This book focuses on mathematical methods in information management, but its structural techniques are applicable in many areas of science and technology.

The intended audience includes upper-division students in a computer science/engineering program. The prerequisites include skill in some programming language and elementary courses in discrete mathematics and probability theory. The book is divided into four parts.

Part I contains brief reviews of the relevant mathematical tools

Part II deals with the original hashing for data storage management

Part III examines the appearance of new applications of hashing ideas

Problems and solutions are included as well as an extensive bibliography which concludes the presentation

0.6 ACKNOWLEDGMENTS

I am in debt to many people who have helped and encouraged me in writing this book.

- My friend of 36 years, Dr. Raymond Pickholtz, Professor Emeritus at George Washington University, who visited UCSB several times, read all the chapters, and provided advice.
- His son, Mr Andrew Pickholtz *Esq.*, who read and added material on his dongle patent in Chapter 19.
- Dr. Michele Covell of Google who was kind enough to read and suggest changes to Chapter 17.
- Mr. Vlado Kitanovski of the Faculty of Electrical Engineering and Information Technologies (Ss Cyril and Methodius University—Skopje) who was kind enough to read and suggest changes to the section on watermarking of images in Chapter 19.
- Dr. Feng Hao of Thales e-Security (Cambridge, United Kingdom) and Professor Ross Anderson of the University of Cambridge Computer Laboratory for their comments on iris identification in Chapter 19.
- Professor Haim Wolfson of Tel-Aviv University who clarified my presentation on geometric hashing.
- My son Keith whose graphic skills were indispensable.

- Finally, to my wife Carol of more than 52 years, who continues to amaze me by her wide-ranging talents. I could not have undertaken this book without her encouragement, assistance, and advice.

It's what you learn after you know it all that counts, attributed to Harry S. Truman.

Anyone who stops learning is old, whether at twenty or eighty, Henry Ford.

Teaching is the highest form of understanding, Aristotle.

PART I

MATHEMATICAL PRELIMINARIES

CHAPTER 1

Counting

Walter Crane (1814–1915), *The Song of Sixpence*: "The Queen was in the parlor, eating bread and honey."

This chapter reviews the basic counting techniques that will be used throughout the book. An excellent reference for this material is [Rosen 2003].

Hashing in Computer Science: Fifty Years of Slicing and Dicing, by Alan G. Konheim

1.1 THE SUM AND PRODUCT RULES

SUM RULE

If the first of two tasks can be performed in any of n_1 ways and the second in any of n_2 ways, and if these tasks cannot be performed at the same time, then there are $n_1 + n_2$ ways of performing either task.

Example 1.1. The formula $|N_1 \cup N_2| = n_1 + n_2$ is valid if

a) Sets N_1 and N_2 contain $n_1 = |N_1|$ and $n_2 = |N_2|$ elements, respectively.
b) The sets have no elements in common $N_1 \cap N_2 = \emptyset$.

GENERALIZED SUM RULE

If the ith of m tasks $1 \leq i \leq m$ can be performed in any of n_i ways, and if these cannot be performed at the same time, then there are $n_1 + n_2 + \cdots + n_m$ ways of performing any of the m tasks.

Example 1.2. The formula $\left|\bigcup_{i=1}^{m} N_i\right| = n_1 + n_2 + \cdots + n_m$ is valid if

a) m sets $N_1, N_2, \cdots, N_m$ contain $n_1 = |N_1|, n_2 = |N_2|, \cdots, n_m = |N_m|$ elements, respectively.
b) No pair of (distinct) sets N_i and N_j $(1 \leq i < j \leq m)$ has an element in common (*pairwise disjoint sets*) $N_i \cap N_j = \emptyset$.

If the m sets $N_1, N_2, \cdots, N_m$ are *not* pairwise disjoint, the sum $n_1 + n_2 + \cdots + n_m$ overcounts the size of $\bigcup_{i=1}^{m} N_i$. The *principle of inclusion-exclusion*, to be discussed in §1.8, provides the corrections.

PRODUCT RULE

If a procedure is composed of two tasks, the first can be performed in any of n_1 ways and thereafter, the second task in any of n_2 ways (perhaps depending on the outcome of the first task), then the total procedure can be performed in any of $n_1 \times n_2$ ways.

GENERALIZED PRODUCT RULE

If a procedure is composed of m tasks, the first can be performed in any of n_1 ways and the ith task can be performed in any of n_i ways (perhaps depending on the outcome of the first i outcomes), then the total procedure can be performed in any of $n_1 \times n_2 \times \cdots \times n_m$ ways.

Example 1.3. A sequence of letters that reads the same forward and backward is a palindrome[1]; for example, `ABADABA` is a palindrome and `ABADABADO` is not.

a) The number of 5- or 6-letter palindromes is (by the product rule) 26^3.
b) The number of 5- or 6-letter palindromes that do *not* contain the letter `R` (by the product rule) is (by the product rule) 25^3.
c) The number of 5- or 6-letter palindromes that do contain the letter `R` is $26^3 - 25^3$.
e) The number of 5- or 6-letter palindromes in which *no* letter is repeated (by the product rule) is (by the product rule) $26 \times 25 \times 24$.

Jenny Craig and Weight Watchers should note the following palindrome (with spaces deleted) created by the distinguished topologist Professor Peter Hilton during World War II:

`DOC NOTE, I DISSENT. A FAST NEVER PREVENTS A FASTNESS, I DIET ON COD.`

Example 1.4. A *bit string* of length n is an n-tuple $\underline{x} = (x_0, x_1, \cdots, x_{n-1})$ with $x_i \in \mathcal{Z}_2 = \{0, 1\}$ for $0 \le i < n$.

a) The number of bit strings of length n is (by the product rule) 2^n.
b) The number of bit strings of length $n \ge 4$ that start with `1100` is (by the product rule) 2^n.
c) The number of bit strings of length $n \ge 4$ that begin or end with `1` is (by the product rule) 2^{n-2}.
d) The number of bit strings of length $n \ge 2$ that begin or end with either `0` or `1` is (by the sum *and* product rules) $4 \times 2^{n-2}$.
e) The number of bit strings of length $n \ge 2$ in which the 4th or 8th bits is equal to 1 is (by the sum *and* product rules) $3 \times 2^{n-2}$.

1.2 MATHEMATICAL INDUCTION

In the sections that follow, we define various *couting functions*, to include permutation, combinations, and so forth. Often, we need to answer the question, "In how many ways can ⋯?" where ⋯ describes some property.

[1]From the Greek *palindromos*, meaning "running backward."

For example, various lotteries require the (hopeful) participant to choose n integers from 0 to 9, perhaps subject to some rules. What is the formula for the number of ways this can be done? Although no universal technique is available for solving such problems, mathematical induction may be successful to prove a formula, if the correct answer can be guessed. Several examples in this chapter will illustrate its usefulness.

Guiseppe Peano (1858–1932) studied mathematics at the University of Turin. He made many contributions to mathematics, and his most celebrated is the Peano axioms, which define the natural numbers $\mathcal{Z} = \{0, 1, \cdots\}$, the letter $\mathcal{Z}$ derived from the German *zahlen* for numbers. Why the adjective *natural*? Are there unnatural numbers? Numbers describing quantity arise naturally when we count things. The Babylonians, Egyptians, and Romans advanced the idea of using symbols to represent quantities. The Roman number system used symbols—X for 10, I for 1, and V for 5, and then incorporated the *place value system* in which XIV was the representation of 14.

Peano defined the natural numbers axiomatically; his fifth axiom

PA#5. A **predicate**[2] $\mathcal{P}$ defined on $n \in \mathcal{Z}$ is true for $n \in \mathcal{Z}$ if

— $\mathcal{P}(0)$ is true—the *base case.*
— the implication $\mathcal{P}(n) \to \mathcal{P}(n+1)$ is true for every $n \geq 0$—the *inductive step.*

This is referred to as the principle of mathematical induction.

Example 1.5. Prove

$$\sum_{i=1}^{n} i = \frac{1}{2} n(n+1) \quad 1 \leq n < \infty \tag{1.1}$$

Solution (by mathematical induction). $\mathcal{P}(n)$ is true for $n = 1$ because

$$\sum_{i=1}^{1} i = 1 = \frac{1}{2}(1 \times 2)$$

Assume $\mathcal{P}(n)$ is true; then equation (1.1) for $(n+1)$ gives

$$\sum_{i=1}^{n+1} i = \sum_{i=1}^{n} i + (n+1) = \frac{1}{2} n(n+1) + (n+1)$$
$$= \frac{1}{2}(n+1)(n2)$$

1.3 FACTORIAL

The factorial of a non-negative integer n is

[2] A *predicate* $\mathcal{P}$ on $\mathcal{Z}$ is a function defined on $\mathcal{Z}$ for which the value $\mathcal{P}(N)$ is either true or false.

$$n! = \begin{cases} 0 & \text{if } n = 0 \\ n \times (n-1) \times (n-2) \times \cdots \times 2 \times 1 & \text{if } n > 0 \end{cases} \tag{1.2a}$$

which may be written as

$$n! = n \times (n-1)! \quad n > 0 \quad 0! \equiv 1 \tag{1.2b}$$

The m-factorial of n is the product

$$(n)_m = \begin{cases} 1 & \text{if } m = 0 \\ n \times (n-1) \times (n-2) \times \cdots \times (n-m+1) & \text{if } 1 \leq m \leq n \\ 0 & \text{if } m > n \end{cases} \tag{1.3a}$$

We have the formula

$$(n)_m = \frac{n!}{(n-m)!} \quad 0 \leq m \leq n \tag{1.3b}$$

Table 1.1 gives the values of $n!$ for[3] $n = 1(1)14$.

Stirling's Formula (1730). Since $n!$ increases very rapidly with n, it is necessary to have a simple formula for large n. A derivation of the following asymptotic formula

$$n! \approx \sqrt{2\pi} n^{n+\frac{1}{2}} e^{-n} \quad n \to \infty \tag{1.4a}$$

is given in [Feller 1957, pp. 50–53]. The meaning of the $\approx$ in equation (1.4*a*) is

$$1 = \lim_{n \to \infty} \frac{n!}{\sqrt{2\pi} n^{n+\frac{1}{2}} e^{-n}} \tag{1.4b}$$

The following correction to equation (1.4*b*) is in [Feller 1957, p. 64].

$$n! \approx \sqrt{2\pi} n^{n+\frac{1}{2}} e^{-n+\frac{1}{12n}-\frac{1}{360n^3}} \quad n \to \infty \tag{1.4c}$$

Table 1.2 compares $n!$ and the approximation in equation (1.4*a*) for $n = 1(1)12$.

TABLE 1.1. n! for n = 1(1)14

n	n!	n	n!	n	n!	n	n!	n	n!
0	1	1	1	2	2	3	6	4	24
5	120	6	720	7	5040	8	40320	9	362880
10	3628800	11	39916800	12	479001700	13	6227020800	14	87178291200

[3]The table maker's notation $n = 1(1)12$ indicates the table contains entries for the parameter values $n = 1$ increasing in steps of 1 until $n = 14$.

TABLE 1.2. Comparison of *n*! and Stirling, Approximation [Equation (1.4*a*)] for *n* = 1(1)12

n	n!	Equation (1.4*a*)	% Error
1	1	0.922137	7.786299
2	2	1.919004	4.049782
3	6	5.836210	2.729840
4	24	23.506175	2.057604
5	120	118.019168	1.650693
6	720	710.078185	1.378030
7	5040	4980.395832	1.182622
8	40320	39902.395453	1.035726
9	362880	359536.872842	0.921276
10	3628800	3598695.618741	0.829596
11	39916800	39615625.050577	0.754507
12	479001600	475687486.472776	0.691879

1.4 BINOMIAL COEFFICIENTS

The binomial coefficient sometimes displayed as $\binom{n}{m}$ and sometimes as $C(n, m)$ is defined for $n \geq 0$ by

$$\binom{n}{m} = \begin{cases} \dfrac{n!}{m!(n-m)!} & \text{if } 0 \leq m \leq n \\ 0 & \text{otherwise} \end{cases} \tag{1.5}$$

If $n \geq 0$ and $0 \leq m \leq n$, the negative binomial coefficient is defined by

$$\binom{-n}{m} = \frac{-n(-n-1)(-n-2)\cdots(-n-(m-1))}{m!} = (-1)^m \binom{n+m-1}{m} \tag{1.6}$$

Table 1.3 lists the binomial coefficients $\binom{n}{m}$ for $m = 0(1)n$ and $n = 0(0)10$.

In addition to his many contributions to mathematics, Blaise Pascal (1623–1662) invented the *Pascaline*, which is a digital calculator using 10-toothed gears to speed on arithmetic. Pascal observed an important recurrence connecting the entries in Table 1.3 and providing a natural way to extend it.

Theorem 1.1 (Pascal's triangle). $\binom{n}{m} + \binom{n}{m+1} = \binom{n+1}{m+1}$ for $0 \leq m \leq n$.

Proof. First, write

$$\binom{n}{m} = \frac{n!}{m!(n-m)!} = \frac{m+1}{n+1}\binom{n+1}{m+1}$$

TABLE 1.3. Binomial Coefficients $\binom{n}{m}$ with $m = 0(1)n$ and $n = 0(1)10$

					$m \rightarrow$						
$\downarrow n$	0	1	2	3	4	5	6	7	8	9	10
0	1										
1	1	1									
2	1	2	1								
3	1	3	3	1							
4	1	4	6	4	1						
5	1	5	10	10	5	1					
6	1	6	15	20	15	6	1				
7	1	7	21	35	35	21	7	1			
8	1	8	28	56	70	56	28	8	1		
9	1	9	36	84	126	126	84	36	9	1	
10	1	10	45	120	210	252	210	120	45	10	1

$$\binom{n}{m+1} = \frac{(n+1)!}{m!(n-m+1)!} = \frac{n-m}{n+1}\binom{n+1}{m+1}$$

Next, adding gives

$$\binom{n}{m} + \binom{n}{m+1} = \binom{n+1}{m+1} \quad \blacksquare$$

Pascal's observation led Isaac Newton (1642–1727) to discover the following theorem.

Theorem 1.2 (the binomial theorem). If $0 \le n < \infty$, then

$$(x+y)^n = \sum_{m=0}^{n} x^m y^{n-m} \binom{n}{m} \tag{1.7}$$

Proof. Using Pascal's triangle

$$\begin{aligned}
\sum_{m=0}^{n} x^m y^{n-m} \binom{n}{m} &= \sum_{m=0}^{n} x^m y^{n-m} \left[\binom{n-1}{m} + \binom{n-1}{m-1}\right] \\
&= y \sum_{m=0}^{n-1} x^m y^{n-1-m} \binom{n-1}{m} + x \sum_{m=1}^{n} x^{m-1} y^{n-1-(m-1)} \binom{n-1}{m-1} \\
&= y(x+y)^{n-1} + y(x+y)^{n-1} \\
&= (x+y)^n. \qquad \blacksquare
\end{aligned}$$

The analysis of many hashing protocols will involve expressions involving the binomial coefficients. A few useful identities are provided in the Appendix. A more extensive collection may be found in [Gould 1972]. Incidentally, a mathematician could make a living just proving binomial coefficient identities. A search on Google on July 9, 2007 for binomial coefficients yielded 37,200 hits.

1.5 MULTINOMIAL COEFFICIENTS

We shall show in §1.6 that the binomial coefficient $\binom{n}{m}$ may be interpreted as

- The number of subsets of size m chosen from a universe $\mathcal{U}$ of size n.
- The number of ways of selecting a sample of m elements (objects) from a universe $\mathcal{U}$ of size n.

However, instead of only selecting a sample consisting of one kind from a universe of size n, we can partition the elements of $\mathcal{U}$ into k subsets $M_0, M_1, \cdots, M_{k-1}$ of sizes $m_0, m_1, \cdots, m_{k-1}$, where

$$m_i \geq 0 \quad 0 \leq i < k \quad n = m_0 + m_1 + \cdots + m_{k-1}$$

The *multinomial coefficient* is an extension of the binomial coefficient defined by

$$\binom{m}{m_0 m_1 \cdots m_{k-1}} = \frac{n!}{m_0!m_1!\cdots m_{k-1}!} \tag{1.8}$$

The analog of the binomial theorem is

Theorem 1.3 (the multinomial theorem). If $0 \leq n < \infty$, then

$$(x_0 + x_1 + \cdots + x_{k-1})^n = \sum_{\substack{m_0, m_1, \cdots, m_{k-1} \\ m_i \geq 0 \quad 0 \leq i \leq k \\ n = m_0 + m_1 + \cdots + m_{k-1}}} \prod_{i=0}^{k-1} x_i^{m_i} \binom{m}{m_0 m_1 \cdots m_{k-1}} \tag{1.9}$$

Proof. By induction on k. ■

1.6 PERMUTATIONS

An m-permutation from the universe $\mathcal{U}$ of n elements is an ordered sample $\underline{x} = (x_0, x_1, \cdots, x_{m-1})$ whose elements $\{x_i\}$ are in $\mathcal{U}$.

There are different flavors of permutations, as follows:

Permutations Without Repetition

$x_i = x_j \Leftrightarrow i = j$

Permutations With Unrestricted Repetition

$x_{i_0} = x_{i_1} = \cdots = x_{i_{s-1}}$ for s distinct indices $i_0, i_1, \cdots, i_{s-1}$ with *no* restriction on s or the indices $\{i_j\}$

Permutations With Restricted Repetition

$x_{i_0} = x_{i_1} = \cdots = x_{i_{s-1}}$ for s distinct indices $i_0, i_1, \cdots, i_{s-1}$ with *some* restrictions on s and/or the indices $\{i_j\}$

Theorem 1.4. The number of m-permutations from the universe $\mathcal{U}$ of n elements is n^m.

Proof. The product rule. ■

Example 1.6. The number of n-bit sequences $\underline{x} = (x_0, x_1, \cdots, x_{n-1})$ with $x_i \in \mathcal{Z}_2 = \{0, 1\}$ for $0 \le i < n$ is 2^n by the product rule.

Example 1.7. The American Standard Code for Information Interchange (ASCII) alphabet is a 7-bit code representing

- Uppercase and lowercase alphabetic characters `a b` ⋯ `z A B` ⋯ `Z`;
- Digits `0 1` ⋯ `9`;
- Blank space, punctuation `. , ! ? ; : -`;

Of the $128 = 2^7$ possible 7-bit sequences, only 95 are *printable*; the remaining 33 are nonprintable characters that consist mostly of control characters; for example,

- `BELL`, which rings a bell when the typewriter carriage returns—I hope everyone remembers what a typewriter is?[4]
- `CR` (or linefeed), which shifts the *cursor* to the next line

plus many communication control characters.

Theorem 1.5. The number $N_{n,m}(\neg\mathcal{R})$ of m-permutations without repetition from the universe $\mathcal{U}$ of n elements is $(n)m$.

Note in the special case $m = n$, the number of n-permutations without repetition from the universe $\mathcal{U}$ of n elements is $n!$.

Let $\mathcal{U} = \{0, 1, \cdots, n-1\}$ and suppose x is an n-permutations without repetition of the elements of $\mathcal{U}$. x can be interpreted as a rearrangement of the elements of $\mathcal{U}$, and the 2-rowed notation $\left(x = \begin{pmatrix} 0 & 1 & 2 & \cdots & n-1 \\ x_0 & x_1 & x_2 & \cdots & x_{n-1} \end{pmatrix}\right)$ is often used to emphasize this interpretation of x.

[4]My former employer used to manufacturer and sell typewriters; I believe they were assembled in Lexington, Kentucky. The *Selectric*, which was introduced in 1961, provided a convenient way to enter data into a computer. Various versions of the Selectric followed during the next 30 years. Finally in 1990, IBM formed a wholly owned subsidiary consolidating the company's typewriter, keyboard, intermediate and personal printers, and supplies business in the United States, including manufacturing and development facilities. IBM also reported that it was working to create an alliance under which Clayton & Dubilier, Inc. would become the majority owner of the new subsidiary and that IBM was studying a plan to include the remainder of its worldwide "information products" business in the alliance in the United States, including manufacturing and development facilities. A year later, IBM and Clayton & Dubilier, Inc. created a new information products company called Lexmark International, Incorporated to develop, manufacture, and sell personal printers, typewriters, keyboards, and related supplies worldwide.

When we interpret a permutation x as a rearrangement of the elements of $\mathcal{U}$, it is natural to compose or multiply them as follows:

$$x=(x_0, x_1, \cdots, x_{n-1}); \quad i \xrightarrow{x} x_i$$

$$y=(y_0, y_1, \cdots, y_{n-1}); \quad i \xrightarrow{y} y_i$$

$$x \times y \equiv z=(z_0, z_1, \cdots, z_{n-1}); \quad i \xrightarrow{x \times y} y_{x_i}$$

then the set of *all* $n!$ permutations of the elements of $\mathcal{U}$ forms a group[5] the symmetric group G_n. A transposition x is a permutation of the elements of $\mathcal{U}$, which leaves all of the elements alone other than elements i and j, which it interchanges. For example, if $i = 1, j = 4$, and $n = 8$, then

$$x=\begin{pmatrix} 0 & 1 & 2 & 3 & 4 & 5 & 6 & 7 \\ 0 & 4 & 2 & 3 & 1 & 5 & 6 & 7 \end{pmatrix}$$

is a transposition.

The transpositions form the building blocks of the symmetric group G_n. The following theorem summarizes the basic properties that we will later use.

Theorem 1.6. For the symmetric group G_n.

a) A transposition x is *idempotent*, meaning $x \times x = 1$, where 1 is the identity permutation

$$1=\begin{pmatrix} 0 & 1 & 2 & \cdots & n-1 \\ 0 & 1 & 2 & \cdots & n-1 \end{pmatrix}$$

b) Every permutation can be written, not necessarily in a unique way, as a product of transpositions; that is, the transpositions of $\mathcal{U}$ *generate* the symmetric group G_n.

c) If $x = y_0 \times y_1 \times \cdots \times y_{m-1} = y'_0 \times y'_1 \times \cdots \times y'_{m'-1}$, then m and m' are either even or odd.

The alternating group A_n of $\mathcal{U}$ consists of all permutations whose representation as a product of transpositions involving an *even* number of transpositions.

An m-permutation with repetition from the universe $\mathcal{U}$ of n elements is an ordered sample $\underline{x} = (x_0, x_1, \cdots, x_{m-1})$, whose elements $\{x_i\}$ are in $\mathcal{U}$ with no restriction on the number of times each element of $\mathcal{U}$ appears.

The term *sampling* in statistics refers to a process by which an m-permutation may be constructed. Imagine an urn that contains n balls bearing the numbers 0, 1,

[5]The elements $\{u\}$ of $\mathcal{U}$ form a group if
1. $u_1, u_2 \in \mathcal{U}$ implies $u_1 \times u_2 \in \mathcal{U}$ (closure).
2. $u_1, u_2, u_3 \in \mathcal{U}$ implies $u_1 \times (u_2 \times u_3) = (u_1 \times u_2) \times u_3$ (associativity law).
3. There exists an element $e \in \mathcal{U}$ such that $u \times e = e \times u = u$ for all $u \in \mathcal{U}$ (identity).
4. for every $u \in \mathcal{U}$, there exists an element $u^{-1} \in \mathcal{U}$ such that $u \times u^{-1} = u^{-1} \times u = e$ (inverse).

$\cdots, n-1$. A sampling process describes how the sample is constructed; two variants of sampling are worth noting, as follows:

Sampling without Replacement
After a ball is drawn from the urn (the person who draws the ball is of course blindfolded), its number is recorded and the ball is not returned to the urn.

Sampling with Replacement
After a ball is drawn from the urn (same security provisions as before), its number is recorded and the ball is returned to the urn.

Frequently, the ball manufacturers insist on business, and the urn contains N replicas of each numbered ball.

Example 1.8 (Powerball). His truck broke down the morning he and his wife of 20 years discovered they had won a \$105.8 million Powerball jackpot from June 27th. Powerball is an American lottery operated by the Multi-State Lottery Association (MUSL), a consortium of lottery commissions in 29 states, the District of Columbia, and the U.S. Virgin Islands. Powerball is licensed as the monopoly provider of multistate lotteries in these jurisdictions.

A player picks 5 numbers from 1 to 55 and one number from 1 to 42.

Every Wednesday and Saturday night at 10:59 p.m. Eastern time, the Powerball management draws 5 white balls out of a drum with 55 balls and 1 red ball out of a drum with 42 red balls. Five balls from 53 plus 1 power ball from a separate group of 42 are selected. First prize is won by matching all 6 balls drawn. There are nine prize levels.

This process demonstrates sampling without replacement.

There are many varieties of permutations with specified repetition, for example, specifying the number of repetitions m_i of the universe $\mathcal{U}$ element a_i subject to the obvious conditions

$$m_i \geq 0 \quad m = m_0 + m_1 + \cdots + m_{n-1}$$

Theorem 1.7. The number $N_{n,\underline{m}}(\mathcal{R})$ of m-permutations from $\mathcal{U} = \{0, 1, \cdots, n-1\}$ with the specified repetition pattern $\underline{m} = (m_0, m_1, \cdots, m_{n-1})$ is

$$N_{n,m}(\mathcal{R}) = \binom{m}{m_0 m_1 \cdots m_{n-1}} = \frac{m!}{m_0!m_1!\cdots m_{n-1}!} \tag{1.10}$$

Example 1.9. How many ways are there of permutating the letters of `CALIFORNIA`?

Answer. $\dfrac{12!}{3!2!} = 11!$ ■

Example 1.10. I gave the following problem during the recall election for the Governor of California when I taught discrete mathematics at the University of California at Santa Barbara in the fall of 2003.

Which of the two names GRAYDAVIS or ARNOLDSCHWARZENEGGER has the most permutations?

Answer. There are

$$N_{GD} = \binom{9}{2} = 181\,440$$

permutations of the letters of GRAYDAVIS because only the letter A is repeated and

$$N_{AS} = \binom{20}{2\,3\,2\,3\,2} = 1\,550\,400 \times 13!$$

of the letters of ARNOLDSCHWARZENEGGER because the letters A and G each occur twice and the letters R and E each occur three times. ■

And the winner was Arnold Schwarzenegger. I am reasonably certain this question did not affect the outcome.

Other types of restrictions on repetitions are possible.

Example 1.11. How many permutations are there of MASSACHUSETTS in which no S's are adjacent?

Answer. There are $\binom{9}{1\,2\,1\,1\,1\,1\,2}$ permutations of MAACHUETT. It remains to place the 4 Ss, one in each of the positions between, before, or after the letters in MAACHUETT, for example, as shown by ↑ in

↑ M ↑ A ↑ A ↑ C ↑ H ↑ U ↑ E ↑ T ↑ T ↑

These four positions may be chosen from the 10 ↑s (without repetition) in $\binom{10}{4}$ ways. ■

1.7 COMBINATIONS

An m-combination from the universe $\mathcal{U}$ of n elements is an unordered sample $\{x_0, x_1, \cdots, x_{m-1}\}$ whose elements $\{x_i\}$ are in $\mathcal{U}$.

Note that we have written $\{x_0, x_1, \cdots, x_{r-1}\}$ rather than $(x_0, x_1, \cdots, x_{r-1})$ because an *ordering* is implicit in the latter.

Theorem 1.8. The number of m-combinations from the universe $\mathcal{U}$ of n elements is $\binom{n}{m}$.

Just as in the case of permutations, there are combinations with additional constraints; two are mentioned here.

Example 1.12. How many permutations are there of the 15 letters of POLYUNSATURATED maintaining the relative order of the vowels A, E, I, O, and U?

Solution.

1. The positions in which the vowels OUAUAE appearing in POLYUNSATURATED can be chosen is $\binom{15}{6}$.
2. Having chosen their positions, their orders are determined.
3. There remain $\binom{9}{1\,1\,1\,1\,1\,2\,1\,1} = 181\,400$ permutations of the remaining letters of PLYNSTRTD. ■

A is a multiset of size m of a universe $\mathcal{U}$ with n elements if it contains m_i copies of the element $a_i \in \mathcal{U}$ for $0 \le i < k$. We can write

$$A = \{(a_0)_{\ell_0}, (a_1)_{\ell_1}, \cdots (a_{m-1})_{\ell_{m-1}}\}$$

where the elements $a_0, a_1, \cdots, a_{m-1}$ of $\mathcal{U}$ are distinct and

$$\ell_i \ge 0 \quad 0 \le i < m \quad n = \ell_0 + \ell_1 + \cdots + \ell_{m-1}$$

A multiset can also be interpreted as an ordered permutation with specified repetition, equivalently, an ordered partition[6] of the integer n into m non-negative parts.

To count the number $C_{n,m}$, we consider a set of $n + m - 1$ positions (horizontal lines) corresponding to the n elements of $\mathcal{U}$ together with $m - 1$ *fictitious* elements. We place a divider (a vertical line) in the following locations:

- Immediately to the left of the leftmost position
- Immediately to the right of the leftmost position
- $m - 1$ dividers through some $m - 1$ of the $n + m - 1$ positions

The partition is determined by dividing the set $\{0, 1, \cdots, n - 1\}$ according to the number of positions between dividers (Figure 1.1).

Theorem 1.9. The number $C_{n,m}$ of partitions of the integer n into m non-negative parts is $\binom{n+m-1}{m-1} = \binom{n+m-1}{n}$.

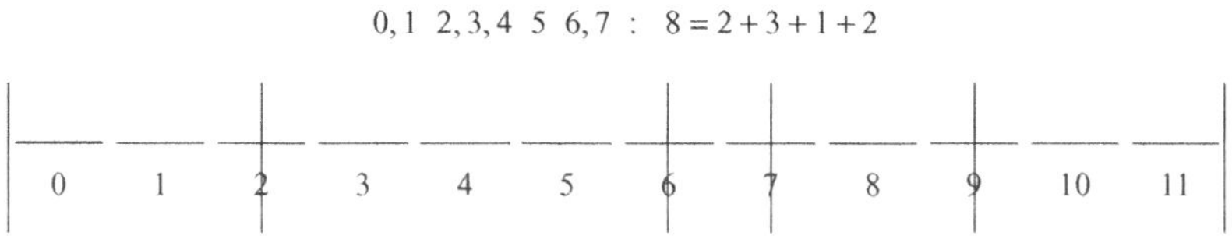

Figure 1.1. An ordered partition of $n = 8$ into $m = 4$ parts.

[6]See *Example 2.6* in Chapter 2 for another type of partition.

Corollary 1.10. The number $C_{n,m}$ of partitions of the integer n into m positive parts is $\binom{n-1}{m-1}$.

Proof. If

$$\ell_i \geq 0 \quad 0 \leq i < m \quad n = \ell_0 + \ell_1 + \cdots \ell_{m-1} \tag{1.11a}$$

then $\ell^* = (\ell_0^*, \ell_1^*, \cdots, \ell_{m-1}^*)$ defined by

$$\ell_i^* = \ell_i + 1 \quad 0 \leq i < m \tag{1.11b}$$

implies (1.11*c*)

$$\ell_i^* > 0 \quad 0 \leq i < m \quad n - m = \ell_0^+ \ell_1^* + \cdots \ell_{m-1}^* \tag{1.11c}$$

and conversely, if $\underline{l}$ and $\underline{l}^*$ are related by equation (1.11*b*), then the conditions of equation (1.11*b*) imply the conditions of equation (1.11*a*). ■

Example 1.13 (donuts). Dunkin' Donuts offers more than 30 varieties of donuts, including the ever popular chocolate creme-filled donut.[7]

How many ways are there of buying 12 donuts from the 30 varieties without any restriction on the number of each kind?

Solutions. $\binom{29+12}{29}$

How many ways are there of buying 12 donuts from the 30 varieties if at least four must be of one specific variety?

Solutions. Suppose the desired donut is of the 0th variety. Equation (1.11*a*) is replaced by

$$\ell_i \begin{cases} \geq 4 & \text{if } i = 0 \\ \geq 9 & \text{if } 0 < i < 30 \end{cases} \qquad 12 = \ell_0 + \ell_1 + \cdots \ell_{29}$$

If we set

$$\ell_i^* = \begin{cases} \ell_i - 4 & \text{if } i = 0 \\ \ell_i & \text{if } 0 < i < 30 \end{cases}$$

[7]I worked at the Puzzle Palace during the summer of 1997. While on this assignment, I violated one of the security rules by failing to turn off my workstation monitor after a logout. The NSA logo remained there for all to see—of course, only those who were cleared to enter the facility could see it! Nevertheless, there was a punishment; I had to buy a dozen donuts for my group after I was told the next day of my infraction, and I was further informed that the *Chief* preferred the chocolate creme-filled variety. I did what I had to do!

then

$$\ell_i^* \geq 0 \quad 0 \leq i < 30 \quad 8 = \ell_0 + \ell_1 + \cdots \ell_{29}$$

yielding the answer is $\binom{29+8}{29}$.

1.8 THE PRINCIPLE OF INCLUSION-EXCLUSION

Let $N_0, N_1, \cdots, N_{n-1}$ be sets in some universe $\mathcal{U}$. Then

$$|N_0 \cup N_1| = |N_0| + |N_1| - |N_0 \cap N_1| \tag{1.12a}$$

$$|N_0 \cup N_1 \cup N_2| = |N_0| + |N_1| + |N_2| - |N_0 \cap N_1| - |N_0 \cap N_2| - |N_1 \cap N_2| + |N_0 \cap N_1 \cap N_2| \tag{1.12b}$$

Note that

- A point $u \in N_0$ is counted once on the right-hand side of equation (1.12*a*) $[1 = 1 + 0 - 0]$ if $u \neq N_1$.
- A point $u \in N_0$ is also counted once on the right-hand side of equation (1.12*a*) $[1 = 1 + 1 - 1]$ if $u = N_1$.

Similarly,

- A point $u \in N_0$ is counted once on the right-hand side of equation (1.12*b*) $[1 = 1 + 0 + 0 - 0 - 0 - 0 + 0]$ if $u \neq N_1$ and $u \neq N_2$.
- A point $u \in N_0$ is also counted once on the right-hand side of equation (1.12*b*) $[1 = 1 + 1 + 0 - 1 - 0 - 0 - 0$ or $1 = 1 + 0 + 1 - 0 - 1 - 0 - 0]$ if $u = N_1$ and $u \neq N_2$ or $u = N_2$ and $u \neq N_1$.
- A point $u \in N_0$ is also counted once on the right-hand side of equation (1.12*b*) $[1 = 1 + 1 + 1 - 1 - 1 - 1 + 1]$ if $u = N_1$ and $u = N_2$.

These equalities are special cases of the following theorem.

Theorem 1.11 (principle of inclusion-exclusion). If $N_0, N_1, \cdots, N_{n-1}$ are sets in a universe $\mathcal{U}$, then

$$\left|\bigcup_{i=0}^{n-1} N_i\right| = \sum_{i=0}^{n-1} |N_i| - \sum_{0 \leq i_0 < i_1 < n} |N_{i_0} \cap N_{i_1}| + \sum_{0 \leq i_0 < i_1 < i_2 < n} |N_{i_0} \cap N_{i_1} \cap N_{i_2}| \cdots (-1)^{n-1} |N_0 \cap N_1 \cap \cdots \cap N_{n-1}| \tag{1.13}$$

Proof. By mathematical induction using the equality in equation (1.12*a*). ■

Sometimes the enumeration combines the use of both ordered permutations and the principle of inclusion-exclusion.

Example 1.14 (more donuts). In how many ways can 27 donuts be chosen from the 30 varieties if *fewer* than 10 of the 0th variety is to be included?

Solution. Equation (1.11*a*) is now replaced by

$$\ell_i \begin{cases} <10 & \text{if } i=0 \\ \geq 9 & \text{if } 0<i<30 \end{cases} \qquad 27 = \ell_0 + \ell_1 + \cdots \ell_{29} \tag{1.14a}$$

The complementary problem is

$$\ell_i \begin{cases} \geq 10 & \text{if } i=0 \\ \geq 9 & \text{if } 0<i<30 \end{cases} \qquad 27 = \ell_0 + \ell_1 + \cdots \ell_{29} \tag{1.14b}$$

If we set

$$\ell_i^* = \begin{cases} \ell_i - 10 & \text{if } i=0 \\ \ell_i & \text{if } 0<i<30 \end{cases}$$

then

$$\ell_i^* \geq 0 \quad 0 \leq i < 30 \quad 18 = \ell_0 + \ell_1 + \cdots \ell_{29}$$

which has $\binom{29+18}{29}$ solutions, which means the original problem equation (1.14*a*) has

$$\binom{29+28}{29} - \binom{29+18}{29}$$

solutions. ■

1.9 PARTITIONS

A partition[8] Π of a set of n elements, say $\mathcal{Z}_n = \{0, 1, \cdots, n-1\}$, is a collection of nonempty sets whose union is $\mathcal{Z}_n$. For example, the five partitions of $\mathcal{Z}_3$ are

$$\Pi_1 : \{0\}\{1\}\{2\} \quad \Pi_2 : \{0,1\}\{2\} \quad \Pi_3 : \{0,2\}\{1\} \quad \Pi_4 : \{1,2\}\{0\} \quad \Pi_6 : \{0,1,2\}$$

The number of partitions of $\mathcal{Z}_n$ is the Bell number B_n. *Example 2.6* in Chapter 2 asks the reader to derive the recursion

$$B_{n+1} = \binom{n}{0} B_0 + \binom{n}{1} B_1 + \cdots + \binom{n}{n} B_n \quad 0 \leq n < \infty; B_0 = 1, B_2 = 1 \tag{1.15}$$

and the generating functions of the $\{B_n\}$.

[8] In §1.6 we defined the ordered partition; without the prefix *order*, the order of the elements in a set in Π and the order in which these sets are listed in Π are both immaterial.

The Stirling number (of the second kind) $S_{n,k}$ is the number of partitions of $\mathcal{Z}_n$ into k (nonempty) sets; thus,

$$S_{3,1} = 1 \quad S_{3,2} = 3 \quad S_{3,3} = 1$$

The Stirling numbers (of the second kind) $\{S_{n,k}\}$ are obviously related to the Bell numbers $\{B_n\}$ by

$$B_n = \sum_{k=1}^{n} S_{n,k} \tag{1.16}$$

A proof of the formula below will be given in Chapter 2.

$$S_{n,k} = \frac{1}{k!} \sum_{s=0}^{k} \binom{k}{s} (-1)^s (k-s)^n \tag{1.17}$$

There is an extensive literature dealing with Stirling numbers, which arise in many applications; see, for example, [Bleick and Wang 1974]. We will encounter the $\{S_{n,k}\}$ in Chapter 10.

1.10 RELATIONS

A relation $\sim$ on a yset X generalizes the notion of function specifying some collection of pairs (x, y), and we write $x \sim y$ for $x, y \in X$ read (elements) *x and y are related* and $x \nsim y$, if they are not related.

A partition $X_0, X_1, \cdots$ of a set X as in §1.9 determines a relation $\sim$ by the rule $x \sim y$ if and only if $x, y \in X_i$ for some i.

Conversely, a relation $\sim$ on a set X determines a partition $X_0, X_1, \cdots$ in which

- X_i consists of $\sim$-related elements.
- If $x \in X_i$, $y \in X_j$ and $i \neq j$, then $x \nsim y$.

The following properties may or not be enjoyed by a relation $\sim$:

1. $\sim$ is *reflexive* if $x \sim x$; $\sim$ is *irreflexive* if $x \nsim x$.
2. $\sim$ is *symmetric* if $x \sim y$ implies $y \sim s$; a reflexive relation $\sim$ is *asymmetric* if $x \sim y$ and $y \sim x$ can only occur if $y = x$.
3. $\sim$ is *transitive* if $x \sim y$ and $y \sim z$ implies $x \sim z$; $\sim$ is *intransitive* if $x \sim y$ and $y \sim z$ implies $x \nsim z$.

In addition to the modifiers *ir*, *anti*, and *in*, there are definitions for the modifier **not** as in *not* reflexive, *not* symmetric, and *not* transitive; we leave to the reader's creativity, their definitions.

$\sim$ is an **equivalence relation** if it is reflexive, symmetric, and transitive.

1.11 INVERSE RELATIONS

The binomial coefficients can be used to define a sort of transformation on sequences; for example,

$$a = (a_0, a_1, \cdots) \to b = (b_0, b_1, \cdots)$$
$$b_n = \sum_l (-1)^k \binom{n}{k} a_k \quad n = 0, 1, \cdots \qquad (1.18a)$$

Theorem 1.12. If equation (1.18*a*) holds, then

$$a_n = \sum_l (-1)^k \binom{n}{k} v_k \quad n = 0, 1, \cdots \qquad (1.18b)$$

Solution: Start with the identity E8 in the Chapter 1 Appendix that follows, setting $b_n = (-1)^m \binom{m}{n}$. ■

Theorem 1.12 is one of many inverse relations (see [Riordan 1968]); we will use it in Chapter 10.

Summations Involving Binomial Coefficients

Equations **E1** through **E8** follow easily from the binomial theorem

$$(x+y)^n = \sum_{m=0}^{n} x^m y^{n-m} \binom{n}{m}$$

and derivatives of it by evaluating for special values of x, y. ■

E1. $\binom{n}{0}+\binom{n}{1}+\binom{n}{2}+\cdots+\binom{n}{n}=2^n$

E2. $\binom{n}{0}-\binom{n}{1}+\binom{n}{2}+\cdots+(-1)^n\binom{n}{n}=0$

E3. $\binom{n}{0}+\binom{n}{2}+\binom{n}{4}+\cdots=2^{n-1}$

E4. $\binom{n}{1}+\binom{n}{3}+\binom{n}{5}+\cdots=2^{n-1}$

E5. $\binom{n}{1}+2\binom{n}{2}+3\binom{n}{3}+\cdots+n\binom{n}{n}=n2^{n-1}$

E6. $\binom{n}{1}-2\binom{n}{2}+3\binom{n}{3}-\cdots+(-1)^n n\binom{n}{n}=0$

E7. $2\binom{n}{2}+6\binom{n}{3}+12\binom{n}{4}+\cdots+n(n-1)\binom{n}{n}=n(n-1)2^{n-2}$

E8. $\sum_k (-1)^{k+m}(-1)^k\binom{n}{k}\binom{k}{m}=\delta_{n,m}=\begin{cases}1 & \text{if } n=m\\ 0 & \text{otherwise}\end{cases}$

E9. $\binom{n+m}{k}=\sum_{j=0}^{n}\binom{n}{j}\binom{m}{k-j}$

Vandermonde's Identity E9 is proved by counting the ways of choosing k elements from the set $A \cup B$, where $A \cap B = \emptyset$, A contains n, and B contains m elements.

E10. $$\sum_{r=1}^{m}\frac{\binom{m}{r}}{\binom{n}{r}}=\frac{m}{n-m+1}$$

E10 is a special case of the formula

$$\sum_{r=j}^{k}\binom{m}{r}\Big/\binom{n}{r}=(n+1)/(n-m+1)\left\{\binom{m}{j}\Big/\binom{n+1}{j}-\binom{m}{k+1}\Big/\binom{n+1}{k+1}\right\}$$

contained in [Gould 1972, p. 46]. It may be proved by recognizing the identity

$$\binom{m}{r}\Big/\binom{n}{r}=m/n\binom{m-1}{r-1}\Big/\binom{n-1}{r-1}$$

E11. $$x^{-1}(x+y+na)^n=\sum_{j=0}^{n}\binom{n}{j}(x+ja)^{j-1}(y+(n-j)a)^{n-j})$$

E11 is a nontrivial generalization of the Binomial theorem from Niel Henrik Abel (1802–1829) published in 1826 (see [Riordan 1968]). Abel is famous for proving the impossibility of representing a solution of a general equation of fifth degree or higher by a radical expression.

REFERENCES

W. W. Bleick and P. C. C. Wang, "Asymptotics of Stirling Numbers of the Second Kind", *Proceedings of the American Mathematical Society*, **42**, #2, pp. 575–580, 1974.

W. Feller, *An Introduction to Probability Theory and Its Applications*, Volume 1 (Second Edition), John Wiley & Sons (New York), 1957; (Third Edition), John Wiley & Sons (New York), 1967.

H. W. Gould, *Combinatorial Identities*, Henry W. Gould (Morgantown, West Virginia), 1972.

J. Riordan, *Combinatorial Identities*, John Wiley & Sons (New York), 1968.

K. H. Rosen, *Discrete Mathematics and Its Applications*, McGraw-Hill (New York), 2003.

CHAPTER 2

Recurrence and Generating Functions

Standard references on the theory and application of generating functions are [Riordan 1968 and 1980].

2.1 RECURSIONS

A *recursion* for a sequence $a_0, a_1, \cdots$ is a rule for computing a_n in terms of quantities depend on n and perhaps some of the previous terms $a_0, a_1, \cdots, a_{n-1}$.

Examples 2.1.

a) The factorial $n!$ of the integer n may be defined recursively by

$$n! = \begin{cases} 1 & \text{if } n = 0 \\ n \times (n-1)! & \text{if } 1 \le n < \infty \end{cases}$$

b) The harmonic numbers $\{H_n : 1 \le n < \infty\}$ may be defined recursively by

$$H_n = \begin{cases} 1 & \text{if } n = 1 \\ H_{n-1} + \dfrac{1}{n} & \end{cases}$$

c) The Fibonacci sequence[1] $\{F_n : 0 \le n < \infty\}$ may be defined recursively by

$$F_n = \begin{cases} 1 & \text{if } n = 0, 1 \\ F_{n-2} + F_{n-1} & \text{if } 2 \le n < \infty \end{cases}$$

[1]Leonardo of Pisa (1170–1250) is better known as Fibonacci (the son of Bonaccio). He grew up in North Africa coming into contact with the mathematical knowledge of the Arab scholars of that period. It is claimed that Fibonacci was the first to recognize that when a financial calculation produced a *negative* result that it represented a *loss*; it might be that a particular calculation was incorrect but could also result from the investor's bad judgment.

Hashing in Computer Science: Fifty Years of Slicing and Dicing, by Alan G. Konheim

d) The Catalan numbers[2] $\{C_n : 0 \le n < \infty\}$ may be defined by the recursion

$$C_n = \begin{cases} 1 & \text{if } n = 0 \\ \sum_{i=0}^{n-1} C_i C_{n-i-1} & \text{if } 1 \le n < \infty \end{cases}$$

2.2 GENERATING FUNCTIONS

The generating function $A(z)$ of the sequence $a_0, a_1, \cdots$ is the power series

$$A(z) = a_0 + a_1 z + \cdots + a_{n-1} z^{n-1} + \cdots \tag{2.1a}$$

Whenever we write an infinite series, there is the question of convergence. The series in equation (2.1a) converges provided the radius of convergence $r = \limsup_n \left| \frac{a_n}{a_{n-1}} \right| < \infty$. In this case, $A(z)$ converges for *all* z in the complex disk $\mathcal{D}_r(0)$ of radius r centered at $z = 0$. If $r > 0$, then we write

$$\{a_n\} \Leftrightarrow A(z) \tag{2.1b}$$

because $\{a_n\}$ determines the generating function $A(z)$, and conversely

$$a_n = \frac{1}{n!} \frac{d^n}{dz^n} A(z) \bigg|_{z=0} \tag{2.1c}$$

Equivalently, the **Cauchy integral theorem** (see [Ahlfors 1953; pp. 92–97]) gives

$$a_n = \frac{n!}{2\pi i} \oint_C \frac{F(\zeta)}{\zeta^{n+1}} d\zeta \tag{2.1d}$$

where C a circle of radius slightly less than r about $\zeta = 0$ so that $A(\zeta)$ is analytic in the closed disk bounded by C. The typographically awkward notation in equation (2.1d) for a_n is often replaced by $a_n = [z^n]A(z)$.

For the sequence $a_n = n!$, the radius of convergence is 0 and something else has to be done. We put this off for the moment and pretend that the sequence $a_0, a_1, \cdots$, whose values we do *not* know converges in a disk of some positive radius.

Examples 2.2.

a) Geometric sequence: $a_n = q^n \ (0 \le n < \infty) \Leftrightarrow A(z) = \frac{1}{1-qz}$.
b) Truncated geometric sequence: $a_n = q^n \ (0 \le n < N) \Leftrightarrow A(z) = \frac{q^N z^N}{1-qz}$.

[2]The Catalan sequence was first described in the 18th century by Leonhard Euler (1707–1783), who was interested in the number of different ways of dividing a polygon into triangles. The sequence was named after the Belgian mathematician Eugene Charles Catalan (1814–1894), who discovered the connection to parenthesized expressions during his exploration of the Towers of Hanoi puzzle.

c) Poisson distribution sequence: $a_n = \frac{q^n}{n!} e^{-n}\ (0 \le n < \infty) \Leftrightarrow A(z) = e^{-q(1-z)}$.

d) Reciprocal factorial sequence: $a_n = \frac{1}{n!}(0 \le n < \infty) \Leftrightarrow A(z) = e^z$.

When two polynomials

$$p(z) = p_0 + p_1 z + \cdots + p_n z^n \tag{2.2a}$$

$$q(z) = q_0 + q_1 z + \cdots + q_m z^m \tag{2.2b}$$

are multiplied $r(z) = p(z)q(z)$

$$r(z) = r_0 + r_1 z + \cdots + r_{m+n} z^{m+n} \tag{2.2c}$$

their coefficients $\{r_k\}$ are given by

$$r_k = \sum_{\substack{0 \le i \le n, 0 \le j \le m \\ i+j=k}} p_i q_j \quad 0 \le k \le m+n \tag{2.2d}$$

The operation $\underline{p} * \underline{q} \to \underline{r}$ is the convolution of the two sequences $\underline{p}$ and $\underline{q}$. It is the *raison d'être* of generating functions since

Theorem 2.1. If $P(z)$ is the generating function of $\underline{p}$ and $Q(z)$ is the generating function of $\underline{q}$, then $P(z)Q(z)$ is the generating function of $\underline{r} = \underline{p} * \underline{q}$.

Convolution is a basic operation on sequences for another reasons; if $\underline{p}$ and $\underline{q}$ are discrete probability distributions of independent random variables X and Y (see Chapter 4), then

$$p_i, q_j \ge 0 \quad (0 \le i, j < \infty)$$
$$1 = \sum_{j=0}^{\infty} p_j = \sum_{j=0}^{\infty} q_j$$

the radii of convergence of each of their generating functions $P(z)$ and $Q(z)$ is 1 and the operation $\underline{r} = \underline{p} * \underline{q}$ is the probability distribution of the random variable $X + Y$.

2.3 LINEAR CONSTANT COEFFICIENT RECURSIONS

A linear recurrence of degree k with constant coefficients (LCCR) is a sequence a_0, a_1, $\cdots$ defined by

$$a_n = C_1 a_{n-1} + C_2 a_{n-2} + \cdots + C_k a_{n-k} + D_n \quad C_k \ne 0 \quad n = k, k+1, \cdots \tag{2.3}$$

where $C_1, C_2, \cdots, C_k$ are constants and D_n is a function *only* of n. The LCCR is homogeneous if $D_n \equiv 0$. To determine completely the sequence $a_0, a_1, \cdots$, it is necessary that the initial conditions $a_0, a_1, \cdots, a_{k-1}$ be given.

Examples 2.1b and *2.1c* are LCCR, but *Example 1c* is a homogeneous LCCR.

Example 2.3 (The Towers of Hanoi). The simplest version is a puzzle invented by E. Lucas in 1883. n real (*not* computer) disks of increasing radii are spindled on a common peg P_1. There are two additional pegs P_2 and P_3 on which the disks may also be spindled. The object of this children's game is to move the n disks from P_1 to P_3, say, perhaps using peg P_2 so that they are spindled with increasing radii. The rules of the moves are as follows:

1. Only one disk may be moved at a time.
2. At no time may a larger radius disk be on top of a smaller radius disk.

Solution (Derive the Recurrence). Let a_n be the total number of moves with n disks (of different radii). Leave the largest disk on P_1 and use a_{n-1} moves to place the other $n-1$ disks on P_2. Then, move the largest disk from P_1 to P_3 and complete the puzzle using a_{n-1} moves to move the disks from P_2 to P_3. This gives

$$a_n = 2a_{n-1} + 1 \quad 1 \le n < \infty \tag{2.4}$$

(*Guessing*). Because a_1 is clearly 1, we must define $a_0 = 0$ if equation (2.4) is to hold as advertised for $n = 1, 2, \cdots$. Evaluating a_n for $n = 2, 3$ gives 3 and 7 and so $a_n = 2^n - 1$ for $n = 0, 2, \cdots$. ■

There are variations of the Tower of Hanoi, for example, more disks [Allouche 1994; Gardner 1957], that are not as susceptible to guessing and solution. A very extensive bibliography is given in [Stockmeyer 2005].

How are LCCRs solved?

2.4 SOLVING HOMOGENEOUS LCCRS USING GENERATING FUNCTIONS

Example 2.4 (The Fibonacci sequence). The recurrence is

$$F_n = F_{n-1} + F_{n-2} \quad 2 \le n < \infty \quad F_0 = F_1 \tag{2.5a}$$

Multiply both sides of equation (2.5*a*) by z^n and sum $n = 2, 3, \cdots$ yielding

$$F(z) - 1 - z = z[F(z) - 1] + z^2 F(z) \tag{2.5b}$$

where

$$F(z) = \sum_{n=0}^{\infty} F_n z^n \tag{2.5c}$$

Rearranging equation (2.5*b*) gives

$$F(z) = \frac{1}{1 - z - z^2} = \frac{1}{(1 - s_0 z)(1 - s_1 z)} \tag{2.5d}$$

Next, by the partial fraction expansion of the right-hand side of equation (2.5*d*)

$$F(z) = \alpha_0 \frac{1}{1 - s_0 z} + \alpha_1 \frac{1}{1 - s_1 z} \tag{2.5e}$$

where

$$r_0 = \frac{1}{s_0} = \frac{-1 + \sqrt{5}}{2} \quad s_0 = \frac{2}{-1 + \sqrt{5}} = \frac{1 + \sqrt{5}}{2} \tag{2.5f}$$

$$r_1 = \frac{1}{s_1} = \frac{-1 - \sqrt{5}}{2} \quad s_1 = \frac{2}{-1 - \sqrt{5}} = \frac{1 - \sqrt{5}}{2} \tag{2.5g}$$

so that

$$F_n = \alpha_0 s_0^n + \alpha_1 s_1^n \quad (0 \leq n < \infty) \tag{2.5h}$$

The values of α_0 and α_1 are found as the solution of linear equations

$$\alpha_0 = \lim_{z \to r_0} \frac{1}{1 - s_1 z} \quad \alpha_1 = \lim_{z \to r_1} \frac{1}{1 - s_0 z} \tag{2.5i}$$

giving

$$\alpha_0 = \frac{1}{1 - r_0 s_1} = \frac{1}{\sqrt{5}} s_0 \quad \alpha_1 = \frac{1}{1 - r_1 s_0} = \frac{1}{\sqrt{5}} s_1 \tag{2.5j}$$

yielding finally

$$F_n = \frac{1}{\sqrt{5}} \left(\frac{1 + \sqrt{5}}{2} \right)^{n+1} - \frac{1}{\sqrt{5}} \left(\frac{1 - \sqrt{5}}{2} \right)^{n+1} \quad (0 \leq n < \infty) \tag{2.5k}$$

Is This Method to Solve LCCR Fail-Safe? We consider only second-order homogeneous LCCRs.

Suppose

$$a_n = C_1 a_{n-1} + C_2 a_{n-2} \quad C_2 \neq 0; a_0, a_1 \text{ given} \quad n = 2, 3, \cdots \tag{2.6a}$$

Suppose we look for a solution of the form $a_n = \alpha x^n$ $(0 \leq n < \infty)$. Then

$$\alpha x^n = C_1 \alpha x^{n-1} + C_2 \alpha x^{n-2}$$

so that

$$0 = \alpha x^{n-2} [x^2 - C_1 x - x^2] \tag{2.6b}$$

which implies that either $\alpha = 0, x = 0$ or

$$0 = x^2 - C_1 x - C_2 \Leftrightarrow 0 = 1 - C_1 x^{-1} - C_2 x^{-2} \tag{2.6c}$$

In this last case, x is a root of the characteristic equation $0 = 1 - C_1 x^{-1} - C_2 x^{-2}$. The two possible cases are as follows:

Case 1.

The characteristic equation has *two distinct* (real) roots $s_0 = \frac{1}{r_0}$ and $s_1 = \frac{1}{r_1}$. Neither s_0 nor s_1 can equal 0 since $C_2 \neq 0$.

The linearity of equation (2.6*a*) implies that for any constants α_0, α_1

$$a_n = \alpha_0 s_0^n + \alpha_1 s_1^n \quad 2 \le n < \infty \tag{2.6d}$$

satisfies equation (2.6*a*) for $0 \le n < \infty$. If they are to satisfy equation (2.6*a*) for *all* $n \ge 0$, then

$$a_0 = \alpha_0 + \alpha_1 \tag{2.6e}$$

$$a_1 = \alpha_0 s_0 + \alpha_1 s_1 \tag{2.6f}$$

which may be written as

$$\begin{pmatrix} a_0 \\ a_1 \end{pmatrix} = V \begin{pmatrix} a_0 \\ a_1 \end{pmatrix} \quad V = \begin{pmatrix} 1 & 1 \\ s_0 & s_1 \end{pmatrix} \tag{2.6g}$$

Note that $det(V) \neq 0$ when the roots are distinct. ■

Case 2.

The characteristic equation has a multiple root $s_0 = s_1 = s$. Still, $a_n = \alpha_0 s^n$ is a solution, but there is a second solution. We claim that $\alpha_1 n s^n$ is also a solution. Why?

If x is a multiple root of the characteristic equation

$$0 = x^2 - C_1 x - C_2 \tag{2.7a}$$

then

$$0 = 2x - C_1 \tag{2.7b}$$

and if $\alpha n x^n$ is a solution for a_n, then equation (2.6*a*) may be written as

$$\alpha n x^n = \alpha C_1 (n-1) x^{n-1} + \alpha C_2 (n-2) x^{n-2} \tag{2.7c}$$

which requires

$$0 = \alpha (n-2) x^{n-2} [x^2 - C_1 x - C_2] + 2\alpha x^{n-1} [2\alpha - C_1] \tag{2.7d}$$

The case of as k^{th} order homogeneous LCCR with k distinct roots is only slightly more complicated; the matrix $V = \begin{pmatrix} 1 & 1 \\ s_0 & s_1 \end{pmatrix}$ is replaced by a Vandermonde matrix[3]

[3]Named after Alexandre-Thophile Vandermonde (1735–1796), even though it did not appear in his collected works.

$$V = \begin{pmatrix} 1 & 1 & \cdots & 1 \\ s_0 & s_1 & \cdots & s_{k-1} \\ s_0^2 & s_1^2 & \cdots & s_{k-1}^2 \\ \vdots & \vdots & \ddots & \vdots \\ s_0^{k-1} & s_1^{k-1} & \cdots & s_{k-1}^{k-1} \end{pmatrix} \tag{2.8a}$$

with determiant

$$det(V) = \prod_{0 \le i < j < n} (s_j - s_i) \tag{2.8b}$$

The analog of equations (2.6*e* and 2.6*f*) is

$$\begin{pmatrix} a_0 \\ a_1 \\ \vdots \\ a_{k-1} \end{pmatrix} = V \begin{pmatrix} \alpha_0 \\ \alpha_1 \\ \vdots \\ \alpha_{k-1} \end{pmatrix} \tag{2.8c}$$

A related interpolation problem points out the nature of the solution method.

Given: ordinates $\underline{y} = (y_0, y_1, \cdots, y_{k-1})$ and abscissas $\underline{x} = (x_0, x_1, \cdots, x_{k-1})$
Find: the minimal degree polynomial $p(x) = p_0 + p_1 x + \cdots + p_{k-1} x^{k-1}$, which interpolates $\underline{y}$ at $\underline{x}$; that is, $y_i = p(x_i)$ for $0 \le i < k$.

The solution was published in 1795 by Joseph Louise Lagrange (1736–1813). First,

$$\begin{pmatrix} y_0 \\ y_1 \\ y_3 \\ \vdots \\ y_{k-1} \end{pmatrix} = \begin{pmatrix} 1 & x_0 & x_0^2 & \cdots & x_0^{k-1} \\ 1 & x_1 & x_1^2 & \cdots & x_1^{k-1} \\ 1 & x_2 & x_2^2 & \cdots & x_2^{k-1} \\ \vdots & \vdots & \vdots & \ddots & \vdots \\ 1 & x_{k-1} & x_{k-1}^2 & \cdots & x_{k-1}^{k-1} \end{pmatrix} \begin{pmatrix} p_0 \\ p_1 \\ p_1 \\ \vdots \\ p_{k-1} \end{pmatrix} \tag{2.8d}$$

and we see that the $k \times k$ matrix in equation (2.8*d*) is the transpose of the Vandermonde matrix in equation (2.8*a*). Lagrange's interpolation theorem states that *if* the abscissas $\underline{x} = (x_0, x_1, \cdots, x_{k-1})$ are distinct, then

$$p(x) = \sum_{i=0}^{k-1} y_i L_{\{i:\underline{x},\underline{y}\}}(x) \tag{2.8e}$$

with

$$L_{\{i:\underline{x},\underline{y}\}}(x) = \prod_{\substack{0 \le j < k \\ j \ne i}}^{k-1} \frac{x - x_j}{x_i - x_j} \tag{2.8f}$$

Note

$$L_{\{i:\underline{x},\underline{y}\}}(x_j) = \begin{cases} 1 & \text{if } j = i \\ 0 & \text{if } j \ne i \end{cases} \tag{2.8g}$$

If there is an inhomogeneous term D_n in equation (2.1a), then the generating function $A(z)$ of $\{a_n\}$ is given by

$$A(z)-\ell(z)=Q(z)A(z)+D(z)-\delta(z)$$

where

- $\ell(z)$ is a polynomial determined by the initial conditions $a_0, a_1, \cdots, a_{k-1}$.
- $Q(z) = C_1 z + C_2 z^2 + \cdots + C_k z^k$.
- $D(z)$ is the generating function of $\{D_n\}$ and $\delta(z)$ is a polynomial of degree at most $k - 1$.

The solution is found in the same way using the roots of $1 - Q(z)$.

2.5 THE CATALAN RECURSION

We start with the recursion

$$C_n = \begin{cases} 1 & \text{if } n=0 \\ \sum_{i=0}^{n-1} C_i C_{n-i-1} & \text{if } 1 \le n < \infty \end{cases} \tag{2.9a}$$

Let $C(z)$ be the generating functions of $\{C_n\}$

$$C(z)=\sum_{n=0}^{\infty} C_n z^n$$

1. Multiply the relationship in equation (2.9a) for $1 \le n < \infty$ by z^n and sum $n = 1, 2, \cdots$

$$C(z)-1=\sum_{n=1}^{\infty} z^n \sum_{i=0}^{n-1} C_i C_{n-i-1}$$

2. Next, interchange the order of summations

$$C(z)-1=\sum_{i=0}^{\infty} C_i \sum_{n=i+1}^{\infty} z^n C_{n-i-1}$$

3. Replace the n-summation variable by $m = n - i - 1$

$$C(z)-1=\sum_{i=0}^{\infty} C_i z^{i+1} \sum_{m=0}^{\infty} z^m C_m = zC^2(z) \tag{2.9b}$$

The quadratic equation (2.9b) has two roots

$$C(z)=\frac{1\pm\sqrt{1-4z}}{2z}$$

Question. Which roots?

Answer. The root with the minus sign, because otherwise, $C(z)$ will *not* be analytic at $z = 0$

$$C(z) = \frac{1-\sqrt{1-4z}}{2z} \tag{2.9c}$$

The Catalan numbers are

$$C_n = \frac{1}{n+1}\binom{2n}{n} \quad 0 \le n < \infty \tag{2.9d}$$

2.6 THE UMBRAL CALCULUS[4]

The umbral calculus originating in the 19th century corresponds to the Jewish kabbala, which deals with mysticism. In the umbral case, strange proofs of theorems involving polynomials were contrived; the result was correct, but the method seemed wrong.

For example, note the structural similarity between the binomial theorem

$$(x+y)^n = \sum_{i=0}^{n}\binom{n}{i} x^{n-i} y^i \tag{2.10a}$$

and

$$B_n(x+y) = \sum_{i=0}^{n}\binom{n}{i} B_{n-i}(x) y^i \tag{2.10b}$$

where $B_n(x)$ is the nth Bernstein polynomial (see [Riordan 1980]). Can the structural similarity between

$$\frac{d}{dx} x^n = n x^{n-1} \tag{2.11a}$$

and

$$\frac{d}{dx}\frac{d}{dx} B_n(x) = n B_{n-1}(x) \tag{2.11b}$$

be proved by replacing x^k in equation (2.11*a*) by $B_k(x)$ in equation (2.11*b*)?

Riordan's book [Riordan 1968] uses umbral-like proofs to establish equations (2.10) and (2.11).

Gian-Carlo Rota (1932–1999) provided an explanation in [Rota, Kahaner and Odlyzko 1973] and [Roman and Rota 1978].

[4] The word umbra in Latin means *shadow*.

2.7 EXPONENTIAL GENERATING FUNCTIONS

We noted in §2.3 that some sequences, for example, the growth of the sequence $f_n = n!$, implies that it does not have a generating function $F(z) = \Sigma_n f_n z^n$; the radius of convergence of $F(z)$ is 0. There is another version of generating functions that remedies this problem, one that we will use in Chapter 13.

The exponential generating function of a sequence $\{a_n\}$ is defined by

$$A(z) = \sum_{n=0}^{\infty} \frac{a_n}{n!} z^n \tag{2.12a}$$

provided the radius of convergence of $A(z)$ is positive.

If the exponential generating function $B(z)$ of the sequence $\{b_n\}$ also exists and

$$c_n = \sum_{i=0}^{n} \binom{n}{i} a_i b_{n-i} \quad 0 \le n < \infty \tag{2.12b}$$

then

$$C(z) = \sum_{n=0}^{\infty} \frac{c_n}{n!} z^n = A(z)B(z) \tag{2.12c}$$

Example 2.5 (Derangements). n guests check their coats and hats at the restaurant's checkroom, perhaps at an upscale McDonald's, Burger King, or Taco Bell. When they leave, their coats and hats are returned in random order; that is, the ith person receives the coat of the $\pi(i)^{\text{th}}$ customer with $\pi(i) \ne i$ for $0 \le i < n$. In how many ways D_n can this happen? For example

n	D$_n$	π
2	1	(1, 0)
3	2	(1, 2, 0) (2, 0, 1)
4	9	(1, 2, 3, 0) ⋯ (3, 0, 1, 2)

Solution. Suppose $\pi(0) = j \ne 0$; there are $n - 1$ choices for j. It follows that $(\pi(1), \pi(2), \cdots, \pi(n-1))$ must be a dearrangement of $(0, 1, \cdots, j-1, j+1, \cdots, n-1)$. The two possibilities are as follows:

1. If $\pi(j) = 0$, then $(\pi(2), \pi(3), \cdots, \pi(n-1))$ is a dearrangement of $(2, 3, \cdots, j-1, j+1, \cdots, n-1)$.
2. If $\pi(j) = k \ne 0$, and $\pi(r) = j$, then $(\pi(2), \pi(3), \cdots, \pi(n-1))$ with $\pi(r)$ replaced by j is a dearrangement of $(2, 3, \cdots, j-1, j+1, \cdots, n-1)$.

Therefore

$$D_n = (n-1)[D_{n-1} + D_{n-2}] \quad 2 \le n < \infty, D_0 = 1, D_1 = 0 \tag{2.13a}$$

If

$$D(z) = \sum_{n=0}^{\infty} \frac{D_n}{n!} z^n \tag{2.13b}$$

then multiplying both sides of equation (2.13*a*) by $\frac{z^{n-1}}{(n-1)!}$ and summing $n = 2$, 3, ⋯ , the result is

$$\sum_{n=2}^{\infty} D_n \frac{z^{n-1}}{(n-1)!} = \frac{d}{dz}[D(z)-1] \tag{2.13b}$$

$$\sum_{n=2}^{\infty} D_{n-1} \frac{(n-1)z^{n-1}}{(n-1)!} = \sum_{n=1}^{\infty} D_n \frac{nz^n}{n!} = z[D(z)-1] \tag{2.13c}$$

$$\sum_{n=2}^{\infty} D_{n-2} \frac{(n-1)z^{n-1}}{(n-1)!} = \sum_{n=0}^{\infty} D_n \frac{(n+1)z^{n+1}}{(n+1)!} = zD(z) \tag{2.13d}$$

yielding

$$D'(z) = \frac{z}{1-z}$$

which is equivalent to

$$\frac{d}{dz} \log D(z) = \frac{z}{1-z} \Rightarrow \log D(z) = \int \frac{\zeta}{1-\zeta} d\zeta = -z + \log \frac{1}{1-z} \tag{2.13e}$$

where log here and elsewhere in this book denotes logarithm to be e. Because $D(0) = D_0 = 1$, we have

$$D(z) = e^{-z+\log\frac{1}{1-z}} = \frac{e^{-z}}{1-z} \tag{2.13f}$$

Writing

$$D(z) = \left(\sum_{j=0}^{\infty} \frac{(-1)^j}{j!} z^j \right) \left(\sum_{k=0}^{\infty} z^k \right) \tag{2.13g}$$

yields

$$D_n = n! \sum_{j=0}^{n} (-1)^j \frac{1}{j!} \tag{2.13h}$$

completing the solution. ■

2.8 PARTITIONS OF A SET: THE BELL AND STIRLING NUMBERS

A partition $\Pi(n)$ of $\mathcal{Z}_n = \{0, 1, \cdots, n-1\}$ is a collection of nonempty sets whose union is $\mathcal{Z}_n$; for example, $\Pi(5) : \{0, 3\}\ \{1, 2, 4\}$.

The number of partitions of $\mathcal{Z}_n$ is denoted by B_n, referred to as the nth-**Bell number**[5]. The five partitions $\Pi(3)$ of $\mathcal{Z}_3$ are

$$\Pi_1(3):\{0\}\ \{1\}\ \{2\}\quad \Pi_2(3):\{0,1\}\ \{2\}\quad \Pi_3(3):\{0,2\}\ \{1\}$$
$$\Pi_4(3):\{1,2\}\ \{0\}\quad \Pi_6(3):\{0,1,2\}$$

We prove

Theorem 2.2

a) The $\{B_n\}$ satisfy the recurrence relation

$$B_{n+1}=\binom{n}{0}B_0+\binom{n}{1}B_1+\cdots+\binom{n}{n}B_n \quad 0\le n<\infty;\ B_0=1,\ B_2=1 \tag{2.14a}$$

b) The exponential generating function of $B(z)$ is

$$B(z)=e^{e^z-1} \tag{2.14b}$$

Solution. The proof is by induction on n, the starting value being $n = 3$. Let the subset of a partition Π_{n+1} that contains $n + 1$ contain k additional elements with $0 \le k \le n$; these elements may be chosen from $\mathcal{Z}_n$ in $\binom{n}{k}$ ways. The remaining $n - k$ elements have B_{n-k} partitions and this proves equation (2.14*a*).

Multiply equation (2.14*a*) by $\frac{z^n}{n!}$ and sum $n = 0, 1, \cdots$ to obtain

$$\sum_{n=1}^{\infty} B_{n+1}\frac{z^n}{n!}=\sum_{n=0}^{\infty}\sum_{k=0}^{n} B_{n-k}\frac{1}{k!(n-k)!} \tag{2.14c}$$

Interchaining the order of integration and changing the variable of summation as in Chapter 2 section 2.5 gives

$$\sum_{n=1}^{\infty} B_{n+1}\frac{z^n}{n!}=e^z D(z) \tag{2.14d}$$

Next, the left-hand side of the previous equation is recognized as $\frac{d}{dz}B(z)$, which gives the differential equation

$$\frac{d}{dz}B(z)=e^z B(z) \tag{2.14e}$$

The initial value $B(0) = B_0 = 1$ completes the induction argument and the solution. ■

James Stirling (1692–1770) was a Scottish mathematician. His asymptotic formula for $n!$, which will be improved on in Chapter 3 *Example 3.5* was introduced in Chapter 1.

[5]In honor of the distinguished mathematician Eric Temple Bell (1883–1960) author of *The Development of Mathematics*, 1945 [Bell 1945].

In his 1930 publication *Methodus Differentialis Sive Tractatus Summations et Interpolations*, Stirling introduced the following two indexed sequences of numbers:

- $\{s_{n,k} : 0 \leq k \leq n < \infty\}$ Stirling numbers of **the first kind**,
- $\{S_{n,k} : 0 \leq k \leq n < \infty\}$ Stirling numbers of **the second kind**

which convert between powers $\{x^n : 0 \leq n < \infty\}$ and the $(x)_k$ the Pochhammer symbol or *falling factorial* [see Chapter 1, equation (1.3*a*)] $\{(x)_n: 0 \leq n < \infty\}$ defined for real x and integer k by

$$(x)_k = \begin{cases} 1 & \text{if } k = 0 \\ x \times (x-1) \cdots \times (x-k+1) & \text{if } k \geq 1 \\ 0 & \text{if } k > n \end{cases} \tag{2.15a}$$

$$x^n = \sum_{k=0}^{n} S_{n,k}(x)_k \tag{2.15b}$$

$$(x)_n = \sum_{k=0}^{n} s_{n,k}(x)_k \tag{2.15c}$$

Gottfried Wilhelm Leibnitz (1646–1716), who along with Sir Isaac Newton (1643–1727), is regarded as the inventor of calculus, was convinced that good mathematical notation was the key to progress. D. Knuth noted [in 1992] Marx's observation [1962] who pointed out the similarity between formulas for the Stirling numbers the binomial coefficients $\left\{\binom{n}{k} : 0 \leq k \leq n < \infty\right\}$. Marx suggested the notations

$$s_{n,k} \equiv \begin{bmatrix} n \\ k \end{bmatrix} \quad S_{n,k} \equiv \begin{Bmatrix} n \\ k \end{Bmatrix} \tag{2.16}$$

The $\{S_{n,k}\}$ will arise in Chapter 10. Whereas $S_{n,k}$ was defined by equation (2.15*b*), these numbers have an important combinatorial significance; $\binom{n}{k}$ is the number of subsets of $\mathcal{Z}_n$ of size k and the Stirling number $S_{n,k}$ is the number of partitions of $\mathcal{Z}_n$ into k (nonempty) sets.

The most easily accessible tables of $\{S_{n,k} : 0 \leq k \leq n \leq N)\}$ for $N = 25$ are found in [Abramowitz and Stegun 1972, p. 835]; more extensive, but less readily accessible tables for $N = 100$ appear in a technical report [Andrews 1965] supported by the Air Force Office of Scientific Research. Copies in 1965 might be purchased for $2 from the University of Illinois (Urbana). Now available from the *Defense Technical Information Center* (DTIC), the price for this text has soared to $69. Furthermore, some of the values of $S_{n,k}$, for example, for $n = 95(1)100$, are missing because of the 108-digit precision of the program.

We begin with

Theorem 2.3

$$S_{n,k} = \frac{1}{k!} \sum_{s=0}^{k} \binom{k}{s} (-1)^s (k-s)^n \tag{2.17}$$

Solution. Suppose n balls are placed into k distinguishable (or ordered urns)[6], where ordering mean the urns are labeled $0, 1, \cdots, k-1$ so that

$$\underbrace{\{1,2,4\}}_{\text{Urn \#1}}\ \underbrace{\{0,3\}}_{\text{Urn \#2}} \qquad \underbrace{\{0,3\}}_{\text{Urn \#1}}\ \underbrace{\{1,2,4\}}_{\text{Urn \#2}}$$

are distinct placements of five balls into two urns. Let $T_{n,k}$ count the distinct number of placements of n balls into k ordered urns and

$$T_k(z) = \sum_n \frac{1}{n!} z^n T_{n,k} \tag{2.18a}$$

If

$$E(z) \equiv e^z - 1 = z + \frac{1}{2!}z^2 + \frac{1}{3!}z^3 + \cdots \tag{2.18b}$$

then

$$E^k(z) = \sum_{\substack{\ell_0,\ell_1,\cdots,\ell_{k-1} \\ \ell_j \geq 1, 0 \leq j < k \\ \ell_0+\ell_1+\cdots+\ell_{k-1}=n}} z^n \prod_{j=0}^{k-1} \frac{1}{\ell_j!} z^{\ell_j} \tag{2.18c}$$

The multinomial coefficient (see Chapter 1, §1.4) implies that $\frac{T_{n,k}}{n!}$ is the coefficient of $\frac{z^n}{n!}$ in $E^k(z)$.

Using the binomial theorem

$$E^k(z) = \sum_{i=0}^{k} \binom{k}{i} (-1)^i e^{(k-i)z} \tag{2.18d}$$

yielding

$$T_{n,k} = \sum_{i=0}^{k} \binom{k}{i} (-1)^i (k-i)^n \tag{2.18e}$$

Finally, because there are $k!$ ways to undistinguish the urns

$$S_{n,k} = \frac{1}{k!} \sum_{i=0}^{k} \binom{k}{i} (-1)^i (k-i)^n \tag{2.18f}$$

completing the proof. ■

The computation of the $\{S_{n,k}\}$ can used [equation (2.17)], although it is easier to make use of

[6]No combinatorialist or probabilist can exist without urns. \ 'arn *noun*: A vessel that typically has the form of a vase on a pedestal and is often used to hold the ashes of the dead. Of course, it could hold balls!

Theorem 2.4. The Stirling numbers of the second kind satisfy the triangular recursion

$$S_{n+1,k} = kS_{n,k} + S_{n,k-1} \tag{2.19}$$

Proof

(=) if the $(n+1)^{st}$ element augments one of the k sets constituting the partition of $\mathcal{Z}_n$ into k sets, a partition of $\mathcal{Z}_{n+1}$ into k sets, each of cardinality ≥2 results;

(+) if the $(n+1)^{st}$ element remains aloof, forming the *singleton* set $\{k\}$, a partition of $\mathcal{Z}_{n+1}$ into k sets results. ■

We conclude this section by drawing the reader's attention to the *One-Line Encyclopedia of Integer Sequences* at www.research.att.com/~njas/sequences/. If the sequence 1, 2035, 74316, 302995, 190575 is entered, the response is A060843: *Triangle of associated Stirling numbers of second kind.* A hard-copy published version is authored by Neil J. A. Sloane and Simon Plouffe. Incidentally, Professor Plouffe broke the world record in 1975 for memorizing digits of π by reciting 4096 digits; alas, he remained the champion until 1977. And, I sometimes forget a TEXcommand!! Perhaps a little *Aricept* might help.

2.9 ROUCHÉ'S THEOREM AND THE LAGRANGE'S INVERSION FORMULA

No discussion of generating functions would be complete without describing and illustrating the following:

- *Rouché's theorem* [Ahlfors 1953, p. 124][7] gives a set of conditions guaranteeing root(s) of an analytic function in some set.
- The *Lagrange inversion formula* [Whittaker and Watson 1952, p. 153][8] often can be used to give an explicit solution for the root.

We begin with a simplified, plain vanilla version of Rouché's theorem.

Theorem 2.5 (Rouché's theorem). If $f(z)$ and $g(z)$ are both analytic in $\mathcal{C}_r(0)$, the closed circle of radius R (about 0) and $|f(z)| < |g(z)/$ for $|z| = R$, then $f(z)$ has exactly as many zeros as $g(z)$ in $\mathcal{C}_r(0)$, the open circle of radius R (about 0).

[7] Rouché's theorem is a consequence of the argument principle in complex function theory. It was published by the French mathematician Eugéne Rouché (1832–1910) in volume 39 of the *Journal of École Polytechique* in 1862 [Rouché 1862].

[8] Joseph-Louis Lagrange (1736–1813) nominally a French mathematician is also claimed by to be Italian in Enciclopedia Italiana XX (Rome, 1933) [Lagrange 1933]. He published his inversion formula in *Mmoires de l'Acadmie Royale des Sciences et Belles-Lettres de Berlin*, **24**, [Lagrange 1770]. It is claimed that Lagrange wrote his papers without requiring any subsequent corrections.

Example 2.6. Suppose $f(z) = \sum_{n=0}^{\infty} p_n z^n$ is a probability generating function (§2.2)

H1. $p_n \geq 0$ $(n \geq 0)$ and $1 = \sum_{n=0}^{\infty} p_n = 1$
H2. $\mu \equiv \sum_{n=0}^{\infty} n p_n < 1.$

The equation $z^m - f(z) = 0$ certainly has a zero at $z = 1$. Setting $g(z) = z^m$, Rouché's theorem implies that $f(z)$ has m zeros (counting multiplicity) in $\mathcal{C}_1(0)$.

Proof. If $R = 1 + \varepsilon > 1$, the triangle inequality[9] implies a constant $c > 0$ exists for which

$$|f(1+\varepsilon)| = \sum_{n=0}^{\infty} p_n(1+\varepsilon)^n = 1 + \mu\varepsilon + c\varepsilon^2 < \left|(1+\varepsilon)^m\right| \quad \text{as } \varepsilon \downarrow 0$$

We conclude $z^m - f(z) = 0$ has m roots (counting multiplicity) in $\mathcal{C}_{1+\varepsilon}(0)$ by Rouché's theorem. Because ε can be made arbitrarily small, $z^m - f(z) = 0$ has m roots (counting multiplicity) in $\mathcal{C}_1(0)$. ■

Theorem 2.7 (Lagrange inversion formula theorem (1770))**.** Let $f(z)$ is analytic in $\mathcal{C}_R(0)$). Suppose w is such that the inequality $w|f(z)| < |z|$ is satisfied on $|z| = R$. Then, the equation $\zeta = w\, f(\zeta)$ with $\zeta \equiv \zeta(w)$ has a root in $\mathcal{C}_R(0)$, is analytic in w and may be expressed as a power series

$$\zeta = \sum_{n=1}^{\infty} \frac{w^n}{n!} \left\{ \frac{d^{n-1}}{dz^{n-1}} f^n(z) \right\} \Bigg|_{z=0} \tag{2.20}$$

converging for $w \in \bar{\mathcal{C}}_R(0)$.

Remark 2.1. If

- The conditions $f(z)$ is analytic about $z = 0$ and the bounds $|f(z)| < |z|$ on $|z| = R$ are replaced by
- the conditions $f(z)$ is analytic about $z = a$ and the bounds $|f(z) - a| < |z - a|$ on $|z - a| = R$.

then the solution $\zeta \equiv \zeta(w)$ of the equation $\zeta = w\, f(\zeta)$ is analytic in w and may be expressed as a power series

$$\zeta = \sum_{n=1}^{\infty} \frac{w^n}{n!} \frac{d^{n-1}}{dz^{n-1}} f^n(z) \Bigg|_{z=a} \tag{2.21}$$

[9]The version of the triangle inequality we use is the generalization of

$$\sum_{n=0}^{N-1} |(a_n + b_n)| \leq \sum_{n=0}^{N-1} |a_n| + \sum_{n=0}^{N-1} |b_n|$$

for vectors $\underline{a} = (a_0, a_1, \cdots, a_{N-1})$ and $\underline{b} = (b_0, b_1, \cdots, b_{N-1})$ with equality, if and only if $\underline{a} = C\underline{b}$ for some constant C.

Example 2.7. The generating function $f(z) = e^{\lambda(z-1)} = e^{-\lambda} \sum_{n=0}^{\infty} \frac{(\lambda z)^n}{n!}$ satisfies the same hypotheses H1 and H2 as in *Example 2.6* with $\mu = \lambda$. If $m = 1$, Rouché's theorem asserts $z - we^{\lambda(z-1)} = 0$ has a unique root, say $\zeta \equiv \zeta(w)$, in $C_1(0)$ for each (complex) w satisfying $|w| < 1$. We apply Lagrange's theorem, because

$$f(z) = e^{\lambda(z-1)} \quad f^n(z) = e^{n\lambda(z-1)}$$
$$\left\{\frac{d^{n-1}}{dz^{n-1}} f^n(z)\right\}\bigg|_{z=0} = (n\lambda)^{n-1} e^{n\lambda(z-1)}$$

we have the expansion

$$\zeta(w) = \sum_{n=1}^{\infty} \frac{w^n}{n!} \left\{\frac{d^{n-1}}{dz^{n-1}} f^n(z)\right\}\bigg|_{z=0} = w \sum_{n=0}^{\infty} \frac{1}{n!} \left((n+1)\lambda e^{-\lambda} w\right)^n \tag{2.22}$$

REFERENCES

M. Abramowitz and I. A. Stegun, (Editors), *Handbook of Mathematical Functions with Formulas, Graphs, and Mathematical Tables*, Dover Publications (New York), 1972.

L. V. Ahlfors, *Complex Analysis*, McGraw-Hill (New York), 1953.

Jean-Paul Allouche, "Note on the Cyclic Towers of Hanoi", *Theoretical Computer Science*, **123**, p. 37, 1994.

Alex M. Andrew, "Table of The Stirling, Numbers of the Second Kind $S(N, K)$ for (N, K) up to 100 and for Values of $S(N, K)$ Equal or Smaller Than $10^{109} - 1$", Air Force Office of Scientific Research, 1965.

E. T. Bell, *The Development of Mathematics*, McGraw Hill (New York), 1945.

Martin Gardner, "Mathematical Games: About the Remarkable Similarity between the Icosian Game and the Tower of Hanoi", *Scientific American*, **196**, #5, pp. 150–156, May 1957.

Donald Knuth, "Two Notes on Notation", *The American Mathematical Monthly*, **99**, #5, pp. 403–422, May 1992.

J. L. Lagrange, *Enciclopedia Italiana* XX (Rome, Italy), 1933.

J. L. Lagrange, *Memoires de l'Academie Royale des Sciences et Belle-Lettres de Berlin*, **24**, 1770.

Imanuel Marx, "Transformation of Series by a Variant of Stirling Numbers", *The American Mathematical Monthly*, **69**, pp. 530–532, 1962.

J. Riordan, *Combinatorial Identities*, John Wiley & Sons (New York), 1968.

J. Riordan, *An Introduction to Combinatorial Analysis*, John Wiley & Sons (New York), 1980.

E. Roúche, *Journal of Ecole Polytechnique*, 1862.

G.-C. Rota, D. Kahaner, and A. Odlyzko. "On the Foundations of Combinatorial Theory VII. Finite Operator Calculus", *Journal Mathematical Analysis Applications*, **42**, pp. 684–760, 1973.

Steven Roman and Gian-Carlo Rota, "The Umbral Calculus", *Advances in Mathematics*, **27**, pp. 95–188, 1978.

Paul K. Stockmeyer, "The Tower of Hanoi: A Bibliography", download from stockmeyer@cs.wm.edu, Version 2.2, 2005.

E. T. Whittaker and G. N. Watson, *A Course of Modern Analysis*, Cambridge University Press (Cambridge, UK), 1952.

CHAPTER 3

Asymptotic Analysis

3.1 GROWTH NOTATION FOR SEQUENCES

The first of the symbols for growth notation O, o, Ω, ω, ~ were introduced in 1894 by number theorist Paul Bachmann [Bachman 1994]. It was popularized in the work of another German number theorist Edmund Landau (1877–1938)) and the symbol O is often referred to as the Landau symbol.

The notation $a_n = O(b_n)$, read "a_n is big-Oh of b_n," specifies that b_n ultimately provides a uniform upper bound on a_n; that is, $C > 0$ and a positive integer N exist such that

$$|a_n| \leq C|b_n| \quad \text{as } n \geq N \tag{3.1}$$

Examples of Sequence Growth

Polynomial growth:	$b_n = n^k$	$k > 0$	as $n \to \infty$
Exponential growth:	$b_n = e^{\alpha n}$	$\alpha > 0$	as $n \to \infty$
Super-exponential growth:	$b_n = n^{\alpha n}$	$\alpha > 0$	as $n \to \infty$
Logarithmic growth:	$b_n = \log \alpha n$	$\alpha > 0$	as $n \to \infty$
Log-log growth:	$b_n = \log \log \alpha n$	$\alpha > 0$	as $n \to \infty$

The notation $a_n = \Omega(b_n)$, read "a_n is big-Omega of b_n, specifies that b_n ultimately provides a corresponding uniform lower bound on a_n; that is, a $C > 0$ and a positive integer N exist such that

$$|a_n| \geq C|b_n| \quad \text{as } n \geq N \tag{3.2}$$

Examples 3.1

a) $a_n = 3n^2 + 4n - 1 = \begin{cases} O(n^2) \\ \Omega(1) \end{cases} = \begin{cases} O(1+\varepsilon)^n & \text{for every } \varepsilon > 0 \\ \Omega(\log n) \end{cases}$ as $n \to \infty$.

b) $a_n = e^{-An} = O(n^{-j})$ for every $j > 0$ as $n \to \infty$.

c) $a_n = (1+\varepsilon)^n = \begin{cases} O(n^n) & \text{for every } j > 0 \\ \Omega(n^j) & \text{for every } \varepsilon > 0 \end{cases}$ as $n \to \infty$.

Hashing in Computer Science: Fifty Years of Slicing and Dicing, by Alan G. Konheim

d) $a_n = \log n = \begin{cases} O(n^j) & \text{for every } j > 0 \\ \Omega(\log\log n) & \end{cases} \quad \text{as } n \to \infty.$

e) $a_n = 2^{\alpha\sqrt{\log n \log\log n}} = \begin{cases} O(n^\alpha) \\ \Omega((\log n)^\alpha) \end{cases} \quad \text{for every real } \alpha > 0 \quad \text{as } n \to \infty.$

We write $a_n = o(b_n)$, read "a_n is little-oh of b_n", if the sequence $\{a_n\}$ ultimately grows much slower than the sequence $\{b_n\}$; that is, for every real $\varepsilon > 0$, a positive integer N exists, such that

$$|a_n| \le \varepsilon |b_n| \quad \text{as } n \ge N \tag{3.3a}$$

Analagous to the o notation, we write $a_n = \omega(b_n)$, read "a_n is little-omega of b_n," reverses the roles of a_n and b_n; that is, for every $C > 0$, a positive integer N exists, such that

$$|a_n| \ge \varepsilon |b_n| \quad \text{as } n \ge N \tag{3.3b}$$

Examples 3.2

a) $a_n = 3n^2 + 4n - 1 = \begin{cases} o(n^3) \\ \omega(n^4) \end{cases} \quad \text{as } n \to \infty$

b) $a_n = 3n^2 + 4n - 1 = \begin{cases} o((1+\varepsilon)^n) & \text{for every } \varepsilon > 0 \\ \omega(\log n) & \end{cases} \quad \text{as } n \to \infty$

c) $a_n = \log n = \begin{cases} o(n^j) & \text{for every } j > 0 \\ \omega(1) & \end{cases} \quad \text{as } n \to \infty$

Finally, the notation $a_n \sim b_n$, read "a_n and b_n are asymptotic to one another," occurs when both sequences ultimately grow at the same rate; that is, for every $\varepsilon > 0$, a positive integer N exists, such that

$$\left|1 - \frac{a_n}{b_n}\right| \le \varepsilon \quad \text{as } n \ge N \tag{3.4}$$

The O, o, Ω, and ω growth notations may also be used to compare functions; for example, the polynomial $p(x) = 1 - 1\,000\,000\,000x + x^2$ is $\begin{cases} O(x^2) & \text{as } x \to \infty \\ O(1) & \text{as } x \to 0 \end{cases}$.

Example 3.3. The number of dearangements D_n of the integers $0, 1, \cdots, n-1$ is asymptotic as $n \to \infty$ to $\frac{n!}{e}$; equivalently, D_n is asymptotic to the nearest integer to $\frac{n!}{e}$.

Proof. The recurrence for $\{D_n\}$

$$D_n = (n-1)[D_{n-1} + D_{n-2}] \quad 2 \le n < \infty \tag{3.5a}$$

was derived in Chapter 2 [equation (2.3)]. Mathematical induction can be used to prove that D_n is given by

$$D_n = n!\sum_{j=0}^{n}(-1)^j\frac{1}{j!} \quad 1 \le n < \infty \tag{3.5b}$$

which is exactly the assertion. ■

3.2 ASYMPTOTIC SEQUENCES AND EXPANSIONS

N. G. de Bruijn [de Bruijn 1958] defines a sequence of functions $\phi_0(x)$, $\phi_1(x)$, $\cdots$ to be an **asymptotic sequence** at $x = x_0 \in \mathcal{X}$ provided $\phi_n(x) = o(\phi_{n-1}(x))$ for $1 \le n < \infty$ as $x \to x_0 \in \mathcal{X}$.

Examples 3.4

a) $\phi_n(x) = (x - x_0)^n$ at x_0
b) $\phi_n(x) = x^{-n}$ at ∞

An asymptotic sequence $\{\phi_n(x)\}$ determines an asymptotic expansion or series for $f(x)$ for $x \in X$ if constants $c_0, c_1, \cdots, c_{n-1}, \cdots$ exist such that

$$R_n \equiv f(x) - [c_0\phi_0(x) + c_1\phi_1(x) + \cdots + c_{n-1}\phi_{n-1}(x)] = O(\phi_n(x)) \quad \begin{cases} x \to x_0 \in \mathcal{X} \\ 1 \le n < \infty \end{cases} \tag{3.6a}$$

We write

$$f(x) \sim c_0\phi_0(x) + c_1\phi_1(x) + \cdots + c_{n-1}\phi_{n-1}(x) \quad \begin{cases} x \to x_0 \in \mathcal{X} \\ 1 \le n < \infty \end{cases} \tag{3.6b}$$

As $n \to \infty$, the asymptotic series at $x \in X$ may either converge to $f(x)$, diverge, or converge to a value different that $f(x)$.

When $\phi_n(x) = x^{-n}$ for $0 \le n < \infty$, an asymptotic power series results.

The coefficients $\{c_n\}$ in an asymptotic expansion of $f(x)$ for a specific asymptotic sequence $\{\phi_n(x)\}$ are uniquely determined; indeed, if for $x = x_0$ and $n = N$, the coefficients $c_0, c_1, \cdots c_{N-1}$ are such that equation (3.6*a*) is satisfied, then c_N defined by

$$c_N = \lim_{x \to x_0} \left| \frac{f(x) - \sum_{n=0}^{N-1} a_n\phi_n(x)}{\phi_N(x)} \right| \tag{3.7}$$

satisfies equation (3.6*a*) for $x = x_0$ and $n = N + 1$.

Different asymptotic sequences $\{\phi_n(x)\}$ and $\{\phi_n'(x)\}$ may give rise to an asymptotic expansion for the same function. Erdélyi [Erdelyi 1956, p. 14] gives the example

$$\frac{1}{1+x} \sim \begin{cases} \sum_n (-1)^{n-1} x^{-n} \\ \sum_n (x-1)x^{-2n} \\ \sum_n (-1)^{n-1}(x^2 - x + 1)x^{-2n} \end{cases} \quad x \to \infty \tag{3.8}$$

of different asymptotic expansions for $\dfrac{1}{1+x}$, which are each convergent for $|x| > 1$.

Remark 3.1. The multiplication or division of asymptotic series

$$f(x) = \sum_n a_n \phi_n(x), \quad g(x) = \sum_n b_n \phi_n(x) \nRightarrow f(x)g(x) = \sum_n \left(\sum_k a_k b_{n-k} \right) \phi_n(x) \tag{3.9}$$

as in equation (3.9) do not generally produce asymptotic series. However, in the special case $\phi_n(x) = x^{-n}$ $(0 \le n < \infty)$ the following occur:

- The formal term-by-term computation of the coefficients of the *product* $f(x)g(x)$ as in equation (3.9) yields an asymptotic power series.
- If $b_0 \neq 0$, the formal term-by-term computation of the coefficients of the *quotient* $\dfrac{f(x)}{g(x)}$ yields an asymptotic power series.

Examples 3.5

a) The gamma function $\Gamma(x) = \int_0^\infty t^{x-1} e^{-t} dt$ $(0 < x < \infty)$ extends the domain of the factorial $n!$ from integers to a subset of the real numbers. It is proved in [Whittaker and Watson 1952, pp. 251–253] that $\Gamma(x) = (x-1)!$ for $x = 1, 2, \cdots$.

The first five terms[1] in the asymptotic expansion of $\Gamma(x)$ as $x \to \infty$ [Whittaker and Watson 1952, pp. 251–253] are

$$\Gamma(x) \sim e^{-x} x^{x+\frac{1}{2}} \sqrt{2\pi}\, \Gamma^*(x) \tag{3.10a}$$

[1]For those with an unbounded lust for details, the following table of γ_n for $n = 1(1)20$ is from [Wrench 1968, p. 619].

n	γ_n	n	γ_n
1	$-\frac{1}{12}$	2	$\frac{1}{288}$
3	$\frac{139}{51\,480}$	4	$-\frac{571}{2\,488\,320}$
5	$-\frac{163\,879}{209\,018\,880}$	6	$\frac{5\,246\,819}{75\,246\,796\,800}$
7	$\frac{534\,703\,531}{909\,961\,561\,600}$	8	$-\frac{4\,483\,131\,259}{86\,684\,309\,913\,600}$
9	$\frac{432\,261\,921\,612\,371}{514\,904\,800\,886\,784\,000}$	10	$\frac{6\,232\,523\,20o2\,521\,089}{86\,504\,006\,548\,979\,712\,000}$
11	$-\frac{25\,834\,629\,665\,134\,204\,969}{12\,494\,625\,021\,640\,835-72\,000}$	12	$-\frac{1\,579\,029\,138\,854\,919\,086\,429}{9\,716\,130\,015\,581\,401\,251\,840\,000}$
13	$\frac{746\,590\,869\,962\,651\,602\,203\,151}{116\,593\,560\,186\,976\,815\,022\,080\,000}$	14	$\frac{1\,511\,513\,601\,028\,097\,903\,631\,961}{2\,798\,245\,444\,487\,443\,560\,529\,920\,000}$
15	$-\frac{8\,849\,272\,268\,392\,873\,147\,705\,987\,190\,261}{299\,692\,087\,104\,605\,205\,332\,754\,432\,000\,000}$	16	$-\frac{142\,801\,712\,490\,607\,530\,608\,130\,701\,097\,261}{57\,540\,880\,724\,084\,199\,423\,888\,850\,944\,000\,000}$
17	$\frac{2\,355\,444\,393\,109\,967\,510\,921\,431\,436\,000\,087\,153}{13\,119\,320\,805\,091\,197\,468\,646\,658\,015\,232\,000\,000}$	18	$\frac{2\,346\,608\,607\,351\,903\,737\,647\,919\,577\,082\,115\,121\,863}{1\,870\,290\,373\,973\,801\,111\,130\,267\,566\,651\,473\,920\,000\,000}$
19	$-\frac{2\,603\,072\,187\,220\,373\,277\,160\,999\,431\,416\,562\,396\,331\,667}{1\,870\,290\,373\,973\,801\,111\,130\,267\,566\,651\,473\,920\,000}$	20	$-\frac{73\,239\,727\,436\,811\,935\,976\,471\,475\,430\,268\,695\,630\,993}{628\,417\,565\,655\,197\,173\,339\,769\,902\,394\,895\,237\,120\,000\,000}$

$$\Gamma^*(x) \sim \sum_{n=0}^{\infty}(-1)^n \gamma_n x^{-n} = \left\{1+\frac{1}{12x}+\frac{1}{288x^2}-\frac{139}{5184x^3}-\frac{571}{2488320x^4}+\cdots\right\} \quad (3.10b)$$

$$\frac{1}{\Gamma^*(x)} \sim \sum_{n=0}^{\infty}\gamma_n x^{-n} \quad (3.10c)$$

$$0=\sum_{j=0}^{n}(-1)^j \gamma_j \gamma_{n-j} \quad (3.10d)$$

Equations (3.10*a* and 3.10*b*) is a refinement of Stirling's formula (Chapter 1, equation (1.4*a*)).

Although there is no closed form expression for the $\{\gamma_n\}$, a recurrence formula is given in [Wrench 1968, p. 618].

b) The (Gaussian) error function Erf(x) and complementary (Gaussian) error function Erfc(x), with domain $-\infty < x < \infty$, are defined by

$$\mathrm{Erf}(x)\frac{2}{\sqrt{\pi}}\int_0^x e^{-t^2}dt \quad \mathrm{Erfc}(x)=\frac{2}{\sqrt{\pi}}\int_x^{\infty} e^{-t^2}dt \quad (3.11a)$$

Because $\mathrm{Erf}(x) \to \frac{\sqrt{\pi}}{2}$ as $x \to \infty$, it follows that

$$1=\mathrm{Erf}(x)+\mathrm{Erfc}(x) \quad (3.11b)$$

The asymptotic expansion of Erfc(x) is

$$\sqrt{\pi}xe^{x^2}\mathrm{Erfc}(x) \sim 1+\sum_{n=1}^{\infty}(-1)^n\frac{(2n)!}{n!(2x)^{2n}} \quad (3.11c)$$

How does one obtain the asymptotic expansions in equations (3.10*a* through 3.10*d*) and (3.11*c*)? For the gamma function, the complicated process is described in great detail in [Whittaker and Watson, pp. 251–253]. In some special cases, and the gamma function is not one of these, when a function $F(x)$ is defined by an integral with x as a parameter, the asymptotic expansion may be obtained using integration by parts. For example, to obtain equation (3.11*c*), start with the defining integral $\int_x^{\infty} e^{-t^2}dt$ and setting

$$u=t^{-1} \quad dv=te^{-t^2}dt \Rightarrow dv=t^{-2}dt \quad v=-\frac{1}{2}e^{-t^2}$$

$$\mathrm{Erfc}(x)=\frac{2}{\sqrt{\pi}}\int_x^{\infty} e^{-t^2}dt \Rightarrow \frac{2}{\sqrt{\pi}}\left[\frac{1}{2x}e^{-x^2}-\frac{1}{2}\int_x^{\infty} t^{-2}e^{-t^2}dt\right]$$

Repeated integration by parts formally proves the expansion. Of course, this does *not* prove that equation (3.11*c*) is the asymptotic expansion. It remains to verify the size of the error when the series is truncated with N terms.

Example 3.6. [Whittaker and Watson 1952, p. 150] shows that integration by parts applied to the function

$$f(x) = \int_x^\infty t^{-1} e^{x-t} dt \tag{3.12a}$$

gives the series

$$f(x) = \frac{1}{x} - \frac{1}{x^2} + \frac{2!}{x^3} - \cdots + \frac{(-1)^{n-1}(n-1!}{x^n} + (-1)^n n! \int_y^\infty \frac{e^{x-t}}{t^{n+1}} dt \tag{3.12b}$$

and yields a series for $x > 0$, which is both divergent and an asymptotic expansion.

3.3 SADDLE POINTS

In introductory calculus courses, we learn that the extremal points of a smooth function $y = f(x)$ including the local minima/maxima of the function defined on the interval $a \leq x \leq b$ can occur at one of the following locations:

i) One of the endpoints $x = a$ or $x = b$ of the interval.
ii) An interior point $a < x^* < b$ of the interval; in this case, x^* is a stationary point of $f(x)$ and the tangent to the curve $y = f(x)$ is horizontal so that $f'(x^*) = 0$.

Of course, there *may* be points $a < x^\# < b$ of $f(x)$ that are stationary points of $f(x)$ without being *either* local minima or maxima. The point $x = x^\#$ is an *inflection point* of $f(x)$ if $f'(x^\#) = f''(x^\#) = 0$ and there is a change of sign of $f''(x)$ at $x = x^\#$.

Example 3.7. The curve $y = x^3$ with $-1 < x < 1$ shown next in Figure 3.1 has no maxima or minima; it has an inflection point at $x = 0$.

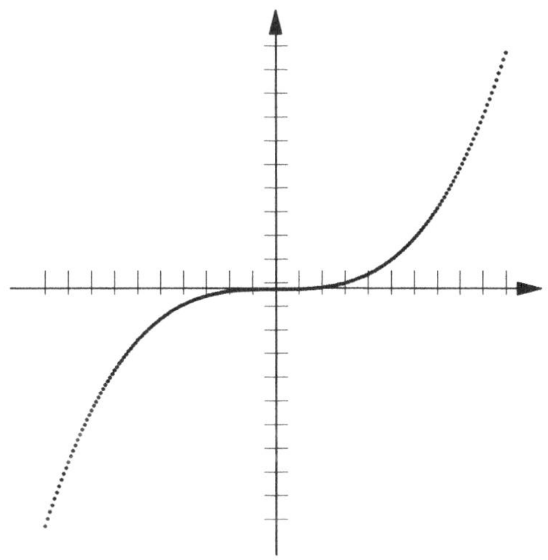

Figure 3.1. The curve $y = x^3$ on the interval $[-1, 1]$.

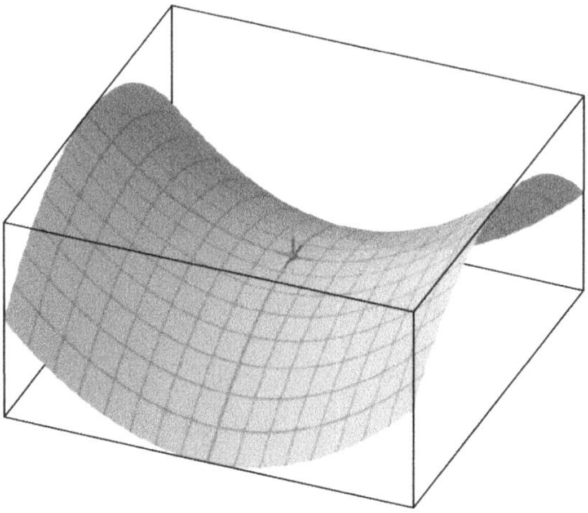

Figure 3.2. Saddle point[2] at $(x, y) = (0, 0)$ of $z = f(x, y) = x^2 - y^2$.

A *saddle point* is the analog of an inflection points in three (or more) dimensions for *smooth* functions $z = f(x, y)$ of two (or more variables); they are points whose neighborhood contains points that lie on different sides of the tangent.

Example 3.7. Figure 3.2 portrays the function $z = f(x, y) = x^2 - y^2$, which has a saddle point at $(x, y) = (0, 0)$.

A saddle point of a function f with continuous second derivatives may be defined in terms of its Hessian matrix $H[f]$

$$H[f] = \begin{pmatrix} \frac{\partial^2 f}{\partial x^2} & \frac{\partial^2 f}{\partial x \partial y} \\ \frac{\partial^2 f}{\partial x \partial y} & \frac{\partial^2 f}{\partial y^2} \end{pmatrix} \tag{3.13a}$$

The point (x^*, y^*) is a saddle point if and only if *neither* of the matrices $H[f]$ or $-H[f]$ are positive-definite at $(x, y) = (x^*, y^*)$. Recall that $H[f]$ is positive-definite at $(x, y) = (x^*, y^*)$ if the quadratic form

$$Q(x, y) = (x, y) \begin{pmatrix} \frac{\partial^2 f}{\partial x^2} & \frac{\partial^2 f}{\partial x \partial y} \\ \frac{\partial^2 f}{\partial x \partial y} & \frac{\partial^2 f}{\partial y^2} \end{pmatrix} \begin{pmatrix} x \\ y \end{pmatrix} \tag{3.13b}$$

satisfies

$$0 \leq Q(x, y) = x^2 \frac{\partial^2 f}{\partial x^2} + y^2 \frac{\partial^2 f}{\partial x^2} + xy \left[\frac{\partial^2 f}{\partial x \partial y} + \frac{\partial^2 f}{\partial x \partial y} \right] \tag{3.13c}$$

[2]Courtesy of Keiff/Wikopedia.com.

in a neighborhood of (x^*, y^*) with $Q(x, y) = 0$ only at $(x, y) = (x^*, y^*)$. Thus, if (x^*, y^*) is a saddle point, the quadratic form $Q(x, y)$ changes sign in a neighborhood of $(x, y) = (x^*, y^*)$.

3.4 LAPLACE'S METHOD

One biography of Pierre-Simon Laplace (1749–1827) attributes the origin of Laplace's method (*aka* steepest descent, saddle point method) in Laplace's "Mémoire sur les probabilités" (1780), although another recent historical paper by S. Petrova and A. Solov'ev claims that it also appears in Cauchy's 1827 paper "Sur divers points danalyse." Laplace's method is an important technique for determining the asymptotic behavior of a function $F(x) = \int_a^b e^{-xg(t)} f(t)\,dt$, where x is large and $-\infty \le a < b \le \infty$. It is assumed that $f(t)$ and $g(t)$ are continuous and real valued, and $g(t)$ is smooth (differentiable). Suppose that $g(t)$ has a unique minimum at $t = t_0$ with $a < t_0 < b$, and $g(t)$ may be represented by $g(t) = g(t_0) + g'(t_0)(t - t_0) + \frac{1}{2} g''(t_0)(t - t_0)^2 + o\left((t - t_0)^2\right)$ for t in a neighborhood $\mathcal{N}$ of t_0. Writing

$$F(x) = \int_a^b e^{-xg(t)} f(t)\,dt = e^{xg(t_0)} f(t_0) \int_{\mathcal{N}} e^{-x\left(\frac{1}{2}g''(t_0)\right)} dt + e^{xg(t_0)} \int_{\bar{\mathcal{N}}} e^{-x(g(t)-g(t_0))} f(t)\,dt \tag{3.14a}$$

Because $g(t) > g(t_0)$ on $\bar{\mathcal{N}}$, it is to be expected that

$$\int_{\bar{\mathcal{N}}} e^{-x(g(t)-g(t_0))} f(t)\,dt \to 0 \quad x \uparrow \infty$$

Because t_0 is a minimum, $g'(t_0) = 0 < g''(t_0)$ and with some optimism and slight of hand we are led to

$$F(x) \sim e^{-xg(t_0)} f(t_0) \int_0^\infty e^{-x\left(\frac{1}{2}g''(t_0)\right)} dt = e^{-xg(t_0)} f(t_0) \sqrt{\frac{2\pi}{xg''(t_0)}} \tag{3.14b}$$

A discussion of the Laplace's method and the conditions that validate the asymptotic result in equation (3.14*b*) is given in [Erdelyi 1956, pp. 56–57].

Remarks 3.2

a) If t_0 is either a or b, the right-hand side of equation (3.14*b*) must be multiplied by $\frac{1}{2}$.
b) If *minimum* is replaced by *maximum*, then $g''(t_0) < 0$ and $\sqrt{\frac{2\pi}{xg''(t_0)}}$ is replaced by $\sqrt{\frac{2\pi}{-xg''(t_0)}}$

3.5 THE SADDLE POINT METHOD

Saddle point analysis is a natural extension of Laplace's method to integrals in which the path is a curve in the complex plane. The problem is to evaluate $F(z,n) \equiv \int_A^B e^{f(z)}dz$ where $f(z) = f(z, n)$ is a smooth function depending on a large parameter n.

If $z_0 \equiv z_0(n) \in (A, B)$ is a saddle point maximum of $f(z)$ on (A, B) (Figure 3.3), the saddle point principle argues that the largest contribution to $I(n) = \frac{1}{2\pi i}\int_A^B e^{f(z)}dz$, which occurs in a neighborhood of $z = z_0 \equiv z_0(n)$, and the integral can be evaluated by replacing $e^{f(z)}$ by $e^{f(z_0)}e^{\frac{1}{2}f''(z_0)(z-z_0)^2}$. This type of problem arises when the Cauchy integral theorem is applied to evaluate the behavior of the coefficient of z^n in the generating function of $F(z)$ as $n \to \infty$. Here

$$F(z) = \sum_{n=0}^{\infty} f_n z^n \quad f_n = \frac{1}{n!}\left\{\frac{d^n}{dz^n}F(z)\right\}_{z=0} \tag{3.15a}$$

$$f_n = \frac{1}{2\pi i}\oint_C \frac{F(w)}{w^{n+1}}d\zeta \tag{3.15b}$$

To prepare for the role Laplace's method will play, we define $f(z) \equiv f(z, n)$ by

$$e^{f(z)} = \frac{F(z)}{z^{n+1}} \Leftrightarrow f(z) = \log F(z) - (n+1)\log z \tag{3.15c}$$

and define $I(n) = \frac{f_n}{n!}$ so that

$$I(n) = \frac{1}{2\pi i}\oint_C e^{f(w)}dw \tag{3.15d}$$

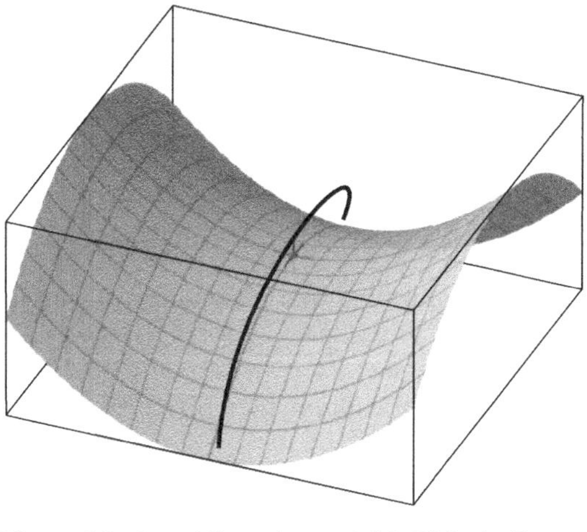

Figure 3.3. A saddle point path [A,B] including z_0.

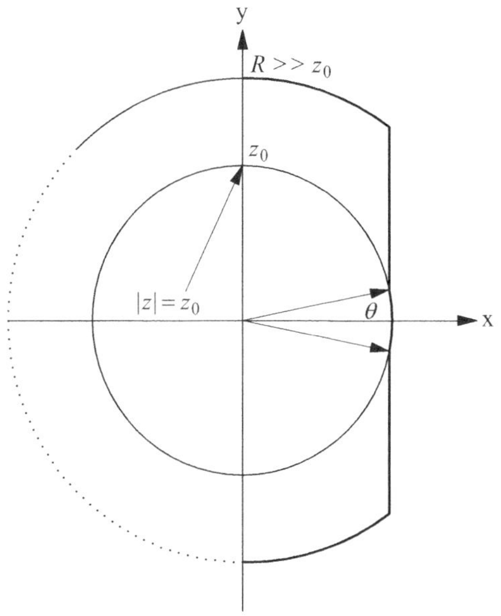

Figure 3.4. Modifying the path of integration.

We suppose that $f(z)$ has a unique real maximum, say at $z_0 \equiv x_0 \equiv x_0(n) > 0$ and that it is a saddle point $f''(x_0) < 0 = f'(x_0)$. Because $F(z)$ is analytic, the path $\mathcal{C}$ is the circle of radius 1 centered at $z = 0$. To apply the saddle point analysis, the path $\mathcal{C}$ is modified to the path of radius x_0 and then deformed $\mathcal{C} \to \mathcal{C}_1(R) \cup \mathcal{C}_2(R)$ to consist of two segments as indicated in Figure 3.4.

These two segments consist of the following:

$\mathcal{C}_1(R)$: the infinitesimal arc of the circle $|z| = x_0$ near x_0
(the limit as $\theta \downarrow 0$) joined by the vertical segment $(x, y) = (x_0, y)$ $(-R \le y \le R)$ to the
$\mathcal{C}_2(R)$: large arc of radius R.

By analyticity

$$I(n) = \frac{1}{2\pi i} \oint_{\mathcal{C}_1(R)} e^{f(w)} dw + \frac{1}{2\pi i} \oint_{\mathcal{C}_2(R)} e^{f(w)} dw \tag{3.15e}$$

The intention of the path deformation is to split the integral $\frac{1}{2\pi i} \oint_{\mathcal{C}} e^{f(w)} dw$ into two integrals; the first

$$\mathcal{C}_1(R): \quad \lim_{R \to \infty} \frac{1}{2\pi i} \oint_{\mathcal{C}_1} e^{f(w)} dw \sim \frac{1}{2\pi} \int_{-\infty}^{\infty} e^{f(x_0 + iy)} dy = \frac{1}{2\pi} \int_{-\infty}^{\infty} e^{f(x_0) + f'(x_0)(iy) - f''(x_0) y^2 + \cdots} dy$$

to which Laplace's method can be applied in the limit as $R \to \infty$, and the second, which will be negligible as $R \to \infty$.

$$\mathcal{C}_2(R):\quad \lim_{R\to\infty} \oint_{\mathcal{C}_2} e^{f(w)}dz = 0$$

Because $F(z) \downarrow 0$ as $z \to \infty$, we might expect that z_0 is the unique maximum of $F(z)$ on $\mathcal{C}_i(R)$ $(i = 1, 2)$ as $R \to \infty$. If this is the case, Laplace's method might then be applies to evaluate

$$I(n) = \frac{1}{2\pi} \int_{-\infty}^{\infty} e^{f(x_0)+f'(x_0)(iy)-\frac{1}{2}f''(x_0)y^2+\cdots} dy \sim \frac{1}{2\pi} e^{f(x_0)} \int_{-\infty}^{\infty} e^{-\frac{1}{2}f''(x_0)y^2} dy \quad n \to \infty \tag{3.15f}$$

In their extensive paper on the analysis of algorithms, Flajolet and Sedgewick [Flajolet and Sedgewick 1994] formulate the following.

Theorem 3.1 (saddle point method). Suppose $F(z) \equiv F(z, n)$ is a smooth function depending on a large parameter n, and $F(z, n) \equiv e^{f(z,n)} \Leftrightarrow f(z, n) = \log F(z, n)$, where $f(z) \equiv f(z, n)$ has a saddle point maximum at $z_0 \in \mathcal{C}$.

The asymptotic behavior of $I(n) = \frac{1}{2\pi} \int_{\mathcal{C}} e^{f(z)}dz$ and is determined by writing

$$I(n) = \frac{1}{2\pi} \int_{\mathcal{C}_1(n)} e^{f(z)}dz + \int_{\mathcal{C}_2(n)} e^{f(z)}dz \tag{3.16a}$$

such that

$$\begin{aligned}\lim_{n\to\infty} \frac{1}{2\pi} \int_{\mathcal{C}_1(n)} e^{f(z)}dz &\sim \frac{1}{2\pi} e^{f(z_0)} \int_{\mathcal{C}^*} e^{\frac{1}{2}f''(z_0)(z-z_0)^2} dz \\ &\to \frac{1}{2\pi} e^{f(z_0)} \int_{-\infty}^{\infty} e^{\frac{1}{2}f''(z_0)(z-z_0)^2} dz = \frac{e^{f(z_0)}}{\sqrt{2\pi|f''(z_0)|}}\end{aligned} \tag{3.16b}$$

$$\lim_{n\to\infty} \frac{1}{2\pi} \int_{\mathcal{C}_2(n)} e^{f(z)}dz = 0 \tag{3.16c}$$

The devil is in the details; to apply the saddle point method, it is necessary to show that the path $\mathcal{C}$ can be modified so that the following will occur:

- Laplace's method (or a modification of it) can be used to show that the central part $(z \approx z_0)$ of $\int_{\mathcal{C}_1(n)} e^{f(z)}dz$ provides the major contribution to $\int_{\mathcal{C}_1(n)} e^{f(z)}dz$ leading to equation (3.16*b*).
- The contribution of the second term, at least as $n \to \infty$, is insignificant.

As we have noted already, when $F(z)$ is analytic in z except for isolated poles, the Cauchy integral theorem can be used to modify the path as long as the poles remain within $\mathcal{C}$ (or outside of $\mathcal{C}$). But there is still work to be done, namely, to show that $\mathcal{C}_2$ can be chosen so that $\int_{\mathcal{C}_2(n)} \cdots$ is negligible. In many cases, this is far from routine or trivial. For example, the papers by Blieck and Yang [Bleick and Wang 1974] and

Temme [Temme 1993], which give an asymptotic analysis by the saddle point method of the Stirling numbers of the second kind (see 3.7), is complicated and represent a triumph of mathematical skills.

3.6 WHEN WILL THE SADDLE POINT METHOD WORK?

Conditions that guarantee the success of the saddle point method when applied to the determination of the coefficients of an analytic function $F(z)$ were discovered by Hayman [Hayman 1956]. We follow the presentation of A. Odlyzko [Odlyzko 1995] when the saddle point method is applied to determine the asymptotics of the coefficients of an analytic function.

$$F(z) = \sum_{n=0}^{\infty} f_n z^n \quad |z| < R \tag{3.17a}$$

$$f_n = \frac{1}{2\pi i} \oint_{|z|=R} \frac{F(z)}{z^{n+1}} dz \tag{3.17b}$$

F is *admissible* if

i) it is analytic in z with $|z| < R$ for $0 < R \leq \infty$.
ii) $F(z)$ is real for z real and $|z| < R$.
iii) There exists an R_0 with $R_0 < R$ such that $\max_{|z|=R} |F(z)| = F(r)$ for $R_0 < r < R$[3].
iv) If $a(r)$ and $b(r)$ are defined by

$$a(r) \equiv r\frac{F'(r)}{F(r)} \tag{3.18a}$$

$$b(r) \equiv ra'(r) = r\frac{F'(r)}{F(r)} + r^2\frac{F''(r)}{F(r)} - r^2\left(\frac{F'(r)}{F(r)}\right)^2 \tag{3.18b}$$

there exists function $\delta(r)$ defined for $R_0 < r < R$ satisfying $0 < \delta(r) < \pi$ for which the following conditions holds:

a) $F(re^{i\theta}) \sim F(r)e^{ia(r)\theta - \frac{1}{2}b(r)\theta^2}$ as $r \to R$ uniformly for $|\theta| < \delta(r)$

b) $f(r) = o\left(F(r)b^{-\frac{1}{2}}(r)\right)$ as $r \to R$ uniformly for $|\theta| < \delta(r)$

c) $b(r) \to \infty$ as $r \to$ R

Hayman proved the following.

[3]This should be compared with the maximum principle, which states that a function $f(z)$ analytic is a circle of radius R attains its maximum modulus on the boundary of the circle; that is, $\max_{|z|\leq R} |f(z)| = \max_{|z|=R} |f(z)|$, (see [Ahlfors 1953, pp. 108–110]).

Theorem 3.2. If F given by equation (3.17a and 3.17b) is admissible in $|z| < R$, then

$$f_n \sim \frac{1}{\sqrt{2\pi b(r)}} F(r) r^{-n} e^{-\frac{(a(r)-n)^2}{b(r)}} + o(1) \tag{3.19a}$$

If r_n as $n \to \infty$ is defined by $a(r_n) = n$, then

$$f_n \sim \frac{1}{\sqrt{2\pi b(r_n)}} F(r_n) r_n^{-n} \tag{3.19b}$$

as $r \to R$, the $o(1)$ error term being uniform in n.

3.7 SADDLE POINT BOUNDS

An alternate approach is to obtain bounds on integrals of the form

$$I(n) = \frac{1}{2\pi i} \int_C \frac{F(z)}{z^{n+1}} dz \tag{3.20a}$$

where $F(z) = F(z, n)$ is analytic and n is large. If the path C can be deformed, then an obvious upper bound for $I(n)$ is

$$I(n) \leq \frac{1}{2\pi} \min_{C^*} \|C^*\| \max_{z \in C^*} \left| \frac{F(z)}{z^{n+1}} \right| \tag{3.20b}$$

where $\|C^*\|$ is the length of the path C^*, the minimum being taken over the family of appropriate paths. The bound in equation (3.20b) might be improved by a judicious choice of the path $C \to C^*$. For example, using the maximum principle and restricting the contour of integration to be circles of radius R, the bound in equation (3.20b) may be improved to

$$I(n) \leq \frac{1}{2\pi} \min_R \left\{ \frac{1}{R^n} \max_{|z|=R} |F(z)| \right\} \tag{3.20c}$$

Additionally, if the power series of $F(z)$ (about $z = 0$) has non-negative coefficients, then the triangle inequality implies the still sharper bound

$$I(n) \leq \min_R \frac{F(R)}{R^n} \tag{3.20d}$$

The bound in equation (3.20d) is particularly nice, because the best value of R should be that which satisfies

$$RF'(R) = nF(R) \tag{3.20e}$$

We now prove[4] a result that appears Flajolet and Sedgewick [Flajolet and Sedgewick 1994, p. 8].

Theorem 3.3. If $F(z)$ is analytic function in a circle about $z = 0$ with radius of convergence $\rho > 0$ with non-negative Taylor coefficients,

$$F(z)=\sum_{n=0}^{\infty} f_n z^n \quad f_n=\frac{1}{2\pi i}\int_{|z|=r}\frac{F(z)}{z^{n+1}}dz \tag{3.21}$$

satisfying

a) $F(\rho) = \infty$
b) $F(z)$ is not a polynomial in z with $f(0) \neq 0$
c) $f_n = \inf_{0 \le r < \rho} \frac{F(r)}{r^n}$

Then

$$f_n \le F(\zeta(n)) \quad (\zeta(n))^{-n} \tag{3.22a}$$

where $\zeta \equiv \zeta(n)$ is the unique stationary point of $\frac{F(z)}{z^{n+1}}$, solution of

$$\frac{d}{dz}\left(\frac{F(z)}{z^{n+1}}\right)=\frac{z^{n+1}F'(z)-nz^n F(z)}{z^{2(n+1)}}=\frac{zF'(z)-nF(z)}{z^{n+2}}=0 \tag{3.22b}$$

Proof. We claim that

- Equation (3.20*e*) has only one solution on the (real) interval $(0, \rho)$.
- The point ζ is a local maximum of $\frac{F(z)}{z^{n+1}}$.

First, $F(0) = f_0 > 0$ so that $F(r)r^{-n}$ increases as $r \downarrow 0$; next, $F(r)$ increases as $r \to \infty$. These two facts prove that $\frac{F(z)}{z^{n+1}}$ has at least one local minimum on $(0, \rho)$. Next

$$\frac{d^2}{dz^2}\left(\frac{F(z)}{z^{n+1}}\right)=\frac{z^2F''(z)-2nzF'(z)+n(n+1)F(z)}{z^{n+2}} \quad z>0 \tag{3.23a}$$

The Taylor series of the numerator in equation (3.23*a*) is

$$\sum_{m=0}^{\infty} f_k(m+1-k)(m-k)z^k \tag{3.23b}$$

from which we observe

$$0 \le f_k(m+1-k)(m-k) \tag{3.23c}$$

[4]Because Part I of this book is an summary of relevant mathematics, proofs have been offered sparsely. Here, we make an exception because the technique is relevant to the asymptotic methodology.

Conclusion. $\frac{F(z)}{z^{n+1}}$ is convex, proving the uniqueness of the minimum. Thus, by hypothesis *c* of Theorem 3.3, $f_n \leq \frac{F(\zeta)}{\zeta^n}$ completing the proof. ■

In §3.8, we illustrate how the saddle point method and saddle point bounds can be applied to the Catalan numbers (Chapter 2 §2.4), the Bell numbers (Chapter 2, *Example 2.6*) and the Stirling numbers of the second kind (Chapter 2, *Example 2.7*).

3.8 EXAMPLES OF SADDLE POINT ANALYSIS

Example 3.8 (Inverse factorial). We apply Theorem 3.1 to the formula

$$\frac{1}{n!} = \frac{1}{2\pi i} \int_{|z|=R} \frac{e^z}{z^{n+1}} dz = \frac{1}{2\pi i} \int_{|z|=R} e^{z-(n+1)\log z} dz \tag{3.24a}$$

$$F(z) = e^{f(z)} \Leftrightarrow f(z) \equiv e^{z-(n+1)\log z} = \log F(z) \tag{3.24b}$$

where R is yet to be specified. The derivatives of $f(z) \equiv z - (n + 1) \log z$ are

$$f'(z) = 1 - \frac{n+1}{z} \quad f''(z) = \frac{n+1}{z^2} \tag{3.24c}$$

so that the unique saddle point is

$$z_0 = n+1 \quad f(z_0) = e^{n+1} \quad f'(z_0) = 0 \quad |f''(z_0)| = \frac{1}{(n+1)^{2n-1}} \tag{3.24d}$$

We now select $R = n + 1$ and modify the path as in Figure 3.4, yielding

$$\frac{1}{n!} \approx \frac{e^{n+1}}{(n+1)^{n-1}\sqrt{2\pi(n+1)}} \approx \frac{e^n n^{-n}}{\sqrt{2\pi n}} \tag{3.24e}$$

Of course, this development is purely *formal*, and there are considerable details to be supplied, which are left to the reader to [Flajolet and Sedgewick 1994, pp. 520–522].

An extension of Theorem 3.1 (see [Flajolet and Sedgewick 1994, p. 522]) shows how the full asymptotic expansion of $\frac{1}{n!}$ can be obtained.

$$\frac{1}{n!} \approx \frac{n^{-n}e^n}{\sqrt{2\pi n}} \left\{ 1 - \frac{1}{12n} + \frac{1}{288n^2} + \frac{139}{51\,840n^3} - \frac{571}{2\,488\,320n^4} + \cdots \right\} \quad n \to \infty \tag{3.25a}$$

or equivalently

$$n! \sim e^{-n} n^n \sqrt{2\pi n} \left\{ 1 - \frac{1}{12n} + \frac{1}{288n^2} + \frac{139}{51\,840n^3} - \frac{571}{2\,488\,320n^4} + \cdots \right\}^{-1} \quad n \to \infty \tag{3.25b}$$

Oops! Equations (3.10*b*) and (3.25*b*) suggest a typographic error because the coefficients of the two asymptotic expansions of $n!$ seemed to be equal except for their sign. Well, there might be typos and maybe even *real* mathematical errors, but this is not one. This is pointed out by Flajolet and Sedgewick who pose the question *Why?* as an exercise.

The formulas in equation (3.10*a* and 3.10*b*) and (3.25*b*) suggest that

$$\begin{aligned}&\left\{1-\frac{1}{12n}+\frac{1}{288n^2}+\frac{139}{51\,840n^3}-\frac{571}{2\,488\,320n^4}\pm\cdots\right\}^{-1}\\&=\left\{1+\frac{1}{12x}+\frac{1}{288x^2}-\frac{139}{5184x^3}-\frac{571}{2\,488\,320x^4}\pm\cdots\right\}\end{aligned} \tag{3.26}$$

where the equals sign in equation (3.26) is an equality of asymptotic expansions. If

$$A(z)=\sum_{n=0}^{\infty}a_n x^{-n}$$

is an asymptotic expansion with $a_0 = 1$, the by *Remark 3.1*,

$$B(z)=A^{-1}(z)=\sum_{n=0}^{\infty}b_n x^{-n}$$

can be computed using the steps in equation (3.9); that is, by imposing the condition

$$0=\sum_j a_j b_{n-j} \qquad n>0$$

Given $\{a_n\}$, the coefficients $\{a_n\}$ are determined. The final step is to use equation (3.10*d*), which shows that equation (3.26) is valid. ■

Example 3.9 (Catalan Numbers). Using generating functions, we proved in Chapter 2 §2.4 the solution of the recursion

$$C_n=\begin{cases}1 & \text{if } n=0\\ \sum_{i=0}^{n-1} C_i C_{n-i-1} & \text{if } 1\le n<\infty\end{cases} \tag{3.27a}$$

was

$$C_n=\frac{1}{n+1}\binom{2n}{n} \qquad 0\le n<\infty \tag{3.27b}$$

$\binom{2n}{n}$ is the central term in the binomial expansion of $(x + y)^n$ (see Theorem 1.2 in Chapter 1 §1.3). Because

$$\binom{2n}{n}=\frac{(2n)!}{n!n!} \tag{3.27c}$$

we have from Stirling's formula

$$n! \approx \frac{1}{\sqrt{2\pi n}} e^{-n} n^n \qquad (3.27d)$$

we have

$$\binom{2n}{n} \approx \frac{1}{\sqrt{\pi n}} 2^{2n} \qquad (3.27e)$$

which gives

$$C_n \approx \frac{1}{\sqrt{\pi n^3}} 2^{2n} \quad \blacksquare \qquad (3.27f)$$

Example 3.10 (The Bell numbers). We apply Theorem 3.3 to determine the growth of the Bell numbers (see Chapter 2 *Example 2.6*) defined by

$$B_{n+1} = \binom{n}{0} B_0 + \binom{n}{1} B_1 + \cdots + \binom{n}{n} B_n \quad 0 \le n < \infty;\ B_0 = 1,\ B_2 = 1 \qquad (3.28a)$$

$$B(z) = \sum_{n=0}^{\infty} \frac{B_n}{n!} z^n = e^{e^z - 1} \qquad (3.28b)$$

$$e^1 \frac{B_n}{n!} = \frac{1}{2\pi i} \int_{|z|=1} \frac{e^{e^z}}{z^{n+1}} dz \qquad (3.28c)$$

To find the stationary point of

$$f(z) = e^z - (n+1)\log z \qquad (3.29a)$$

we compute

$$f'(z) = e^z - \frac{n+1}{z} \qquad (3.29b)$$

whose solution $\zeta = \zeta(n)$ satisfies

$$\zeta e^{\zeta} = n+1 \qquad (3.29c)$$

This equation and variations of it arises in a variety of problems (see, for example Chapter 13 §13.2); in [Wright 59, p. 89], it is shown that

$$\zeta(n) = \log(n+1) - \log\log(n+1) + \frac{\log\log(n+1)}{\log(n+1)} + O\left(\frac{\ln^2 \log(n+1)}{\ln^2(n+1)}\right) \quad n \to \infty \qquad (3..29d)$$

which gives the upper bound

$$B_n \leq n!\frac{e^n}{(\log(n+1))^{(n+1)}} \tag{3.29e}$$

Flajolet and Sedwick point out that equation (3.29*e*) shows that there are much fewer partitions of a set than permutations. ■

To obtain the asymptotic formula for $\frac{B_n}{n!}$ from Theorem 3.1, we proceed formally writing

$$e^1\frac{B_n}{n!} = \frac{1}{2\pi i}\int_{|z|=1}\frac{e^{e^z}}{z^{n+1}}dz \tag{3.30a}$$

requires that we study

$$f(z) \equiv e^z - (n+1)\log z \tag{3.30b}$$

and its the derivatives

$$f'(z) = e^z - \frac{n+1}{z} \quad f''(z) = e^z + \frac{n+1}{z^2} \tag{3.30c}$$

The maximum of $f(z)$ occurs at $z_0 = \zeta \equiv \zeta(n)$ satisfying

$$f'(\zeta) = 0 \Rightarrow \zeta e^\zeta = n+1 \tag{3.31a}$$

Next

$$f(\zeta) = e^\zeta - (n+1)\log\zeta \tag{3.31b}$$

Because $\zeta e^\zeta = n + 1$

$$\log\zeta = \log(n+1) - \zeta \tag{3.31c}$$

giving

$$f(\zeta) = e^\zeta - \log(n+1)^{(n+1)} + (n+1)\zeta \tag{3.31d}$$

so that

$$e^{f(\zeta)} = e^{\frac{n+1}{\zeta}}\zeta^{-(n+1)} \tag{3.31e}$$

Next, we need $f''(\zeta)$

$$f''(\zeta) = e^\zeta + \frac{n+1}{\zeta^2} = (n+1)\left(\frac{1}{\zeta} + \frac{1}{\zeta^2}\right) \tag{3.32a}$$

Using equation (3.31e), we have $\zeta \approx \log(n+1)$, which gives

$$e^{f(\zeta)} \approx \frac{e^{\frac{n+1}{\log(n+1)}}}{(\log(n+1))^{n+1}} \tag{3.32b}$$

$$f''(\zeta) \approx \frac{n+1}{\log(n+1)} \tag{3.32c}$$

yielding the following asymptotic formula

$$B_n \approx n! \frac{e^{\frac{n+1}{\log(n+1)}}}{(\log(n+1))^{n+1}} \frac{1}{\sqrt{2\pi \frac{n+1}{\log(n+1)}}} \tag{3.32d}$$

Example 3.11 (Stirling numbers of the second kind). Theorem 2.3 (Chapter 2) provides the formula for $S_{n,m}$

$$S_{n,m} = \frac{1}{r!} \sum_{s=0}^{m} \binom{m}{s} (-1)^s (m-s)^n \tag{3.33}$$

the Stirling numbers of the second kind. For their asymptotic formula, we start with the Cauchy integral formula writing

$$S_{n,m} = \frac{n!}{m!} \frac{1}{2\pi i} \oint_{|z|=R} \frac{(e^z-1)^m}{z^{n+1}} dz = \frac{n!}{m!} \frac{1}{2\pi i} \oint_{|z|=R} \frac{e^{f(z)}}{z} dz \tag{3.34a}$$

with $R > 0$ a circle of radius centered at $z = 0$ and

$$F(z,n,m) \equiv \frac{(e^z-1)^m}{z^n} = e^{f(z)} \quad f_{n,m}(z) \equiv f(z) = m\log(e^z-1) - n\log z \tag{3.34b}$$

We follow the presentation in N. Temme [Temme 1993] and the subsequent paper [Chelluri, Richmond and Temme 1999]. Let ζ be the positive solution of

$$f'(z) = \frac{me^z}{e^z-1} - \frac{n}{z} = \frac{mze^z - n(e^z-1)}{z(e^z-1)} = 0 \tag{3.35a}$$

$$\alpha\zeta = 1 - e^{-\zeta} \Leftrightarrow \alpha\zeta e^{\zeta} = e^{\zeta} - 1 \quad \alpha \equiv \frac{m}{n} \tag{3.35b}$$

While $\zeta = 0$ is clearly a solution of equation (3.35b), $\lim_{z \to \zeta} F(z,n,m) = 0$ when $m < n$ so that it is not related to the growth of $S_{n,m}$; however,

- $1 - e^{-z}$ is convex and increasing on $[0, \infty)$ with values $\begin{cases} 0 & \text{at } z = 0 \\ 1 - e^{-1} & \text{at } z = 1 \\ 1 & \text{at } z \to \infty \end{cases}$.

- $\alpha\zeta$ is increasing on $[0, \infty)$ with values $\begin{cases} 0 & \text{at } z=0 \\ \alpha & \text{at } z=1 \\ 1 & \text{as } z \to \infty \end{cases}$.

Conclusions

1. If α satisfies $0.632120559 \cdots \approx 1-e^{-1} < \alpha < 1$, then equation (3.35*b*) has a second solution ζ in the interval (0, 1]
   - If $\alpha \uparrow 1$, then $\zeta \downarrow 0$.
   - If $\alpha \downarrow 1-e^{-1}$, then $\zeta \uparrow 1$.
2. If α satisfies $0 < \alpha < 1-e^{-1}$, then equation (3.35*b*) has a second solution in the interval $(1, \infty]$
   - If $\alpha \downarrow 0$, then $\zeta \uparrow \infty$.

The solution of equation (3.35*b*) can be obtained directly by iteration or an explicit power series for ζ, as a function of α can be found using Lagrange's interpolation theorem (Chapter 2). Such a solution appears in [Bleick and Wang 1974, equation (5), p. 576]. Their result was obtained directly using the solution of the functional equation $we^w = \sigma$, which is analytic for $|\sigma| < 1$ given in [Wright 59, p. 89] as

$$w = w(\sigma) = \sum_{k=0}^{\infty} \frac{k^{k-1}}{k!} \sigma^k \quad |\sigma| < 1 \tag{3.36a}$$

To apply Wright's solution of $we^w = \sigma$ to solve equation (3.35*b*), note that when we define

$$\tau = \frac{1}{\alpha} \quad \sigma = \tau e^{-\tau}$$

then equation (3.35*b*) may be written as

$$z = \tau - \tau e^{-z} \tag{3.36b}$$

so that setting $w = \tau - z$

$$(\tau - z)e^{z-\tau} = \tau e^{-\tau} \Leftrightarrow w = w(\sigma) \tag{3.36c}$$

Unfortunately, the series in equation (3.36*a*) converges slowly as $\alpha \uparrow 1 \Leftrightarrow \sigma \uparrow 1$; the size of the kth term may be estimated using Stirling's formula [equation (3.25*a*)] as

$$\frac{(k+1)^{k-1}}{k!} \sigma^k \sim \frac{(e\sigma)^k}{\sqrt{2\pi k} k} \tag{3.37a}$$

If the summation is truncated after $K - 1$ terms, the remainder R_K is

$$R_K = \sum_{j=K}^{\infty} \frac{(k+1)^{k-1}}{k!} \sigma^k \sim \frac{(e\sigma)^k}{\sqrt{2\pi k} k} \tag{3.37b}$$

We are led to bound R_K by

$$R_K \overset{?}{\leq} \frac{(e\sigma)^K}{\sqrt{2\pi K}K}\sum_{j=0}^{\infty}(e\sigma)^j \leq \frac{(e\sigma)^K}{\sqrt{2\pi K}K(1-e\sigma)} \tag{3.37c}$$

To obtain a solution with error $\leq \varepsilon$, we need to be conservative and truncate the series in equation (3.36a) after $K^* \equiv K^*(\varepsilon)$ terms where

$$\frac{(e\sigma)^{K^*}}{\sqrt{2\pi K^*}K^*(1-e\sigma)} \leq \varepsilon \tag{3.37d}$$

Table 3.1 lists the solution ζ obtained by means equation (3.36a) obtained using truncation with $\varepsilon = 10^{-8}$ and $n = 25$ for $m = 2(1)21$. The program continues summing the terms in equation (3.36a) until either a maximum number ($K^* = 2000$) is reached or $K^* = \min_{k \leq K^*} \frac{(k+1)^{k-1}}{k!}\sigma^k \frac{1}{1-e\sigma} < \varepsilon$. The last two columns compare ζ with $\frac{1-e^{-\zeta}}{\alpha}$.

Temme [Temme 1993] remarks that when the saddle point method is used to evaluate the asymptotics of $\oint e^{h(z)}dz$ by expanding $h(z) \equiv h(z, n)$ in a power series $h(z) = h(z_0) + (z - z_0)h'(z_0) + (z - z_0)^2h''(z_0) + \cdots$ about the saddle point z_0, it leads to a formula for the $S_{n,m}$ with insufficient accuracy as $n, m \to \infty$. To remedy this, Temme considers $\oint \frac{e^{f(z)}}{z}dz$ with $f(z)$ given by equation (3.35b) and to which he applies the transformation

TABLE 3.1. The Computation of ζ for $n = 25$, $m = 2(1)21$

m	α	σ	K^*	R_{K^*}	ζ	$\frac{1-e^{-\zeta}}{\alpha}$
2	0.080000	0.00012663	3	0.00000000	12.49995341	12.49995341
3	0.120000	0.00544493	5	0.00000000	8.33132623	8.33132623
4	0.160000	0.03279699	6	0.00000000	6.23778640	6.23778640
5	0.200000	0.09157819	8	0.00000000	4.96511423	4.96511423
6	0.240000	0.17559935	10	0.00000000	4.09743662	4.09743662
7	0.280000	0.27295102	12	0.00000001	3.45907561	3.45907561
8	0.320000	0.37322803	16	0.00000000	2.96365700	2.96365699
9	0.360000	0.46948143	20	0.00000001	2.56387230	2.56387230
10	0.400000	0.55782540	25	0.00000001	2.23161189	2.23161188
11	0.440000	0.63651537	31	0.00000001	1.94907925	1.94907925
12	0.480000	0.70513630	39	0.00000001	1.70442187	1.70442186
13	0.520000	0.76402829	50	0.00000001	1.48941386	1.48941386
14	0.560000	0.81391789	64	0.00000001	1.29814846	1.29814846
15	0.600000	0.85569520	84	0.00000001	1.12626123	1.12626123
16	0.640000	0.89028566	111	0.00000001	0.97044813	0.97044813
17	0.680000	0.91858048	149	0.00000001	0.82815526	0.82815525
18	0.720000	0.94140219	206	0.00000001	0.69737252	0.69737252
19	0.760000	0.95949073	296	0.00000001	0.57649272	0.57649272
20	0.800000	0.97350098	448	0.00000001	0.46421276	0.46421276
21	0.840000	0.98400647	730	0.00000001	0.35946302	0.35946302

$$z \to t(z) \Leftrightarrow t(z) = \log(e^z - 1) - \log z \quad z \in \mathcal{C}_1(0) \tag{3.38a}$$

yielding

$$f(z) = m\log(e^z - 1) - n\log z \Rightarrow mt + (m-n)\log t + A \tag{3.38b}$$

We evaluate

$$f(\zeta) = m\log(e^\zeta - 1) - n\log\zeta = n[\alpha\log(e^\zeta - 1) - \log\zeta]$$

and use equation (3.35*b*) replacing $e^\zeta - 1$ by $\alpha\zeta e^\zeta$ to obtain

$$f(\zeta) = n[-\log\zeta + \alpha\log(\alpha\zeta e^\zeta)] = m[\zeta + \log\zeta + \log\alpha] \tag{3.38c}$$

$$A = f(\zeta) - mt_0 + (n-m)\log t_0 \quad t_0 \equiv \frac{n-m}{m} = \frac{1}{\alpha} - 1 \tag{3.38d}$$

A detailed analysis is given in [Chelluri, Richmond and Temme 1999] using the results of Moser and Wyman [Moser and Wyman 1958a,b][5]. This gives the following formula result appearing in Temme [Temme 1993] and [Chelluri, Richmond and Temme 1999, Theorem 1, p. 3].

Theorem 3.4

a) Uniformly as $n \to \infty$ for $\delta < m \le n$ (equation (2.9) in [Temme 1993, p. 236])

$$S_{n,m} \sim e^A m^{n-m} f(t_0)\binom{n}{m} \tag{3.39a}$$

$$f(t_0) = \sqrt{\frac{t_0}{(1+t_0)(\zeta - t_0)}} \tag{3.39b}$$

b) If $\delta < m \le n \le n - \sqrt[5]{n}$, then ([Chelluri, Richmond and Temme 1999, Theorem 1, p. 3])

$$S_{n,m} = \frac{n!}{m!}\frac{(e^\zeta - 1)^m}{2\zeta^n}\frac{1}{\sqrt{\pi\zeta m H_0(\zeta)}}[1 + O(n^{-1})] \tag{3.40a}$$

$$H_0(\zeta) \equiv \frac{1}{2}\left[\zeta(e^\zeta - 1)^{-1} - \zeta e^\zeta(e^\zeta - 1)^{-2}\right] = \frac{1}{2}\frac{e^\zeta - 1 - \zeta}{\left(e(^\zeta - 1\right)^2} \tag{3.40b}$$

The implicit constant in the $O(n^{-1})$ error estimate in equation (3.40*b*) depends only on δ.

c) The restriction $n - \sqrt[3]{n} \le m \le n$ results in the simplification of equations (3.40*a* and 3.40*b*) ([Chelluri, Richmond and Temme 1999])

[5]Their analysis that applies to more than the $\{S_{n,m}\}$ shows the advantage of dealing with problems arising from analytic functions as in equations (3.17*a* and 3.17*b*). They deform the contour in equation (3.34*a*) from a circle of radius $R > 0$ to one which allows better estimates.

TABLE 3.2. Comparison of $\tilde{S}_{25,m}$ and $S_{25,m}$ for m = 2(1)21

m	α	$\tilde{S}_{25,m}$	$S_{25,m}$
2	0.080000	1.683350×10^{7}	1.677721×10^{7}
3	0.120000	1.417263×10^{11}	1.411980×10^{11}
4	0.160000	4.699236×10^{13}	4.677129×10^{13}
5	0.200000	2.450692×10^{15}	2.436685×10^{15}
6	0.240000	3.726761×10^{16}	3.702641×10^{16}
7	0.280000	2.294365×10^{17}	2.227632×10^{17}
8	0.320000	6.953396×10^{17}	6.902237×10^{17}
9	0.360000	1.176917×10^{18}	1.167921×10^{18}
10	0.400000	1.212745×10^{18}	1.203163×10^{18}
11	0.440000	8.089620×10^{17}	8.023559×10^{17}
12	0.480000	3.653564×10^{17}	3.622626×10^{17}
13	0.520000	1.155052×10^{17}	1.148507×10^{17}
14	0.560000	2.620141×10^{16}	2.595811×10^{16}
15	0.600000	4.342218×10^{15}	4.299395×10^{15}
16	0.640000	5.323001×10^{14}	5.266552×10^{14}
17	0.680000	4.862630×10^{13}	4.806333×10^{13}
18	0.720000	3.318413×10^{12}	3.322716×10^{12}
19	0.760000	1.686932×10^{11}	1.662190×10^{11}
20	0.800000	6.329270×10^{9}	6.220195×10^{9}
21	0.840000	1.721601×10^{8}	1.685195×10^{8}

$$S_{n,m} = \frac{1}{2^{n-m}} \frac{n^{2(n-m)}}{(n-m)!}\left[1+O\left(n^{-\frac{1}{3}}\right)\right] \tag{3.41}$$

Table 3.2 compares the [equations (3.40*a* and 3.40*b*)] approximation $\tilde{S}_{25,m}$ with the *exact* values $S_{25,m}$ from [Abramowitz and Stegun 1972, Table 24.3, p. 835] for n = 25 and m = 2(1)21.

The growth rate of $S_{n,m}$ if $n, m \to \infty$ and $m = \alpha n$ can be estimated from equations (3.40*a* and 3.40*b*) as follows:

Remarks 3.3. with $0 < \alpha < 1$.

a) Using Stirling's formula for the factorial, we have $\frac{n!}{m!} \sim n^{(1-\alpha)n}(\alpha)^{-\frac{1}{2}}\alpha^{-\alpha n}e^{-n(1-\alpha)}$.

b) Remarks 3.3a) and b) give $\frac{\left(e^{\zeta}-1\right)^{m}}{\zeta^{n}} \sim \left(\zeta^{\alpha-1}e^{\alpha\zeta}\right)^{n}\alpha^{\alpha n}$.

Combining Remarks 3.3a) and b), we have the $S_{n,m}$ growth rate based on equations (3.40*a* and 3.40*b*)

Corollary 3.5. If $\delta < m \le n \le n-\sqrt[5]{n}$, then uniformly as $n \to \infty$ for $\delta < m \le n$

$$S_{n,m} \sim C_1 C_2^n n^{(1-\alpha)n-\frac{1}{2}} \tag{3.42a}$$

where

$$C_1 = \frac{1}{2\alpha} \frac{1}{\sqrt{\pi \zeta H_0(\zeta)}} \quad C_2 = \zeta^{\alpha-1} e^{\alpha\zeta} e^{-(1-\alpha)} \tag{3.42b}$$

REFERENCES

M. Abramowitz and I. A. Stegun (Editors), *Handbook of Mathematical Functions with Formulas, Graphs, and Mathematical Tables*, Dover Publications (New York), 1972.

L. V. Ahlfors, *Complex Analysis*, McGraw-Hill (New York), 1953.

P. Bachman, *Analytische Zahlentheorie*, University of Michigan Library (Ann Arbor, MI), [2006].

W. W. Bleick and P. C. C. Wang, "Asymptotics of Stirling Numbers of the Second Kind", *Proceedings of the American Mathematical Society*, **42**, #2, pp. 575–580, 1974.

R. Chelluri, L. B. Richmond, and N. Temme, "Asymptotic Estimates for Generalized Stirling Numbers", *Analysis*, **20**, p. 113, 2000.

N. G. de Bruijn, *Asymptotic Methods in Analysis*, North Holland Publishing (Amsterdam, The Netherlands), 1958.

A. Erdelyi, *Asymptotic Expansions*, Dover Publications (New York), 1956.

P. Flajolet and R. Sedgewick, "The Average Case Analysis of Algorithms: Saddle Point Asymptotics", Research report 2376, Institut National de Recherch en Informatique et en Automatique, 1994.

W. Hayman, "A Generalization of Stirlings Formula", *Journal fur die Reine und Angewandte Mathematik*, **196**, pp. 67–95, 1956.

L. Moser and M. Wyman, "Asymptotic Development of the Stirling Numbers of the First Kind", *Journal of the London Mathematical Society*, **33**, pp. 133–146, 1958.

L. Moser and M. Wyman, "Stirling Numbers of the Second Kind", *Duke Mathematics Journal*, **25**, pp. 29–43, 1958.

A. M. Odlyzko, "Asymptotic Enumeration Methods", in *Handbook of Combinatorics*, Volume 2, R. L. Graham, M. Groetschel, and L. Lovasz (Editors), Elsevier (New York), 1995, pp. 1063–1229.

N. M. Temme, "Asymptotic Estimates of Stirling Numbers", *Studies in Applied Mathematics*, **89**, pp. 233–243, 1993.

E. T. Whittaker and G. N. Watson, *A Course of Modern Analysis*, Cambridge University Press (Cambridge, UK), 1952.

E. M. Wright, "Solution of the Equation $ze^z = a$", *Bulletin of the American Mathematical Society*, **65**, pp. 89–93, 1959.

John W. Wrench, Jr., "Concerning Two Series for the Gamma Function", *Mathematics of Computation*, **22**, #103, pp. 617–626, July 1968.

CHAPTER 4

Discrete Probability Theory

My favorite reference book in probability is *An Introduction to Probability Theory and Its Applications* by William Feller [Feller 1957].

Another more recent excellent choice is Grimmett and Stirzaker's *Probability and Random Processes* [1992].

4.1 THE ORIGINS OF PROBABILITY THEORY

Probability theory formalizes the notion of randomness, just as Euclid axiomatized geometry by abstracting the intrinsic properties of lines and points and their interrelations. Fermat and Pascal introduced probabilistic reasoning more than 300 years ago to explain the outcomes in games of chance.

Jacob Bernoulli (1654–1705) was a member of a large family of scientists and mathematicians. His book *Ars Conjectandi* published in 1713 proved the *law of large numbers*, certainly the most fundamental result in probability. Abraham de Moivre (1667–1754) published a proof in 1733 of the *central limit theorem*, the second cornerstone of probability. De Moivre's 1718 book *Doctrine of Chance* used probability theory to explain the nature of games of chance to gamblers.

Still, the development of probability theory as an independent mathematical discipline begsn in 1812 in *Théorie Analytique des Probablités* by Pierre Simon Laplace (1749–1817).

Questions in statistical mechanics motivated the subsequent development of the subject; the two major publications were the following:

1933 Andrey Nikolaevich Kolmogorov (1903–1987), *Grundbegriffe des Warcheinlichkeitrechnung* (*Foundations of Probability Theory*) [Kolmogorov 1933], and

1936 Maurice René Fréchet (1878–1973), *Recherches Théorie modernses en calcul des probablités* [Fréchet 1936].

4.2 CHANCE EXPERIMENTS, SAMPLE POINTS, SPACES, AND EVENTS

The mathematical theory of probability theorem begins with a chance experiment ε, which has possible outcomes denoted by sample points $\omega_0, \omega_1, \cdots$. The sample space Ω consists of the set of all outcomes $\Omega = \{\omega_i\}$.

Hashing in Computer Science: Fifty Years of Slicing and Dicing, by Alan G. Konheim

The simplest environment—more than satisfactory for *most* but *not all* of computer science—is *discrete probability theory* in which the sample space Ω consists of either of the following:

- A *finite* number $\omega_0, \omega_1, \cdots, \omega_{N-1}$ of sample points
- A *countably infinite* number $\omega_0, \omega_1, \cdots, \omega_{N-1}, \cdots$ of sample points

An event E is any subset of sample points.

In this case, a probabilistic model for Ω assumes given a **probability mass function** $p(\omega_i)$ satisfying

P1. $p(\omega_i) \geq 0$ for $i = 0, 1, \cdots,$

P2. $1 = \sum_i p(\omega_i)$,

P3. The probability of the event E is $Pr\{E\} = \sum_{\omega_i \in E} p(\omega_i)$.

The probabilities of events satisfy

E1. $Pr\{E\} \geq 0$, $Pr\{\emptyset\} = 0$, $Pr\{\Omega\} = 1$,

E2. If the events E and F are disjoint, $E \cap F = \emptyset$, then $Pr\{E \cup F\} = Pr\{E\} + Pr\{F\}$.

The pair (Ω, p) is a *probability space.*

Example 4.1. The chance experiment ε is the tossing the same coin m times; an outcome of a toss of a coin is either heads (1) (with probability α) or tails (0) (with probability $1 - \alpha$) .

The ith outcome of ε is $\omega_i = (\omega_{i,0}, \omega_{i,1}, \cdots, \omega_{i,m-1})$. The number of sample points in the sample space Ω is $N = 2^m$; the number of possible events is 2^{2^m}.

The coin tosses are *independent* if

$$p(\omega_i) = \prod_{j=0}^{m-1} \alpha^{\omega_{i,j}}(1-\alpha)^{1-\omega_{i,j}} = \alpha^{N_1(\omega_i)}(1-\alpha)^{N_0(\omega_i)}$$

where $N_1(\omega_i)$ (*resp.* $N_0(\omega_i)$) is the total number of heads (*resp.* tails) in the outcome ω_i

The probability of the event $E_{m,k} = k$ heads in m tosses is $Pr\{E_{m,k}\} = \binom{m}{k}\alpha^k(1-\alpha)^{m-k}$ [see Chapter 1, Theorem 1.7].

The coin is *fair* wh en $\alpha = \frac{1}{2}$.

Example 4.2. The chance experiment ε is the tossing the same coin an infinite number of times; an outcome of a toss of a coin is either heads (1) (with probability α) or tails (0) (with probability $1 - \alpha$).

An outcome of ε is $\omega = (\omega_0, \omega_1, \cdots, \omega_{m-1}, \cdots)$. The number of sample points in the sample space Ω is ∞; the number of possible events is ∞.

Just as in *Example 4.1*, the outcomes of the first m tosses is $\omega_i = (\omega_{i,0}, \omega_{i,1}, \cdots, \omega_{i,m-1})$ and if the first m tosses coin tosses are independent, then

$$p(\omega_i) = \prod_{j=0}^{m-1} \alpha^{\omega_{i,j}}(1-\alpha)^{1-\omega_{i,j}} = \alpha^{N_1(\omega_i)}(1-\alpha)^{N_0(\omega_i)}$$

Example 4.2 has a much larger number of sample points and events than in *Example 4.1*. If a set A has k elements, the cardinality of the set of all subsets of A, called the power set of A, is 2^k.

- In *Example 4.1*, the cardinality of Ω is 2^m; the set of all events has cardinality 2^{2^m}.
- In *Example 4.2*, the cardinality of Ω is $2^{\aleph_0}$ (where $\mathcal{N}_0$ is read *aleph0*). By analogy, with *Example 4.1*, the cardinality of the set of all events is $2^{2^{\aleph_0}}$. If the continuum hypothesis is accepted (see [Halmos 1960]), the cardinality of the set of all events in *Example 4.2* is $\aleph_2 = 2^{\aleph_1}$ where $\aleph_1 = 2^{\aleph_0}$.

When probability theory is enriched by moving from *discrete* to *continuous* probability theory, some strange things happen. For additional clarification or mild indigestion, see [Loéve 1981, Chapter 1].

4.3 RANDOM VARIABLES

A (discrete valued) random variable is any function X on a (discrete probability) sample space. There are many possible choices for the range; in the simplest case, the range is the integers $\mathcal{Z}$

$$X \quad \Omega \rightarrow \mathcal{Z} \tag{4.1a}$$

or some subset of them, for example, the non-negative integers $\mathcal{Z}_+$

$$X \quad \Omega \rightarrow \mathcal{Z}_+ \tag{4.1b}$$

or perhaps the integers modulo m $\mathcal{Z}_m = \{0,1, \cdots, m-1\}$

$$X \quad \Omega \rightarrow \mathcal{Z}_m \tag{4.1c}$$

The probability distribution of X is defined by

$$p_X(k) = Pr\{E_X(k)\} \tag{4.2a}$$

where

$$E_X(k) = \{\omega_i \in \Omega : X(\omega_i) = k\}$$

The distribution function of X is

$$F_X(k) = \sum_{j \le k} p_X(j) \tag{4.2b}$$

Example 4.1 (continued). The number of heads N_1 and the number of tails N_0 are both random variables.

- N_1 has the binomial distribution BIN(m, α); that is, $p_{N_1}(k) = \binom{m}{k}\alpha^k(1-\alpha)^{m-k}$.
- N_0 has the binomial distribution BIN(m,1 − α); that is, $p_{N_0}(k) = \binom{m}{k}\alpha^{m-k}(1-\alpha)^k$.

The definition of a random variable can be greatly extended; first, its *range* need not be the integers $\mathcal{Z}$ but pairs or triples of r tuples of integers; for example

$$\underline{X} \quad \Omega \to \mathcal{Z}^r$$

where (Ω, p) is a (discrete probability) sample space. The probability distribution of $\underline{X}$ is defined by

$$p_{\underline{X}}(\underline{k}) = Pr\{E_{\underline{X}}(\underline{k})\} \quad \underline{k} \in \mathcal{Z}^r$$

where

$$E_{\underline{X}}(\underline{k}) = \{\omega_i \in \Omega : \underline{X}(\omega_i) = \underline{k}\} \quad \underline{k} \in \mathcal{Z}^r$$

Of course, $\underline{X}$ as defined above really consists of r random variables $\underline{X} = (X_0, X_1, \cdots, X_{r-1})$.

The *joint distribution function* of $\underline{X}$ is

$$F_{\underline{X}}(\underline{k}) \equiv Pr\{X_i \leq k_i, 0 \text{ eqi} < r\}$$

In fact, in probability theory allows—just as California does—that (almost) anything goes; so when we visit the San Diego Zoo we might consider the random species X of the first animal we encounter.

4.4 MOMENTS—EXPECTATION AND VARIANCE

The kth moment of a random variable X on a (discrete probability) sample space (Ω, p), often denoted by μ_k, is

$$E\{X^k\} = \sum_{j=0}^{\infty} j^k p_X(j) \tag{4.3}$$

If Ω contains a (countable) infinite number of points and X is *nice* function, then the series presented previously converges if X is *bounded*.

The expected value $E\{X^k\}$ with $k = 1$ is the *expectation* of X, is often denoted by $\mu \equiv \mu_X$, and is often referred to as the mean or average. The variance of X, often denoted by σ_X^2, is defined by

$$\text{Var}\{X\} = \sum_{j=0}^{\infty} (j - \mu_X)^2 p_X(j) \tag{4.4}$$

The expectation operation is linear; if X and Y are random variables, then

$$E\{aX+bY\}=aE\{X\}+bE\{Y\} \tag{4.5}$$

where a and b are constants.

Theorem 4.1. (Cauchy–Schwarz inequality).

$$E^2\{XY\} \le E\{X^2\}E\{Y^2\} \tag{4.6}$$

The covariance of X and Y is

$$\text{Cov}\{X,Y\}=E\{(X-\mu_X)(Y-\mu_Y)\} \tag{4.7}$$

and the correlation coefficient of X and Y is

$$\rho\{X,Y\}=\frac{\text{cov}\{X,Y\}}{\sqrt{\text{Var}\{X\}\text{Var}\{Y\}}} \tag{4.8}$$

Example 4.1 (continued). The expectation and variance of the random variable N_1 with the binomial distribution BIN(m, α) are

$$E\{N_1\}=m\alpha$$
$$\text{Var}\{N_1\}=m\alpha(1-\alpha)$$

4.5 THE BIRTHDAY PARADOX[1]

Consider the chance experiment ε of teaching a *random* class with m students. Your choice of class determines an outcome, one of the 365^m sample points $\omega = (\omega_0, \omega_1, \cdots, \omega_{m-1})$, which lists the birthdays (0,1, $\cdots$, 364) of the m students in the class. We calculate the probability of the event E_m^*: Two or more students have the same birthday.

Our model of randomness assumes that all the 365^m sample points ω are equally likely to occur; that is, $p(\omega) = 365^{-m}$. It follows that

$$1-Pr\{E_m(i)\}=Pr\{\overline{E_m(i)}\}=365\alpha(1-\alpha)^{m-1}$$

where

$$\alpha=\frac{1}{365}$$

while

$$Pr\{E_m^*\}=\frac{365\times 364\times\cdots\times(366-(m-1))}{365^m}$$
$$=\left(1-\frac{1}{365}\right)\left(1-\frac{2}{365}\right)\cdots\cdots\left(1-\frac{m-1}{365}\right)$$

[1]A paradox is a tenet contrary to believe opinion; it is also an assertion or sentiment seemingly contradictory or opposed to common sense.

Using the approximation

$$1 - x \simeq e^{-x}$$

we have

$$Pr\{E_m^*\} \simeq Pr\{\tilde{E}_m^*\} = e^{-\frac{m(m-1)}{720}}$$

The values of $Pr\{E_m^*\}$ and $Pr\{\tilde{E}_m^*\}$ are tabulated in Table 4.1 for $10 \leq n \leq 53$.

Using the approximation, $Pr\{\tilde{E}_m^*\} = 0.5$ requires $m \approx \sqrt{2 \ln 2 \times 365}$ so that $1 - P_{23} \geq 0.5$.

If there were d instead of 365 birthdays in a year, then $Pr\{\tilde{E}_m^*\} = 0.5$ requires $n \approx 1.7741\sqrt{d}$.

The term paradox is applied because $Pr\{E_m\} \leq \frac{1}{2}$ for $m \geq 23$ for $m \geq 23$. In a class of 53 students, the probability of two or more students having the *same* birthday is greater than 0.97.

There are many applications in cryptography; suppose we have a (block cipher) (see Chapter 6) with an example of ciphertext $y = E\{k, x\}$, where k is an unknown key and x is a known plaintext. If there are N keys, then k could be discovered by testing *all* N keys. Instead, memory of size M might be allocated to hold the results of some precomputation. The two possible cases are as follows:

Exhaustive search: $(T, M) = (N, 1)$;
Dictionary search: $(T, M) = (1, N)$.

M. Hellman [Hellman 1980] used the birthday paradox and analyzed the case $(T, N) = \left(N^{\frac{2}{3}}, \left(N^{\frac{2}{3}}\right)\right)$.

Recently, with the use of *hashed values* (aka message digests) in Web-based transactions using the Secure Socket Layer (see Chapter 18), the birthday paradox has been used again.

4.6 CONDITIONAL PROBABILITY AND INDEPENDENCE

To some extent, our presentation of probability theory parallels the ideas of area. The probability of an event E is the (relative) area of E compared with that of the sample space Ω. The distinguishing feature of probability theory, which causes it to diverge from the theory of area (aka measure theory) (see [Loéve 1981]) is the concept of *independence*.

The conditional probability of the event E *given* or *conditioned by* the event F is defined when $Pr\{F\} > 0$ by

$$Pr\{E/F\} = \frac{Pr\{E \cap F\}}{Pr\{F\}} \tag{4.9}$$

TABLE 4.1. Probability of Distinct Birthdays in a Class of *m* Students

m	$Pr\{E_m^*\}$	$Pr\{\tilde{E}_m^*\}$	m	$Pr\{E_m^*\}$	$Pr\{\tilde{E}_m^*\}$	m	$Pr\{E_m^*\}$	$Pr\{\tilde{E}_m^*\}$	m	$Pr\{E_m^*\}$	$Pr\{\tilde{E}_m^*\}$
10	0.883052	0.884009	11	0.858859	0.860119	12	0.832975	0.834584	13	0.805590	0.807592
14	0.776897	0.779334	15	0.747099	0.750008	16	0.716396	0.719811	17	0.684992	0.688939
18	0.653089	0.657587	19	0.620881	0.625945	20	0.588562	0.594195	21	0.556312	0.562512
22	0.524305	0.531062	23	0.492703	0.499998	24	0.461656	0.469464	25	0.431300	0.439588
26	0.401759	0.410487	27	0.373141	0.382264	28	0.345539	0.355007	29	0.319031	0.328792
30	0.293684	0.303680	31	0.269545	0.279718	32	0.246652	0.256942	33	0.225028	0.235375
34	0.204683	0.215028	35	0.185617	0.195903	36	0.167818	0.177990	37	0.151266	0.161273
38	0.135932	0.145726	39	0.121780	0.131318	40	0.108768	0.118010	41	0.096848	0.105761
42	0.085970	0.094524	43	0.076077	0.084250	44	0.067115	0.074887	45	0.059024	0.066382
46	0.051747	0.058682	47	0.045226	0.051734	48	0.039402	0.045483	49	0.034220	0.039879
50	0.029626	0.034869	51	0.025568	0.030405	52	0.021995	0.026440	53	0.018862	0.022929

Conditional probability relates to the information contained in the knowledge that an event has occurred. For example, the relationship in equation (4.9) can be restated as follows:

$Pr\{E \cap F\}$ is the probability of observing both the events E and F.
$Pr\{F\}$ is the unconditional (or *a prior*) probability of observing the event F.
$Pr\{E/F\}$ is the conditional (or *a posteriori*) probability of observing the event E given (or after) observing the occurrence of the event F.

Bayes' law relates the conditional probability $Pr\{E/F\}$ and $Pr\{F/E\}$

$$Pr\{E/F\} = \frac{Pr\{F/E\}Pr\{E\}}{Pr\{F\}} \tag{4.10}$$

and plays a central role in cryptanalysis (see Chapter 6). If the following occurs:

- $E(Y)$ is the event Y is ciphertext
- $F(K)$ is the event that the ciphertext Y has resulted by encipherment $y = E\{K, X\}$ using the *key* K, then $Pr\{F(k)/E(y)\}$ is the conditional probability that the key $K = k$ was used to encipher given the observation of the ciphertext $Y = y$, so that equation (4.10) can be used to *infer* the secret information (the identity of the key) from the observed information (the intercepted ciphertext). More information is contained in [Konheim 2007].

E and F are independent events if $Pr\{E/F\} = Pr\{E\}$ or equivalently $Pr\{F/E\} = Pr\{F\}$; that is, the information that the event E has occurred, does not alter the likelihood $Pr\{F/E\}$ that the event F will occur. If a cryptographic system (see Chapter 6) has this property for all ciphertext Y and keys K, then the cryptographic system offers perfect secrecy.

Finally, X and Y are independent random variables if

$$Pr\{X = i/Y = j\} = Pr\{X = i\} \tag{4.11}$$

or equivalently, in light of equation (4.9),

$$Pr\{X = i, Y = j\} = Pr\{X = i\} \times Pr\{X = j\} \tag{4.12}$$

More generally, $X_0, X_1, \cdots, X_{n-1}$ are independent random variables if'

$$\begin{aligned} &Pr\{X_0 = i_0, X_1 = i_1, \cdots, X_{n-1} = i_{n-1}\} \\ &= Pr\{X_0 = i_0\} \times Pr\{X_1 = i_1\} \times \cdots \times Pr\{X_{n-1} = i_{n-1}\} \end{aligned} \tag{4.13}$$

4.7 THE LAW OF LARGE NUMBERS (LLN)

The law of large numbers is the connection of probability theory to games of chance (gambling). It is claimed that Jacob Bernoulli believed the LLN was so simple that even the stupidest man instinctively knows it is true. Be that as it may, it is reported that his proof appeared in print in 1773 only 20 years after discovery by Bernoulli.

The LLN comes in two flavors: If $X_0, X_1, \cdots, X_{n-1}$ are independent random variables, then their sample mean (or sample average) is

$$\hat{\mu}_n = \frac{X_0 + X_1 + \cdots + X_{n-1}}{n} \tag{4.14}$$

What happens as $n \to \infty$?

Theorem 4.2 (The weak law of large numbers). If the $X_0, X_1, \cdots, X_{n-1}, \cdots$ are **iid**, meaning independent random variables with the same distribution and $\mu = E\{X_i\}$, then for every real number $\varepsilon > 0$

$$1 = \lim_{n\to\infty} Pr\{|\hat{\mu}_n - \mu| < c\} \tag{4.15}$$

Although we will not offer proofs for any of the theorems in this chapter, we note that the proof of the weak LLN follows immediately from

Theorem 4.3 (Chebychev's inequality). If the random variable X has mean μ and variance σ^2, then

$$Pr\{|X - \mu| \geq a\} \leq \frac{\sigma^2}{a^2} \tag{4.16}$$

The second version of the LLN appears in A. N. Kolmogorov's fundamental 1933 memoir [Kolmogorov 1933].

Theorem 4.4 (The strong law of large numbers). With the same hypothsis as in Theorem 4.2 and a new and stronger conclusion, the sequence of random variables $\{\hat{\mu}_n = \frac{X_0 + X_1 + \cdots + X_{n-1}}{n} : 1 \leq n < \infty\}$ converges in probability

$$1 = Pr\left\{\lim_{n\to\infty} \hat{\mu}_n = \mu\right\} \tag{4.17}$$

There is a subtle and technical difference between and weak and strong laws, which we will not discuss in this text. For more details, see [Loéve 1981].

If I foolishly go to Las Vegas with my first royalty check and play n games of blackjack with the payoff (winnings) X_i on the ith game, then $\hat{\mu}_n$ is my net good or bad fortune.

Why do most gamblers fail to smile while the casino owners in Las Vegas and the Internal Revenue Service (IRS) are grinning ear to ear? Well, the rules of games of chance are arranged so that $E\{X_i\} < 0$, and therefore, $\hat{\mu}_n \to -\infty$. Therefore, I will exhaust my meager resources, unless the publisher is more generous.

Even if the game is fair, meaning $E\{X_i\} = 0$, I am not much better off. While $0 = Pr\left\{\lim_{n\to\infty} \hat{\mu}_n\right\}$, the fluctuations of $\hat{\mu}_n$ are very large. The law of the iterated logarithm (see [Feller 1957, pp. 191–192] and [Grimmettand and Stirzaker 1992, p. 301]) is

$$1 = \limsup_{n\to\infty} \frac{\hat{\mu}_n}{\sqrt{2 ln\, ln(n)}} \tag{4.18}$$

Of course, with $\mu = 0$, there might be a way to beat the house; see [Grimmettand and Stirzaker 1992, p. 302–304].

4.8 THE CENTRAL LIMIT THEOREM (CLT)

The CLT appears in the works of de Moivre (1718) and Laplace (1812). It is, like the LLN and law of the iterated logarithm, an example of a limit theorem stating some property of a random process $\{X_n\}$ that does *not* depend on the formula of the individual distributions,

Theorem 4.5 (The central limit theorem). If the $X_0, X_1, \cdots, X_{n-1}, \cdots$ iid random variables with $\mu = E\{X_i\}$ and $\sigma^2 = Var\{Xi\}$, then for every real number $\beta > 0$

$$\lim_{n\to\infty} Pr\left\{\left|\frac{\hat{\mu} - n\mu}{\sigma\sqrt{n}}\right| \leq \beta\right\} = \Phi(\beta) \tag{4.19}$$

where

$$\Phi(x) = \frac{1}{2\pi}\int_{-\infty}^{x} e^{-\frac{1}{2}y^2}\,dy \tag{4.20}$$

4.9 RANDOM PROCESSES AND MARKOV CHAINS

A (discrete-time) random process (*aka* a stochastic process) $\mathcal{X}$ is a sequence of random variables

$$\mathcal{X} = \{X_n : n = 0, 1, 2, \cdots\} \tag{4.21}$$

A standard interpretation of the index n is time.

To define $\mathcal{X}$, it is necessary to specify the joint distribution function

$$F_{X_n, X_{n+1}, \cdots, X_m}(k_n, k_{n+1}, \cdots, k_m)) \equiv Pr\{X_i \leq k_i, n \leq i \leq m\} \tag{4.22}$$

for every pair (n, m) with $0 \leq n \leq m < \infty$.

The simplest example of a random process occurs when the $\{X_n\}$ are independent and identically distributed random variables. In this case, the conditional probability $Pr\{X_{n+1}/X_n\} = Pr\{X_{n+1}\}$ since the $\{X_n\}$ are independent. Such a random process is usually denoted by iid.

Example 4.3. The iid random process model for English language is derived from a large sample of text by *i*) changing uppercase letters to lowercase letters, and *ii*) ignoring all symbols other than letters. A count of the resulting sample of lowercase letters yields the probabilities $\{\pi(i)\}$ of the symbol "*i*" in Table 4.2.

TABLE 4.2. English Language Letter Probabilities

i	$\pi(i)$	i	$\pi(i)$	i	$\pi(i)$
A	0.0856	J	0.0013	S	0.0607
B	0.0139	K	0.0042	T	0.1045
C	0.0279	L	0.0339	U	0.0249
D	0.0378	M	0.0249	V	0.0092
E	0.1304	N	0.0707	W	0.0149
F	0.0289	O	0.0797	X	0.0017
G	0.0199	P	0.0199	Y	0.0199
H	0.0528	Q	0.0012	Z	0.0008
I	0.0627	R	0.0677		

An iid model of English language text does not realistically model English language text; for example, it would assign to both `Goodmorning` and `Gdmoogninr` the same probability.

The more interesting examples assume some time-dependence, meaning that the value of X_{n+1} depends on the value of X_n so that the *absolute distribution* of X_{n+1} is not equal to the *conditional distribution* $Pr\{X_{n+1}/X_n\}$.

The simplest example of such a random process is a stationary discrete-time Markov chain $\mathcal{X}$ (see [Feller 1957, Chapter 15] and [Grimmettand and Stirzaker 1992, Chapter 6]). If the range of each X_n is $\mathcal{Z}_m$, then $\mathcal{X}$ is determined by the following two functions:

i) A probability distribution $\pi(i)$ on the initial state X_0

$$Pr\{X_0 = i\} = \pi(i) \geq 0 \quad 0 \leq i < m \qquad (4.23a)$$
$$1 = \sum_{i=0}^{m-1} \pi(i)$$

ii) A (one-step) transition function $P(j/i)$ on pairs of states

$$Pr\{X_n = j/X_{n-1} = i\} = P(j/i) \geq 0 \quad 0 \leq i, j < m \quad n > 0 \qquad (2.1b)$$

$$1 = \sum_{j=0}^{m-1} P(j/i) \quad 0 \leq i < m \qquad (4.23b)$$

The ℓth order transition function $P^{(\ell)}(j/i)$ is defined by

$$Pr\{X_{n+\ell-1} = j/X_{n-1} = i\} = P^{(\ell)}(j/i) = \begin{cases} P(j/i) & \text{if } \ell = 1 \\ \sum_{k=0}^{m-1} P^{(\ell-1)}(j/k)P(k/i) & \text{if } \ell > 1 \end{cases} \quad \begin{cases} 0 \leq i, j < m \\ n, \ell > 0 \end{cases} \qquad (4.23c)$$

A state j is periodic with period p if $P^{(e)}(j/j) = 0$ unless ℓ is a multiple of p. The Markov chain is aperiodic if has no periodic states.

Remark 4.1. If $P^{(\ell)}(j/i) > 0$ for all i, j for some $\ell > 0$, then the Markov chain is aperiodic. The probability that the Markov chain generates the r-vector of plaintext $(k_0, k_1, \cdots, k_{r-1})$ is

$$Pr\{(X_0, X_1, \cdots, X_{r-1}) = (k_0, k_1, \cdots, k_{r-1})\} = \pi(k_0)\prod_{t=1}^{r-1} P(k_t/k_{t-1}) \qquad (4.23d)$$

The Markov chain $\mathcal{X}$ is stationary or homogeneous (in time) if

$$\pi(j) = \sum_{i=0}^{m-1} \pi(i)P(j/i); \quad 0 \le i < m \qquad (4.23e)$$

Equation (4.23*e*) implies the probability $Pr\{(X_n, X_{n+1}, \cdots, X_m) = (k_n, k_{n+1}, \cdots, k_m)\}$ is the same for each pair of times $0 \le n \le m$. In particular, it implies the following:

- The probability of observing $\{X_n = i\}$ is $\pi(i)$ at each *time n*.
- The probability of observing $\{X_n = i, X_{n+1} = j\}$ is $\pi(i)P(j/i)$ for each pair of times $0 \le n \le m$.

Suppose (π, P) are given, which satisfy the conditions in equations (4.23*a* and 4.23*b*) but *not* necessarily equation (4.23*d*). In this case, $\pi \equiv \pi_0$ is the distribution of the Markov chain's initial state, and the formula in equation (4.23*c*) permits the calculation of the *path* $k_0 \to k_1 \to \cdots \to k_{r-1}$.

The Markov chain $\mathcal{X}$ has a stationary distribution $\pi \equiv \pi_\infty$ if

$$\pi_\infty(i) = \lim_{n\to\infty} Pr\{X_n = i\} \qquad (4.24)$$

In this case, pi_∞ satisfies

$$\pi_\infty(j) = \sum_{i=0}^{m-1} \pi_\infty(i)P(j/i); \quad 0 \le i < m \qquad (4.25)$$

That is, the long-term behavior of the random process forgets its humble origins.

When does a Markov Chain have a stationary distribution and how can it be computed? The full details are given in [Feller 1957] and [Grimmettand and Stirzaker 1992], but briefly details are presented in the next section.

Theorem 4.6. If

a) The states of the chain are irreducible; that is, it is possible to make a transition from i to j in one or more steps $P^{(\ell)}(j/i) > 0$ for *some* $\ell \ge 1$ and
b) If the 1states are aperiodic

then, the matrix equation

$$\underline{\pi} = \mathbf{P}\underline{\pi} \qquad (4.26a)$$

$$\mathbf{P} = \mathbf{P}(j/i) \qquad \mathbf{P}(j/i) \equiv (P(j/i)) \qquad (4.26b)$$

has the unique solution $\underline{\pi} = \underline{\pi_\infty}$ and

$$\underline{\pi_\infty} = \lim_{n\to\infty} \mathbf{P}^n \underline{\pi_\infty} \qquad (4.26c)$$

REFERENCES

W. Feller, *An Introduction to Probability Theory and Its Applications*, Volume 1 (Second Edition), John Wiley & Sons (New York), 1957; (Third Edition), John Wiley & Sons (New York), 1967.

M. R. Fréchet, "Recherches Theorie Modernes en Calcul des Probabilites", 1936.

G. R. Grimmett and D. R. Stirzaker, *Probability and Random Processes*, Oxford University Press (Oxford, UK), 1992.

Paul R. Halmos, *Naive Set Theory*, Springer-Verlag (New York), 1960.

Martin E. Hellman, "A Cryptanalytic Time—Memory Trade-Off", *IEEE Transactions on Information Theory*, **IT-26**, #4, July 1980, pp. 401–406.

A. N. Kolmogorov, *Foundations of the Theory of Probability*, Chelsea Publishing Company (New York), 1933.

A. G. Konheim, *Computer Security and Cryptography*, John Wiley & Sons (New York), 2007.

Michel Loéve, *Probability Theory*, D. Van Nostrand (New York), 1981.

CHAPTER 5

Number Theory and Modern Algebra

5.1 PRIME NUMBERS

The integer p is a *prime* integer, if it is divisible only by 1 and itself; otherwise, p is a *composite* integer. Although there are infinitely many primes, like honest politicians, primes are rare; the number of primes $\pi(n)$ less than or equal n is $o(n)$. Rosser and Schoenfeld [Rosser et al. 1962, p. 64] wrote that first Legendre (1808) and later Gauss (1849) conjectured that the density of primes should be logarithmic; that is

$$\pi(n) \equiv \sum_{p \le n} 1 \approx \int_2^n \frac{dy}{\log y}$$

Rosser and Schoenfeld start with this observation to obtain approximations for sums of the form

$$\sum_{p \le n} f(p)$$

The special case $f(p) = 1$ is the celebrated theorem as follows.

Theorem 5.1. (The prime number theorem).[1] The number $\pi(n)$ of primes less than or equal to n is asymptotic to $\dfrac{n}{\log n}$ as $n \to \infty$; that is,

$$1 = \lim_{n \to \infty} \frac{\pi(n)}{\dfrac{n}{\log n}}.$$

The distance between consecutive primes increases much faster than n as $n \to \infty$ so that the density of primes is 0.

We will state without proof several refinements of the the prime number theorem from [Rosser and Schoenfeld 1962], which we will use in Chapter 16.

Lemma 5.2. Let $\mathcal{P}_n$ denote the set of primes $\le n$.

a) If $n \ge 10^8$, then $\prod_{p \in \mathcal{P}_n} p > e^{n - 2.05282\sqrt{n}}$.

[1] Proofs from Jacques Hadamard and Charles Jean de la Vallée Poussin were published in 1896.

Hashing in Computer Science: Fifty Years of Slicing and Dicing, by Alan G. Konheim

b) If $n \geq 17$, then $\frac{n}{\log n} \leq \pi(n) \leq 1.255506 \frac{n}{\log n}$.

c) There exists a constant B such that

$$\left| \sum_{p \in \mathcal{P}, p \leq n} \frac{1}{p} - \log \log n - B \right| < \frac{1}{2 \log^2 n}$$

for all n.

The two central number-theoretic issues in its application to cryptography are as follows:

- Efficient methods for finding the prime factors of an integer n
- Efficient methods for generating prime number

Although these two problems in number theory existed before the invention of public-key cryptography (see Chapter 6), it is the scale of the numbers involved that sets these problems apart from those in "classic" number theory and has invigorated this ancient branch of mathematics.

5.2 MODULAR ARITHMETIC AND THE EUCLIDEAN ALGORITHM

Our starting point is the following theorem.

Theorem 5.2. (The division algorithm for integers). If $a, b \in \mathcal{Z} \equiv \{0, \pm 1, \pm 2, \cdots\}$ with $b > 0$, there exist unique integers $q, r \in \mathcal{Z}$ such that $a = qb + r$ with $0 \leq r < b$; we write $a = r(\text{mod } b)$.

For every integer $n \geq 2$, addition, subtraction, and multiplication can be defined on the set of residues mod n, that is, on the set of integers $\mathcal{Z}_n = \{0, 1, 2 \cdots, n - 1\}$. If $a, b \in \mathcal{Z}_n$, the division algorithm can be used to define addition, subtraction, and multiplication as follows:

1. Addition + is defined : $a + b = qn + r; 0 \leq r < n \quad r = (a + b) \text{ (mod } n)$
2. Subtraction – is defined; : $a - b = sn + t; 0 \leq t < n \quad t = (a - b) \text{ (mod } n)$
3. Multiplication × is defined; : $a \times b = un + v; 0 \leq v < n \quad v = (a \times b) \text{ (mod } n)$
4. The *distributive law* is satisfied : $a \times (b + c) = (a \times b) + (a \times c)$
5. The *associative law* is satisfied : $a \times (b \times c) = (a \times b)$ timesc
6. The *commutative laws* are satisfied : $a \times b = b \times a; a + b = b + a$
7. An *additive identity* exists : the integer 0 satisfies $x + 0 = 0 + x = x \text{ (mod } n)$
8. A *multiplicative identity* exists : the integer 1 satisfies $x \times 1 = 1 \times x = x \text{ (mod } n)$

The previous list demonstrates the following:

- $\mathcal{Z}_n$ with the arithmetic operation + is an example of a commutative or *Abelian* group; the sum and difference of integers in $\mathcal{Z}_n$ are integers in $\mathcal{Z}_n$ and an additive identity (=0) exists.

- $\mathcal{Z}_n$ with the arithmetic operations + and × is an example of a ring; the sum, difference, and product of integers in $\mathcal{Z}_n$ are integers in $\mathcal{Z}_n$ and both an additive identity (=0) and multiplicative identity (=1) exist.

It is not always possible to solve the equation $ax = b \pmod{n}$ for x given a, b. For example, if $n = 6$, a solution exists for $a = 1, 5$ and all b; for $a = 4, 5$, a solution exists only for $b = 0, 2$ and 4. The formal solution $x = a^{-1}b \pmod{n}$ requires a to have a multiplicative inverse (modulo n); that is, an integer $c \equiv a^{-1}$ in $\mathcal{Z}_n$ such that $1 = (a \times c) \pmod{n} = (c \times a) \pmod{n}$. When n is a prime, every $a \neq 0$ has a multiplicative inverse and $\mathcal{Z}_n$ is a field. Conversely, $\mathcal{Z}_n$ is a prime only if n is a prime.

If $a, b \in \mathcal{Z}^+$, the greatest common divisor of a and b, denoted by $d = gcd\{a, b\}$, is the unique integer d satisfying the following requirements:

1. d divides *both* a and b.
2. If c divides *both* a and b, then c divides d.

Theorem 5.3. $d = gcd\{a, b\}$ is uniquely determined.
$a, b \in \mathcal{Z}^+$ are relatively prime if $1 = gcd\{a, b\}$. According to the proof of Theorem 5.2, if $1 = gcd\{a, b\}$, there exist integers x, y such that $1 = ax + by$. Moreover,

- If $x > 0$, then $y < 0$ and $xa = 1 + (-y \times b) \Rightarrow xa = 1 \pmod{b} \Rightarrow a^{-1} = x \pmod{b}$,
- If $x < 0$, then $y > 0$; if r is such that $rb + x > 0$, then $ra - y > 0$ and $(rb + x)a = 1 + (-y + ra)b \Rightarrow (rb + x)a = 1 \pmod{b} \Rightarrow a^{-1} = (rb + x) \pmod{b}$

so that if $1 = gcd\{a, b\}$ and $0 < a < b$, the multiplicative inverse of a modulo b, denoted by a^{-1}, exists and it satisfies $1 = (a \times a^{-1}) \pmod{b} = (a^{-1} \times a) \pmod{b}$.

The computation of x and y is provided by an algorithm appearing in *Euclid's Elements* (300 BC).

Theorem 5.4. (The Euclidean algorithm).

If $a, b \in \mathcal{Z}^+$, the sequence $r_0, r_1, \cdots, r_s, r_{s+1}$ defined by

$a, b \in \mathcal{Z}^+; r_0 = a; r_1 = b$		
$r_0 = c_1 r_1 + r_2$	$0 < r_2 < r_1$	$r_2 = r_0 \pmod{r_1}$
$r_1 = c_2 r_2 + r_3$	$0 < r_3 < r_2$	$r_3 = r_1 \pmod{r_2}$
$r_2 = c_3 r_3 + r_4$	$0 < r_4 < r_3$	$r_4 = r_2 \pmod{r_3}$
$\vdots$	$\vdots$	$\vdots$
$r_{s-2} = c_{s-1} r_{s-1} + r_s$	$0 < r_s < r_{s-1}$	$r_s = r_{s-2} \pmod{r_{s-1}}$
$r_{s-1} = c_s r_s + r_{s+1}$	$0 < r_{s+1} < r_s$	$r_{s+1} = r_{s-1} \pmod{r_s}$

satisfies

a) For some value of s, $r_j \begin{cases} \neq 0 & \text{if } 0 \leq j \leq s \\ = 0 & \text{if } j = s+1 \end{cases}$.
b) $r_s = gcd\{a, b\}$.
c) $\exists x, y \in \mathcal{Z}, r_s = xa + yb$.

TABLE 5.1. The Euler Totient Function $\phi(n)$ for n = 2(1)13

n	2	3	4	5	6	7	8	9	10	11	12	13
$\phi(n)$	1	2	2	4	2	6	4	6	4	10	4	12

Example 5.1. a = 654, b = 1807 ??? = $gcd\{645, 1807\}$

$$\begin{array}{rl} 1807 = 2\times 654+499 & 499 = 1807-(2\times 654) \\ 654 = 1\times 499+155 & 155 = 654-499 = -1807+(3\times 654) \\ 499 = 3\times 155+34 & 34 = 499-(3\times 155) = (4\times 1807)-(11\times 654) \\ 155 = 4\times 34+19 & 19 = 155-(4\times 34) = (-17\times 1807)+(47\times 654) \\ 34 = 1\times 19+15 & 15 = 34-19 = (21\times 1807)-(58\times 654) \\ 19 = 1\times 15+4 & 4 = 19-15 = (-38\times 1807)+(105\times 654) \\ 15 = 3\times 4+3 & 3 = 15-(3\times 4) = (135\times 1807)-(373\times 654) \\ 4 = 1\times 3+1 & 1 = 4-3 = (-173\times 1807)+(478\times 654) \\ 3 = 3\times 1+0 & \end{array}$$

$$1 = gcd\{560, 1547\} = (-173\times 1807)+(478\times 654) \quad x = 478,\ y = 173$$
$$477 = 654^{-1}(\text{mod } 1807) \Leftrightarrow 1 = (478\times 654)(\text{mod } 1807)$$

The Euler totient function $\phi(n)$ for the positive integer n is the number of positive integers less than n, which are relatively prime to n.

The values of $\phi(n)$ for n = 2(1) 13 are listed in Table 5.1.

Theorem 5.5. If the prime factorization of n is $n = p_1^{n1} p_2^{n2} \cdots p_k^{nk}$, then $\phi(n) = \prod_{i=1}^{k} p_i^{n_i-1}(p_i - 1)$.

5.3 MODULAR MULTIPLICATION

Theorem 5.6. If a, k, n are positive integers, the complexity of modular exponentiation a^k (mod n) is $O((\log_2 k)(\log_2 n)^2)$.

The complexity of the multiplication two s-bit numbers is $O(s^2)$; thus, if we write

$$k = k_0 + k_1 2 + k_2 2^2 + \cdots + k_{s-1} 2^{s-1}$$

each of the $O(\log_2 k)$ powers a^{2^j} (mod n) (j = 1, 2, ···) can be computed in time $O((\log_2 n)^2)$.

```
MOD (a, k, n)

d :=1;
aa :=a;
while (k > 0) do begin
if (1 = (k mod 2)) then
d :=(d * a) mod n;
k :=(k – (k mod 2)) div 2;
aa :=(aa * aa) mod n;
end;
```

Example 5.2. Evaluate $y = 1311^{134}$ (mod 39979); first, the base-2 expansion of the exponent $134 = 128 + 4 + 2 = 2^7 + 2^2 + 2^1$ is determined. Next, $T_j = 1311^{2^j}$ (mod 39979) for $1 \le j \le 7$ is computed by repeated squaring

$$T_1 = T_0^2 \pmod{39\,979} = \left[1311^2 \pmod{39\,979}\right]$$
$$T_2 = T_1^2 \pmod{39\,979} = \left[1311^4 \pmod{39\,979}\right]$$
$$\ddots$$
$$T_7 = Y_6^2 \pmod{39\,979} = \left[1311^{128} \pmod{39\,979}\right]$$

Finally, y is expressed as a product

$$\begin{aligned} y &= 1311^{134} \pmod{39\,979} \\ &= \left[1311^2 \pmod{39\,979}\right] \times \left[1311^4 \pmod{39\,979}\right] \times \left[1311^{128} \pmod{39\,979}\right] \end{aligned}$$

multiplying all the terms $\{T_j\}$ for which 2^j appears in the base-2 expansion of 134 to obtain the value of $y = 17236$.

5.4 THE THEOREMS OF FERMAT[2] AND EULER

Theorem 5.7. (Fermat's little theorem). If p is a prime number

a) $a^p = a \pmod{p}$ for any integer a.
b) $a^{p-1} = 1 \pmod{p}$ if a is not divisible by p.

Proof. We claim $(b + 1)^p = (b^p + 1) \pmod{p}$ for any integer b and prime p; the proof uses the binomial theorem $(b+1)^p = \sum_{i=0}^{p} \binom{p}{i} b^i$. But each binomial coefficient $\binom{p}{i}$ is divisible by p for $1 \le i < p$ giving $(b + 1)^p = (b^p + 1) \pmod{p}$ proving Theorem 5.7a by induction on b. But

$$0 = \left[a^p - a\right] \pmod{p} = a\left[a^{p-1} - 1\right] \pmod{p}$$

so that if $1 = gcd\{a, p\}$, Theorem 5.7b is proved. ■

Remark. Is there a converse to Fermat's theorem? If $a^{n-1} = 1 \pmod{n}$ for every integer a with $1 = gcd\{a, n\}$, does it follow that n is a prime? The answer is no; for example, $561 = 3 \times 11 \times 17$, although it might not be obvious that $a^{560} = 1 \pmod{561}$ for every integer a with $1 = gcd\{a, 561\}$. Moreover, there is an infinite number of such *Carmichael numbers*. We return to the "false" converse of Fermat's theorem in §5.7 when we examine the testing of numbers to determine whether they are prime.

[2]Pierre de Fermat (1601–1665) was both a French lawyer and a mathematician who is credited with the early developments leading to modern calculus.

The Euler totient function $\phi(n)$ of an integer n was defined in §5.4 as the number of integers less than n that are relative prime to n. Theorem 5.5 gave the formula

$$\phi(n)=\prod_{i=1}^{k} p_i^{n_i-1}(p_i-1)$$

when the prime factorization of n is $n=p_1^{n_1}p_2^{n_2}\cdots p_k^{n_k}$. Here, we need only the special case where n is the product of two (distinct) primes $n = p_1p_2$; in this case, $\phi(n) = (p_1 - 1)(p_2 - 1)$.

We need an important generalization of Fermat's little theorem.

Theorem 5.8. (Euler's theorem). If integers n and m are *relatively prime*, meaning their greatest common divisor is 1, $1 = gcd\{m, n\}$, then $m^{\phi(n)} = 1(\text{mod } n)$.

Theorem 5.9. If $1 = gcd\{m, n\}$, then m has a multiplicative inverse in $\mathcal{Z}_n$, which may be computed in time $O((\log_2 n)^2)$.

5.5 FIELDS AND EXTENSION FIELDS

A *field* $\mathcal{F}$ is a mathematical system in which addition (+) and multiplication (×) are defined with the following properties:

- $\mathcal{F}$ is a group under the operation addition + with (additive) identity element 0.
- $\mathcal{F}^* \equiv \mathcal{F} - \{0\}$ is a *cyclic group*,[3] which is an element under the operation multiplication × with (multiplicative) identity 1.

The real $\Re$ and complex numbers C systems are examples of *infinite* fields; we noted in §5.2 that $\mathcal{Z}_p$ is a field for p a prime. There are two possibilities in a field $\mathcal{F}$ when repeated copies $1 + 1 + \cdots$ of the (multiplicative) identity element 1 are added. These possibilities are as follows:

1. if $1 + 1 + \cdots$ is *never* equal to 0, then $\mathcal{F}$ is a field of characteristic 0; $\Re$ and C are examples.
2. $\underbrace{1+1+\cdots+1_n}\begin{cases}\neq 0 & \text{if } 1\leq n<q\\ =0 & \text{if } n=q\end{cases}$.

In the second case, q must be a prime, for in a field $q_1, q_2 \neq 0$ and $q_1q_2 = 0$ implies $q_1 = 0$ or $q_2 = 0$.

When p is a prime number, $\mathcal{Z}_p$ is a field of characteristic p and its nonzero elements $\mathcal{Z}_p^*=\{1,2,\cdots,p-1\}$ form a cyclic group of the order $p - 1$.

$\mathcal{Z}_p$ is not the only field of characteristic p. It is easy to prove that a field $\mathcal{F}$ of characteristic p (a prime) must contain p^m elements for some integer m. Moreover, there is a very simple description of such field, which we now discuss.

The problem

[3]A group is a cyclic group if every element x is s power g^r of a generator g of the group.

Given: $a, y \in \Re$

Find: $x \in \Re$ such that $y = ax$

has a unique solution $x = a^{-1}y$ provided a^{-1} exists.

The same conclusion fails for the equation $y = ax^2 + bx$; it may not always have a root (solution) in the field $\Re$. However, if the field $\Re$ is augmented by including complex numbers, the equation $y = ax^2 + bx$ always has two roots. Defining $\iota = \sqrt{-1}$ as a solution of the equation $0 = x^2 + 1$ and adjoining $\iota = \sqrt{-1}$ to the field $\Re$ produces the complex number system $\mathcal{C}$.

The complex number system $\mathcal{C}$ consisting of all numbers of the form $x = u + \iota v$ forms a field is a field in which the operations addition, subtraction, and division are defined by the following:

Addition: $(u_1 + \iota v_1) + (u_2 + \iota v_2) = (u_1 + u_2) + \iota(v_1 + v_2)$

Subtraction: $(u_1 + \iota v_1) - (u_2 + \iota v_2) = (u_1 - u_2) + \iota(v_1 - v_2)$

Multiplication: $(u_1 + \iota v_1) \times (u_2 + \iota v_2) = (u_1 u_2 - v_1 v_2) + \iota(u_1 v_2 + u_2 v_1)$

Division: $(u_1 + \iota v_1) \div (u_2 + \iota v_2) = \dfrac{(u_1 u_2 + v_1 v_2) + \iota(u_2 v_1 - u_1 v_2)}{D}$ if $D = (u_2^2 + v_2^2) \neq 0$

Moreover, every polynomial $p(x) = p_0 + p_1 x + \cdots + p_n x^n$ with coefficients in $\mathcal{C}$ and $p_n \neq 0$

- Has precisely n roots
- $p(x)$ splits into the product $p(x) = p_n(x - x_1)(x - x_2)\cdots(x - x_n)$ of linear factors where each of the $x_j = u_j + \iota v_j$ are roots of $p(x) = 0$

This same process can be defined for polynomials $p(x) = p_0 + p_1 x + \cdots + p_m x^m$ whose coefficients $\{p_i\}$ are in $\mathcal{Z}_2 = \{0, 1\}$; we write $p \in \mathcal{P}_{\mathcal{Z}_2}[x]$ and, if $p_m = 1$, then $\deg(p) = m$.

1. A polynomial $p \in \mathcal{P}_{\mathcal{Z}_2}[x]$ is *reducible* if it may be factored, that is, written as a product $p(x) = r(x)s(x)$ of polynomials in $\mathcal{P}_{\mathcal{Z}_2}[x]$ each of positive degree. $r(x)$ and $s(x)$ are factors of $p(x)$.
2. If $p \in \mathcal{P}_{\mathcal{Z}_2}[x]$, then
   –$p(1) = 0$ implies $p(x) = (1 + x)q(x)$.
   –$p(0) = 0$ implies $p(x) = xq(x)$.
3. A polynomial $p \in \mathcal{P}_{\mathcal{Z}_2}[x]$ of degree ≥ 1 is *irrreducible* if $p(x) = r(x)s(x)$ implies either $\deg(p) = 0$ or $\deg(q) = 0$.
4. If $p, q \in \mathcal{P}_{\mathcal{Z}_2}[x]$, there exist polynomials $a, b \in \mathcal{P}_{\mathcal{Z}_2}[x]$ such that $p(x) = a(x)q(x) + b(x)$.
5. *Modular arithmetic* for $\mathcal{P}[x]$ is defined by the rule if $p(x) = a(x)q(x) + r(x)$; $r(x)$, then we write $r(x) = p(x) \pmod{q(x)}$ for the residue of $p(x)$ mod $q(x)$.
6. **Theorem 5.10.** (see [Menezes et al. 1996, Chapter 6]). If the polynomial $p \in \mathcal{P}_{\mathcal{Z}_2}[x]$ satisfies $p(0) = 1$, then $1 + x^d$ for some d with $d \geq 1$ is a factor of $p(x)$. The exponent of p is the smallest integer e for which this occurs; if e is the exponent of p, then $0 = (1 + x^e) \pmod{p(x)}$ or $1 = x^e \pmod{p(x)}$.

7. The exponent e of $p \in \mathcal{P}_{\mathcal{Z}_2}[x]$ of degree m satisfies $1 \le e \le 2^m - 1$ [Menezes, Van Oorschot and Vanstone 1996].
8. A polynomial $p \in \mathcal{P}_{\mathcal{Z}_2}[x]$ of degree m is **primitive** if its exponent is $e = 2^m - 1$.
9. Let $p \in \mathcal{P}_{\mathcal{Z}_2}[x]$ of $\deg(p) = m$ with $p(x) = p_0 + p_1x + \cdots + p_mx^m$ be irreducible. The set of residues of $\mathcal{P}_{\mathcal{Z}_2}[x]$ modulo $p(x)$, denoted by $\mathcal{P}_{\mathcal{Z}_2}[x]/p(x)$, consists of all polynomials $a(x) \equiv a_0 + a_1x + \cdots + a_{m-1}x^{m-1}$ of degree $< m$.

9a. *Uniqueness*

$$\text{If } a(x) \equiv a_0 + a_1x + \cdots + a_{m-1}x^{m-1} \in \mathcal{P}_{\mathcal{Z}_2}[x]/p(x) \text{ and}$$
$$b(x) \equiv b_0 + b_1x + \cdots + b_{m-1}x^{m-1} \in \mathcal{P}_{\mathcal{Z}_2}[x]/p(x),$$

then

$$a(x) \overset{p(x)}{=} b(x) \Rightarrow a_i = b_i \text{ for } 0 \le i < m.$$

The addition, multiplication, and division on the set of residues can be defined as follows;

9b. *Addition*

$$\text{If } a(x) \equiv a_0 + a_1x + \cdots + a_{m-1}x^{m-1} \in \mathcal{P}_{\mathcal{Z}_2}[x]/p(x) \text{ and}$$
$$b(x) \equiv b_0 + b_1x + \cdots + b_{m-1}x^{m-1} \in \mathcal{P}_{\mathcal{Z}_2}[x]/p(x),$$

then

$$a(x) \overset{p(x)}{+} b(x) = (a_0 + b_0) + (a_1 + b_1)x + \cdots + (a_{m-1} + b_{m-1})x^{m-1} \in \mathcal{P}_{\mathcal{Z}_2}[x]/p(x).$$

9c. *Multiplication*

$$\text{If } a(x) \equiv a_0 + a_1x + \cdots + a_{m-1}x^{m-1} \in \mathcal{P}_{\mathcal{Z}_2}[x]/p(x) \text{ and}$$
$$b(x) \equiv b_0 + b_1x + \cdots + b_{m-1}x^{m-1} \in \mathcal{P}_{\mathcal{Z}_2}[x]/p(x),$$

then

$$c(x) \equiv a(x) \overset{p(x)}{\times} b(x) \text{ is defined by } c(x) = (a(x)b(x)(\text{mod } p(x)) \in \mathcal{P}_{\mathcal{Z}_2}[x]/p(x).$$

We claim that for every $a(x) \in \mathcal{P}_{\mathcal{Z}_2}[x]/p(x)$ with $a(x) \neq 0 \in \mathcal{P}_{\mathcal{Z}_2}[x]/p(x)$, there exists a polynimial $b(x) \in \mathcal{P}_{\mathcal{Z}_2}[x]/p(x)$ such that $1 = a(x)b(x)(\text{mod } p(x)) \in \mathcal{P}_{\mathcal{Z}_2}[x]/p(x)$ so that $1 \overset{p(x)}{=} \left(a(x) \overset{p(x)}{\times} b(x) \right)$.

Proof. The statement $a(x)b(x) \overset{p(x)}{=} 0$ contradicts the irreducibility of $p(x)$. Because there are $2^m - 1$ nonzero residues $b(x) \in \mathcal{P}_{\mathcal{Z}_2}[x]/p(x)$, there must be *exactly* one residue $b(x) \in \mathcal{P}_{\mathcal{Z}_2}[x]/p(x)$ for which $1 \overset{p(x)}{=} \left(a(x) \overset{p(x)}{\times} b(x) \right)$. ■

Now, we abandon the notations $\overset{p(x)}{=}$, $\overset{p(x)}{+}$ and $\overset{p(x)}{\times}$ returning to our old favorites $=$, $+$, and $\times$ and conclude

Theorem 5.11. $\mathcal{P}_{\mathcal{Z}_2}[x]/p(x)$ is a field obtained by the *adjunction* of the root, say ϑ, to the field $\mathcal{Z}_2$. Because $\mathcal{P}_{\mathcal{Z}_2}[x]/p(x)$ consists of all polynomials of degree $< m$, we may replace this notation with $\mathcal{Z}_{m,2} = \{(a_0, a_1, \cdots, a_{m-1}) : a_i \in \mathcal{Z}_2, 0 \le i < m\}$.

Example 5.3. Table 5.2 lists x^j (mod $p(x)$) together with its *polynomial coding*; the 4-bit representation for $0 \le j \le 5$ for the irreducible polynomial $p(x) = 1 + x + x^2 + x^3 + x^4$ with exponent $e = 5$.

The nonzero elements of $\mathcal{Z}_{m,2}$ form a (multiplicative) cyclic group of order $2^m - 1$ [Menezes, Van Oorschot and Vanstone 1996, p. 81].

When $p \in \mathcal{P}_{\mathcal{Z}_2}[x]$ of degree m is irreducible, the $2^m - 1$ elements of the extension field $\mathcal{Z}_{m,2}$ may be represented in two different ways, as follows:

1. The polynomial coding, the $2^m - 1$ nonzero m-bit binary sequences $\underline{y} = (y_0, y_1, \cdots, y_{m-1})$
2. As the $2^m - 1$ powers of a generator of $\mathcal{Z}_{m,2} - \{0\}$

When $p \in \mathcal{P}_{\mathcal{Z}_2}[x]$ is primitive, these representations are the same; when $p \in \mathcal{P}_{\mathcal{Z}_2}[x]$ is irreducible but not primitive, these representations are different.

In *Example 5.3*, the 16 nonzero elements of the field $\mathcal{Z}_{4,2}$ form a cyclic group of order $15 = 2^4 - 1$; it contains generators of orders 3, 5 and 15. Tables 5.3L, R below give two coding of the elements of the extension field $\mathcal{Z}_{2,2^4}$.

- The generator ϑ in Table 5.3L is of order 15.
- The element $\eta = (0, 1, 1, 0)$ in Table 5.3R is of order 15.

Example 5.3 illustrates the following two important ideas:

1. There are different ways of generating the extension field $\mathcal{Z}_{m,2}$:
   - Adjoining a root of a primitive m^{th} degree polynomial to $\mathcal{Z}_2$
   - Adjoining a root of an irreducible but *non*primitive m^{th}-degree polynomial to $\mathcal{Z}_2$.

TABLE 5.2. x^j (mod $p(x)$) $0 \le j \le 5$ for $p(x) = 1 + x + x^2 + x^3 + x^4$

$1 = x^0$ (mod $p(x)$) $\leftrightarrow$ (0, 0, 0, 1)	$x = x^1$ (mod $p(x)$) $\leftrightarrow$ (0, 0, 1, 0)
$x^2 = x^2$ (mod $p(x)$) $\leftrightarrow$ (0, 1, 0, 0)	$x^3 = x^3$ (mod $p(x)$) $\leftrightarrow$ (1, 0, 0, 0)
$x^3 + x^2 + x + 1 = x^4$ (mod $p(x)$) $\leftrightarrow$ (1, 1, 1, 1)	$1 = x^5$ (mod $p(x)$) $\leftrightarrow$ (0, 0, 0, 1)

TABLE 5.3L. Polynomial Coding of ϑ^j $0 = 1 + \vartheta + \vartheta^4$

$p(x) = 1 + x + x^4$				$p(\vartheta) = 0$	
(0, 0, 0, 1)	1	(0, 0, 1, 0)	ϑ	(0, 1, 0, 0)	ϑ^2
(1, 0, 0, 0)	ϑ^3	(0, 0, 1, 1)	ϑ^4	(0, 1, 1, 0)	ϑ^5
(1, 1, 0, 0)	ϑ^6	(1, 0, 1, 1)	ϑ^7	(0, 1, 0, 1)	ϑ^8
(1, 0, 1, 0)	ϑ^9	(0, 1, 1, 1)	ϑ^{10}	(1, 1, 1, 0)	ϑ^{11}
(1, 1, 1, 1)	ϑ^{12}	(1, 1, 0, 1)	ϑ^{13}	(1, 0, 0, 1)	ϑ^{14}

TABLE 5.3R. Polynomial Coding of η^j 0 = $1 + \eta + \eta^2 + \eta^3 + \eta^4$

$p(x) = 1 + x + x^2 + x^3 + x^4$					
(0, 1, 1, 0)	η	(1, 0, 1, 1)	η^2	(0, 1, 0, 0)	η^3
(0, 1, 1, 1)	η^4	(1, 1, 0, 1)	η^5	(1, 1, 1, 1)	η^6
(1, 0, 0, 1)	η^7	(1, 0, 1, 0)	η^8	(0, 0, 1, 0)	η^9
(1, 1, 0, 0)	η^{10}	(1, 0, 0, 1)	η^{11}	(1, 0, 0, 0)	η^{12}
(1, 1, 1, 0)	η^{13}	(0, 1, 0, 1)	η^{14}	(0, 0, 0, 1)	η^{15}

In the first instance, the root, say ϑ, is a generator of the multiplicative (or *Galois*) group of nonzero elements of the extension field; in the second, it is no a generator.

2. In both instances
   - The polynomial coding forms a **basis** for the extension field $\mathcal{Z}_{m,2}$; every element in the field is a linear combination of one of the vectors

$$\underline{u}_0 = ((0)_{n-1}, 1), \quad \underline{u}_1 = ((0)_{n-2}, 1, 0) \quad \underline{u}_{n-1} = \left(1, (0)_{n-j-1}\right).$$

   - There is an element $\eta = (0, l, 1, 0) \in \mathcal{Z}_{m,2}$ whose powers $1, \eta, \eta^2, \cdots, \eta^{2^m-2}$ generate *all* elements of $\mathcal{Z}_{m,2}$.
   - There is a normal basis for the extension field $\mathcal{Z}_{m,2}$; an element $\lambda \in \mathcal{Z}_{m,2}$ whose powers $\lambda, \lambda^2, \lambda^{2^2}, \cdots, \lambda^{2^m-1}$ form a basis for $\mathcal{Z}_{m,2}$

$$z \in \mathcal{Z}_{m,2} \Rightarrow z = \sum_{j=0}^{2^m-1} a_j \lambda^{2^j} \quad a_j \in \{0, 1\}\ 0 \le j < m.$$

Example 5.3 (continued). If $\lambda = \eta^j$ with $1 = gcd\{j, 15\}$, then $\lambda, \lambda^2, \lambda^{2^2}, \cdots, \lambda^{2^m-1}$ form a normal basis for $\mathcal{Z}_{m,2}$.

5.6 FACTORIZATION OF INTEGERS

We may assume that the integer n is odd; several factorization methods are based on a simple idea attributed to Dixon [Dixon 1981]. If integers x and y can be found so that $x^2 = y^2 \pmod n$, then $(x - y)(x + y) = 0 \pmod n$. If $x \neq \pm y \pmod n$, then either $gcd\{n, x - y\}$ or $gcd\{n, x + y\}$ is a nontrivial factor of n. In fact, the factorizations $n = ab$ are in 1:1 correspondence with pairs of integers s, t such that $0 = (t^2 - s^2) \pmod n$ in the sense that

$$t = \frac{a+b}{2} \text{ and } s = \frac{a-b}{2} \text{ [Koblitz 1987, Proposition V.3.1].}$$

Example 5.4. $37^2 = 7^2 \pmod{55}$.

$$(37-7)\times(37+7) = 30\times 44 = 0 \pmod{55}$$
$$5 = gcd\{55, 30\} \quad 11 = gcd\{55, 44\}$$

To find pairs (x, y), random values of s in $\mathcal{Z}_n$ are chosen and $u = s^2 \pmod{n}$ is computed. If u is a perfect square (mod n), say $u = t^2 \pmod{n}$ and both $0 \neq (s - t) \pmod{n}$ and $0 \neq (s + t) \pmod{n}$, then we find a factor of N.

For example, if $n = 55$ and $s = 13$, then

$$13^2 \pmod{55} = 4 = 2^2 \pmod{55}$$

leading to then factorization

$$11 = gcd\{55, 13-2\} \quad 5 = gcd\{55, 13+2\}$$

Of course, if we have chosen $s = 12$, then

$$12^2 \pmod{55} = 34$$

which is not a perfect square. However, we may lessen the effect of bad choice of s by randomly choosing r integers $\{s_i\}$ and computing their squares (mod n).

$$u_i = s_i^2 \pmod{n}; \quad 1 \leq i \leq r$$

Write the prime factorizations of the $\{u_i\}$

$$u_1 = \prod_k p_k^{e1,k} \quad u_2 = \prod_k p_k^{e2,k} \quad \cdots \quad u_r = \prod_k p_k^{e_r,k}$$

The strategy is to combine by multiplication *some* of the $\{u_i\}$ so that the total exponent of the terms included is even. In this way, the product of the terms included is a perfect square.

Example 5.5. $n = 77$

$$\begin{aligned} 15 &= 13^2 \pmod{77}; \quad 15 = 3 \times 5 \\ 56 &= 21^2 \pmod{77}; \quad 56 = 2^3 \times 7 \\ 60 &= 37^2 \pmod{77}; \quad 60 = 2^2 \times 3 \times 5 \\ 70 &= 42^2 \pmod{77}; \quad 2 \times 5 \times 7 \end{aligned}$$

yielding

$$15 \times 60 = 2^2 \times 3^3 \times 5^2 = 30^2 = (13 \times 37)^2 \pmod{77}$$

leading to the factorization

$$11 = gcd\{77, 481-30\} \quad 7 = gcd\{77, 481+30\}$$

Combining the $\{s_i : 1 \leq i \leq r\}$ can be performed systematically as follows:

1. Find the prime factorization of $u_i = s_i^2 \pmod{n} = \prod_j p_j^{e_{j,i}}$.

2. If $p_1 < p_2 < \cdots < p_m$ denote the set of primes arising in the factorization of the $\{u_i\}$, then $r \times m$ form the r array of exponents $\Gamma = \begin{pmatrix} e_{1,1} & e_{1,2} & \cdots & e_{1,m} \\ e_{2,1} & e_{2,2} & \cdots & e_{2,m} \\ \cdots & \cdots & \ddots & \cdots \\ e_{r,1} & e_{r,2} & \cdots & e_{r,m} \end{pmatrix}$ and try to find an r vector $\underline{x} = (x_1, x_2, \cdots, x_r) \in \mathcal{Z}_{r,2}$ such that $\underbrace{(0, 0, 0, \cdots, 0)}_{\text{length } m} = (x_1, x_2, \cdots, x_r)\Gamma \pmod 2$.

Using Gaussian elimination [Riesel 1994, p. 195], calculate

$$v^2 \equiv (s_{i_1} \times s_{i_2} \cdots s_{i_M}) \pmod n \quad u^2 \equiv (u_{i_1}^2 \times u_{i_2}^2 \cdots u_{i_M}^2) \pmod n$$
$$u^2 = v^2 \pmod n$$

The process is successful if $u \neq \pm v \pmod n$; if it is unsuccessful, then choose another set $\{u_i\}$.

Many improvements to Dixon's method are described in [Koblitz 1987], [Pomerance 1980], and [Riesel 1994].

5.7 TESTING PRIMALITY

Various cryptographic protocols (Diffie-Hellman, RSA, and elliptic curve) described in Chapter 6 require prime numbers for their implementation. The testing whether an integer n is a prime comes in two flavors-deterministic and probabilistic. A deterministic test makes an unequivocal answer, either yes or no, to the question "*Is n a prime*?". Trial division of the integers $\leq \sqrt{n}$ (as in The Sieve of Eratosthenes discussed in §5.2) is deterministic, but its running time is too large for the primes required in cryptography. Agrawal, Kayal, and Saxena [Agrawal, Kayal and Saxena 2004] gave a simple deterministic algorithm to test the primality of an integer. Their discovery, more correctly their paper, was the winner of the 2006 Gödel prize.[4]

The AKS algorithm uses the same idea as in the proof of Fermat's little theorem.

Lemma 5.11. If $gcd\{a, n\} = 1$, then n is prime if and only if the two polynomials $(X + a)^n$ and $(X^n + a)$ are equal modulo n (in X).

Proof. If n is a prime

$$(X+a)^n = X^n + \sum_{i=1}^{n} \binom{n}{i} x^i a^{n-i} + a^n$$

For $1 \leq i < n$, we have $\binom{n}{i} = 0 \pmod n$ and $a^n = a \pmod n$ so that

[4]This prize for outstanding papers in theoretical computer science is named after Kurt Gödel. It is awarded annually jointly by the European Association for Theoretical Computer Science (EATCS) and the Association for Computing Machinery (ACM).

$$(X+a)^n = [X^n + a] (\text{mod } n)$$

If n is not a prime, suppose q is a factor of n but the integer k with $1 \leq k \leq n$ satisfies $0 \neq n$ (mod q^k). It follows that

i) q^k does not divide $\binom{n}{q}$.
ii) $1 = gcd\{a^{n-q}, q^k\}$.

It follows that the two polynomials $(X + a)^n$ and $(X^n + a)$ are not identical completing the proof. ■

The drawback of Lemma 5.11 is the running time of the polynomial comparison, requiring $O(n)$ (comparison) steps. A simplification is achieved in AKS by comparing these two polynomials modulo $gcd\{X^r - 1, n\}$. The AKS algorithm is extremely simple.

AKS algorithm

Input: integer $n > 1$

1. If $n = ab$, for integers $a > 0$ and $b > 1$, output composite.
2. Find the smallest r such that $o_r(n) > \log^2 n$.
3. If $1 < gcd\{a, n\} < n$ for some $a \leq r$, output composite.
4. If $n \leq r$, output prime.
5. For $a = 1$ to $\lfloor \sqrt{\phi(r)} \log n \rfloor$ do ϕ [Euler totient function] if $(X + a)^n \neq (X^n + a)$ mod $gcd\{X^r - 1, n\}$, output composite.
6. Output prime.

It is proved that the running time of the AKS Algorithm is a polynomial if *log n*; the timing estimate in the original paper was $O(\log^{12} n)$. Subsequently in 2005, H. W. Lenstra Jr. and Carl Pomerance modified AKS to improve its running time to $O(\log^6 n)$.

The bad news is that $O(\log^6 n)$ might take a long time to test the primality of

- The 384-bit prime $p = 2^{384} - 2^{128} - 2^{96} + 2^{32} - 1$, which is cited in [FIPS 186-3].
- The 15071-digit prime $4405^{2638} + 2638^{4405}$ cited by R. Brent [Brent 2006].

Probabilistic tests of primality are based on Fermat's little theorem; $a^{n-1} = 1$ (mod n) for every a for which $1 = gcd\{a, n\}$ provided n is a prime number. The test of the primality of n will randomly choose a and evaluate a^{n-1} (mod n) until either of the following occur:

i) An a is found for which $a^{n-1} \neq 1$ (mod n), proving n is composite.
ii) We lose patience.
   - If n is an odd composite number and $1 \leq a < n$ such that $a^{n-1} \neq 1$ (mod n), then a is a Fermat witness to the compositeness of n.
   - If n is an odd composite number and $1 \leq a < n$ such that $a^{n-1} = 1$ (mod n), then n is a pseudoprime to base a an a is a Fermat liar to the primality of n.

Fermat's primality test

Input : n, odd integer; $t \geq 1$.
For $i = 1$ to t do
a) Choose a random integer a with $2 \leq a \leq n - 2$
b) Compute $r = a^{n-1} \pmod{n}$
c) If $r \neq 1$, Return ("Composite").
Return("Prime")

If n is a prime, then Fermat's primality test will *Return* ("*Prime*"), but it may *falsely* conclude n is a prime when it is composite as a result of a single Fermat primality test if $a^{n-1} = 1 \pmod{n}$.

Indeed, it is possible that $a^{n-1} = 1 \pmod{n}$ for every a for which $1 \leq a \leq n - 1$. Such a composite integer n is a Carmichael number.[5]

- The smallest Carmichael number is $n = 561 = 3 \times 11 \times 17$.
- The so-called *taxicab* number 1729[6]
- And ··· for triple jeopardy, there are only 105 212 Carmichael numbers $\leq 10^{15}$.

Theorem 5.12.

a) n is a Carmichael number if and only if
   - n is square free.
   - $p - 1$ divides $n - 1$ for every prime divisor of n.

b) Each Carmichael number has at least three distinct prime factors.
c) There are an infinite number of Carmichael numbers.

Suppose n is composite, a primality test that computes the value $a^{n-1} \pmod{n}$ until an a is found yielding a residue $\neq 1$ may fail for the following two reasons:

- n might be a Carmichael number.
- The computation could be infeasible if the first a for which computing a^{n-1} (mod n) is *not* satisfied is too large.

Theorem 5.13. (Miller-Rabin) [Rabin 1976]. Let n be an odd prime, $n - 1 = 2^r s$ where r is odd and let a be any integer such that $1 = gcd\{a, n\}$. Then, either $a^r = 1 \pmod{n}$ or $a^{2^j r} = -1 \pmod{n}$ for some j with $0 \leq j \leq s - 1$.

Let n be an odd composite number, $n - 1 = 2^r s$ where r is odd and a is integer in $[1, N - 1]$.

- If $a^r \neq 1 \pmod{n}$ or $a^{2^j r} \neq -1 \pmod{n}$ for it every j with $0 \leq j \leq s - 1$, then a is a strong witness to the compositeness of n.

[5]Discovered by R. D. Carmichael in 1910.

[6]The number Carmichael $1729 = 7 \times 13 \times 29 = 1^3 + 12^3 = 9^3 + 10^3$ appears in a story about the Indian genius Srinivasa Ramanujan as related by the great English mathematician G. H. Hardy and cited in his book *A Mathematician's Apology* [Hardy 1992].

- Otherwise, if $a^r = 1 \pmod{n}$ or $a^{2^j r} = -1 \pmod{n}$ for some j with $0 \le j \le s-1$, then n is a strong pseudoprime to the base a to the compositeness of a.

Miller-Rabin Primality Test

MR1. Write $n - 1 = 2^r s$, where r is odd.
MR2. Choose a random integer a with $2 \le a \le n - 2$.
MR3. Compute $b = a^r \pmod{n}$.
MR4. If $b = 1 \pmod{n}$, then RETURN("prime") and END.
MR5. For $i = 0$ to $s - 1$ do
MR6a. If $b = -1 \pmod{n}$, then RETURN("prime") and END.
MR6b. Otherwise, compute $b^2 \pmod{n}$.
MR7. RETURN("composite").

END.

Theorem 5.14.

a) If n is an odd prime, the output of the Miller-Rabin test is RETURN("prime").
b) If N is an odd composite number, the probability that the Miller-Rabin test fails RETURN("prime") for t independent values of a is less than $\left(\frac{1}{4}\right)^t$.

Example 5.6. Miller-Rabin test for $n = 229$.

$n = 229$;	$n - 1 = 2^2 \times 57$	
$a = 225$	$y = a^r \pmod{n} = 1$	prime
$a = 47$	$y = a^r \pmod{n} = 107$	prime
$a = 151$	$y = a^r \pmod{n} = 1$	prime
$a = 101$	$y = a^r \pmod{n} = 122$	prime
$a = 52$	$y = a^r \pmod{n} = 107$	prime
$a = 21$	$y = a^r \pmod{n} = 107$	prime
$a = 180$	$y = a^r \pmod{n} = 1$	prime
$a = 189$	$y = a^r \pmod{n} = 107$	prime
$a = 79$	$y = a^r \pmod{n} = 107$	prime
$a = 126$	$y = a^r \pmod{n} = 1$	prime

Example 5.7. Miller-Rabin test for $231 = 3 \times 77$.

$n = 231$	$n - 1 = 2^1 \times 115$	
$a = 227$	$y = a^r \pmod{n} = 164$	composite
$a = 47$	$y = a^r \pmod{n} = 89$	composite
$a = 152$	$y = a^r \pmod{n} = 89$	composite
$a = 101$	$y = a^r \pmod{n} = 164$	composite
$a = 53$	$y = a^r \pmod{n} = 221$	composite
$a = 21$	$y = a^r \pmod{n} = 21$	composite

$a = 182$	$y = a^r \pmod{n} = 98$	composite
$a = 190$	$y = a^r \pmod{n} = 1$	prime
$a = 79$	$y = a^r \pmod{n} = 142$	composite
$a = 127$	$y = a^r \pmod{n} = 43$	composite

Let n be odd and $n - 1 = 2^s r$ with r odd. An integer a with $2 \leq a \leq n - 2$ is a witness to the compositeness of n if the Miller-Rabin test fails to RETURN("prime"); that is,

1. $a^r \pmod{n} \neq 1$, and
2. $a^{2^j r} \pmod{n} \neq 1$ for all j with $1 \leq j < s$.

Remarks. The chances that a the Miller-Rabin test will fail for a composite n is estimated using Theorem 2.1 in [Schoof 2008].

Theorem 5.15. The number of composite integers of the form $n = 1 + 2^k m$ for which Miller-Rabin fails is bounded above by $\frac{1}{4}\phi(n) \leq \frac{n-2}{4}$ where ϕ is the Euler-totient function.

Corollary 5.16. The probability that a composite n passes the Miller-Rabin test is less than $\frac{1}{4}$.

A deterministic version of Miller-Rabin from G. Miller exists provided the Generalized Riemann Hypothesis holds.

Theorem 5.17. (Gary Miller) [Miller 1976][7]: If n is composite and the Generalized Riemann Hypothesis holds, then a constant c exists such that there exists an a which is a witness to the compositeness of n with $1 < a \leq c(\log_2 n)^2$.

It has been shown that c may be taken equal to 2 thus showing the existence of a *deterministic* test for the primality of n whose complexity is a polynomial in the bit length of n.

There is also an earlier primality test created by Solovay and Strassen [Solovay and Strassen 1977].

There are other algorithms [Goldwasser and Kilian 1986], [Atkin and Morian 1993], and [Adleman and Huang 1992] formulated over elliptic curves (see Part IV).

But more important, the errors in [Adleman and Huang 1992] return *false prime numbers* as opposed to *false composite numbers* as in Miller-Rabin or Solovay-Strassen. This suggests the use of Las Vegas algorithms, which use say, Miller-Rabin and alternating with Adelman-Huang.

[7]The *Riemann zeta function* is defined for complex z by $\zeta(z) = \sum_{n=1}^{\infty} n^{-z}$. The series converges when $Re(z) > 1$. The zeta function may be continued analytically to a domain in the complex plane including the region $0 < Re(z) < 1$.
The Riemann Hypothesis : All zeros of $\zeta(z)$ lie on the line $Re(z) = \frac{1}{2}$.
The Generalized Riemann Hypothesis replaces $\zeta(z)$ by the Dirichlet L-series and makes the same assertion about zeros.

REFERENCES

L. Adleman and M. D. A. Huang, *Primality Testing and Two Dimensional Abelian Varieties over Finite Fields*, Springer-Verlag Notes in Mathematics, **1512**, 1992.

M. Agrawal, N. Kayal, and N. Saxena, "PRIMES Is in P", *Annals of Mathematics*, **160**, #2, pp. 781–793, 2004.

A. O. L. Atkin and F. Morian, "Elliptic Curves and Primality Proving", *Mathematics of Computation*, **61**, pp. 29–68, 1993.

R. P. Brent, "Uncertainty Can Be Better Than Certainty", presented at the Australian National University, November 27, 2006.

J. D. Dixon, "Asymptotically Fast Factorization of Integers", *Mathematics of Computation*, **36**, pp. 255–260, 1980.

Federal Information Processing Standards Publication, Digital Signature Standard (DSS), November 20, 2008.

S. Goldwasser and J. Kilian, "Almost All Primes Can Be Quickly Certified", Proceedings of the Eighteenth Annual ACM Symposium on Theory of Computing, pp. 316–329, 1986.

G. H. Hardy, *A Mathematician's Apology*, Cambridge University Press (Cambridge, UK), 1992.

N. Koblitz, *A Course in Number Theory and Cryptography*, Springer-Verlag (Berlin, Germany), 1987.

A. J. Menezes, P. C. Van Oorschot, and S. A. Vanstone, *Handbook of Applied Cryptography*, CRC Press (Boca Raton, FL), 1996.

G. L. Miller, "Riemann's Hypothesis and Tests for Primality", *Journal of Computer and System Sciences*, **13**, pp. 300–317, 1976.

C. Pomerance, "Cryptology and Computational Number Theory", *Proceedings of Symposia in Applied Mathematics*, **142**, American Mathematical Society (Providence, RI), 1980.

M. Rabin, "Probabilistic Algorithms", in J. F. Traub, editor, *Algorithms and Complexity*, Academic Press, San Diego, California, 1976, pp. 21–38.

H. Riesel, *Prime Numbers and Computer Methods for Factorization* (Second Edition), Birkhauser (Berlin, Germany), 1994.

J. B. Rosser and L. Schoenfeld, "Approximate Formulas for Some Functions of Prime Numbers", *Illinois Journal of Mathematics*, **6**, pp. 64–94, 1962.

R. Schoof, "Four Primality Algorithms", in *Algorithmic Number Theory*, Cambridge University Press (Cambridge, UK), 2008, pp. 101–126.

R. Solovay and V. Strassen, "A Fast Monte-Carlo Test for Primality", *SIAM Journal of Computing*, **6**, pp. 84–85, 1977.

CHAPTER 6

Basic Concepts of Cryptography

6.1 THE LEXICON OF CRYPTOGRAPHY

The word cryptography is derived from the Greek word kryptos, which means hidden, and graphien, which means to write. David Kahn's book [Kahn 1967] provide extensive accounts of cryptography and its influence on history.

Every scientific discipline develops its own lexicon and cryptography is no exception. We begin with a brief summary of the principal terms used in cryptography.

An alphabet $\mathcal{A} = \{a_0, a_1, \cdots, a_{m-1}\}$ is a finite set of **letters**; for our purposes, the most natural alphabets are

Binary ($m = 2^r$): (0,1)-sequences of fixed length r: $Z_{r,2} = \{\underline{x} = (x_0, x_1, \cdots, x_{r-1})\}$.

American Standard Code for Information Interchange (**ASCII**)/Extended Binary Coded Decimal Interchange (**EBCDIC**) ($m = 2^7/m = 2^8$).

Text is formed by concatenating letters of $\mathcal{A}$; an n-gram $(a_0\ a_1 \cdots a_{n-1})$ is the concatenation of n letters. We do not require that the text be *understandable* nor that it be grammatically correct relative to a *natural language*; thus, examples of ASCII text include both `GoodMorning` and `vUI*_9Uiing8`.

Encipherment (or encryption) is a transformation process T the plaintext $\underline{x} = (x_0, x_1, \cdots, x_{n-1})$ is enciphered to the ciphertext $\underline{y} = (y_0, y_1, \cdots, y_{m-1})$.

$$T : \texttt{Good\_Morning} \rightarrow \texttt{Kssh\_Qsvrmrk}$$

is an example of encipherment introduced nearly 2000 years ago by Julius Caesar during the Gallic Wars to communicate with his friend and lawyer Marcus Tullius Cicero. The following are not necessary:

i) The plaintext and ciphertext alphabets be identical

ii) Encipherment leave the number of letters unchanged.

The only requirement on T is the obvious one; it must be possible to reverse the process of encipherment. *Decipherment* (or decryption) is the transformation T^{-1} which recovers the plaintext $\underline{x}$ from the ciphertext $\underline{y}$.

Hashing in Computer Science: Fifty Years of Slicing and Dicing, by Alan G. Konheim

$$T^{-1}: \texttt{Kssh\_Qsvrmrk} \rightarrow \texttt{Good\_Morning}$$

Additional properties are sometimes imposed on T; for example, encipherment does not change the number of letters.

6.2 STREAM CIPHERS

Stream ciphers combine ASCII character plaintext $x_0, x_1, \cdots, x_{n-1}$ letter by letter with a key stream of 0s and 1s. Each plaintext letter is first coded into its 7-bit ASCII ordinal value

$$x_0, x_1, \cdots, x_{n-1} \rightarrow \underline{x}_0, \underline{x}_1, \underline{x}_2, \cdots, \underline{x}_{n-1}$$

and then enciphered by the exclusive-OR (XOR) with the key stream.

Various methods for generating the key stream are possible. Good references for this material are [Beker and Piper 1982] and [Lidl and Niederrieter 1997]. The original research on linear recurring (periodic) sequences is contained in [Selmer 1966] and [Zierler 1959].

6.3 BLOCK CIPHERS

A block cipher is a special form of encipherment in which the plaintext data is divided into blocks of some fixed length, for example 64, and each block is transformed by T. *Chaining* is a mode of operation for block ciphers that causes the encipherment of the ith block to depend on the $(i-1)^{\text{st}}$ block. Chaining generally enciphers identical blocks differently, further hiding the plaintext. The Data Encryption Standard (DES) [FIPS 146-3, 1999] (also pp. 288–302 in [Konheim 2007]) and the Advanced Encryption Standard (AES) [Daemen 1995] and [FIPS 197, 201] are examples of block ciphers.

6.4 SECRECY SYSTEMS AND CRYPTANALYSIS

When a pair of users encipher the data to be exchanged over a network, the cryptographic transformation they use must be specific to the users. A cryptographic system is a family $\mathcal{T} = \{T_k : k \in \mathcal{K}\}$ of cryptographic transformations. A key k is an identifier specifying a transformation T_k in the family $\mathcal{T}$. The key space $\mathcal{K}$ is the totality of all key values. The sender and receiver agree in some way on a particular k and encipher their data with the enciphering transformation T_k.

Encipherment originally involved pen-and-pencil calculations. Mechanical devices were introduced to speed-up encipherment in the 18th century, and they in turn were replaced by electromechanical devices a century later. Encipherment today is often implemented in software: i) T_k is an algorithm, ii) T_ks input consists of both plaintext $\underline{x}$ and key k, and iii) T_k's output is the ciphertext $\underline{y}$.

Will encipherment provide secrecy? Cryptography is a contest between the following two adversaries:

- The *designer* of the system (algorithm, key space, and protocol implementation)
- The *opponent*, who attempts to circumvent the effect of encipherment

Can an opponent recover all or part of the plaintext $\underline{x}$ from the ciphertext $\underline{y} = T_{k0}(\underline{x})$ with knowledge of the cryptographic system $\mathcal{T}$ but without the key k_0? **Cryptanalysis** encompasses all of the techniques to recover the plaintext and/or key from the ciphertext.

The ground rules of this contest were set forth in the 19th century by Jean-Guiullaume-Hubert-Victor-Francois-Alexandre-Auguste-Kerckhoffs von Niuewen of(1835–1903).[1] Kerckhoffs formulated six properties (see [Konheim 2007, pp. 4–5], pp. 4–5) in his book *La Cryptographie Militare* that a cryptographic system should enjoy for the designer to triumph in the struggle.

In assessing the strength of an encipherment system, it must be assumed that the cryptographic system $\mathcal{T} = \{T_k : k \in \mathcal{K}\}$ is known, but the key k_0 producing the ciphertext $T_{k_0} : \underline{x} \to \underline{y}$ is not. Three environments in which cryptanalysis may be attempted are as follows:

Ciphertext Only

The ciphertext $\underline{y} = T_{k0}(\underline{x})$ is known by the opponent; $\underline{x}$ and k_0 are unknown.

Corresponding Plain- and Ciphertext

The plaintext $\underline{x}$ and ciphertext $\underline{y} = T_{k0}(\underline{x})$ are both known by the opponent; k_0 is unknown.

Chosen Plaintext and the Corresponding Plain- and Ciphertext

The plaintext $\underline{x}$ and ciphertext $\underline{y}_i = T_{k0}(\underline{x} + i)$ are both known for some set of *chosen* plaintext $\{\underline{x}_i\}$; k_0 is unknown.

The term *unbreakable* is used colloquially to mean that *no* technique exists to determine the key k or plaintext $\underline{x}$ from the ciphertext $\underline{y} = T_k(\underline{x})$. It is possible to design an unbreakable system, but it is impractical to use in Web-based transactions and is viable only when a modest amount of traffic is exchanged and an alternate secure path for exchanging the key is available.

More relevant is the amount of computational effort—as measured by time and memory—needed to produce k and/or $\underline{x}$. Claude Shannon's paper [Shannon 1949] developed a theory of secrecy systems and defined the work function, which is a quantitative measure (computational time/memory) of the strength of encipherment. The larger the work function, the more secrecy that results from encipherment. The minimum work function required is application dependent; for example, a patient's medical records may require protection for the patient's lifetime, while military plans, need be secreted for a shorter time. Alas, the work function is not generally computable. It may be possible to bound the work function from above and thereby to show often that secrecy is not achieved. It is much more difficult to obtain a lower bound needed to conclude that *no* methods exist that will break the system with an effort less than the lower bound.

[1] Born in in Nuth (Netherlands), he was professor of languages in Paris in the late 19th century.

Because plaintext usually can be found by deciphering the ciphertext with *evey* possible key—a process referred to as key trial—the size of the key space $\mathcal{K}$ must be large enough to discourage this attack.

6.5 SYMMETRIC AND TWO-KEY CRYPTOGRAPHIC SYSTEMS

For nearly two millennia, encipherment was provided exclusively by *conventional* or *single-key* cryptosystems, $\mathcal{T} = \{T_k : k \in \mathcal{K}\}$ with $y = T_k(x)$ denoting the ciphertext y, which results from the encipherment of plaintext x when using the key k. Knowledge of k permitted the computation of T_k^{-1} and the recovery of the plaintext $x = T_k^{-1}(y)$. Each party to an enciphered communication either agreed in advance to key k or a third party delivered the key over an alternate secure path. The secrecy proffered by the encipherment data depended on whether the cryptosystem $\mathcal{T}$ would resist cryptanalysis. Could the key k or plaintext be recovered from $\{y_i = T_k(x_i): 1 \le i \le N\}$ until suitable conditions?

The traditional role of cryptography is *secrecy*, which means to hide the data in communications. In the 20th century, the emergence of private computer systems and public data networks meant that large amounts of data might be i) stored in file systems or ii) transmitted over potentially insecure channels. Methods were needed to protect the privacy of such information while at the same time providing relatively open access for users with a need to obtain the information. When our government uses cryptography, it employs couriers to distribute the keys over an alternate secure path.

When N users are connected on a computer network, typically, the network links are insecure because they might be wiretapped by an opponent.

If a single-key cryptosystem is used to encipher data, it is necessary that a key $k_{i,j}$ $(i \neq j)$ be specified and available for each pair of networked users. It is not feasible for each user in a network of N users to maintain a table of $\approx N^2$ keys $\{k_{i,j}\}$. The problem of key exchange or key distribution is to implement a secure mechanism to make the keys available for each pair of the users.

Needham and Schroeder [Needham and Schroeder 1978] proposed a simple solution using a trusted authority or key server together with a protocol for exchanging keys; the ith user has a network-unique (user) *identifier* User_ID[i] and a secret key k_(ID[i]) known only by the user and the key server. To enable secure communications between the ith and i^{th} users, the network key server generates a session key k_{SK}, which is delivered to User_ID[i] and User_ID[j] enciphered in each of their secret keys. For more information, see [Konheim 2007, Chapter 16].

The Needham-Schroeder solution using a key server suffers from the need to maintain a table, adding/deleting entries as users join/leave the network. This might have been feasible when the Internet consisted of a few thousand users, but it is difficult to manage a network with millions of users today. Moreover, independent public networks may have different operating systems, and it is not feasible for a single key server to provide network-wide service. There must be a hierarchy of key servers with different domains used to exchange the keys. The issue of *consistency* and the complexity of the information transferred between the key servers if User_ID[i] and User_ID[j] are in different domains and are important issues.

In addition, in the 1980s the use of networks for electronic commerce (or e-commerce) to be discussed in Chapter 19 brought about the birth of a new paradigm for cryptography.

6.6 THE APPEARANCE OF PUBLIC KEY CRYPTOGRAPHIC SYSTEMS

Cryptography changed in 1976 with the appearance of papers by Whitfield Diffie (then a graduate student) and Martin Hellman [Diffie and Hellman 1976a], [Diffie and Hellman 1976b]. They invented public key cryptography (PKC) in response to an expanded role of information processing technology in our society coupled with access to public data networks. Encipherment would not only be needed by governments, but also for the following reasons:

- To protect the confidentiality of medical record
- By the participants in commercial transactions carried out over a public data network

In traditional commercial transactions, the parties meet and sign in each others presence a document (contract) specifying the rules of their transaction. Often, a notary is needed to authenticate the parties. In e-commerce, an electronic transaction requires a digital signature to be appended to the transaction data.

Diffie and Hellman proposed a new type of cryptosystem that would alleviate but not eliminate the problem of key distribution and also provide a mechanism for digital signatures.

The characteristic property of conventional cryptosystems $\mathcal{T} = \{T_k : k \in \mathcal{K}\}$ is that T_k determines the inverse transformation T_k^{-1}. Often, the key k determines a second key k^{-1} so that $T_k^{-1} = T_{k-1}$.

Diffie and Hellman proposed (public key) cryptosystems that use two keys: a public key PuK for encipherment and a private key PrK for decipherment.

$$\text{Encipher}: \underline{x} \rightarrow \underline{y} = E_{\text{PuK}}\{\underline{x}\} \quad \text{Decipher}: \underline{y} \rightarrow \underline{x} = E_{\text{PrK}}\{\underline{x}\}$$

In addition to the usual properties required of a strong cryptosystem, it was crucial that PrK could not be determined from PuK. User_ID[i] would publish its public key PuK(ID[i]) thereby enabling any user to encipher information for User_ID[i]. Knowledge of PrK(ID[i]), which is presumptively known only by User_ID[i], would permit the decipherment of PuK(ID[i])-enciphered messages. How can such pairs (PuK(ID[i]),PrK(ID[i]) be found?

Diffie and Hellman argued that there are complex mathematical functions $f(x)$ for which the problem

Given: x

Find: $y = f(x)$

is easy to solve, but for which the problem

Given: $y = f(x)$

Find: x

is hard to solve.

A solution to the easy problem would be computation of the ciphertext, which is the encipherment $y = E_{\text{PuK(ID}[i])}\{x\}$ of the plaintext x using User_ID[i]'s public key PuK(ID[i]). A solution to the the hard problem would be computation of the plaintext, which is the decipherment $x = E_{\text{PrK(ID}[i])}\{y\}$ of the ciphertext y using User_ID[i]'s private key PrK(ID[i]).

Easy and hard refer to the complexity class of the problem. A problem is considered *computationally infeasible* if the cost of finding a solution, as measured by either the amount of memory used or the computing time, while finite, is extraordinarily large; it is much greater than the value of the solution and/or is beyond the computing resources.

The *execution time* of an algorithm A with n inputs is the number of times some basic operation is performed.

Algorithm A with n inputs executes in *polynomial time* or is an $O(n^d)$ algorithm if there is a constant C such that the execution time is no larger than Cn^d.

Many problems admit such a description; two examples are as follows:

Addition

- *Given* n-bits $(x_0, x_1, \cdots, x_{n-1})$ and n integers $(b_0, b_1, \cdots, b_{n-1})$ each expressed with n bits
- *Compute* the sum $S = \sum_{i=0}^{n-1} b_i x_i$.

  The sum may be computed by an $O(n^2)$ algorithm.

Modular Exponentiation

- *Given* M, e and N, each an n-bit integer
- *Compute* $C = M^e \pmod{N}$.

  C may be computed using a $O(n^3)$ algorithm.

For some problems, either a polynomial time algorithm $O(n^d)$ for the solution is unknown or the running time of the best known algorithm is exponential like.

Knapsack problem

- *Given* the sum S
- *Compute* n-bits $(x_0, x_1, \cdots, x_{n-1})$ to satisfy $S = \sum_{i=0}^{n-1} b_i x_i$.

  $(x_0, x_1, \cdots, x_{n-1})$ may be computed by a $O\left(2^{\frac{n}{2}}\right)$ algorithm.

  No polynomial time algorithm is known.

Logarithm problem (mod N)

- *Given* $C = M^e \pmod{N}$, M and N where C, M, and N are each n-bit integers.
- *Calculate* the (discrete) logarithm $e = \log_M C \pmod{N}$.

  $\log_M C \pmod{N}$ may be calculated using a $O\left(2^{\beta\sqrt{\log n \log\log n}}\right)$ algorithm.

  No $O(n^d)$ algorithm to compute $\log_M C \pmod{N}$ is known.

Generally speaking a problem is

- Easy, if a $O(n^c)$ algorithm is known/to find a solution, and
- Hard, if no $O(n^d)$ algorithm to find a solution is known.

Complexity theory stemming from the work of Alan Turing [Turing 1936] classifies algorithms (or problems) depending on their execution times; for more information see [Aho, Ullman and Hopcroft 1973]] or [Papdimitrious 1995].

P Polynomial-time problems with n inputs.
An $O(n^d)$ algorithm to solve the problem exists.
NP Nondeterministic polynomial-time problems with n inputs.
An $O(n^d)$ algorithm to check a possible solution to the problem exists.

Complexity theory identifies a distinguished subclass of NP consisting of problems that are equivalent in the sense that a solution to any one NP complete problem can be transformed to a solution to another problem in this class.

The relationship between the classes is not known; in particular, the truth of the equality P = NP or proper inclusion $P \subset NP$ remains unsettled. If the second statement is true, there are problems for which *no* $O(n^d)$ algorithmic solution exists.

Examples of corresponding easy (f) and hard (f^{-1}) problems include the following:

E *Addition (of knapsack weights)*

- *Given* a knapsack vector $\underline{b} = (b_1, b_1, \cdots, b_n)$ and a selection vector $\underline{x} = (x_0, x_1, \cdots, x_{n-1})$; $(x_i = 0, 1)$
- *Compute* $S = b_0x_0 + b_1x_1 + \cdots + b_{n-1}x_{n-1}$

  Addition (of knapsack weights) is in the complexity class P.

H *Knapsack problem (subset sum problem)*

- *Given* a knapsack vector $\underline{b} = (b_0, b_1, \cdots, b_{n-1})$ and a sum S
- *Determine* a vector $\underline{x} = (x_0, x_1, \cdots, x_{n-1})$ with components 0 and 1 such that $S = b_0x_0 + b_1x_1 + \cdots + b_{n-1}x_{n-1}$

  The knapsack problem is in the complexity class NP-complete.
  No P-algorithm to solve the knapsack problem is *known*.
  The fastest algorithm to solve the knapsack problem runs in time $O\left(2^{\frac{n}{2}}\right)$.

E *Multiplication of integers*

- *Given* integers p and q whose lengths are each n-bits
- *Calculate* the product $N = pq$

  Multiplication of integers is in the complexity class P.

H *Integer factorization*

- *Given* an n-bit integer which is the product of two primes p, q
- *Calculate* the factors p and q

  Factorization is in the complexity class NP; it is not believed to be NP-complete.

  No P-algorithm to factor is known.

  There is a $O\left(2^{\alpha\sqrt{\log n \log\log n}}\right)$ algorithm to factor.

E *Modular exponentiation* (mod p)

- *Given* p a prime, q a primitive root of p[2], and e an exponent each number requiring n-bits
- *Calculate* $N = q^e \pmod p$

  Exponentiation modulo p is in the complexity class P.

  Exponentiation modulo p is an $O(n^3)$ algorithm.

H *Logarithm problem* (mod p)

- *Given* p a prime, q a primitive root of p, and $N = q^e \pmod p$
- *Calculate* the exponent $e = \log_q N \pmod p$

  Taking logarithms modulo p is in the complexity class NP; it is not believed to be NP-complete. No P-algorithm to calculate logarithms modulo p is known.

  There is a $O\left(2^{\beta\sqrt{\log n \log\log n}}\right)$ algorithm to calculate logarithms modulo p.

Diffie and Hellman suggested that encipherment be based on an easy problem, whereas decipherment requires the solution of the corresponding hard problem.

But there is a defect! If computing $y = f(x)$ is easy but computing $x = f^{-1}(y)$ is infeasible for a third party, it must also be so for the creator of the (easy, hard) pair.

Diffie and Hellman called f a trap door one-way function if it satisfies the following three properties:

1. *Given*: A description of $f(x)$ and x
   It is computationally feasible to compute $y = f(x)$
2. *Given*: A description of $f(x)$ and $y = f(x)$
   It is computationally infeasible to compute $x = f^{-1}(x)$
3. *Given*: A description of $f(x)$ and $y = f(x)$ and parameters z
   It is computationally feasible to compute $x = f^{-1}(y)$.

If

E. The computation of $y = f(x)$ is the encipherment of plaintext $y - E_{\text{PuK}}\{x\}$ with the public key PuK, and

D. The computation of $x = f^{-1}(y)$ is the decipherment of ciphertext $y - E_{\text{PrK}}\{x\}$ with the public key PrK, then

[2] q is a *primitive root* of p if the powers $q^i \pmod p$ are distinct for $0 \le i \le p - 1$ and therefore a rearrangement (permutation) of the integers $1, 2, 3, \cdots, p - 1$.

knowledge of the trap door z for a trap door one-way function f permits the construction of a pair (PuK,PrK) of public key cryptosystem keys. Without the trap door z, a user is not in a position to find PrK from PuK.

What functions f are one way and which of them have trap doors? Diffie and Hellman could not in [Diffie and Hellman 1976a] or [Diffie and Hellman 1976b] provide any example of a trap door one-way function.

Merkle and Hellman [Merkle and Hellman 1978] described a PKS that satisfied some but not the most crucial of the requirements of a trap door PKS.

Shortly thereafter, Ronald Rivest, Adi Shamir, and Len Adelman [Rivest, Shamir and Adelman 1978] defined RSA, which was the first example of a public key cryptosystem. To the best of our current knowledge, RSA meets all the desiderata of a strong PKS system. The strength of the RSA cryptosystem, defined in §6.8 that follows, seems to depend on the difficulty of factoring large numbers.

6.7 A MULTITUDE OF KEYS

Cryptography makes use of a variety of keys, as follows:

- *Symmetric keys*, perhaps in pairs k and k' that that $T_{k\text{prime}}(T_k(x)) = x$ for plaintext x
- (Asymmetric) *public/private key* (pairs) Puk and Prk such that $T_{\text{Pu}k}(T_{\text{Pr}k}(x)) = x$ for plaintext x
- (Symmetric) *session keys* k_{SK} generated by a "system" entity to be used by two parties to encipher information during a session

6.8 THE RSA CRYPTOSYSTEM

The very basic number theory is required to understand the RSA system and indeed most of modern cryptography is contained in Chapter 5.

A four-tuple (p, q, e, d) is an RSA parameter set if

a) p and q are prime numbers; $N = pq$ is the RSA modulus with $\phi(N) = (p - 1)(q - 1)$
b) the RSA enciphering exponent e satisfies $1 = gcd\{e, \phi(N)\}$
c) the RSA deciphering exponent d satisfies $1 = gcd\{d, \phi(N)\}$
d) e and d are multiplicative inverses modulo $\phi(N)$ of one another, $ed = 1 \pmod{\phi(N)}$

Theorem 6.1. When (p, q, e, d) is an RSA parameter set, the RSA encipherment transformation E_e is modular exponentiation, which is defined for integers n in $\mathcal{Z}_N \equiv \{0, 1, 2, \cdots, N - 1\}$ by

$$\mathbf{E}_e : n \to \mathbf{E}_e(n) = n^e \pmod{N} \tag{6.1a}$$

$\mathbf{E}_e$ is a one-to-one mapping on $\mathcal{Z}_N$ onto $\mathcal{Z}_N$. Its inverse, the RSA decipherment transformation D_d, is also modular exponentiation, which is defined for integers n in $\mathcal{Z}_N$ by

$$\mathbf{D}_d : n \to \mathbf{D}_d(n) = n^d \pmod N \tag{6.1b}$$

$$I = \mathbf{E}_e\mathbf{D}_d = \mathbf{D}_d\mathbf{E}_e \tag{6.1c}$$

Proof. The second assertion clearly implies the first; if (p, q, e, d) is an RSA parameter set, then

$$(n^e)^d = n^{1+C\phi(N)} = n^{1+C(p-1)(q-1)}$$

because $ed = 1$ (*modulo* $\phi(N)$). If $n = pm$ with $1 = gcd\{m, q\}$, then p divides n and q does not divide n, so that

$$(n^e)^d - n = n\left[\left(n^{C(p-1)}\right)^{q-1} - 1\right]$$

By Fermat's theorem, $y^{q-1} - 1 = 0 \pmod q$ with $y = n^{(p-1)}$.

A similar argument shows that if $n = qm$ with $1 = gcd\{m, p\}$ that $(n^e)^d - n = 0 \pmod N$.

Finally, if n is relatively prime to both p and q

$$(n^e)^d - n = n\left[\left(n^{p-1}\right)^{C(q-1)} - 1\right] \quad (n^e)^d - n = n\left[\left(n^{q-1}\right)^{C(p-1)} - 1\right] \quad \blacksquare$$

Equations (6.1*a* through 6.1*c*) state that *both* encipherment and decipherment in the RSA system are modular exponentiation, differing only by the exponent.

Example 6.1. $p = 31, q = 5, e = 7, d = 103$

$$N = 155 = 31 \times 5$$

$$\phi(N) = 120 = (31-1)\times(5-1)$$

$$ed = 721 = 1 \pmod{120}$$

$$67 = 98^7 \pmod{155}$$

$$98 = 67^{103} \pmod{155}$$

Theorem 6.2. If e, p, and q are given, then d can be computed in time $O(\log_2 \phi(N))$.

Although factorization *Example 6.1* is misleading, the numbers used are small in the sense that they are expressible as 4-byte words. However, to achieve real cryptographic strength for secrecy and authentication systems using RSA, this size limitation must be considerably relaxed. We need to perform arithmetic operations *modulo m* with large integers m that have thousands of digits. The necessary size may in fact increase as more refined factorization and cryptanalytic techniques are introduced; for more information, see [Riesel 1994].

As we have defined RSA, it enciphers and deciphers integers in $\mathcal{Z}_N$. When cryptography is applied to protect transactions on the Web, the data are expressed using the ASCII character alphabet. Thus, RSA and other modern encipherment and authentication systems must first process the raw plaintext data transforming it into a form to which RSA may be applied; see [Konheim 2007, pp. 379–382]. An Internet standard is described in [RFC 1113].

How strong is RSA? The strength of RSA seems to depend on the complexity of factorization of $n = pq$.

Collaborative computing describes a variety of activities involving the iteration of people using desktops (or laptops) and sophisticated digital cellular phones. There is no better example than in the factorization of large integers.

RSA Data Security Incorporated (Redwood City, CA) is now a subsidiary of EMC^2 where it is claimed information lives, supplies encryption protocols using the RSA algorithm. Because the strength of RSA seems to depend on the intractability of factoring $n = pq$ for suitably large prime numbers, the RSA (Factoring) Challenge was set up in March 1991; it consists of a list of numbers, each the product of two primes of roughly comparable size. There are 42 numbers in the challenge; the smallest length is 100 digits and they increase in steps of 10 digits to the 500 digits. Table 6.1 gives some results in the RSA Challenge.

Martin Gardner's article [Gardner 1979], which appeared a year before the publication of the paper [Rivest, Shamir and Adelman 1978] that defined the RSA cryptosystem, contained the first of many factoring RSA challenges. RSA-129 is a 129-digit integer that is the product of two primes.

```
RSA-129 = 1143816257578888676692357799761466120102182967212
          4236256256184293570693524573389783059712356395870
          5058989075147599290026879543541

N = pq

p = 3490529510847650949147849619903898133417764638493387843990820577

q = 32769132993266709549961988190834461413177642967992942539798288533
```

As a way of testing a possible factorization, the message `THE MAGIC WORDS ARE SQUEAMISH OSSIFRAGE`[3] was enciphered with RSA-129 using the public key $e = 9007$ and the private key

TABLE 6.1. The RSA Challenge

Number	Date Factored
RSA-100	April 1991
RSA-110	April 1992
RSA-120	June 1993
RSA-129	April 1994
RSA-130	April 1996
RSA-140	February 1999
RSA-155	August 1999

[3]ossifrage *n.*, bone breaking.

```
d=1066986143685780244428687713289201547807099066339378628
  01226224496631063125911774470873340168597462306553968544
  513277109053606095
```

RSA-129 was factored in 8 months (April 1991) and as Gardner's article suggests, it does not "··· take millions of years ···" to factor and to claim the prize of $100 for the first solution.

To achieve the desired strength, the sizes of the primes p and q must be large, say 512 bits. Although efficient implementations of modular exponentiation exist, the Web world is impatient and fast must be very fast. The current usage of RSA, as described in Chapter 18, is not for encipherment/decipherment but for authentication.

6.9 DOES PKC SOLVE THE PROBLEM OF KEY DISTRIBUTION?

In a PKC system, User_ID[i] enciphers data for User_ID[j] using User_ID[j]'s public key PuK(ID[j]). How does User_ID[i] learn the value of PuK(ID[j])? Several possible solutions include the following:

i) The existence of a network-wide table of pairs (ID[···],PuK(ID[···])) maintained by some entity that User_ID[i] accesses,
ii) User_ID[j] delivers PuK(ID[j]) to User_ID[i] on demand.
iii) User_ID[i] receives PuK(ID[j]) at the time of a transaction from some third entity.

We seem to be faced with the same problem of key exchange considered in §6.4. In a second, less burdensome solution, User_ID[i] asks a presumptive User_ID[j] to transmit a copy of PuK(ID[j]). Communications enciphered with PuK(ID[j]) would then be able to be read by someone with knowledge of PrK(ID[j]). Of course, User_ID[A] has be certain that the communication was with User_ID[j]. It is necessary for User_ID[i] to have some way of verifying the link ID[j] $\leftrightarrow$ PuK(ID[j]).

The need for a provable link between the public key and identifier of a user was recognized in 1978 by Adelman's student Kohnfelder. In Part I, Section D *Weaknesses in Public-Key Cryptosystems* of [Kohnfelder 1978], Kohnfelder writes

> Although the enemy may eavesdrop on the key transmission system, the key must be sent via a channel in such a way that the originator of the transmission is reliably known.

Kohnfelder observed that all public-key cryptosystems are vulnerable to a *spoofing* attack[4] if the public keys are not certified; User_ID[?] pretending to be User_ID[i] to User_ID[j] by providing User_ID[?]'s public-key (in place of User_ID[i]'s public key) *to* User_ID[j]. Unless User_ID[j] has some way of checking the correspondence between ID[i] and PuK(ID[i]), this type of spoofing attack is possible.

[4]To spoof is to cause a deception or hoax.

Kohnfelder wrote

> ··· each user who wishes to receive private communications must place his enciphering algorithm (his public key) in the public file.

Kohnfelder proposed a method to make spoofing more difficult in Part III of [Kohnfelder 1978]. He postulated the existence of a *public file* $\mathcal{F}$, which contains (in my notation) pairs {(ID[*i*]), PuK(ID[*i*])} for each user in the system. Although it might be possible for User_ID[A] to contact $\mathcal{F}$ to ask for a copy of User_ID[*j*]'s public key, this solution suffers from the same operational defect as a network-wide key server.

- What entity will maintain and certify of a large database that is continually changing?
- The public file will need to be replicated to prevent severe access times to obtain information.

Kohnfelder proposed a certificate as a data set providing an authentication between the user's identifier and its public key. We take this up in Chapter 18.

6.10 ELLIPTIC GROUPS OVER THE REALS

In the 1980s, N. Koblitz [Koblitz 1987b] and V. Miller [Miller 1986] proposed the use of elliptic curves to generate cryptographic systems. Elliptic curve cryptosystems seem to offer a considerable efficiency with respect to key size. We give a brief description of elliptic groups beginning with the case of real elliptic groups. We briefly touch on their generalization to modular elliptic groups in §6.11. There is an extensive literature including the wonderful survey paper [Menezes and Vanstone 1993].

A *plane curve* is the locus of points (x, y) in the plane which are the solutions of $f(x, y) = 0$ with $f(x, y)$ a polynomial in two variables with rational coefficients. The study of plane curves has fascinated mathematicians for nearly two millennia.

By changing variables, the general cubic equation $y^2 + b_1xy + b_2y = x^3 + a_1x^2 + a_2x + a_3$ yields the normal form for an elliptic curve $y^2 = x^3 + ax + b$. An elliptic curve may have one or three real roots; it does not have multiple roots provided the discriminant[5] $D = 4a^3 + 27b^2$ is not 0.

An elliptic curve $y^2 = x^3 + ax + b$ with one real root is shown in Figure 6.1.

If $x^3 + ax + b$ has three (distinct) real roots, the curve consists of two sections. An example of is shown in Figure 6.2.

The curve $y^2 = x^3 - x = x(x - 1)\ (x + 1)$ enjoys a property characteristic of all elliptic curves called the chordtangent group law discovered by Carl Gustav Jacob Jacobi (1804–1851) in the 19th century.

[5]The discriminant of the polynomial $f(x) = \prod_i a(x - r_i)$ of degree n is $D = a^{n-1} \prod_{i<j} (e_i - e_j)^2$. The roots of $f(x)$ are distinct if and only if $D \neq 0$.

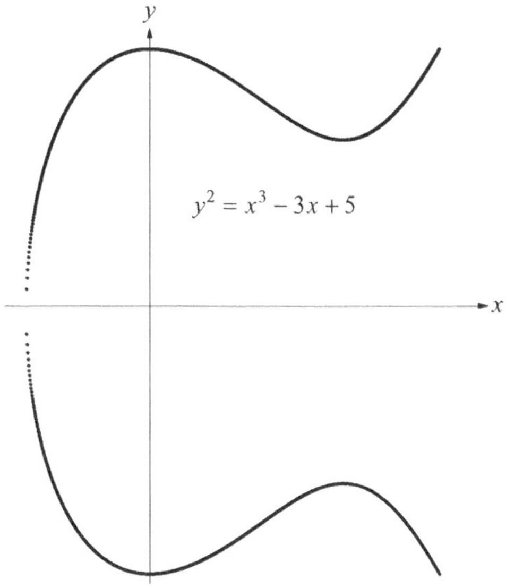

Figure 6.1. Elliptic curve with one real root.

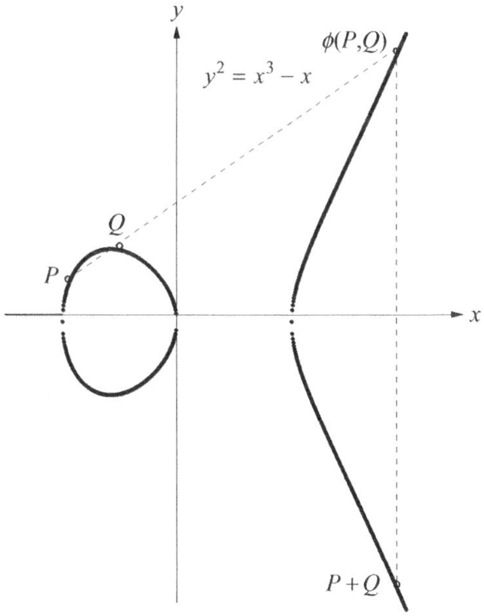

Figure 6.2. Elliptic curve with three real distinct roots.

Theorem 6.3 (*Bezout's theorem*). If $P = (x_1, y_1)$ and $Q = (x_2, y_2)$ are two points on the elliptic curve $y^2 = x^3 + ax + b$ with $4a^3 + 27b^2 \neq 0$ and if the line PQ joining these points is not vertical, then the line PQ will intersect the curve in a third place $\phi(P, Q) = R = (x_3, -y_3)$.

Bezout's Theorem implies the points on an elliptic curve form a group; if P and Q are points of $\mathcal{E}(a, b)$, then define the point $\phi(P, Q)$ at the intersection of the line PQ and the elliptic curve.

1. If the line PQ is vertical ($x_1 = x_2$), the vertical line *meets* the curve at ∞ so that $\phi(P, Q) = \infty$.
2. The value of $\phi(\infty, P)$ is the reflection about the x-axis of P so that $\phi(\infty, Q) = P$ if $Q = \phi(\infty, P)$.

Define the sum of P and Q by

$$P + Q \equiv \phi(\infty, \phi(P, Q))$$

The third point on the line PQ is determined and reflected about the x-axis.

3. The point at ∞ satisfies $\phi(P, \infty) = \phi(\infty, P) = P$ so that $\mathcal{O} = \infty$ acts as an identity element under $+$.
4. The point $\phi(\infty, P)$ satisfies $\phi(P, \phi(\infty, P)) = \infty$ so that $\phi(\infty, P)$ is the inverse of P under $+$ satisfying $P + \phi(\infty, P) = \mathcal{O}$.

Although it is surprisingly difficult, it can be proved that addition + satisfies the associativity law [Husemoller 1986].

$$P + (Q + R) = (P + Q) + R$$

Theorem 6.4. $\mathcal{O}$ together with the points on the elliptic curve $y^2 = x^3 + ax + b$ ($4a^3 + 27b^2 \neq 0$) form the Abelian elliptic group $\mathcal{E}(a, b)$ with group operation + and identity element $\mathcal{O}$.

6.11 ELLIPTIC GROUPS OVER THE FIELD $\mathcal{Z}_{m,2}$

The elliptic group described in Theorem 6.4 consists of points in the plane; elliptic groups can be defined over any finite field. For applications to the encipherment of text, it is more natural to consider elliptic groups whose points can be associated with ASCII-coded text; for example, the extension field $\mathcal{Z}_{m,2}$.

Let $p(x)$ be a primitive polynomial over $\mathcal{Z}_2$ of degree m. The elliptic group $\mathcal{E}_{\mathcal{Z}_{m,2}}(a, b)$ consists of the point at infinity $\mathcal{O}$ together with all pairs $(x, y) \in \mathcal{Z}_{m,2}$, which satisfy the equation

$$y^2 + xy = x^3 + ax^2 + b \quad a, b \in \mathcal{Z}_{m,2}, \quad b \neq 0$$

Because $b \neq 0$, the point (0, 0) is not a solution; (0,0) is used as the representation of $\mathcal{O}$.

Theorem 6.5. If $y \in \mathcal{Z}_{m,2}$ is a solution of $y^2 + xy = x^3 + ax^2 + b$, then $y + x \in \mathcal{Z}_{m,2}$ is also a solution.

Proof. If

$$y^2 + xy = x^3 + ax^2 + b$$

then

$$(y+x)^2 + x(y+x) = y^2 + x^2 + xy + x^2 = y^2 + xy$$

because the field $\mathcal{Z}_{m,2}$ is of characteristic 2.

Theorem 6.6. $\varepsilon_{\mathcal{Z}_{m,2}}(a, b)$ is a group under addition where

a) $(x, y) + (0, 0) = (0, 0) + (x, y) = (x, y)$.
b) $(x, y) + (x, y + x) = (0, 0)$.
c) If $x_1 \neq x_2$, then $(x_1, y_1) + (x_2, y_2) = (x_3, y_3)$ where (x_3, y_3) is the reflection of the y-value $x_3 + y_3$ of the point (x, y) on the linear form $y = \lambda x + \mu$ at x_3.

$$\lambda = \frac{y_1 + y_2}{x_1 + x_2}$$

$$x_3 = \lambda^2 + \lambda + x_1 + x_2 + a$$

$$y_3 = \lambda(x_1 + x_3) + x_3 + y_1$$

d) $2(x, y) = (x_2, y_2)$ where

$$\lambda = x + \frac{y}{x}$$

$$x_2 = \lambda^2 + \lambda + a$$

$$y_2 = x_2 + (\lambda + 1)x_2$$

Proof. There are two cases to be examined; if there are two distinct points of intersection (x_1, y_1) and (x_2, y_2) of the linear form $y = \lambda x + \mu$ and the curve $y^2 + xy + x^3 + ax^2 + b$, then

$$\begin{aligned} 0 &= x^3 - (\lambda^2 + \lambda + a) - x\mu + b - \mu^2 \\ &= (x - x_1)(x - x_2)(x_3) \end{aligned}$$

then

$$\lambda + \lambda^2 + a = x_1 + x_2 + x_3$$

which gives

$$x_3 = \lambda^2 + \lambda + x_1 + x_2 + a$$

$$y_3 = \lambda(x_1 + x_3) + x_3 + y_1$$

In the second case, the linear form $y = \lambda x + \mu$ is tangent to the curve $y^2 + xy = x^3 + ax^2 + b$, which requires (x, y) to be a solution to the pair of equations

$$x\frac{dy}{dx} + y = x^2 \Leftrightarrow 2y\frac{dy}{dx} + x\frac{dy}{dx} + y = 3x^2 + 2ax$$

$$y^2 + xy = x^3 + ax^2 + b$$

This gives

$$\lambda = x + \frac{y}{x}$$

$$x_2 = \lambda^2 + \lambda + a$$

$$y_2 = x^2 + (\lambda + 1)x_2$$

The number of points in $\varepsilon_{\mathcal{Z}_{m,2}}(a, b)$ is given by the next theorem.

Theorem 6.7 (Helmut Hasse (1898–1979)). The order $\#\varepsilon_{\mathcal{Z}_{m,2}}(a, b)$ of the elliptic group $\varepsilon_{\mathcal{Z}_{m,2}}(a, b)$ satisfies

$$1 + 2^m - 2\sqrt{2^m} \le \#\varepsilon_{\mathcal{Z}_{m,2}}(a, b) \le 1 + 2^m + 2\sqrt{2^m} \quad \#\varepsilon_{\mathcal{Z}_{m,2}}(a, b) - (1 + 2^m) \le 2\sqrt{2^m}$$

It follows that for large m, we have $|\varepsilon_{\mathcal{Z}_{m,2}}(a, b)| \approx O(2^m)$.

Example 6.2. Using the primitive polynomial $p(x) = 1 + x + x^6$, construct the elliptic group $\varepsilon_{\mathcal{Z}_{6,2}}(a, b)$ with $a = 0_6$, $b = (0, 0, 1, 0)$. Table 6.2 lists the 31 elements $(x, y) \neq (0_6, 0_6)$, giving both the polynomial coding and power of the adjoined root ϑ.

6.12 ELLIPTIC GROUPS CRYPTOSYSTEMS

A variety of publications describes the use of elliptic curve cryptosystems including [Menezes and Vanstone 2001], [ANSI X9.62], and [FIPS 186-3, 208] Start with a primitive polynomial $p(x)$ over $\mathcal{Z}_2$ of degree m; the elliptic group $\varepsilon_{\mathcal{Z}_{m,2}}(a, b)$ consists of the point at infinity $\mathcal{O}$ together with all pairs $(x, y) \in \mathcal{Z}_{m,2}$, which are on the elliptic curve $y^2 + xy = x^3 + ax^2 + b$ with $a, b \in \mathcal{Z}_{m,2}$, $b \neq 0$.

1. The base point P of the elliptic curve cryptosystem is a nonsecret point on the elliptic group $\varepsilon_{\mathcal{Z}_{m,2}}(a, b)$, which has a prime order $n > 2^{160}$.
2. All quantities of interest, i.e., keys, are elements of the cyclic group consisting of all multiples of xP with $x \in \mathcal{Z}_{m,2}$.
3. The ith user with User_ID[i] has both a secret key $\mathrm{PrK}_{\mathrm{User\_ID[i]}} = \mathrm{Pr}k_{\mathrm{User\_ID}[i]}P$ and a public key $\mathrm{PuK}_{\mathrm{User\_ID[i]}} = \mathrm{Pu}k_{\mathrm{User\_ID}[i]}P$.

   The correct correspondence of the pair of keys ($\mathrm{PrK}_{\mathrm{User\_ID[i]}}$, $\mathrm{PuK}_{\mathrm{User\_ID[i]}}$) is attested to by a *certificate* to be discussed in Chapter 18.

TABLE 6.2. The Elliptic Group $\mathcal{E}_{\mathcal{Z}_{6,4}}(a, b)$; $p(x) = 1 + x + x^6$; $a = 0$ and $b = (0, 0, 1, 0)$

(x, y_j)	$y_j = \vartheta^{k_j}$ $(j = 1, 2)$	
$x = (0,0,0,0,0,1) = \vartheta^0$	$y_1 = (1,1,1,0,1,1) = \vartheta^{21}$	$y_2 = (1,1,1,0,1,0) = \vartheta^{42}$
$x = (0,0,0,0,1,0) = \vartheta^1$	$y_1 = (0,0,1,1,1,1) = \vartheta^{18}$	$y_2 = (0,0,1,1,0,1) = \vartheta^{48}$
$x = (0,0,0,1,0,0) = \vartheta^2$	$y_1 = (0,1,0,0,1,0) = \vartheta^{33}$	$y_2 = (0,1,0,1,1,0) = \vartheta^{36}$
$x = (0,0,1,0,0,0) = \vartheta^3$	$y_1 = (1,0,0,1,1,0) = \vartheta^{17}$	$y_2 = (1,0,1,1,1,0) = \vartheta^{55}$
$x = (0,1,0,0,0,0) = \vartheta^4$	$y_1 = (0,0,1,0,0,0) = \vartheta^3$	$y_2 = (0,1,1,0,0,0) = \vartheta^9$
$x = (0,0,0,0,1,1) = \vartheta^6$	$y_1 = (1,0,0,1,0,0) = \vartheta^{34}$	$y_2 = (1,0,0,1,1,1) = \vartheta^{47}$
$x = (0,0,0,1,1,0) = \vartheta^7$	$y_1 = (0,0,1,1,0,0) = \vartheta^8$	$y_2 = (0,0,1,0,1,0) = \vartheta^{13}$
$x = (0,0,1,1,0,0) = \vartheta^8$	$y_1 = (0,0,0,0,1,1) = \vartheta^6$	$y_2 = (0,0,1,1,1,1) = \vartheta^{18}$
$x = (0,1,1,0,0,0) = \vartheta^9$	$y_1 = (1,0,0,1,0,1) = \vartheta^{31}$	$y_2 = (1,1,1,1,0,1) = \vartheta^{59}$
$x = (0,0,0,1,0,1) = \vartheta^{12}$	$y_1 = (1,0,0,0,0,0) = \vartheta^5$	$y_2 = (1,0,0,1,0,1) = \vartheta^{31}$
$x = (0,0,1,0,1,0) = \vartheta^{13}$	$y_1 = (1,1,0,1,1,1) = \vartheta^{43}$	$y_2 = (1,1,1,1,0,1) = \vartheta^{59}$
$x = (0,1,0,1,0,0) = \vartheta^{14}$	$y_1 = (0,1,0,0,1,1) = \vartheta^{16}$	$y_2 = (0,0,0,1,1,1) = \vartheta^{26}$
$x = (0,1,0,0,1,1) = \vartheta^{16}$	$y_1 = (0,0,0,1,0,1) = \vartheta^{12}$	$y_2 = (0,1,0,1,1,0) = \vartheta^{36}$
$x = (0,0,1,1,1,1) = \vartheta^{18}$	$y_1 = (1,0,1,1,1,0) = \vartheta^{55}$	$y_2 = (1,0,0,0,0,1) = \vartheta^{62}$
$x = (0,1,1,1,1,0) = \vartheta^{19}$	$y_1 = (1,1,1,1,1,1) = \vartheta^{58}$	$y_2 = (1,0,0,0,0,1) = \vartheta^{62}$
$x = (0,1,0,0,0,1) = \vartheta^{24}$	$y_1 = (1,1,0,0,0,0) = \vartheta^{10}$	$y_2 = (1,0,0,0,0,1) = \vartheta^{62}$
$x = (0,0,0,1,1,1) = \vartheta^{26}$	$y_1 = (1,0,1,0,0,1) = \vartheta^{23}$	$y_2 = (1,0,1,1,1,0) = \vartheta^{55}$
$x = (0,0,1,1,1,0) = \vartheta^{27}$	$y_1 = (0,0,0,0,0,1) = \vartheta^0$	$y_2 = (0,0,1,1,1,1) = \vartheta^{18}$
$x = (0,1,1,1,0,0) = \vartheta^{28}$	$y_1 = (0,0,1,0,0,1) = \vartheta^{32}$	$y_2 = (0,1,0,1,0,1) = \vartheta^{52}$
$x = (0,0,1,0,0,1) = \vartheta^{32}$	$y_1 = (0,1,1,0,0,0) = \vartheta^9$	$y_2 = (0,1,0,0,0,1) = \vartheta^{24}$
$x = (0,1,0,0,1,0) = \vartheta^{33}$	$y_1 = (1,0,1,1,1,1) = \vartheta^{40}$	$y_2 = (1,1,1,1,0,1) = \vartheta^{59}$
$x = (0,0,1,0,1,1) = \vartheta^{35}$	$y_1 = (0,1,0,0,0,0) = \vartheta^4$	$y_2 = (0,1,1,0,1,1) = \vartheta^{38}$
$x = (0,1,0,1,1,0) = \vartheta^{36}$	$y_1 = (1,0,0,1,1,1) = \vartheta^{47}$	$y_2 = (1,1,0,0,0,1) = \vartheta^{61}$
$x = (0,1,1,0,1,1) = \vartheta^{38}$	$y_1 = (1,0,1,0,1,0) = \vartheta^{53}$	$y_2 = (1,1,0,0,0,1) = \vartheta^{61}$
$x = (0,1,1,1,0,1) = \vartheta^{41}$	$y_1 = (1,1,1,0,0,0) = \vartheta^{29}$	$y_2 = (1,0,0,1,0,1) = \vartheta^{31}$
$x = (0,1,1,0,0,1) = \vartheta^{45}$	$y_1 = (0,0,0,0,0,1) = \vartheta^0$	$y_2 = (0,1,1,0,0,0) = \vartheta^9$
$x = (0,0,1,1,0,1) = \vartheta^{48}$	$y_1 = (1,1,1,1,0,0) = \vartheta^{20}$	$y_2 = (1,1,0,0,0,1) = \vartheta^{61}$
$x = (0,1,1,0,1,0) = \vartheta^{49}$	$y_1 = (0,0,0,1,0,0) = \vartheta^2$	$y_2 = (0,1,1,1,1,0) = \vartheta^{19}$
$x = (0,1,0,1,0,1) = \vartheta^{52}$	$y_1 = (1,1,0,0,1,0) = \vartheta^{46}$	$y_2 = (1,0,0,1,1,1) = \vartheta^{47}$
$x = (0,1,0,1,1,1) = \vartheta^{54}$	$y_1 = (0,0,0,0,0,1) = \vartheta^0$	$y_2 = (0,1,0,1,1,0) = \vartheta^{36}$
$x = (0,1,1,1,1,1) = \vartheta^{56}$	$y_1 = (0,0,0,0,1,0) = \vartheta^1$	$y_2 = (0,1,1,1,0,1) = \vartheta^{41}$

4. The public and private keys satisfy $\text{PuK}_{\text{User\_ID}[i]} + \text{PrK}_{\text{User\_ID}[i]} = \mathcal{O}$, equivalently, $\text{Pu}k_{\text{User\_ID}[i]} + \text{Pr}lk_{\text{User\_ID}[i]} = 0$.
5. When session key is required for the signing of a document, it is some secret multiple SK P of the base point.

6.13 THE MENEZES-VANSTONE ELLIPTIC CURVE CRYPTOSYSTEM

The Menezes-Vanstone (public key) elliptic curve cryptosystem [Menezes and Vanstone 1993] is a variant of El Gamal's encipherment system ([ElGamal 1985a] and [ElGamal 1985b]). The steps given in the subsequent list are followed by User_ID[i] to send an encrypted message (x_1, x_2) with $x_1, x_2 \in \mathcal{Z}_{m,2}$ to User_ID[j] with public key $\text{PuK}_{\text{User\_ID}[j]} = \text{Pu}k_{\text{User\_ID}[j]}P$.

1. User_ID[i] chooses a random session key SK $\in \mathcal{Z}_{m,2}$ and compute SK $P \equiv (\kappa_1, \kappa_2)$ and SK $Puk_{\text{User\_ID}[j]}P \equiv (\kappa_1', \kappa_2')$ where $\kappa_1, \kappa_2, \kappa_1', \kappa_2' \in \mathcal{Z}_{m,2}$.
2. *Test*: If $\kappa_1' = \kappa_2' = 0 \in \mathcal{Z}_{m,2}$, then choose another random session key and return to E1.
3. Compute $y_1 \equiv x_1\kappa_1' \in \mathcal{Z}_{m,2}$ and $y_2 \equiv x_2\kappa_2' \in \mathcal{Z}_{m,2}$.
4. The ciphertext sent to User_ID[j] consists of $(\kappa_1, \kappa_2, y_1, y_2)$.

User_ID[j] deciphers as follows:

1. User_ID[j] uses the private key to compute $(\kappa_1, \kappa_2)P + Prk_{\text{User\_ID}[j]} = (\kappa_1', \kappa_2')$.
2. User_ID[j] may then compute $x_1' \equiv y_1(\kappa_1')^{-1} \in \mathcal{Z}_{m,2}$ and $x_2' \equiv y_2(\kappa_2')^{-1} \in \mathcal{Z}_{m,2}$.
3. The plaintext is $(x_1, x_2) = (x_1', x_2')$.

All's well that ends well

William Shakespeare (circa 1623).

6.14 SUPER SINGULAR ELLIPTIC CURVES

We continue with a primitive polynomial $p(x)$ over $\mathcal{Z}_2$ of degree m; the elliptic group $\varepsilon_{\mathcal{Z}_{m,2}}(a, b)$ consists of the point at infinity $\mathcal{O}$ together with all pairs $(x, y) \in \mathcal{Z}_{m,2}$, which are on the elliptic curve $y^2 + xy = x^3 + ax^2 + b$ with $a, b \in \mathcal{Z}_{m,2}$, $b \neq 0$.

The papers [Okamoto, Fujisaki and Morita 1999] and [Okamoto and Pointcheval 2000] also devise an El-Gamal like elliptic curve encipherment.

Notwithstanding the titles of [Okamoto, Fujisaki and Morita 1999] and [Okamoto and Pointcheval 2000], are elliptic curve cryptosystems really secure? Are all elliptic curves equivalent in strength?

The discrete logarithm problem for the field $\mathcal{Z}_p$ is

$\text{DLP}(\mathcal{Z}_p)$	
Given:	$x \in \mathcal{Z}_p$ and $x^k \in \mathcal{Z}_p$
Find:	$k \in \mathcal{Z}_p$

By analogy, the discrete logarithm problem for elliptic curves is

$\text{DLP}(\varepsilon_{\mathcal{Z}_{m,2}})$	
Given:	kP
Find:	$k \in \mathcal{Z}_{m,2}$

In the §6.13 cryptosystem, the public and private keys are multiples of the base point. It is known that the DLP over a group is complicated. Indeed, the complexity of the DLP is the basis of all the secure data exchange/signature schemes today. The purveyors of elliptic curve cryptography point out with some rationale that $\text{DLP}(\varepsilon_{\mathcal{Z}_{m,2}})$ is harder to solve than DLP($\mathcal{Z}_p$). Of course, cryptography has been around for at least two millennia; RSA has been available for more than 30 years and elliptic curve cryptography has existed for more than 20 years.

TABLE 6.3. Comparison of the Required Key Bit-Size (Symmetric, RSA, and ECC)

Key Size (in bits)			
Sym.	RSA	Elliptic	E-Ratio
80	1024	160	3:1
112	2048	224	6:1
128	3072	256	10:1
192	7680	384	32:1
256	15360	521	64:1

Even though the fat lady may have finished the overture, it is clear that not all elliptic curves yield strong encipherment systems.

- The elliptic group $\varepsilon_{\mathcal{Z}_{m,2}}(a,b)$ is super singular if $\#\varepsilon_{\mathcal{Z}_{m,2}}(a,b)=2^m+1-t$ and 2 divides t.
- The anomalous condition for $\varepsilon_{\mathcal{Z}_{m,2}}(a,b)$ holds if $\#\varepsilon_{\mathcal{Z}_{m,2}}(a,b)\neq 2^m$.

[Certicom 1997] contains a summary of the difficulty of the $\mathrm{DLP}(\varepsilon_{\mathcal{Z}_{m,2}})$ citing the paper [Menezes, Okamoto and Vanstone 1993], in which $\mathrm{DLP}(\varepsilon_{\mathcal{Z}_{m,2}})$ can be reduced to solving DLP for s simpler group and then only efficient for super singular elliptic curves. The standards [ANSI X9.62] and [FIPS 186-3, 208] explicitly exclude such curves.

Finally, [Satoh and Araki 1998] and [Smart 1999] have shown that elliptic curves that do not satisfy the anomalous condition should be avoided for the same reason as those that are super singular. The standards [ANSI X9.62] and [FIPS 186-3, 208] require checking that the anomalous condition is satisfied.

In their review "The Case for Elliptic Curve Cryptography" of elliptic curve cryptography,[6] several points are made; elliptic curve cryptosystems seem to offer a considerable efficiency with respect to key size. In Table 6.3, the following exist:

- The first three columns contain a comparison of National Institute of Standards and Technology (NIST)-recommended key sizes for a symmetric, RSA, and elliptic curve cryptosystem (ECC).
- The last column gives the ratio of computational efficiency of RSA to elliptic curve cryptography.

The article also points out that intellectual property rights are a major roadblocks to the further adoption of elliptic curve cryptography. The authors cite Certicom, which holds more than 130 patents in this area.

REFERENCES

A. Aho, J. E. Hopcroft, and J. D. Ullman, *The Design and Analysis of Computer Algorithms*, Addison-Wesley (Boston, MA), 1973.

[6]http://www.nsa.gov/business/programs/elliptic_curve.shtml

American National Standard X9.62-1998, Public Key Cryptography for the Financial Services Industry: The Elliptic Curve Digital Signature Algorithm (ECDSA), September 20, 1998.

H. Beker and F. Piper, *Cipher Systems: The Protection of Communications*, John Wiley & Sons (New York), 1982.

Certicom, Remarks on the Security of the Elliptic Curve Cryptosystem, Certicom Technical Publication, 1997, revised 2001.

J. Daemen, "Cipher and Hash Function Design Strategies Based on Linear and Differential Cryptanalysis", PhD Dissertation, Katholieke Universiteit Leuven, 1995.

W. Diffie and M. Hellman, "Multiuser Cryptographic Techniques", National Computer Conference, 1976, pp. 109–112.

W. Diffie and M. Hellman, "New Directions in Cryptography", *IEEE Transactions Information Theory*, **IT-22**, #6, pp. 644–654, 1976.

T. ElGamal, "A Public-Key Cryptosystem and a Signature Scheme Based on Discrete Logarithms", in *Proceedings of CRYPTO 84*—Advances in Cryptology, Springer-Verlag (Berlin, Germany), 1985, pp. 101–108.

T. ElGamal, "A Public-Key Cryptosystem and a Signature Scheme Based on Discrete Logarithms", *IEEE Transactions on Information Theory*, **IT-31**, #4, pp. 469–472, 1985.

Federal Information Processing Standards Publication 46-1, Data Encryption Standard (DES), National Bureau of Standards, January 22, 1988; superceded by Federal Information Processing Standards Publication 46-2, December 30, 1993 and reaffirmed as FIPS PUB 46-2, October 25, 1999.

Federal Information Processing Standards Publication 186-3, Digital Signature Standard, November 2008.

Federal Information Processing Standards Publication 197, Advanced Encryption Standard (AES), November 26, 2001.

M. Gardner, "A New Kind of Cipher That Would Take Millions of Years to Break", *Scientific American*, **237**, pp. 120–124, 1977.

D. Husemoller, *Elliptic Curves*, Springer-Verlag (Berlin, Germany), 1986.

L. M. Kohnfelder, "Towards a Practical Public-Key Cryptosystem", Bachelor's Thesis, MIT, 1978.

N. Koblitz, "Elliptic Curve Cryptosystems", *Mathematics of Computation*, **48**, pp. 203–209, 1987.

A. G. Konheim, *Computer Security and Cryptography*, John Wiley & Sons (New York), 2007.

R. Lidl and H. Niederrieter, *Finite Fields*, Cambridge University Press (Cambridge, United Kingdom), 1997.

R. C. Merkle and M. E. Hellman, "Hiding Information and Signatures in Trapdoor Knapsacks", *IEEE Transactions on Information Theory*, **IT-24**, #5, pp. 525–530, 1978.

A. J. Menezes, T. Okamoto, and S. Vanstone, "Reducing Elliptic Curve Logarithms to Logarithms in a Finite Field", *IEEE Transactions on Information Theory*, **IT-39**, pp. 1639–1646, 1993.

A. J. Menezes and S. A. Vanstone, "Elliptic Curve Cryptosystems and Their Implementation", *Journal of Cryptology*, **6**, pp. 209–224, 1993.

A. J. Menezes and S. Vanstone, "The Elliptic Curve Digital Signature Algorithm (ECDSA)", Certicom Technical Publication, 2001.

V. Miller, "Use of Elliptic Curves in Cryptography", in *Proceedings of CRYPTO 85 Advances in Cryptology*, Springer-Verlag (Berlin, Germany), 1986, pp. 417–426.

R. M. Needham and M. D. Schroeder, "Using Authentication for Authentication in Large Networks of Computers", *Communications of the ACM*, **21**, #12, pp. 993–999, 1978.

T. Okamoto, E. Fujisaki, and H. Morita, "PSEC: Provably Secure Elliptic Curve Ecryption Scheme", presented at IEEE P1363 Meeting on Asymmetric Encryption, 1999.

T. Okamoto and D. Pointcheval, "PSEC-3: Provably Secure Elliptic Curve Encryption Scheme (Version 2)", submitted to IEEE P1363 Meeting on Asymmetric Encryption, 2000.

C. H. Papdimitrious, *Computational Complexity*, Addison Wesley (Boston, MA), 1995.

H. Riesel, *Prime Numbers and Computer Methods for Factorization* (Second Edition), Birkhauser (Berlin, Germany), 1994.

RFC 1113, Internet Engineering Task Force, J. Linn, "Privacy Enhancement for Internet Electronic Mail", August 1989.

R. L. Rivest, A. Shamir, and L. Adelman, "A Method for Obtaining Digital Signatures and Public-Key Cryptosystems", *Communications of the ACM*, **21**, #2, pp. 120–126, 1978.

T. Satoh and K. Araki, "Fermat Quotients and the Polynomial Time Discrete Log Algorithm for Anomalous Elliptic Curves", *Commentari Mathematici University St. Pauli*, **47**, pp. 81–92, 1998.

E. S. Selmer, *Linear Recurrence Relations Over Finite Fields*, Department of Mathematics, University of Bergen (Norway), 1966.

C. E. Shannon, "Communication Theory of Secrecy Systems", *Bell System Technical Journal*, **28**, pp. 656–715, 1949.

N. P. Smart, "The Discrete Logarithm Problem on Elliptic Curves of Trace One", *Journal of Cryptology*, **2**, pp. 193–196, 1999.

A. M. Turing, "On Computable Numbers with an Application to the Entscheidungs Problem (Decision Problem)", *Proceedings of the London Mathematical Society*, **42**, #2, pp. 230–267, 1936.

N. Zierler, "Linear Recurring Sequences", *Journal of the Society of Industrial and Applied Mathematics*, **7**, pp. 31–38, 1959.

PART II

HASHING FOR STORAGE: DATA MANAGEMENT

CHAPTER 7

Basic Concepts

7.1 OVERVIEW OF THE RECORDS MANAGEMENT PROBLEM

The mechanics of using hashing in storage management is examined in this chapter. Storage can mean either the computer main random-access storage or file systems, which is also known as pervasive storage.

We view information in storage as composed of records $\{R_i\}$. The records may be of variable sizes and are located in the system storage media (typically a disk) in a variety of regions, especially in dynamic file systems, where additions and deletions of records make it difficult and impractical to insist on contiguity. Using the information means gaining access to the records and performing operations on records.

A unique key is associated with and stored in each record. This key could be a name or a suitable number, but in general, any string can serve as a unique identifier. It is used by the programs processing the data to locate (address) the desired record. When a telephone directory is searched for a telephone number, the process is somewhat easier than the general instance of file management because the records composed of the triple (`Name_Address_TelephoneNumber`) are stored sorted on the `Name` *field*.

File management is complicated because most data sets are *dynamic* rather than *static*, meaning existing records are updated or deleted and new records are added.

The set of program keywords recognized by a compiler is fixed; the list of variable names it needs to deal with in any compilation is also fixed until there are changes in the program. The set of valid social security numbers, students enrolled in a university, and telephone directories are examples of dynamic data sets.

The menu of possible file operations includes the following:

`INSERT`: Creation of new records

`MODIFY`: Modification of existing records

`DELETE`: Deletion of existing records

All of these processes depend on a procedure

`SEARCH`: Find the address of the record with key KEY

Hashing in Computer Science: Fifty Years of Slicing and Dicing, by Alan G. Konheim

When a program needs to access a specific record with a key K, how does it find its location in storage? This problem is solved by the abstract data type (ADT) dictionary.

Several designs have been proposed for the dictionary, such as arrays, search trees, and associative tables.[1]

The implementation of a dictionary by hashing solves both difficulties of inefficient search and the inflexibility of ordered arrays. It calls for the creation of a table, which is often called a (hashing) address table declared as an array of the suitable type. As a rule, the address table is kept entirely and contiguously in main storage. Although the records may be of variable sizes and scattered in chunks all over the file system with variable-length keys, the address table entries are of fixed size. An entry might consist of the two components (k, ADD[k]) where ADD[k] is the address (a pointer) of the record with key k. As we shall discuss, an address table allows a trade-off of processing time to find the record with table space.

It is assumed that a block of memory is allocated by the system to store variable- or fixed-length records. A simple, albeit inefficient, scheme stores records contiguously, in a scheme referred to as chaining.

7.2 A SIMPLE STORAGE MANAGEMENT PROTOCOL: PLAIN VANILLA CHAINING

Assume m records $R_0, R_1, \cdots, R_{m-1}$ are to be stored contiguously in the block of memory beginning at address A. Each record is prefixed by its key.

Figure 7.1 is meant to illustrate this in the special case when each of the m records has the same length L (bytes); when variable-length records are stored, each record must contain a length field.

To access the record with key KEY, the keys of the records at addresses $A, A + L$, $\cdots$ are tested by making the comparison $k_i \stackrel{?}{=} \text{KEY}$ until either there is

Case 1: a match, $k_i = \text{KEY}$, or

Case 2: an empty cell is encountered $k_i = \emptyset$ implying that *no* record with key $k = \text{KEY}$ has been stored.

In *Case 2*, a cell has been found into which a *new* record with key $k = \text{KEY}$ can be stored.

Figure 7.1 is only a logical representation of a chain of records. Often, a chain of records is implemented by a linked list, which is a data structure[2] composed to an ordered sequence of nodes, each consisting of one or more data fields and a link pointing to the next nodes. The first node in the linked list is pointed or linked to by head of the list. Figure 7.2 depicts a possible structure for a node with a field for the key k, the address of the data corresponding to the record with key k and a link

[1]Special-purpose dictionaries, for example, stacks and heaps, have been designed to provide for other needs, such as finding the record most recently inserted, or `max`, locating the record with the maximal key.

[2]Starting in §6.1, [Knuth 1973] contains a discussion of various data structures. Chapter 4 in [Aho, Hopcroft and Ullman 1973] contains a brief introduction to data structures.

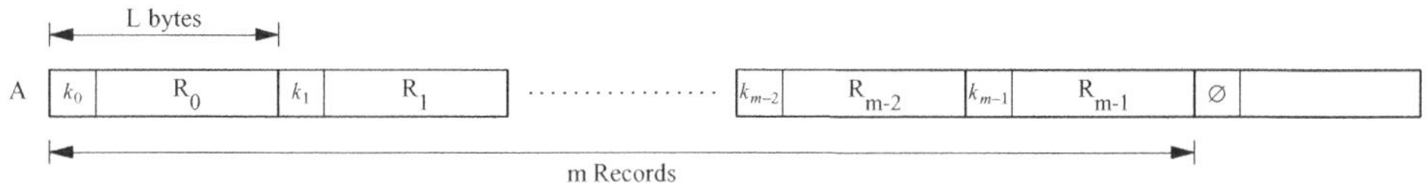

Figure 7.1. Plain vanilla chaining.

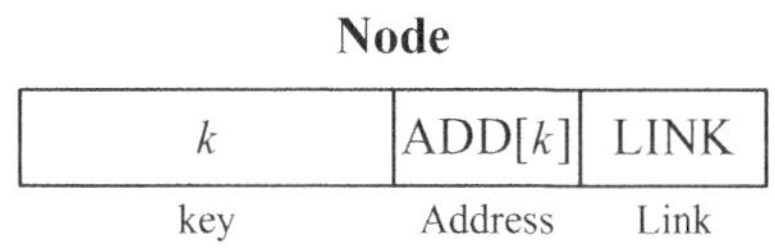

Figure 7.2. The fields in a node.

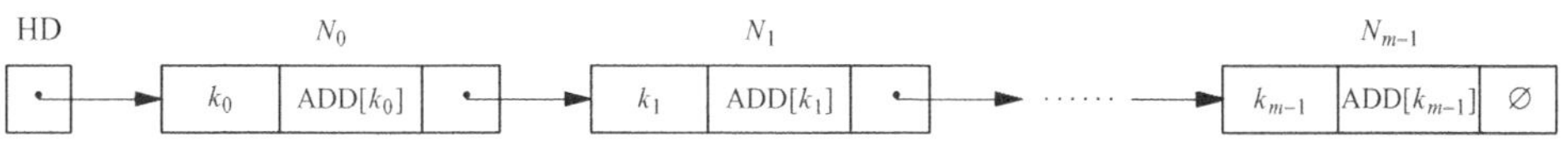

Figure 7.3. Strawberry chaining.

field (LINK) to the next node. Figure 7.3 illustrates an implementation of a chain as a linked list.

Note that LINK = ∅, which indicates that it is the tail of the linked list, that is, the final node in the chain.

Exercise 7.1. (Model for performance evaluation of simple storage): Assume the following:

- m records $R_0, R_1, \cdots, R_{m-1}$, each requiring for storage a single cell (1 unit), have been stored in a block of n contiguous cells with starting address A and
- If an existing record is selected for access, then each of the m records is chosen with equal probability.

`SEARCH(KEY)` is the procedure to locate the record with key = KEY requires a C comparison.

7.1a) Calculate the probability distribution and average number of C needed to execute `SEARCH(KEY)` if such the record with key = KEY is currently in storage.

7.1b) Assuming that no record with key k = KEY has been stored, calculate the average number of comparisons needed to execute `SEARCH(KEY)`.

Remark 7.1. Although the `SEARCH` with plain vanilla chaining or strawberry chaining is certainly simple, it is not efficient, as formulated above. Finding the record with key `XYLOPHONE` among the 50 musical instrument types arranged alphabetically might be time consuming; it is even more so for `ZEBRA` among the 1.75 million animal species.

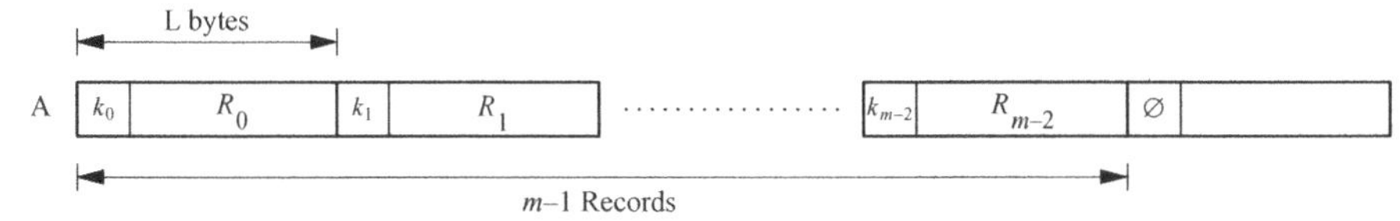

Figure 7.4. Simple storage after record R_{m-1} is deleted.

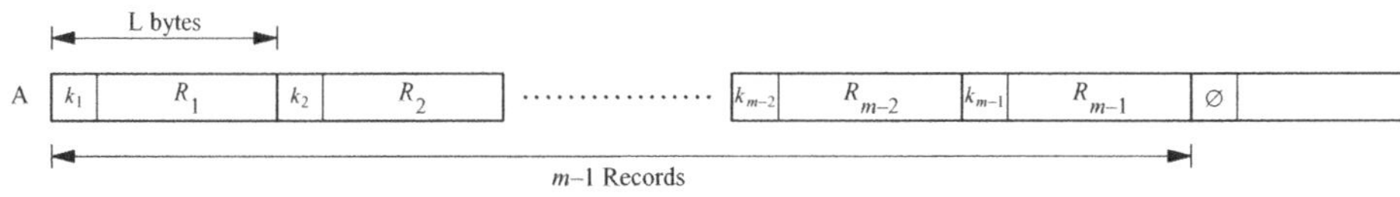

Figure 7.5. Simple storage after record R_0 is deleted.

Deletion is simple when records are chained as in Figure 7.1. To delete the last record R_{m-1} in a chain, it must be purged from the system and LINK must be set to $\emptyset$ in a linked list implementation, as shown in Figure 7.4.

However, if one of the other records is deleted, it is necessary to move several records to preserve contiguity. The worst case occurs when record R_0 is deleted requiring $m - 1$ records to be moved as depicted in Figure 7.5.

The deletion process is easier if the chain is implemented as a linked list because the offending node is merely deleted from the list and the LINKs adjusted. We take this up later in Chapter 10.

7.3 RECORD-MANAGEMENT WITH SORTED KEYS

Searching fixed length records is simplified if the address of the record with key = KEY can be approximately determined. This occurs, for example, when a telephone directory is searched for the telephone number of key = `Name`. Searching is easier in this instance, because the records (`Name_Address_TelephoneNumber`) are stored sorted on the `Name` *field*. In Figure 7.6, we depict a segment of the Santa Barbara telephone directory assuming, the following is true:

- Listings appear in the Santa Barbara telephone directory for both `Konda, Kandarp & Daksha` and `Konheim, Alan & Carol.`
- No names intervene between these two listings.[3]

Even though searching a system table is easy because there is a lexicographic ordering of the keys, alphabetic storage has a serious disadvantage; the insertion of a new record may require rearrangement of existing records. For example, when the `Kong`'s move into Santa Barbara, the insertion of their entry

`Kong, Michael & Michelle`	`110 State Street Santa`	`Santa Barbara`	`967-2234`

[3]Assuming the sexist practice of listing male names *before* the corresponding female name (and the author lives in California).

⋱

$a-2$	Kolib, Herbert & Ronald	123 Palo Alto Street	Isla Vista	685-2330
$a-1$	Kominsky, Thelma & Louise	23 Anapamu Street	Santa Barbara	563-1165
a	Konda, Kandarp & Daksha	512 Canon Perdido	Santa Barbara	543-2321
$a+1$	Konheim, Alan & Carol	3735 Essex Street	Santa Barbara	687-1178
$a+2$	Konheim, Joshua	5320 Traci Drive	Santa Barbara	683-0485
$a+3$	Unoccupied Address Cells			

⋱

Figure 7.6. Telephone directory.

preserving the alphabetic ordering requires providing space between the records of `Konda, Kandarp & Dak-sha` and `Konheim, Alan & Carol`, and thus moving two records. We trust the American civil liberties Union (ACLU) does not file suit!

It might be possible to construct a perfect key-mapping function $h: k \rightarrow h(k) = k^*$, which would effectively alphabetize a known and given set of keys $k_0, k_1, \cdots, k_{m-1}$.

$$k_0, k_1, \cdots, k_{m-1} \rightarrow h(k_{\pi_0}) \prec h(k_{\pi_1}) \prec \cdots \prec h(k_{\pi_{m-1}})$$

where $\prec$ is a lexicographic-like ordering on the keys $k_{\pi_0}, k_{\pi_1}, \cdots, k_{\pi_{m-1}}$.

The construction of such an alphabetizer is possible, and this perfection is discussed in Chapter 11. However, if the keys are unknown or if records are dynamically added to or deleted from the file, then the ordering is likely to change. What is really needed is something less than perfection, namely, a key-mapping function h which produces distinct values $h(k_0), h(k_1), \cdots, h(k_{m-1})$ together with a procedure to handle less than perfectly chosen functions h.

Such a scheme was discovered in 1953 and improved on in the next half century. We will discuss this topic in the next chapters.

REFERENCES

A. Aho, J. E. Hopcroft, and J. D. Ullman, *The Design and Analysis of Computer Algorithms*, Addison-Wesley (Boston, MA), 1973.

D. Knuth, *The Art of Computer Programming: Volume 3/Sorting and Searching*, Addison-Wesley (Boston, MA), 1973.

CHAPTER 8

Hash Functions

8.1 THE ORIGIN OF HASHING

The history of the hashing concept is described in both [Knuth 1973], the third volume of Knuth's seminal work, *The Art of Computer Programming*, and in the paper "Hashing" by D. G. Knott [Knott 1975].

The website `http://www-03.ibm.com/ibm/history/` provides access to various documents relating to IBM's role in the development of computers. The following material in derived in part from the transcription of a December 10, 1970 *Oral History of IBM Technology*[1] interview of Elaine M. McGraw (née Boehme).

Nat Rochester was the software project manager for the IBM 701 machine; together with Gene M. Amdahl, Elaine M. McGraw (née Boehme), and Arthur L. Samuel, he considered a key-value-to-address machine for the 701 assembler.

Samuel claims (in a private communication to Knott cited in [Knott 1975]) that when confronted with the problem of collisions, Amdahl invented linear open addressing (see Chapter 13).

8.1.1 Biographical Notes

- Nat Rochester studied electrical engineering at the Massachusetts Institute of Technology (MIT) and was working in acoustics at the outset of World War II. During the war, he worked on radar at the MIT Radiation Laboratory and at Sylvania building equipment for the radiation laboratory. After the war, he worked on the arithmetic unit for Whirlwind and on cryptanalysis equipment for the National Security Agency (NSA). Deciding in 1948 that computers would be a "major thing," Rochester joined IBM where he urged the development of computers.
- Gene M. Amdahl was born in South Dakota in 1922; he studied electronics and computer programming while in the U.S. Navy during World War II. He received a bachelors degree in engineering physics at South Dakota State University in 1948. In 1952, Amdahl completed his doctorate in theoretical physics at the

[1]Provided courtesy of Ms. Stacy L. Fortner of the IBM Corporate Archives (Somers, New York).

Hashing in Computer Science: Fifty Years of Slicing and Dicing, by Alan G. Konheim

University of Wisconsin, where he designed his first computer, the Wisconsin Integrally Synchronized Computer (WISC).

His IBM career began in 1952. Amdahl gained recognition during the 1960s as the principle architect of IBMs System 360 series of mainframe computers, which were based on the *Stretch*, which Amdahl had worked on in 1955. The System 360 series was one of the greatest success stories in the computer industry and became the main contributor to IBM's enormous profitability in the late 1960s Amdahl became an IBM Fellow, able to pursue his own research projects.

Amdahl left IBM in 1970 and formed Amdahl Corporation in Sunnyvale, California. The company's strategy was to compete head-to-head with IBM in the mainframe market.

- Elaine M. McGraw (née Boehme) was an economics major who learned to program a UNIVAC after joinging Prudential Insurance in the 1950s. She was one of two employees selected by Prudential to attend a class for the first IBM 701 customers. Ms. McGraw believed that Prudential's business computing needs were for a decimal machine—Knuth's `MIX` assembler—and consequently believed that Prudential was not going to receive one of the first IBM 701 machines. Because she enjoyed programming, Ms. McGraw left the rock and applied for a job with IBM. She was hired as a programmer-technician by Gene Amdahl in 1953.

  Ms. McGraw recalls that either Rochester or Amdahl suggested assigning a symbol to instructions and data and "··· having the computer actually figure out the addresses that were appropriate to be assigned to the various instruction and data." Ms. McGraw was given the assignment to implement this symbolic assembler without being told by Rochester how to do it.

  After the 701 project, Ms. McGraw continued to work with Gene Amdahl on the IBM's System 360 series and a predecessor machine; the Stretch was a supercomputer with high-speed memory and performance, designed for the two prospective clients—NSA and the Los Alamos Scientific Laboratory (LASL).

  Little more detail is available on her IBM career other than she was working at the IBM facility in Gaithersburg (Maryland) in December 1970.

- Arthur Samuel (1901–1990) was a pioneer of artificial intelligence research; his research contributions date back to the 1950s. Samuel joined IBM's Poughkeepsie Laboratory in 1949 and worked on the 701, which was IBM's first stored program computer.

  Samuels' Checkers program made history by demonstrating that a computer program could play checkers well enough to beat human experts, and it could learn from its experience in playing against humans to become a better player.

  Samuel retired from IBM in 1966 and came to Stanford University as a research professor.

Both Buckholz [Buckholz 1963; pp. 109–110] and Knuth [Knuth 1973, p. 540] note that the concepts of hashing (aka scatter storage) as well as record chaining described in Chapter 10 first appeared in in 1953 internal IBM documents[2] by Hans Peter Luhn

[2]Provided courtesy of Ms. Dawn M. Stanford of the IBM Corporate Archives (Somers, New York).

(1896–1964) and several years later in an article by Arnold I. Dumey [Dumey 1956]. In these memoranda, Luhn was a mathematician and a prolific inventor on 20 patents; he was hired by IBM in 1941. Luhn worked on the reservation system in 1950 for American Airlines, which was to become the Sabre System. He retired in 1961.

An article by Robert Morris [Morris 1968] used the word *hashing* for the first time.[3]

8.2 HASH TABLES

It is assumed that a block of memory $\mathcal{M}$ is allocated by the system to store variable- or fixed-length records. Memory management is responsible for maintaining, allocating, and deallocating memory for the records. Additional memory is provided for the hash table of n cells, which is a concierge providing the address in $\mathcal{M}$ of the record with KEY = k. The hash table provides the connection between the record identifier = key and the address of the record in this memory block.

The address in the hash table of the entry for the record with KEY = k at address table is determined as the result of a computation

$$h : k \to h(k) \in \mathcal{Z}_n \equiv \{0, 1, \cdots, n-1\} \tag{8.1}$$

Figure 8.1 depicts an instance in which there are no collisions for the three keys k_1, k_2, k_3 in the hashing address table; that is, $h(k_1)$, $h(k_2)$, and $h(k_3)$ are distinct.

The hash table address $h(k)$ in Figure 8.1 contains pairs $(k, \text{ADD}[k])$, where k is a key and ADD[k] is the address of a record with the key k.

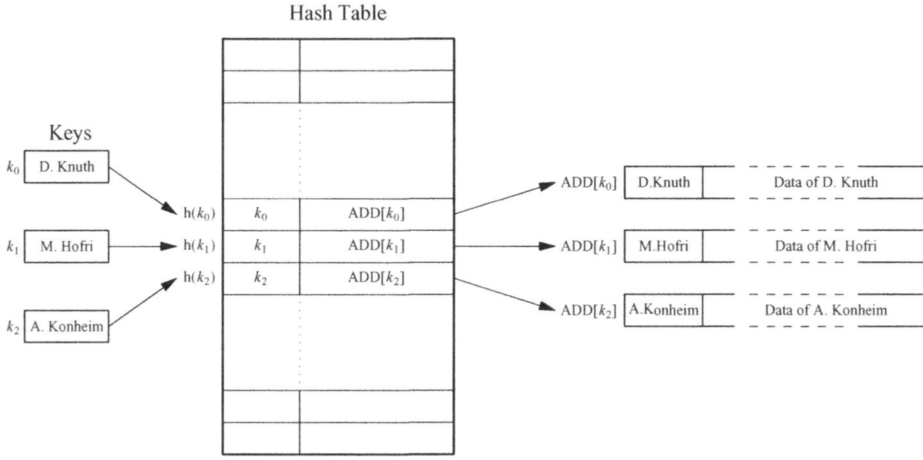

Figure 8.1. Hash table: No collisions for records R_0, R_1, R_2.

[3]The website `www.recipesource.com` lists 95 recipes for hash, which is truly a part of the American cuisine. An additional 7 recipes use the adjective *hashed*, but this list fails to include the Swiss all-time maximum-arterial-clogger Berner rösti. Its preparation, which requires considerable manual dexterity, combines bacon, cheese (optional), onion, black pepper, garlic, and potatoes. Hashing represents the confluence of mathematics, computer science, and gastronomy. We should be grateful, however, that Robert Morris' article did not appear in the Zeitsschrift für, or we might be writing about pfannkuchening.

To access the record with key k = KEY, the following must occur:

- The hashing address table is accessed at address $h(k)$, and the key at this address is compared with KEY.
- The address ADD[k] of the record with key k is read.
- The record at ADD[k] is then accessed.

When there are collisions, say $h(k_0) = h(k_1) = h(k_2)$, some mechanism must be provided to resolve the matter.

Figure 8.2 depicts a resolution in which the records with these three keys have been inserted in the order R_0, then R_1, then R_2.

When R_1 is to be accessed, the hash table at $h(k_1) = h(k_0)$ reveals that there is a hashing collision, and it is necessary to look at another location or locations in the hashing to find it.

Missing from Figure 8.2 is any explanation as to how the system discovers the following:

where R_1's values (k_1, ADD[k_1]) are stored
where R_2's values (k_2, ADD[k_2]) are stored in the hash address table

Of course, Figures 8.1 and 8.2 are again only logical representations of a hash table. These figures suggest equal length keys that may be partially beyond the control of the system.

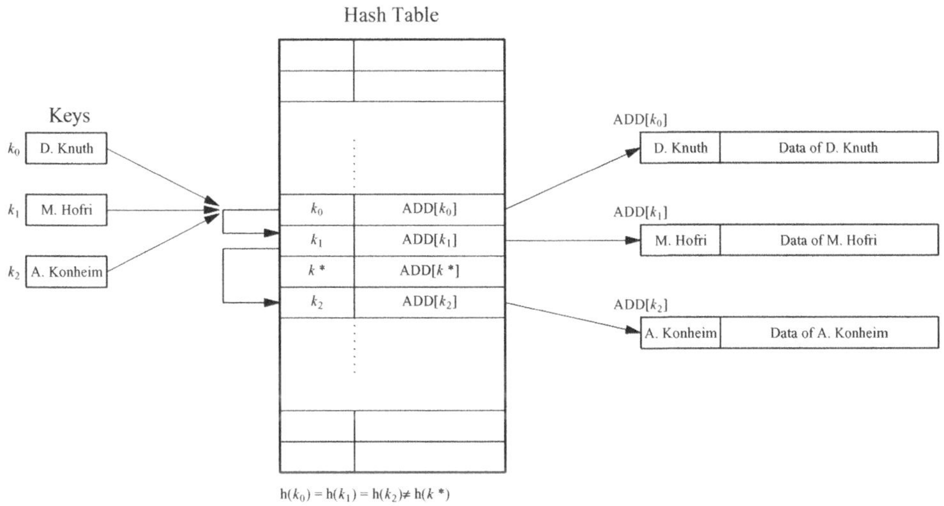

Figure 8.2. Hash table: Instances of hashing collision.

I have always wanted to write a program using the word `antidisestablishmentarianism` as a variable name.[4] How can hashing succeed with such perverse behavior? Linked lists described in Chapter 10 can be used to tolerate variable-length keys, with modern marketing techniques by either imposing a key-length surcharge or by using linked lists.

Figure 8.3 offers a possible solution to handle this deviant behavior. A hash table might consist of the heads of $m \leq n$ linked lists. Figure 8.3 is the analog of Figure 8.1; it displays only $m = 3$ entries that, in this instance, do not result in collisions. The heads HEAD[$h(k_i)$] with $i = 0, 1, 2$ point to the first node in three different linked lists.

Figure 8.4 depicts a possible format for a linked list node which permits variable-length keys.

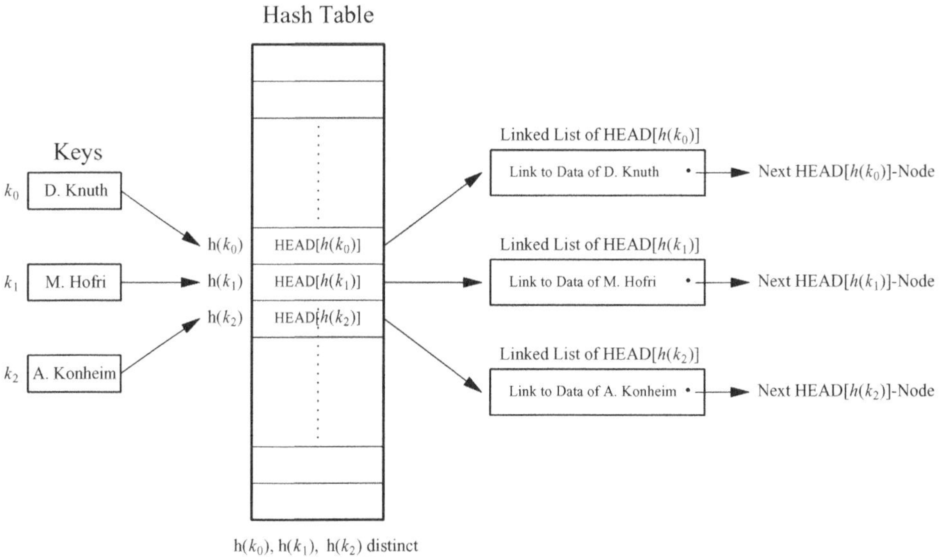

Figure 8.3. Implementation of a hash table using linked lists.

Node

LEN[k]	k	ADD[k]	LINK
key length	key	Address	Link

Figure 8.4. Possible format for a linked list node.

[4]The website `http://www.atomicmpc.com` claims this is the longest word in the dictionary. The website `http://dictionary.reference.com` offers the following definition:

— *noun*, opposition to the withdrawal of state support or recognition from an established church, especially the Anglican Church in 19th-century England

The website `http://www.mail-archive.com` observes that `dichlorodiphenyltrichloroethane` (DDT) contains two more letters. This website also claims that `Pneumonoultramicroscopicsilicovolcanokoniosis` is believed to be the winner. Isn't the Web wonderful?

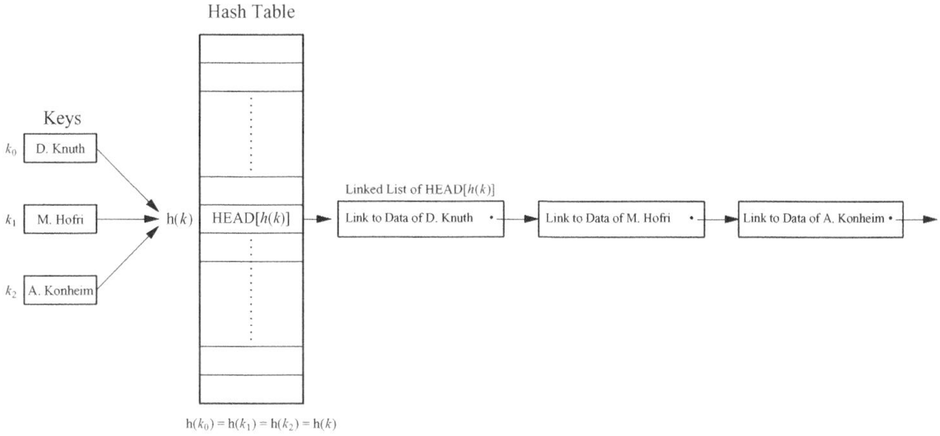

Figure 8.5. Implementation of a hash table using linked lists with three-way collision.

Each of the three linked lists terminates, by pointing to a *terminal node*, omitted above.

We discuss in Chapter 10 how this implementation of hash-tables might be modified to handle the situation depicted in Figure 8.2 in which there is a 3-way collision. One extension, referred to as ***separate chaining*** is shown in Figure 8.5.

The keys of the records of D. Knuth, M. Hofri, and A. Konheim have the same hash value and result in a linked list now consisting of three nodes.

8.3 A STATISTICAL MODEL FOR HASHING

We define a statistical model for hashing that we will be used throughout this book. When hashing is used, the system determines a hashing function h for a hash table of n cells. In the simplest scheme, h is a mapping $h: \mathcal{K} \to \mathcal{Z}_n$ from the set of all possible keys $k \in \mathcal{K}$ to the identifiers $\mathcal{Z}_n$ of cells in the hash table. The pair (h, k) determines a hash-table cell, although in more elaborate schema, for example, double hashing [Chapter 14] and uniform hashing [Chapter 15] it determines a sequence of hash-table cells. Although $\mathcal{K}$ is certainly large, it is finite, because we may assume that a key is an alphanumeric string of some length ≤ 50.

The architect of a hashing scheme has no knowledge of the specific keys that will arise in some application. A statistical model (see [Loéve 1981, pp. 3–7] and [Feller 1967, pp. 1–6]) is used to replace this ignorance by assuming that keys arise randomly. Indeed, this paradigm is often the basis when mathematics is applied to evaluate problems in computer science and engineering. Accordingly, the statistical model supposes a chance experiment $\mathcal{E}$ whose possible outcomes are sample points in the sample space Ω. The selection of a key from $\mathcal{K}$ is an outcome $\mathcal{E}$ the phrase "selecting a key" translates to K is a random variable $K: \omega \to \mathcal{K}$ on the sample space with value in $\mathcal{K}$.

Our hashing model combines the system's choice of h and the random selection of a key making the following two assumptions:

Uniform Distribution: $Pr\{h(K)=i\}=\frac{1}{n}$, $(0 \le i < n)$;
Each possible cell is likely to be the target for a randomly chosen key K.

Independence: $Pr\{h(K_0), h(K_1), \cdots, h(K_{m-1}) = (i_0, i_1, \cdots, i_{m-1})\} = \frac{1}{n^m}$ for each of the n^m possible vectors $\underline{i} = (i_0, i_1, \cdots, i_{m-1})$ of hash table cells.

The hashing sequence $(h(K_0), h(K_1), \cdots, h(K_{m-1}))$ for m randomly chosen keys $(K_0, K_1, \cdots, K_{m-1})$ result from independent experiments.

Note that we are not assuming independence and uniform distribution of keys, but a somewhat weaker assumption that h specified by the system interacts with the keys to result in the previously stated properties.

Most performance analysts question the assumption of a uniform distribution as a matter of faith. However, there is little prognosis for success if the uniform distribution were replaced by another specific distribution, whose appropriateness might also be questioned. The results of McKenzie et al. [Mckenzie, Harries and Bell 1990] described in Chapter 9, §9.6 suggests the uniform distribution assumption is not entirely unreasonable. Similarly, the assumption of independence might also be questioned, but it seems farfetched conceptually to argue for a more general dependence, for example, Markov dependence.

8.4 THE LIKELIHOOD OF COLLISIONS

Section 8.2 illustrates that locating a record with key k using a hash table involves the following two steps:

Step #1: Computing the hash value $h(k)$ and examining the hash table $\mathcal{T}$ at address $h(k)$ to determine whether the key stored there is k.

Step #2A: If so, the pointer ADD[k] stored at address $h(k)$ allows direct access the record with key k.

Step #2B: If not, then either not record with key k has been stored or there is a hash-table *collision*. In the latter case, some other process must be invoked to find the record with key k.

The hashing literature contains different protocols describing how collisions are resolved and how the protocol's performance for `SEARCH` (cost = time/space) is effected.

It is natural to ask how likely is a collision; for example, if the hash-table contains the following:

- Locations for n-records
- Already has entries for m records

What is the probability that hashing for a $(m + 1)^{st}$ record will be either of the following:

- Collision free
- Will result in a k-way collision

It is not possible to answer either of these questions without specifying the collision resolution procedure (CRP). However, the first question can be addressed we assume that the CRP results in a uniform distribution of occupied hash-table addresses. By this, we mean that the CRP causes all m-element subsets of the n-address hash table to be equally likely to occur. Chapter 12 considers this in greater detail.

Exercise 8.1. Assume the following:

H1. The information of m records has been entered into a hash table of size n-entries with $n \gg m$

H2. All possible hash values $h(K) \in \{0, 1, \cdots, n - 1\}$ are equally likely

H3. The CRP causes all m-element subsets of the n-address hash table to be equally likely to occur

Calculate the probability that the insertion into the hash table of the information of an $(m + 1)^{st}$ record is collision free.

The hashing literature contains, among other things, protocols describing specific hash function h, methods by which collisions are resolved, and how the performance (cost = time/space) is affected.

When hashing is used to maintain a file system, there are several technical issues:

1. What functions can serve as hash functions?
2. What is the most efficient way to resolve collisions?
3. What is the performance of hashing schemes? Because `SEARCH` requires a hash computation and one or more comparisons of key (probes), performance might be measured by the distribution, which includes the average or maximum number of comparisons.
4. How should records be deleted from files to best support the file-management scheme?

The hashing literature overflows with schema for the choice of hashing functions h. It also discusses collision-resolution protocols and their performance.

REFERENCES

Werner Buckholz, "File Organization and Addressing", *IBM Systems Journal*, **2** #2, pp. 66–111, June 1963.

Arnold I. Dumey, "Indexing for Rapid Random-access Memory", *Computers and Automation*, **5**, #12, pp. 6–9, 1956.

W. Feller, *An Introduction to Probability Theory and Its Applications*, Volume 1 (Second Edition), John Wiley & Sons (New York), 1957; (Third Edition), John Wiley & Sons (New York), 1967.

G. D. Knott, "Hashing Functions", *The Computer Journal*, **18**, #3, pp. 265–278, 1975.

D. Knuth, *The Art of Computer Programming: Volume 3/Sorting and Searching*, Addison-Wesley (Boston, MA), 1973.

Michel Loéve, *Probability Theory*, D. Van Nostrand (New York), 1981.

B. J. Mckenzie, R. Harries, and T. Bell, "Selecting a Hashing Algorithm", *Software-Practice and Experience*, **20**, #2, pp. 209–224, 1990.

Robert Morris, "Scatter Storage Techniques", *Communications of the ACM*, **11**, #1, pp. 38–44, 1968.

CHAPTER 9

Hashing Functions: Examples and Evaluation

9.1 OVERVIEW: THE TRADEOFF OF RANDOMIZATION VERSUS COMPUTATIONAL SIMPLICITY

The models of collision-resolution protocols to be described in the remaining chapters of Part II assume that the hashing function $h:\mathcal{K} \to \mathcal{Z}_n$ is uniformly distributed; that is, $Pr\{h(K)=i\}=\frac{1}{N}$ for $0 \le iN$. It seems appropriate before embarking on this examination to provide some examples. Because the rationale of hashing is the time/space-trade-off, a hashing function must be easy to compute; this limits the possible choices.

We assume each record requires a unit cell in the hash table that contains space for N record entries. The hash table provides the connection between the m-bit key and the true address in memory of the record. The cell $h(k)$ in hash table contains the pair (k, ADD[k]), where ADD[k] is the true address of the record with key $k \in \mathcal{K}$ and h is a hashing function

$$h:\mathcal{K} \to \mathcal{Z}_N \tag{9.1}$$

$\mathcal{K}$ is the key space and $\mathcal{Z}_N = \{0, 1, \cdots, N-1\}$ identify the cells in the hash table. Often, $\mathcal{K} = \{0, 1\}^\ell$ is the set of all $m = 2^\ell$ (0,1) sequences of length ℓ We assume $2^\ell < n$

If hashing is to simplify the procedure SEARCH, the computation in equation (9.1) must be easy to effect.

9.2 SOME EXAMPLES OF HASHING FUNCTIONS

Example 9.1. [Sedgewick 1998] Assume $s = (s_0, s_1, \cdots, s_{m-1})$ is a string of m American Standard Code for Information Interchange (ASCII) characters and where A_{s_i} is the 7-bit ASCII ordinal of s_i. With a slight abuse of notation, the hash function of the key k associated with the string s is $h(k)=\left(\sum_{i=0}^{m-1} A_{s_i} 127^i\right) \pmod N$

Hashing in Computer Science: Fifty Years of Slicing and Dicing, by Alan G. Konheim

Figure 9.1. The golden ratio.

Example 9.2. [Weiss 1996]: The same parameters as in *Example 9.1.*

Weiss' Hash

```
int Hash (char s; int N)
{int h = 0;
for (s = 0; s + +)
(h ≪ 5) ˜s˜h;        ≪ 5 bitwise left-shift; ˜ bitwise XOR
if (h < 0) h = -h
return h (mod N) }
```

Example 9.3. Donald Knuth [Knuth 1973, §6.4] describes various hashing schema. He points out that hash functions can be constructed using modular arithmetic. According to the Greeks, the golden section ϕ is the ideal proportion in which an interval [A,B,C] should be divided (Figure 9.1).

$$\frac{\overline{AB}+\overline{BC}}{\overline{AB}}=\frac{\overline{AB}}{\overline{BC}}:\phi=\frac{\sqrt{5}+1}{2}\approx 1.6180339887\cdots$$

The reciprocal

$$\phi^{-1}=\frac{\sqrt{5}-1}{2}\approx 0.6180339887\cdots$$

is used to define the Fibonacci hashing function by

$$h(k)=B\times\left(k\phi^{-1}(\mathrm{mod}\ 1)\right)$$

Knuth points out [Knuth 1973, pp. 510–11] the nearly uniformly distribution properties of the hash values with the Fibonacci multiplier.

Example 9.4. Jenkins [Jenkins 97] describes a 32-bit hash Newhash, a 64-bit hash Hash64 and a 96-bit hash; the first two use the operations exclusive-OR and left and right shifts, the third also requires subtraction.

Example 9.5. (Fowler/Noll/Vo)[1]: An unimpeachable source asserts that the FNV hashing, which is defined for $n = 32, 64, 128$, and 256 bit arithmetic, is widely favored. The basic operation deriving a hash value is

[1]The FNV hash algorithm results from an idea sent as reviewer comments to the IEEE POSIX P1003.2 committee by Glen Fowler and Phong Vo and improved on by Landon Curt Noll. For further details www.isthe.com/chongo/tech/comp/fnv.

FNV-1

Input: n, n-offset_basis, n-FNV prime
hash = n-offset_basis;
hash = hash XOR octet_of_data; [XOR-folding]
hash = (hash *n-FNV prime) (mod 2^n); [modular multiplication]

An evaluation of FNV-1 given ar [Mulvey 2007] using the χ^2-test (see §9.4) states the following:

- Nonuniformity in the high-order bits of the hash value when $n = 14,15$, nonsupportive of the desired uniform distribution properties of a good hashing function
- Poor diffusion of the two low-order bits under the avalanche effect,[2] which is not indicative of the desired mixing properties of a good hashing function (see [Konheim 1981, pp. 262–263]).

Example 9.6. Pearson constructed an 8-bit hash of a string [Pearson 1990] $x = (x[0], x[1], \cdots, x[n-1])$ of (8-bit) characters. It uses an auxiliary table (TAB) containing a permutation of all 8-bit integers ($\mathcal{Z}_{256}$).

Pearson's Hash

Parameters
TAB : array[0..255] of integers;
TAB : array[0..255] of integers;
hash : 8-bit integer;
Initialization
$hash[-1] := (n + C)\ (mod\ 256)$;
for i := 0 to n-1 do
{
$hash[i] := (hash[i-1] + x[i])\ (mod\ 256)$;
$hash[i] :=$ TAB[$hash[i]$];
}
Output
$h_x = (h[0], h[1], \cdots, h[n-1])$

A simple argument shows that if $x = (x[0], x[1], \cdots, x[n-1]) \neq y = (y[0], y[1], \cdots, y[n-1])$, then $h_x \neq h_y$. The final byte $h[n-1]$ of $h_x = (h[0], h[1], \cdots, h[n-1])$ may also be used to define the *hash of x*.

Example 9.7. Bell and Kaman [Bell and Kaman 1970] suggested the linear quotient hash code defined by

LQ-1. (R, Q) computed by $R = K \pmod{n}$ and $Q = \dfrac{K-R}{n}$,
LQ-2. $h(K) = R$.

[2]The avalanche effect of a mapping f measures the effect of iteration of f on two input values x and y, which have a small (Hamming) distance. It is desired that the Hamming distance between $f^{(n)}(x)$ and $f^{(n)}(y)$ not remain close to one another but increase with n.

Collision resolution for the key K visits the sequence of cells in the order $h(K)$, $h(K) + Q, h(K) + 2Q, \cdots$

Because the goal of hashing is to map a key into an address randomly using pseudo-random number generators and modular arithmetic. Early candidates considered for such generators were linear congruential generators (LCGs), which is an algorithm producing a sequence $x_0, x_1, \cdots, x_{r-1}, \cdots$ defined by $x_j = (ax_{j-1} + b)$ $(modulo\ m)$ for $j = 1, 2, \cdots$; where a is the multiplier, b is the increment, m is the modulus, and x_0 is the initial seed.

An LCG is either periodic $x_j = x_{j+P}$ for all j or ultimately periodic, $x_j = x_{j+P}$ for all $j > J$. If the LCG is periodic, its period P is at most m, and in most cases it is less than P. The LCG will have a full period $P = m$ if and only if the following is true:

1. b and m are relatively prime.
2. $a - 1$ is divisible by all prime factors of m.
3. $a - 1$ is a multiple of 4 if m is a multiple of 4.

Well-known LCGs include the following:

- $x_j = 279\,470\,273 x_{j-1}$ (mod 4394967291) ([Park and Miller 1988]) is a multiplicative random number generator in which the modulus is $n = 2^{32} - 5 = 4\,394\,967\,291$ and the multiplier is $g = 279\,470\,273$ is a primitive root of n.
- $x_j = 16\,807 x_{j-1}$ (mod $2^{31} - 1$) : IBM 360-random number generator [Lewis, Goodman and Miller 1969] has been used extensively.

A bibliography of references on random number generators has been prepared by Entacher [Entacher 1998]

Of course, the renowned mathematician [von Neumann 1951] once wrote

> *Anyone who considers arithmetical methods of producing random digits is, of course, in a state of sin.*

9.3 PERFORMANCE OF HASHING FUNCTIONS: FORMULATION

McKenzie et al. [Mckenzie, Harries and Bell 1990] analyze the distributional characteristics of several hashing functions including the following:

GNU-cpp:	GNU[3] C preprocessor;
GNU-cc1:	GNU C compiler front end;
PCC:	Portable C Compiler[4] front end;
CPP:	UNIX 4.3 BSD C preprocessor;
C++:	AT&T C++ compiler.

[3]GNU, for GNU's Not UNIX, is a computer operating system composed entirely of free software. Founded in 1983 by Richard Stallman, GNU was the original focus of the Free Software Foundation (FSF). Stallman's goal was to bring a wholly free software operating system into existence.

[4]Chapter 10 describes linking a records to facilitate the operation `SEARCH`.

They introduce the following notations:

- Integers are stored as 32-bit values.
- A key is a sequence of n ASCII characters stored as signed bytes.
- BITS(v,n) is defined to be the integer whose least significant (rightmost) bits are the n least significant bits of v with other bits (if $n < 32$) set to 0.
- The operators AND, OR, XOR on bit strings are performed bitwise.
- n (*modulo* N) is the residue of the integer n mod N.
- The range of a hashing function—the size of the hash table—is the set $\mathcal{Z}_N \equiv \{0, 1, 2, \cdots, N-1\}$.

The evaluation strategy of McKenzie et al. is to use a large collection $\mathcal{K}$ of keys, which are referred to by the authors as identifiers, to assess the randomness of a hashing function. They use the following two such collections;

- $\mathcal{K}_1 : |\mathcal{K}_1| = 36736$ keys collected from programs in C;
- $\mathcal{K}_2 : |\mathcal{K}_2| = 24463$ keys collected from a UNIX dictionary.

The most natural characteristic of a hashing algorithm to evaluate is the distribution of values of the mapping h of $\mathcal{K}$ into $\mathcal{Z}_N$. It is desired that h map the key space $\mathcal{K}$ uniformly onto $\mathcal{Z}_N$. This can be studied experimentally; a large typical set of test keys is chosen and the frequency f_i with which $h(k) = i$ is measured. They measure randomness of the hashing function h in two ways.

Uniform Distribution (UD)

Calculate the frequencies $\{f_i : 0 \le i < N\}$ and compare with the uniform distribution $\left\{q_i \equiv \frac{1}{N} : 0 \le i < N\right\}$

Goodness of Fit (GF)

The number of probes needed by exhaustive matching of keys in a chain[4] of c records needed to find a record with a given key is $\frac{1}{2}c(c+1)$.

McKenzie et al. define the goodness of fit of h on $\mathcal{K}_i$ by measuring the quantities

$$E_N \equiv \frac{|\mathcal{K}|}{N}$$

$$U_N = N\frac{E_N(E_N+1)}{2}$$

$$S_N = \frac{1}{2}\sum_{i=0}^{N-1} f_i(f_i+1)$$

and evaluating

$$R_N \equiv \frac{S_N}{U_N}$$

Because the cost in

The evaluation of UD or GF uses the χ^2 test, to which we now digress.

9.4 THE χ^2-TEST

Suppose T of independent trials of a chance experiment $\mathcal{E}$ are performed; each trial has one of N possible outcomes $O_0, O_1, \cdots, O_{N-1}$, The number of times the outcome O_i occurs is C_i and $f_i \equiv \frac{C_i}{T}$.

Is it likely that the observed outcome counts $\{C_i\}$ are consistent with the

Hypothesis : q_i is the probability of occurrence of O_i $(0 \le i < N)$.

If the hypothesis is true, then for each possible outcome O_i the law of large numbers asserts

$$\lim_{T\to\infty} \frac{C_i}{T} = q_i \quad (0 \le i < N) \tag{9.2}$$

The χ^2 statistic is the quantity defined by

$$\chi_N^2 = \sum_{i=0}^{N-1} \frac{(C_i - Tq_i)^2}{Tq_i} = \sum_{i=0}^{N-1} \frac{T}{q_i}\left(\frac{C_i}{T} - q_i\right)^2 \tag{9.3a}$$

The ith term in (9.1) is the sum the product of two factors; the first

$$\infty = \lim_{T\to\infty} \frac{T}{q_i} \tag{9.3b}$$

increases without bound with N, whereas the second has one of two limiting values

$$\lim_{T\to\infty}\left(\frac{C_i}{T} - q_i\right) = \begin{cases} 0 & \text{if the hypothesis is } \textit{true} \\ \infty & \text{if the hypothesis is } \textit{false} \end{cases} \tag{9.3c}$$

The statistician Karl Pearson [Pearson 1900] proved the limiting distribution of χ_N^2 exists and is independent of the distribution $\{q_i\}$. Moreover, the outcome counts $\{C_i\}$ have $N - 1$ ***degrees of freedom***[5]

Theorem 9.1. If $\{q_i\}$ is the common distribution of the $\{C_i : 0 \le i < N\}$, then

$$\lim_{T\to\infty} Pr\{\chi_N^2 \le x\} = \frac{2^{-(N-1)}}{\Gamma\left(\frac{N-1}{2}\right)} \int_0^x y^{\frac{N-3}{2}} e^{-\frac{y}{2}} dy = \int_0^x k_{r-1}(y)\,dy \tag{9.4a}$$

where $\Gamma(k)$ defined for real positive k is the Gamma function defined by

$$\Gamma(k) = \int_0^\infty x^{k-1} e^{-x} dx \quad 0 < k < \infty \tag{9.4b}$$

[5]The components of the N-vector of counts $\underline{C} = (C_0, C_1, \cdots, C_{N-1})$ are not independent since $T = \sum_{i=0}^{N-1} C_i$.

$\Gamma(k)$ is an extension of the factorial function

$$\Gamma(k) = (k-1)! \quad k = 1, 2, \cdots \tag{9.4c}$$

Given a value of $p \leq 100$, a value $x(p, N-1)$ exists such that χ_N^2 should exceed $x(p, N-1)$ with probability $0.01p$ if the sample size is large enough

$$\frac{p}{100} = \int_{x(p,N-1)}^{\infty} k_{N-1}(y)\,dy \tag{9.5a}$$

when the hypothesis is true. A large χ_N^2-value for $p \approx 99$—in excess of $x(99, N-1)$—therefore casting doubt on the validity of the hypothesis. Tables of the χ_N^2 limits can be found in [Abramowitz and Stegun 1972; pp. 981])

$$x^2(p, N) \simeq N\left(1 - \frac{2}{9N} + x(p, N)\sqrt{\frac{2}{9N}}\right) \simeq N - \frac{2}{3} + \sqrt{2N}\,x(p, N) + \frac{2}{3}x(p, N)^2 + \cdots \tag{9.5b}$$

from [Abramowitz and Stegun 1972; pp. 981] are included in [Knuth 1971; pp. 39].

When the χ^2 test is applied to evaluate the distribution of values of a hashing function and the number of degrees of freedom N, soon to be identified with the size of the hash table, is very large, then a very good approximation exists (see [Abramowitz and Stegun 1972, p. 941])

$$Q_N(\chi^2) \approx Q(x^*) \tag{9.6a}$$

where

$$x^* = \sqrt{2\chi_N^2} - \sqrt{2N-1} \tag{9.6b}$$

$$Q(y) = \frac{1}{\sqrt{2\pi}} \int_y^{\infty} e^{-\frac{u^2}{2}}\,du \tag{9.6c}$$

9.5 TESTING A HASH FUNCTION

Determining whether the hashing function h produces either a UD or GF involves the same process; a count vector $C = (C_0, C_1, \cdots, C_{N-1})$ is calculated as follows if

$$h: \mathcal{K} \to \mathcal{Z}_N$$

on a set $\mathcal{K}$ of $|\mathcal{K}|$ identifiers (keys), initialize the count vector

$$\underline{C} = (C_0, C_1, \cdots, C_{N-1}) = \underbrace{(0, 0, \cdots, 0)}_{N \text{ copies}}$$

then, hash each $k \in \mathcal{K}$

$$h(k) = i \equiv i(k) \in \mathcal{Z}_N$$

and augment the count C_i of the hash table address

$$C_{i(k)} \rightarrow C_{i(k)} + 1$$

Finally, evaluate

$$\chi_N^2 = \sum_{i=0}^{N-1} \frac{(C_i - |\mathcal{K}| q_i)^2}{|\mathcal{K}| q_i} = \sum_{i=0}^{N-1} \frac{|\mathcal{K}|}{q_i} \left(\frac{C_i}{|\mathcal{K}|} - q_i \right)^2$$

Validating the Uniform Distribution

1. χ_N^2 is then computed using equation (3.3*a*).
2. x^* is calculated using equation (3.5*b*).
3. The desired measure of uniformity $Q_N(\chi^2)$ is found using the approximation $Q_N(\chi^2) \approx Q(x^*)$ by reference to a table of the tail of the normal distribution

$$Q(x^*) = \frac{1}{\sqrt{2\pi}} \int_{x^*}^{\infty} e^{-\frac{u^2}{2}} du$$

It is desired that $Q(x^*)$, the approximate probability of deviation, be very small. ■

Validating the Goodness of Fit

1. Calculate $f_i = \frac{C_i}{|\mathcal{K}|}$ for $0 \le i < N$.
2. Calculate $S_N = \frac{1}{2} \sum_{i=0}^{N-1} f_i (f_i + 1)$.
3. The desired measure of goodness of fit is $R_N = \frac{S_N}{U_N}$ where $U_N = \frac{|\mathcal{K}|(|\mathcal{K}|+1)}{2N}$.

An ideal hashing function should produce a value of $R_N \approx 1$. ■

9.6 THE MCKENZIE ET AL. RESULTS

Table 9.1, which was derived from [Mckenzie, Harries and Bell 1990], contains the results of the UD and GF tests.

TABLE 9.1. McKenzie et al. UD and GF Test Results

Algorithm	N	$\frac{\lvert\mathcal{K}\rvert}{N}$	χ_N^2	$Q_N(\chi^2)$	x*	R_N
C++	257	142	334	7.70×10^{-4}	3.23	1.009
GNU-cc1	1008	36	1043	2.09×10^{-1}	0.81	1.028
GNU-cpp	1403	26	1464	8.88×10^{-2}	1.35	1.039
PCC	1013	36	1900	8.34×10^{-57}	16.66	1.051
CPP	2000	18	5449	0.00	41.18	1.142

REFERENCES

M. Abramowitz and I. A. Stegun (Editors), *Handbook of Mathematical Functions with Formulas, Graphs, and Mathematical Tables*, Dover Publications (New York), 1972.

J. R. Bell and C. H. Kaman, "The Linear Quotient Hash Code", *Communications of the ACM*, **13**, #11, pp. 675–676, November 1970.

K. Entacher, "A Collection of Selected Pseudorandom Number Generators With Linear Structures", January 13, 1998, at http://random.mat.sbg.ac.at/-charly/server/node10.html.

R. Jenkins, 1997, at http://www.burtleburtle.net/bob/hash/evahash.html.

D. Knuth, *The Art of Computer Programming: Volume 3/Sorting and Searching*, Addison Wesley (Boston, MA), 1973.

A. G. Konheim, *Cryptography: A Primer*, John Wiley & Sons (New York), 1981.

P. A. Lewis, A. S. Goodman, and J. M. Miller, "A Pseudo-Random Number Generator for the System/360", *IBM Systems Journal*, **8**, pp. 136–146, 1969.

B. J. Mckenzie, R. Harries, and T. Bell, "Selecting a Hashing Algorithm", *Software-Practice and Experience*, **20**, #2, pp. 209–224, 1990.

B. Mulvey, Hash Functions, 2007, http://www.bretmulvey.com/hash/6.html.

S. K. Park and K. W. Miller, "Random Number Generators: Good Ones Are Hard to Find", *Communications of the ACM*, **31**, #10, pp. 1192–1201, 1988.

P. K. Pearson, "Fast Hashing of Variable-Length Text Strings", *Communications of the ACM*, **33**, #6, p. 677, 1990.

K. Pearson, "On the Criterion That a Given System of Deviations from the Probable in the Case of Correlated System of Variables Is Such That It Can Be Reasonably Supposed to Have Arisen from Random Sampling", *Philosophical Magazine*, **50**, pp. 157–175, 1900.

R. Sedgewick, *Algorithms in C++, Parts 14: Fundamentals, Data Structure, Sorting, Searching*, Addison Wesley (Boston, MA), 1998.

J. von Neumann, "Various Techniques Used In Connection With Random Digits", *Applied Mathematics Series*, **12**, pp. 36–38, 1951.

M. A. Weiss, *Algorithms, Data Structures, and Problem Solving With C++*, Addison Wesley (Boston, MA), 1996.

CHAPTER 10

Record Chaining with Hash Tables

10.1 SEPARATE CHAINING OF RECORDS

The cost of SEARCH, as measured by the number of comparisons, when N keys are stored as in *Example 7.1*, in a single **chain** $\mathcal{C}$ is $O(N)$. A significant improvement might be expected by combining record chaining with hashing, which results in chains rooted at hash-table cells. When a single chain of N records is replaced by r chains $\mathcal{C}_0, \mathcal{C}_1, \cdots, \mathcal{C}_{r-1}$ of lengths $L_0, L_1, \cdots, L_{r-1}$ with $N = L_0 + L_1 + \cdots + L_{r-1}$, the number of comparisons required in SEARCH or INSERT is reduced.

Knuth [Knuth 1973, pp. 513–14] referred to the merge of hashing and chaining as *separate chaining*; the idea was invented by H.P. Luhn in 1953 (see Chapter 8) and described in memoranda dated 1/3/53 and 2/20/53 to Mr. J. C. McPherson, who then an IBM Vice-President, Luhn wrote[1]

> When in loading the system, an address is found to be occupied by a previous entry, the address number of the next auxilliary address is recorded behind the original address and the new item is recorded at this auxilliary address. A third item belonging to the same address would similarly be indexed by by appending an auxillary address to the second item

One implementation stores the parameters of the key k (key and address pointer) in a linked list, which is a chain of of nodes whose structure is shown in Figure 10.1.

A node might contain the following:

- The key k.
- A pointer ADD[k] giving the address the record associated with key k.
- A pointer LINK• to the next node in the linked list; LINK• $= \emptyset$, if it is the terminal node in the linked list.

The hash-table (HT) cell $h(k)$ depicted in Figure 10.2 contains the following two entries:

[1]Provided couresty of Ms. Dawn M. Stanford of the IBM Corporate Archives (Somers, NY).

Hashing in Computer Science: Fifty Years of Slicing and Dicing, by Alan G. Konheim

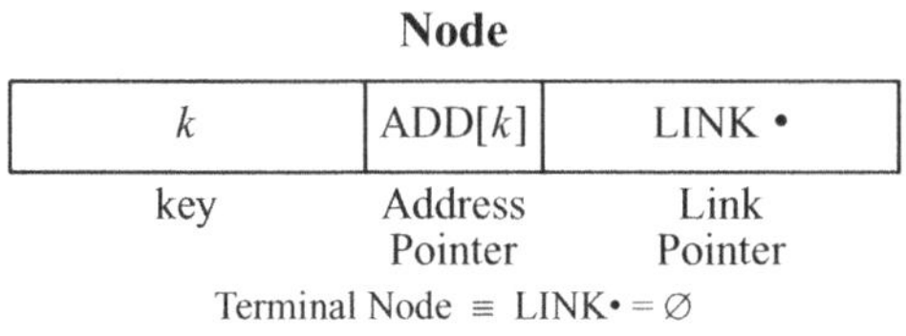

Figure 10.1. Node in a linked list.

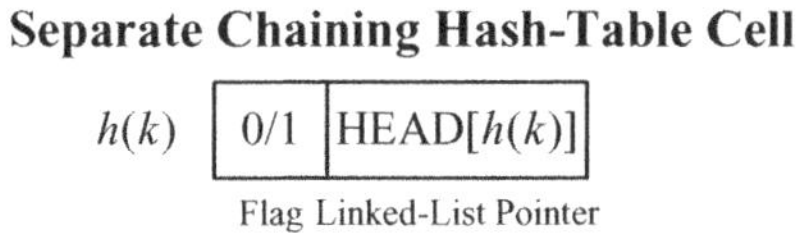

Figure 10.2. Separate chaining hash-table cell.

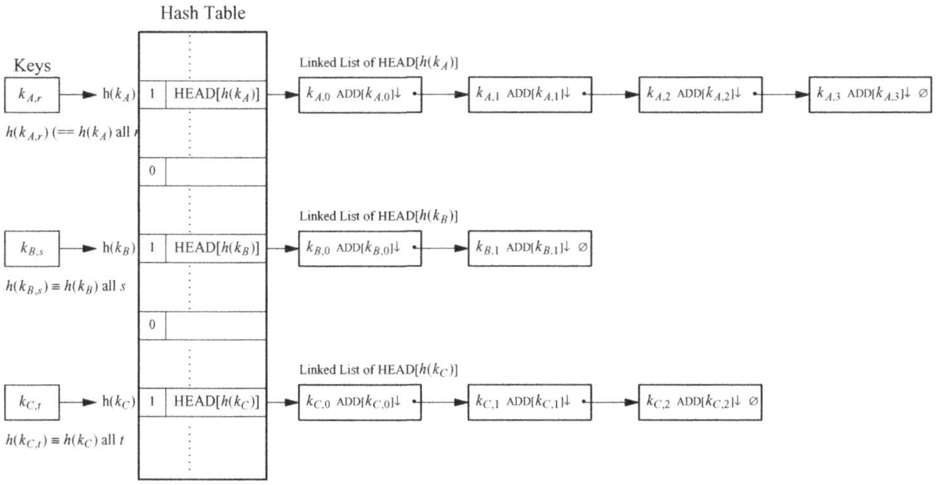

Figure 10.3. Instance of separate chaining with a hash table.

1. A bit signaling cell occupancy (F = 0/1).
2. If the cell is occupied (F = 1), it contains the pointer HEAD[$h(k)$] to the first of node in a chain of nodes forming a linked list.

Figure 10.3 depicts a hash table showing the entries for the following three chains:

1. Chain C_A stores the parameters of four keys $k_{A,0}$, $k_{A,1}$, $k_{A,2}$, $k_{A,3}$ in a linked list pointed to by the hash-table entry HEAD[$h(k_A)$] at hash-table address $h(k_A) = h(k_{A,r})$ $(0 \leq r < 4)$.
2. Chain C_B stores the parameters of two keys $k_{B,0}$, $k_{B,1}$ in a linked list pointed to by the hash-table entry HEAD[$h(k_B)$] at hash-table address $h(k_B) = h(k_{B,s})$ $(0 \leq s < 2)$.

3. Chain C_C stores the parameters of three keys $k_{C,0}$, $k_{C,1}$, $k_{C,2}$ in a linked list pointed to by the hash-table entry HEAD[$h(k_C)$] at hash-table address $h(k_C) = h(k_{C,t})$ $(0 \le t < 3)$.

The SEARCH(k)/INSERT(k) Processes: The entries at the hash-table cell $h(k)$ are examined;

0. If F = 0, no linked list is rooted at this hash-table address
   a) If SEARCH(k) is the preparatory step for INSERT(k), a linked list with a new single node is created; the pointer HEAD[$h(k)$] to the node is stored at hash-table cell $h(k)$.
1. If F = 1, HEAD[$h(k)$] directs access to the L (say) nodes of this linked list; the key k_i at the ith node in this linked list is compared with k for $i = 0, 1, \cdots, L - 1$ until either of the following occur:
   a) $k_i = k$ providing the address ADD[k] of the record associated with the key k.
   b) $k_i \ne k$ for $i = 0, 1, \cdots, L - 1$, signaling that no record associated with key k is currently stored.
      - When SEARCH(k) is the preparatory step for INSERT(k), the linked list can be extended by adjoining an $(L + 1)^{\text{st}}$ node containing the necessary parameters of the key k.

It follows that

- The expected cost of a successful SEARCH(k) is $\frac{1}{2}(L+1)$ comparisons.
- The cost of an unsuccessful SEARCH(k) is L comparisons.

Exercise 10.1. (Deletions with separate chaining): Write pseudocode describing the deletion of the hash-table entries of KEY = k if separate chaining as in Figure 10.1 is used. Assume the following:

a) $h(k) = h(k_i)$ $(0 \le i < L)$ for the L keys $k_0, k_1, \cdots, k_{L-1}$ in the linked list pointed to by HEAD[$h(k)$];
b) LINK[$h(k_i)$] is in the link-field in the node of the key k_i $(0 \le i < L)$.
c) LINK-ADD[i] is in the link-field of record R_i $(0 \le i < L)$.

10.2 ANALYSIS OF SEPARATE CHAINING HASHING SEQUENCES AND THE CHAINS THEY CREATE

The running time of a program implementing separate chaining is analyzed in a publication by Knuth [Knuth 1973]. It is clear that the cost of SEARCH(k) depends on the length of the linked list rooted at HEAD[$h(k)$], and we begin our analysis by using the statistical model defined in Chapter 8, §8.3, which postulates that the combined effect of the hash function h and m keys results in n^m equally likely hash sequences $(h(k_0), h(k_1), \cdots, h(k_{m-1}))$.

The steps in which the insertion of m keys results in a set of r chains $C_0, C_1, \cdots, C_{r-1}$ are as follows:

1. Selecting a partition $\{K_i\}$ of the set K of m keys $K = \{k_0, k_1, \cdots, k_{m-1}\}$ into r nonempty subsets determining the keys contained in the chain C_i

$$K_i = \{k_{i,0} < k_{i,1} < \cdots < k_{i,\ell_i-1}\} \subset K \quad 0 \le i < r$$

$$|K_i| = \ell_i \quad 0 \le i < r$$

$$K_i \cap K_j = \emptyset \quad i \ne j \quad K = K_0 \cup K_1 \cup \cdots, K_{r-1}$$

   The number of partitions of $\{0, 1, \cdots m-1\}$ into r nonempty sets in 1. above is the Stirling number of the second kind denoted by $S_{m,r}$; see §1.9 in Chapter 1 and *Example 2.7* and Theorem 2.3 in Chapter 2.
   With a slight abuse of notation, we use the precedence symbol $\prec$ to signal precedence in time; that is, keys $k_0, k_1, \cdots, k_{m-1}$ in K were inserted in order of increasing subscripts into the hash table.
   The lengths $\underline{\ell} \equiv (\ell_0, \ell_1, \cdot\cdot\cdot, \ell_{r-1})$ of r chains; $\underline{\ell}$ satisfies $\ell_i > 0$ $(0 \le i < r)$ and $m = \sum_{i=0}^{r-1} \ell_i$;
   there are $\binom{m-1}{r-1}$ such length vectors $\underline{\ell}$ (Chapter 1, Corollary 1.10).
2. Selecting the set of r hash-table cells $0 \le j_0 < j_1 < \cdots < j_{r-1} < n$ at which the head of chains are located.
   There are $\binom{n}{r}$ possible selections of r hash-table cells in 2. above.
3. Selecting an assignment $\underline{\pi} = (\pi_0, \pi_1, \cdots, \pi_{r-1})$ determining which chain is rooted at which hash-table cell.

$$j_0 = h(k_{\pi_0,0}),\ j_1 = h(k_{\pi_1,0}), \cdots, j_{r-1} = h(k_{\pi_{r-1},0})$$

   There are $r!$ possible assignments of partitions to cells in **C3**.

Example 10.1. For the parameters $(n, m, r) = (8, 9, 3)$

a) There are $\binom{8}{3} = 56$ possible sets of 3 hash-table cells (j_0, j_1, j_2) at which the chains may be rooted.
b) There are $S_{9,3} = 3025$ partitions $K = K_0 \cup K_1 \cup K_2$ of $K = \{k_0, k_1, \cdots, k_8\}$ into 3 nonempty subsets ([Abramowitz and Stegun 1972, p. 835]).
c) There are $6 = 3!$ ways of assigning the sets K_0, K_1, K_2 to the hash-table cells (j_0, j_1, j_2).

The keys were entered into the hash table by our precedence convention in the order $k_0 \prec k_1 \prec \cdot\cdot\cdot \prec k_8$. Three possible outcome of $\mathcal{E}$ are as follows:

1. The set choice of hash-table cells $(j_0, j_1, j_2) = (3, 5, 6)$.
2. Partition $K = K_0 \cup K_1 \cup K_2$ of K is $K_0 = \{k_0, k_1, k_6, k_8\}$, $K_1 = \{k_1, k_3\}$, $K_2 = \{k_4, k_5\}$

3. The 6 = 3! permutations which are ways of assigning the sets of keys K_0, K_1, K_2 to the hash-table cells (j_0, j_1, j_2) are

$$\underline{\pi}=(0,1,2):\begin{pmatrix}\text{HT-Cell}\,\#\,j_0 \ni \text{HEAD}[h(k_0)] \to K_0,\\ \text{HT-Cell}\,\#\,j_1 \ni \text{HEAD}[h(k_1)] \to K_1,\\ \text{HT-Cell}\,\#\,j_2 \ni \text{HEAD}[h(k_4)] \to K_2\end{pmatrix}$$

$$\underline{\pi}=(0,2,1):\begin{pmatrix}\text{HT-Cell}\,\#\,j_0 \ni \text{HEAD}[h(k_0)] \to K_0,\\ \text{HT-Cell}\,\#\,j_1 \ni \text{HEAD}[h(k_4)] \to K_2,\\ \text{HT-Cell}\,\#\,j_2 \ni \text{HEAD}[h(k_1)] \to K_1\end{pmatrix}$$

$$\underline{\pi}=(1,0,2):\begin{pmatrix}\text{HT-Cell}\,\#\,j_0 \ni \text{HEAD}[h(k_1)] \to K_1,\\ \text{HT-Cell}\,\#\,j_1 \ni \text{HEAD}[h(k_0)] \to K_0,\\ \text{HT-Cell}\,\#\,j_2 \ni \text{HEAD}[h(k_4-)] \to K_2\end{pmatrix}$$

$$\underline{\pi}=(2,1,0):\begin{pmatrix}\text{HT-Cell}\,\#\,j_0 \ni \text{HEAD}[h(k_4)] \to K_2,\\ \text{HT-Cell}\,\#\,j_1 \ni \text{HEAD}[h(k_1)] \to K_1,\\ \text{HT-Cell}\,\#\,j_2 \ni \text{HEAD}[h(k_0)] \to K_0\end{pmatrix}$$

The permutation $\underline{\pi}$ = (0, 1, 2) is consistent with the chains illustrated in Figure 10.3 in which chain $\mathcal{C}_0$ contains 4 keys, chain $\mathcal{C}_1$ 2 keys, and chain $\mathcal{C}_2$ 3 keys.

Theorem 10.1. Let $N\mathcal{C}$ and $L_{\mathcal{C}_0}$, $L_{\mathcal{C}_1}$, ⋯ denote the number and lengths of chains resulting from the insertion of the m keys $k_0, k_1, \cdots, k_{m-1}$ into the hash table of n cells using seperate chaining. The record-chaining model yields the following:

a) The probability that hashing sequences $h(k_0)$, $h(k_1)$, ⋯ , $h(k_{m-1})$ produce r chains is

$$Pr\{N_{\mathcal{C}}=r\}=\frac{1}{n^m}\binom{n}{r}r!S_{m,r} \tag{10.1a}$$

b) The expected number of chains is

$$E\{N_{\mathcal{C}}\}=n\left[1-\left(1-\frac{1}{n}\right)^m\right]\sim n(1-e^{-\mu}) \quad n,m\to\infty, m=\mu n \quad 0<\mu<1 \tag{10.1b}$$

c) The expected value of the average length of a chain $E\{L_{\mathcal{C}}\}\equiv\frac{1}{N_{\mathcal{C}}}\sum_{i=0}^{N_{\mathcal{C}}-1}L_{\mathcal{C}_i}$ converges to

$$\lim_{\substack{n,m\to\infty\\ m=\mu n}} E\{L_{\mathcal{C}}\}=\frac{\mu}{1-e^{-\mu}} \tag{10.1c}$$

Proof of Theorem 10.1

1. There are $S_{m,r}$ partitions $\{K_0, K_1, \cdots, K_{r-1}\}$ of $K=\{k_0, k_1, \cdots, k_{m-1}\}$ into r nonempty subsets $K_i=\{k_{i,0}<k_{i,1}<\cdots<k_{i,\ell_i-1}\}$ $0\le i<r$ as in **S1**.

2. The r occupied hash-table cells $(j_0, j_1, \cdots, j_{r-1})$ containing the heads of linked lists as in **S2** may be chosen in $\binom{n}{r}$ ways.
3. The keys in $K_0, K_1, \cdots, K_{r-1}$ may be assigned to the chains $\mathcal{C}_0, \mathcal{C}_1, \cdots, \mathcal{C}_{r-1}$ in $r!$ ways.

Because the probability of each hashing sequence $h(k_0), h(k_1), \cdots, h(k_{m-1})$ is $\frac{1}{n^m}$, Theorem 10.1a) is proved.

To evaluate

$$E\{N_{\mathcal{C}}\} = \frac{1}{n^m}\sum_r \binom{n}{r} r!\, S_{m,r} \tag{10.2a}$$

we write $\binom{n}{r} r! = (n)_r r$ and observe that $r(n)_r = (n)_r[n - (n - r)]$; we obtain

$$r(n)_r = n[(n)_r - n(n-1)_r] \tag{10.2b}$$

$$r^2(n)_r = n^2(n)_r - n(2n-1)(n-1)_r + n(n-1)(n-2)_r \tag{10.2c}$$

so that

$$E\{N_{\mathcal{C}}\} = \frac{1}{n^{m-1}}\sum_r [(n)_r - n(n-1)_r]!\, S_{m,r} \tag{10.2d}$$

The formula for the Stirling numbers of the second kind

$$S_{m,r} = \frac{1}{r!}\sum_{s=0}^{r}\binom{r}{s}(-1)^s(r-s)^m$$

was given in Chapter 2 (Theorem 2.3). However, it will be more useful to define the $\{S_{m,k}\}$ as in Chapter 2, equations (2.15*a* through 2.15*c*), as a doubly indexed sequence of integers with boundary values

$$S_{m,0} = \begin{cases} 1 & \text{if } m = 0 \\ 0 & \text{if } m > 1 \end{cases}$$

which satisfy the generating functions

$$x^m = \sum_{k=0}^{m} S_{m,k}(x)_k$$

with x real and $0 \le m < \infty$. The *Pochhammer symbol* $(x)_k$ is the generalization of $(n)_k$ from integers n to reals x and is defined by

$$(x)_k = \begin{cases} 1 & \text{if } k = 0 \\ x \times (x-1) \times \cdots \times (x-k+1) & \text{if } k \ge 1 \\ 0 & \text{if } k > n \end{cases}$$

Using equations (16.2*b* and 16.2*c*), the generating function relationship gives

$$x^{m+1} - x(x-1)^m = \sum_{k=0}^{m} S_{m,k} k (x)_k \quad x \text{ real} \quad m \geq 0 \qquad (10.3a)$$

$$x^{m+2} - x(2x-1)(x-1)^m + x(x-1)(x-2)^m = \sum_{k=0}^{m} S_{m,k} k^2 (x)_k \quad x \text{ real} \quad m \geq 0 \qquad (10.3b)$$

Replacing x in equations (10.3*a* and 10.3*b*) by n gives the formulae

$$E\{N_C\} = n\left[1 - \left(1 - \frac{1}{n}\right)^m\right] \sim n\left(1 - e^{-\mu}\right) \quad n, m \to \infty, m = \mu n \quad 0 < \mu < 1 \qquad (10.4a)$$

$$\text{Var}\{N_C\} \sim O(n) \qquad (10.4b)$$

from which it follows that

$$\text{Var}\left\{\frac{1}{n} N_C\right\} \sim O\left(\frac{1}{n}\right) \qquad (10.4c)$$

Finally

$$E\{L_C\} = E\left\{\frac{1}{N_C} \sum_{i=0}^{N_C - 1} L_{C_i}\right\} = \frac{m}{n} E\left\{\frac{n}{N_C}\right\} \qquad (10.4d)$$

We complete the proof by noting that

- $\frac{m}{n} \to \mu$ as $n \to \infty$
- $\lim_{n\to\infty} \frac{E\{N_C\}}{n}$ generally differs from

$$\lim_{n\to\infty} \frac{n}{E\{N_C\}} = \frac{1}{1 - e^{-\mu}},$$

in this instance, the equality is found to be true either by

i) Applying Tchebychev's inequality (Chapter 4, Theorem 5.4) and equation (10.4*c*)

ii) Observing that equation (10.4*c*) states that $\frac{1}{n} N_C$ converges to a constant. ■

Example 10.2. The Stirling numbers $\{S_{m,r}\}$ grow rapidly with m; for example, $S_{25,10} \approx 1.2032 \times 10^{18}$ [Abramowitz and Stegun 1972, p. 835]. The numerical values for $Pr\{NC = r\}$ and $E\{NC\}$ given in Table 10.1 for $m = 25$, $n = 30$ were derived using the asymptotic formula for $S_{m,r}$ (Chapter 3, Theorem 3.4b [Chelluri, Richmond and Temme 2000]. Using this asymptotic formula with an error term $O(m^{-1})$, the expected number of chains $E\{N_C\}$ is

- 16.988 determined from the Table 10.1 data
- 16.962 calculated using the asymptotic approximation in equation (10.4*a*).

TABLE 10.1. Asymptotic Evaluation of $Pr\{N_C = r\}$

$m = 25, n = 30$

r	$Pr\{N_C = r\}$	r	$Pr\{N_C = r\}$
1–9	0.000000	15	0.103954
10	0.000016	16	0.191151
11	0.000208	17	0.244467
12	0.001787	18	0.216881
13	0.010166	19	0.132303
14	0.039204	20	0.054603

10.3 A COMBINATORIAL ANALYSIS OF SEPARATE CHAINING

The random outcome $(h(K_0), h(K_1), \cdots, h(K_{m-1}))$ of the chance experiment $\mathcal{E}$ is an instance an (m,n) sequence; it is defined as a sequence $\underline{h} = (h_0, h_1, \cdots, h_{m-1})$ of length m whose entries $\{h_i\!:\!0 \le i < m\}$ are elements of $\mathcal{Z}_n = \{0, 1, \cdots, n-1\}$. The number of possible (m, n)-sequences $\underline{h}$ is n^m.

An alternative description of the (m, n)-sequence $\underline{h}$ is specified by three quantities

Q1. The number r of different indices of $\mathcal{Z}_n$ that appear (at least once) in the sequence $\underline{h}$; r corresponds to the number of chains resulting from separate chaining.

Q2. The names $0 \le i_0 < i_1 < \cdots < i_{r-1} < n$ of the indices in $\mathcal{Z}_n$ appearing in the sequence $\underline{h}$; the vector $\underline{i} = (i_0, i_1, \cdots, i_{r-1})$ corresponds to those hash-table cells {HT-Cell #i_j : $0 \le j < r$} that contain a set flag F(F = 1).

Q3. The multiplicity vector $\underline{\ell} = (\ell_0, \ell_1, \cdots, \ell_{r-1})$ whose j^{th} component counts the number of times ℓ_j a hashing value j occurs for each name i_j with $0 \le j < r$; ℓ_j corresponds to the length of the jth-chain $\mathcal{C}_1$.

The multiplicities $\underline{\ell} = (\ell_0, \ell_1, \cdots, \ell_{r-1})$ are an ordered partition of the integer m into r positive integer parts (see Chapter 1, §1.9).

If $1 \le r \le m$, an (n, m, r) h sequence is a sequence of hashing-values $\underline{h} = (h_0, h_1, \cdots, h_{m-1})$ that contains exactly r distinct entries from $\mathcal{Z}_n$. The hashing values $\underline{h}$ of an (n,m,r) h sequence produces r chains of total length m.

An alternative description of an (n,m,r) h-sequence specifies the following:

- The number r of indices and their values $0 \le i_0 < i_1 < \cdots < i_{r-1} < n$ occurring in $\underline{h}$,
- The multiplicity vector $\underline{\ell} = (\ell_0, \ell_1, \cdots, \ell_{r-1})$ of $\underline{h}$ where

$$\underline{\ell} = (\ell_0, \ell_1, \cdots, \ell_{r-1})$$

$$\ell_j \ge 0 \quad (0 \le j < r) \quad m = \ell_0 + \ell_1 + \cdots + \ell_{r-1}$$

in which ℓ_j now counts the number of times the value i_j occurs in $\underline{h}$ for $0 \le j < r$. The multiplicity vector $\underline{\ell} = (\ell_0, \ell_1, \cdots, \ell_{r-1})$ is an ordered partition of the integer m into r positive parts.

This exercise develops the formulas counting the number of (n, m, r) h sequences.

The following sets of multiplicity vectors $\{\underline{\ell}\}$ are defined for (m, r) with $1 \leq r \leq m$ and $m \geq 1$.

- $\Upsilon_{m,r}$ consists of multiplicities $\underline{\ell} = (\ell_0, \ell_1, \cdots, \ell_{r-1})$, which are non-negative integers with sum m.

$$\upsilon_{m,r} = |\Upsilon_{m,r}| \tag{10.5a}$$

$$\Upsilon_{m,r} = \begin{cases} \left\{ \underline{\ell} = (\ell_0, \ell_1, \ldots, \ell_{r-1}) : \begin{Bmatrix} 0 \leq \ell_i \\ 1 \leq i < r \end{Bmatrix}, \ell_0 + \ell_1 + \cdots + \ell_{r-1} = m \right\} \\ 1 \leq r \leq m \end{cases} \tag{10.5b}$$

- $\Upsilon_{m,r}^{(s)}$ is the subset of $\Upsilon_{m,r}$ consisting of those multiplicity vectors $\underline{\ell}$ that have exactly s components $\{\ell_j\}$ equal to 0.

$$\upsilon_{m,r}^{(s)} = |\Upsilon_{m,r}^{(s)}| \tag{10.6a}$$

$$\Upsilon_{m,r}^{(s)} = \begin{cases} \left\{ \underline{\ell} = (\ell_0, \ell_1, \ldots, \ell_{r-1}) : \begin{Bmatrix} 0 \leq \ell_i \\ 1 \leq i < r \end{Bmatrix}, s = \sum_{i=0}^{r-1} \chi_{\{\ell_i = 0\}}, \quad \ell_0 + \ell_1 + \cdots + \ell_{r-1} = m \right\} \\ 0 \leq s < r \leq m \end{cases} \tag{10.6b}$$

χ_E, appearing in equation (10.6*a*), is the indicator function of the event (or condition) E;

$$\chi_E = \begin{cases} 1 & \text{if } E \text{ is satisfied} \\ 0 & \text{otherwise} \end{cases} \tag{10.7}$$

Exercise 10.2. (Technical details for separate chaining):

a) Prove

$$\upsilon_{m,r}^{(0)} = \begin{cases} \sum_{\underline{\ell} \in \Upsilon_{m,r}^{(0)}} \binom{m}{\ell_0 \ell_1 \cdots \ell_{r-1}} & 1 \leq r \leq m \\ \upsilon_{m,0}^{(0)} = \begin{cases} 1 & \text{if } m = 0 \\ 0 & \text{if } m > 0 \end{cases} \end{cases} \tag{10.8a}$$

$$\upsilon_{m,r} = \sum_{s=0}^{r} \upsilon_{m,r}^{(s)} \quad \upsilon_{m,r}^{(r)} \equiv 0 \tag{10.8b}$$

$$\upsilon_{m,r}^{(s)} = \binom{r}{s} \upsilon_{m,r-s}^{(0)} \quad 0 \leq s < r \tag{10.8c}$$

and therefore conclude

$$n^m = \sum_{r=0}^{m} \binom{n}{r} \upsilon_{m,r}^{(0)} \quad \upsilon_{m,0}^{(0)} = 0 \tag{10.8d}$$

b) Use equations (10.8*a* through 10.8*c*) to prove

$$\upsilon_{m,r}^{(0)} = \sum_{s=0}^{r} \binom{r}{s} (-1)^s (r-s)^m = r!S_{m,r} \tag{10.9}$$

c) Write a program to evaluate $\upsilon_{m,r}^{(0)}$ for $m = 1(1)10$, $r = 1(1)m$ and tabulate[2] the results.

Note the difference between the following:

- The formula for the number

$$\upsilon_{m,r} = \binom{m}{\ell_0 \ell_1 \cdots, \ell_{r-1}} = \frac{m!}{\ell_0! \ell_1! \cdots, \ell_{r-1}!}$$

  of ordered partitions of the integer m into r non-negative integer parts involving the multinomial coefficient defined in Chapter 1, §1.4
- The formula $\upsilon_{m,r}^{(0)}$ in equation (10.9) for the number of ordered partitions of the integer m into r *positive* integer parts.

Theorem 10.2. (Insertion/search costs): If m keys $k_0, k_1, \cdots, k_{m-1}$ are inserted by separate chaining in a hash table with n cells in the order $k_0 \prec k_1 \prec \cdots \prec k_{m-1}$, then the following must occur:

a) The random number of comparisons[3] $C_{m,n}^{(I)}$ required for the insertion of the $(m+1)^{\text{st}}$ key k_m has the binomial distribution BIN$(m, \frac{1}{n})$

$$Pr\{C_{m,n}^{(I)} = \ell\} = \beta_{m,n}(\ell) \equiv \binom{m}{\ell} \frac{1}{n^\ell} \left(1 - \frac{1}{n}\right)^{m-\ell} \quad 0 \le \ell \le m \tag{10.10a}$$

with expected insertion cost of

$$E\{C_{m,n}^{(I)}\} = \frac{m}{n} \to \mu \quad n, m \to \infty, \mu n \tag{10.10b}$$

b) The probability $Pr\{C_{m,n}^{(S)}(i) = \ell\}$ that ℓ $(0 \le \ell \le i)$ comparisons are required to search for the ith key k_i $(0 \le i < m)$ is

$$Pr\{C_{m,n}^{(S)}(i) = \ell\} = \begin{cases} 1 & \text{if } i = 0 \\ \beta_{i,n}(\ell) & \text{if } 0 < i < m, 0 \le \ell \le i \\ 0 & \text{otherwise} \end{cases} \tag{10.11a}$$

[2] $a = b(1)c$ is the tablemaker's notation indicating that the parameters of the value being tabulated range from $a = b$ to $a = c$ in steps of 1.

[3] We ignore the cost of discovering LINK $= \emptyset$ for a terminal node.

with expected search cost

$$E\{C_{m,n}^{(S)}(i)\} = \frac{i}{n} \quad 0 \leq i < m \tag{10.11b}$$

The probability distribution $Pr\{\hat{C}_{m,n}^{(S)} = \ell\}$ of the average search cost for a randomly chosen one of the keys $k_0, k_1, \cdots, k_{m-1}$ is

$$Pr\{\hat{C}_{m,n}^{(S)} = \ell\} = \frac{1}{m}\sum_{i=\ell}^{m-1} Pr\{C_{m,n}^{(S)}(i) = \ell\} = \frac{1}{m}\sum_{i=\ell}^{m-1} \beta_{i,n}(\ell) \quad 0 \leq i \leq m \tag{10.11c}$$

with *averaged* expected search cost

$$E\{\hat{C}_{m,n}^{(S)}\} = \frac{m+1}{n} \sim \mu \quad n, m \to \infty, m = \mu n \tag{10.11d}$$

Proof. Suppose the insertion of the first m keys resulted in r chains $\mathcal{C}_0, \mathcal{C}_1, \cdots, \mathcal{C}_{r-1}$. When $\ell > 0$, key k_m must have been inserted at the end of a chain of length ℓ. The probability of this event occurring is calculated as follows:

1. There are $\binom{n}{r}$ ways in which the hash-table cells $\mathrm{HT}_{j0}, \mathrm{HT}_{j1}, \cdots, \mathrm{HT}_{jr-1}$ at which these linked lists are rooted could have been chosen.
2. The probability that key k_m hashes to cell HT_{j_s} is

$$\frac{1}{n} = \frac{r}{n}\frac{1}{r} \text{ for } 0 \leq s < r$$

3. $\binom{m}{\ell}$ ways in which the ℓ keys hashed to cell HT_{j_s} could have been chosen.
4. There are $S_{m-\ell,r-1}$ possible partitions of the remaining $m - \ell$ already inserted keys.
5. There are $(r - 1)$ ways in which the chains $\{\mathcal{C}_t : 0 \leq t < r, t \neq s\}$ can be assigned to the hash-table cells $\{j_t : 0 \leq t < r, t \neq s\}$.

This gives

$$\begin{aligned} Pr\{C_{m,n}^{(I)}(m) = \ell\} &= \frac{1}{n^m}\binom{m}{\ell}\sum_{r=1}^{m-(\ell-1)}\binom{n}{r}\frac{r}{n}\frac{1}{r}(r-1)!S_{m-\ell,r-1} \\ &= \frac{1}{n^{m+1}}\binom{m}{\ell}\sum_{r=1}^{m-(\ell-1)} r(n)_r S_{m-\ell,r-1} \end{aligned} \tag{10.12a}$$

Using equations (10.2*b*) and (10.3*a*), we conclude that $C_{m,n}^{(I)}$ has the binomial distribution

$$Pr\{C_{m,n}^{(I)} = \ell\} = \beta_{m,n}(\ell) \equiv \binom{m}{\ell}\frac{1}{n^\ell}\left(1 - \frac{1}{n}\right)^{m-\ell} \quad 1 \leq \ell \leq m \tag{10.12b}$$

If $\ell = 0$, the key k_m must have been inserted at an unoccupied hash-table cell; a slight modification of the argument for $\ell > 0$ gives

$$Pr\{C_{m,n}^{(I)} = 0\} = \frac{1}{n^m}\sum_{r=1}^{m}\binom{n}{r}\left(1-\frac{r}{n}\right)r!S_{m,r} = \left(1-\frac{1}{n}\right)^m \tag{10.12c}$$

so that the binomial distribution in equation (10.12*b*) also is valid for $i = 0$, completing the proof of part a) of Theorem 10.2c. For part b) of Theorem 10.2, note that *no* key comparisons are required when $i = 0$, as we have assumed no cost to determine a vacuous link.

If $0 < i < m$ and $h(k_i) = j$, the event $\{C_{m,n}^{(S)}(i) = \ell\}$ requires that $h(k) = j$ for ℓ of the keys $k \in \{k_0, k_1, \cdots, k_{\ell-1}\}$. Because $0 \le j < n$, equation (10.11*a*) has been proved as well as the expectation formula of equation (10.11*b*).

Finally, the event

$$\{\hat{C}_{m,n}^{(S)} = \ell\}$$

requires

$$\{C_{m,n}^{(S)}(i) = \ell\}$$

for some i with $\ell \le i < m$. Averaging over the values of i gives equation (10.11*c*) and the average expectation formula equation (10.11*d*) completes the proof. ■

Remark 10.1. In principle, there is no restriction on the ratio of m (the number of records) to n (the range of the hashing function h). m is limited by the memory allocated for ADD[k] in the linked-list nodes.

Exercise 10.3. Prove

$$r\,S_{m,r} = \sum_{k=1}^{m-(r-1)}\binom{m}{k}(n)_r S_{m-k,r-1} \quad 1 \le k \le m-(r-1) \tag{10.14}$$

10.4 COALESCED CHAINING

A modification of separate chaining was suggested by [Williams 1959] instead of a chain-of key parameters implemented as a linked list, by a sequence of linked hash-table cells. An implementation of coalesced chaining uses a hash table of $n_0 + n_1$ cells consisting of the following two contiguous parts:

- Hash-table cells $0, 1, \cdots, n_0 - 1$ cells—the *address region*,
- Hash-table cells $n_0, n_0 + 1, \cdots, n_0 + n_1 - 1$ cells—the *cellar*.

A cellar pointer (CP)

Coalesced Chaining HT Cell

0/1	k	ADD[k]	LINK •
Flag	key	Address Pointer	HT Cell Link

Figure 10.4. Coalesced hashing cell.

- is initialized to $n - 1$, the bottom of a hash-table cellar
- points to the first unoccupied hash-table cell in the cellar
- is updated CP $\rightarrow$ CP $- 1$ whenever an existing chain of hash-table cells is extended by the insertion of a key.

CP is allowed to point to a vacant address above the top of cellar; thus, if a hash-table cell is filled

- In the *extended cellar* as a result of extending a sequence of links or
- By the insertion address region, then CP may be decremented by more than one address.

To provide the links, a hash-table cell is augmented by including a link field as shown in Figure 10.4. The link field contains the following:

- A flag bit F = 0/1 indicating whether the hash-table cell is occupied;
- A field for the key;
- A field containing the address pointer ADD[k] to the location at which the record with key k is stored,
- A new field containing a hash-table link LINK[$h(k)$] equal to $\emptyset$ if there is exactly one key stored k^* satisfies $h(k^*) = h(k)$.

Overflow occurs when the insertion of a key causes CP $\rightarrow$ CP $- 1 < 0$.

Figure 10.5 displays an instance of coalesced hashing with three linked chains of hash-table cells for six keys displayed. Figure 10.5 shows the top of cellar but does not indicate the location of the CP.

A. Chain C_A consisting of three keys $k_{A,0}$, $k_{A,1}$, $k_{A,2}$;
   $k_{A,0}$: stored at hash-table cell $h(k_{A,0})$;
   $k_{A,1}$: stored at hash-table cell pointed to by LINK[$h(k_{A,0})$];
   $k_{A,2}$: stored at hash-table cell pointed to by LINK[$h(k_{A,1})$]; the link at this hash-table cell is $\emptyset$ indicating an end of chain.
B. Chain C_B consisting of one key k_B;
   k_B: stored at hash-table cell $h(k_B)$; the link at this hash-table cell is $\emptyset$ indicating an end of chain.
C. Chain C_C consisting of two keys $k_{C,0}$, $k_{C,1}$;
   $k_{C,0}$: stored at hash-table cell h($k_{C,0}$);
   $k_{C,1}$: stored at hash-table cell pointed to by LINK[$h(k_{C,0})$]; the link at this hash-table cell is $\emptyset$ indicating an end of chain.

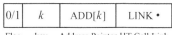

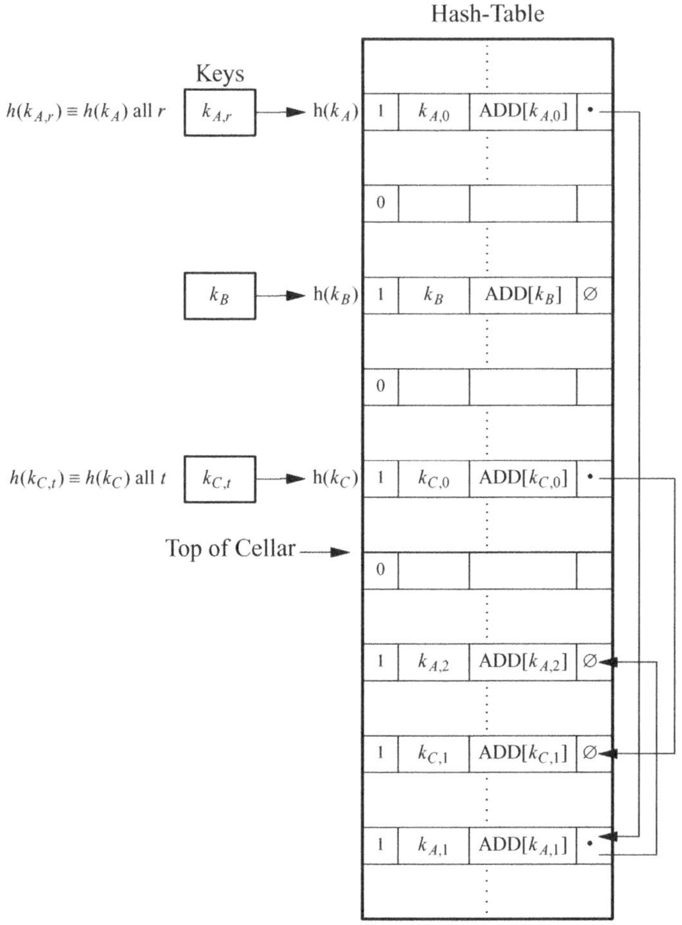

Figure 10.5. An instance of coalesced hashing with a hash table.

Exercise 10.4. (Deletions with coalesced chaining): Write a pseudocode describing a deletion of the hash-table entries of KEY = k if coalesced chaining as in Figures 10.4 and 10.5 is used. Assume the current CP value points to $j \geq 0$ cells bottom of cellar. Distinguish the cases in which CP is above and below the top of cellar.

The term **coalesced hashing**[4] refers to the *coalescing* or combining of chains of keys;

SC. The nodes containing k_i and k_j with $i \neq j$ are in the same linked list or chain, if and only if $h(k_i) = h(k_j)$.

CC. The HT cells containing k_i and k_j with $i \neq j$ are in the same linked list or chain, if $h(k_i) = h(k_j)$, but this condition is not necessary.

Two or more SC chains may coalesce into one CC chain of linked HT cells.

[4]co-a-lesced *intr.v.* 1. To grow together; fuse. 2. To come together so as to form one whole; unite.

The "plain vanilla" version of coalesced hashing proposed by [Williams 1959] with an empty celler ($n_0 = n, n_1 = 0$), referred to by Knuth as a *chained scatter table* is analyzed in [Knuth 1973]. Even if the economy of memory is not virtuous enough, CC slightly decreases the `SEARCH` time for an existing record.

Theorem 10.3. [Knuth 1973] If m keys $k_0, k_1, \cdots, k_{m-1}$ are inserted by coalesced chaining in a hash table with n cells, the expected cost $E\{\hat{C}_{m,n}^{(S)}\}$ to search for a randomly chosen one of the m records is

$$E\{\hat{C}_{m,n}^{(CC)}\} = 1 + \frac{1}{8}\frac{n}{m}\left(\left(1+\frac{2}{n}\right)^m - 1 - \frac{2m}{n}\right) + \frac{1}{4}\frac{m-1}{n} \tag{10.15a}$$

$$\sim 1 + \frac{1}{8\mu}(e^{2\mu} - 1 - 2\mu) + \frac{1}{4}\mu \quad n, m \to \infty;\ m = \mu n \tag{10.15b}$$

Knuth observes that

$$E\{\hat{C}_{m,n}^{(CC)}\} \le 1.80 \quad m = n \to \infty \tag{10.15c}$$

A general form of coalesced hashing was analyzed by J. Vitter in his PhD dissertation and described in [Vitter 1982], [Vitter 1983], and it subsequently was discussed by Chen and Vitter [Chen and Vitter 1986]. Several possible implementations of coalesced hashing are described in [Knuth 1973]; they all locate the parameter of a key k, which encounters a conflict, that is, cell $h(k)$ is occupied in the same manner, but they differ in how these overflow hash table cells are linked.

The types of coalesced chaining are as follows:

- Late-insertion coalesced chaining (LICH)—the conflicting cell for k is linked at-the-end of the current chain of cells;
- early insertion coalesced chaining (EICH)—the most recent conflicting cell k is linked between cell $h(k)$ and the current next-in-chain cell.
- varied-insertion coalesced chaining (VICH).

10.5 THE PITTEL-YU ANALYSIS OF EICH COALESCED CHAINING

Pittel [Pittel 1987] and [Pittel and Yu 1988] determine the distribution of search time in the with CC; the first paper analyzed LICH and the second EICH.

Notation for linked hash-table cells: HT_i will denote the ith hash-table cell; we write $k \in \text{HT}_i$ to indicate the key in this cell and $\text{HT}_i \xrightarrow{\text{L}} \text{HT}_j$ to indicate the link between the ith and j^{th} cells.

The EICH variant of coalesced chaining is defined by the following rule:

If cell $\text{HT}_{h(k_i)}$ is already occupied when the insertion of k_i is attempted, the link $\text{HT}_{h(k_i)} \xrightarrow{\text{L}} \text{HT}_{j0}$ is created to the cell HT_{j0} into which k_i is inserted.

Pittel and Yu introduce several EICH chaining measurements, as follows:

- $T_{1,m,n}(i)$ is the length of the chain C_i of linked hash-table cells $\mathrm{HT}_{h(k_i)}$ to the cell HT_{j0}

$$C_i : \mathrm{HT}_{h(k_i)} \xrightarrow{\mathrm{L}} \mathrm{HT}_{j_{k-2}} \xrightarrow{\mathrm{L}} \cdots \xrightarrow{\mathrm{L}} \mathrm{HT}_{j_2} \xrightarrow{\mathrm{L}} \mathrm{HT}_{j_1} \xrightarrow{\mathrm{L}} \mathrm{HT}_{j_0}$$

  which contains the key k_i; $(0 \le i < m)$ after all m keys have been inserted.
- $U_{1,m} = \max_{0 \le i < m} U_{1,m}(i)$.
- $T_{2,m,n}(i) \equiv T_2(i)$ is the number of extra cells in the chain C_i searched to find k_i after the m keys have been inserted when $\mathrm{HT}_{h(k_i)}$ is occupied by a key other than k_i and 0, otherwise.
- $U_{2,m,n} = \max_{0 \le i < m} U_{2,m,n}(i)$.

$T_{1,m,n}(i)$ measures the `SEARCH`-time for k_i while $T_{2,m,n}(i)$ reflects the effect of coalescing.

We use the statistical model defined in previous chapter; the addresses of the m records identified by the keys $\underline{k} = (k_0, k_1, \cdots, k_{m-1})$ are inserted by EICH CC into a hash table of size n cells using a hash function h. The model postulates the independence and uniform distribution of hashing sequences; that is, the pair $(h, \underline{k})$ produces n^m equally likely hash sequences $h(\underline{k}) = (h(k_0), h(k_1), \cdots, h(k_{m-1}))$. It will be helpful to view insertion as a synchronous process, so that key k_i is inserted into the hash table at *time i*.

Example 10.3. Figure 10.6 is a modification of *Example 1* in [Pittell 1987, p. 1181]; it shows the links in the cells for both LICH and EICH. The hashing values that give rise to these configurations are listed in Table 10.2.

Table 10.3 displays the chains of linked hash-table cells for both separate and EICH coalesced chaining.

The first result in [Pittel and Yu 1988] is presented in the following theorem.

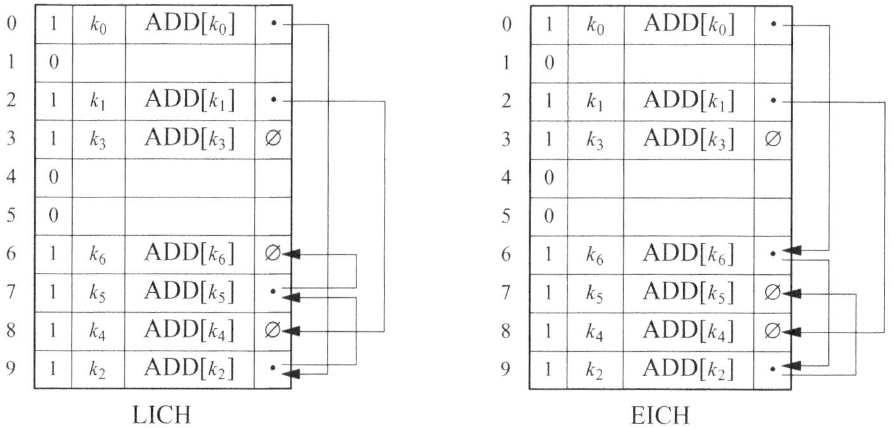

Figure 10.6. Coalesced chaining with LICH and EICH.

TABLE 10.2. The Hashing Values $\{h(k_j) : 0 \le j < 6\}$

$j \to$	0	1	2	3	4	5	6
$h(k_j)$	0	2	0	3	2	9	0

TABLE 10.3. Linked Hash-Table Cells

LICH	EICH
$HT_0 \xrightarrow{L} HT_9 \xrightarrow{L} HT_7 \xrightarrow{L} HT_6 \longrightarrow \emptyset$	$HT_0 \xrightarrow{L} HT_6 \xrightarrow{L} HT_9 \xrightarrow{L} HT_5 \longrightarrow \emptyset$
$HT_2 \xrightarrow{L} HT_4 \longrightarrow \emptyset$	$HT_2 \xrightarrow{L} HT_4 \longrightarrow \emptyset$
$HT_3 \xrightarrow{L} \emptyset$	$HT_3 \xrightarrow{L} \emptyset$

Theorem 10.4. [Pittel and Yu 1988]

$$Pr\{T_{1,m,n}(i) \ge k\} = \begin{cases} 1 & \text{if } k = 1 \\ \dfrac{i}{n}\displaystyle\sum_{j=0}^{k-2} (-1)^j \binom{k-2}{j}\left(1-\frac{j}{n}\right)^{m-i-1} & \text{if } 2 \le k \le n-i+1 \end{cases} \tag{10.16a}$$

If $n, m - i \to \infty$ and $k^2 = o(m - i)$, then

$$Pr\{T_{1,m,n}(i) \ge k\} \sim \frac{i}{n}\left[1 - e^{-\mu + \frac{i}{n}}\right]^{k-2} \qquad m = n\mu \tag{10.16b}$$

The distribution of the SEARCH time averaged over the already inserted keys $\hat{C}_{m,n}$ converges

$$\lim_{\substack{m,n\to\infty\\ m=\mu n}} Pr\left\{\hat{C}_{m,n} = k\right\} = \begin{cases} 1 - \dfrac{\mu}{2} & \text{if } k = 1 \\ \dfrac{1}{\mu(k-1)}\displaystyle\int_0^{\mu}\left[1 - e^{-\mu+y}\right]^{k-1} dy & \text{if } k \ge 2 \end{cases} \tag{10.16c}$$

$$\lim_{\substack{m,n\to\infty\\ m=\mu n}} E\left\{\hat{C}_{m,n}\right\} = \frac{1}{\mu}\left[e^{\mu} - \mu\right] \tag{10.16d}$$

$$\lim_{\substack{m,n\to\infty\\ m=\mu n}} \mathrm{Var}\left\{\hat{C}_{m,n}\right\} = \frac{1}{\mu}\left(\frac{1}{2}e^{2\mu} + e^{\mu} - \mu - \frac{3}{2}\right) - \frac{1}{\mu^2}\left(e^{\mu} - 1\right)^2 \tag{10.16e}$$

Except for the asymptotic results, the size n of the hash table will be fixed; to simplify the notation, the subscript n is suppressed and $T_{1,m}(i)$ is written in place of $T_{1,m,n}(i)$.

Proof. We give the proof only for equation (10.16*a*); we may write equation (10.16*a*) using Pascal's triangle (Chapter 1, §1.4)

$$\binom{k-2}{j} + \binom{k-2}{j-1} = \binom{k-1}{j}$$

as

$$Pr\{T_{1,m}(i)=1\}=1-\frac{i}{n} \tag{10.17a}$$

$$Pr\{T_{1,m}(i)=k\}=Pr\{T_{1,m}(i)\geq k\}-Pr\{T_{1,m}(i)\geq k+1\}$$
$$=\frac{i}{n}\sum_{j=0}^{k-2}(-1)^j\binom{k-2}{j}\left(1-\frac{j+1}{n}\right)^{m-i-1} \quad k=2,3,\cdots \tag{10.17b}$$

The proof consists of providing a purely combinatorial interpretation of equations (10.17*a* and 10.17*b*); for example,

$T_{1,m}(i) = 1$ implies k_i was hashed to a cell not occupied by any of the i keys $k_0, k_1, \cdots, k_{i-1}$, an event of probability $1-\frac{i}{n}$; therefore, at time i

- EICH inserted cell k_i in $\mathrm{HT}_{h(k_i)}$ so that $T_{1,i}(i) = 1$.
- $k_i \in \mathrm{HT}_{h(k_i)}$ implies the length of the chain to the cell containing k_i does not change with the insertion of the final $m - i - 1$ keys $k_{i+1}, k_{i+2}, \cdots, k_{m-1}$ so that $T_{1,m}(i) = 1$.

This yields equation (10.17*a*). ■

To prove the formula for $Pr\{T_{1,m}(i) = k\}$ in equation (10.17*b*) with $k \geq 2$, we start as we did with the case $k = 1$.

$T_{1,m}(i) = k \geq 2$ implies that $\mathrm{HT}_{h(k_i)}$ was occupied already by one of the i keys $k_0, k_1, \cdots, k_{i-1}$ whose hashing value differs from $h(k_i)$, and thus, a new cell HR_{j0} is found into which k_i is inserted.

Although k_i will never be evicted from its cell, additional links between $\mathrm{HT}_{h(k_i)}$ and HT_{j0} may be inserted between these two cells. Indeed, if $k > 2$, then $T_{1,m}(i) = k$ requires the insertion of the keys $k_{i+1}, k_{i+2}, \cdots, k_{m-1}$ to create exactly $k - 2$ additional new links between $\mathrm{HT}_{h(k_i)}$ and HT_{j0}; that is

$$\mathrm{HT}_{h(k_i)} \xrightarrow{\mathrm{L}} \mathrm{HT}_{jk-2} \xrightarrow{\mathrm{L}} \cdots \xrightarrow{\mathrm{L}} \mathrm{HT}_{j2} \xrightarrow{\mathrm{L}} \mathrm{HT}_{j1} \xrightarrow{\mathrm{L}} \mathrm{HT}_{j0} \tag{10.18}$$

The proof is an accounting of all hashing sequences $h(k_{i+1}), h(k_{i+2}), \cdots, h(k_{m-1})$ that will result in the links in equation (10.18). We make this accounting by specifying a sequence of events such as

$$\underbrace{\varepsilon_2(+)}_{i}, \varepsilon_3(-), \underbrace{\varepsilon_3(+)}_{i+R_2}, \varepsilon_4(-), \underbrace{\varepsilon_4(+)}_{i+R_3}, \cdots, \varepsilon_k(-), \underbrace{\varepsilon_k(+)}_{i+R_{k-1}}, \varepsilon_{k+1}(-)$$

at times $i, i + R_2, i + R_3, \cdots, i + R + k - 1$ describing their effects on the links between $\mathrm{HR}_{h(k_i)}$ and HT_{j0} and calculating the probabilities of their occurrence. These events are defined as follows:

$\mathcal{E}_2(+)$

At time i, the insertion of $k_i \in \mathrm{HT}_{j0}$ resulted in the creation of a link between cells $\mathrm{HT}_{h(k_i)}$ and HT_{j0}.

$\mathcal{E}_2(+)$ requires $h(k_i)$ to differ from the cells occupied by keys $k_0, k_1, \cdots, k_{i-1}$, an event of probability $\frac{i}{n}$; Therefore, at time i, the following occurs:

- EICH inserted cell k_i in HT_{j0}.
- EICH created the link $\mathrm{HT}_{h(k_i)} \xrightarrow{\mathrm{L}} \mathrm{HT}_{j0}$ so that[5] $T_{1,i}(i) = 2$.

$\mathcal{E}_3(-)$

The insertion of the next $r_2 \geq 0$ keys starting immediately after time i at which $T_{1,i}(i) = 2$ did not result in the creation of any links between cells $\mathrm{HT}_{h(k_i)}$ and HT_{j0}.

$\mathcal{E}_3(-)$ requires $h(k_{i+j}) \neq h(k_i)$ for $0 \leq j < r_2$, an event of probability $\left(1-\frac{1}{n}\right)^{r_2}$.

$\mathcal{E}_3(+)$

At time $i + R_2 \equiv i + r_2 + 1$, the insertion of k_{i+R_2} resulted in the creation of an additional link between cells $\mathrm{HT}_{h(k_i)}$ and HT_{j0}.

$\mathcal{E}_3(+)$ if and only if $h(k_{i+R_2}) = h(k_i)$, an event of probability $\frac{1}{n}$.

- EICH inserts k_{i+R_2} into a previously unoccupied cell HT_{j1}.
- EICH creates a new link between $\mathrm{HT}_{h(k_i)}$ and HT_{j0} by inserting cell HT_{j1}, which changes $T_{1,i}(i) = 2 \rightarrow T_{1,i+R_2} = 3$.

$\mathcal{E}_4(-)$

The insertion of the next $r_3 \geq 0$ keys starting immediately after time $i + R_2$ at which $T_{1,i+R_2}(i) = 3$ did not result in the insertion of any links from $\mathrm{HT}_{h(k_i)}$ to cell HT_{j0}.

$\mathcal{E}_4(-)$ requires $h(k_{i+R_2+j}) \notin \{h(k_i), h(k_{j1})\}$ for $0 \leq j < r_3$, an event of probability $\left(1-\frac{2}{n}\right)^{r_2}$.

$\mathcal{E}_4(+)$

At time $i + R_3 \equiv i + R_2 + r_3 + 1$, the insertion of k_{i+R_3} resulted in the creation of an additional link between cells $\mathrm{HT}_{h(k_i)}$ and HT_{j0}.

$\mathcal{E}_4(+)$ requires $h(k_{i+R_3}) \in \{h(k_i), h(k_{j1})\} (R_3 = R_2 + r_3 + 1)$, an event of probability $\frac{2}{n}$.

- EICH inserts k_{i+R_3} into a previously unoccupied cell HT_{j2}.
- EICH creates a new link between $\mathrm{HT}_{h(k_i)}$ and HT_{j0} by inserting cell HT_{j2}, which changes $T_{1,i+R_2}(i) = 3 \rightarrow T_{1,i+R_3} = 4$.

$\varepsilon_k(-)$:

The insertion of the next $r_{k-1} \geq 0$ keys starting immediately after time $i + R_{k-2}$ did not result in the insertion of any links from $\mathrm{HT}_{h(k_i)}$ to cell HT_{j0}.

$\varepsilon_k(-)$ requires $h(k_{i+R_{k-2}+j}) \notin \{h(k_i), h(k_{j1}), \cdots, h(k_{jk-3})\}$ for $0 \leq j < r_{k-1}$, which is an event of probability $\left(1-\frac{k-2}{n}\right)^{r_{k-1}}$.

[5]We write $T_{1,j}(i)$ with $i \leq j \leq m$ for the length of the chain from $\mathrm{HT}_{h(k_i)}$ to the cell HT_{j0} at time $j \geq i$.

$\varepsilon_k(+)$

At time $i + R_{k-1} \equiv i + R_{k-2} + r_{k-1} + 1$, the insertion of $k_{i+R_{k-1}}$ resulted in the insertion of a link between $\mathrm{HT}_{h(k_i)}$ to cell HT_{j0}.

- EICH inserts $k_{i+R_{k-1}}$ into a previously unoccupied cell HT_{j2}.
- EICH creates a new link between $\mathrm{HT}_{h(k_i)}$ and HT_{j0} by inserting cell HT_{j2}, which changes $T_{1,i+R_{k-1}}(i) = 3 \to T_{1,i+R_k} = k$.

$\varepsilon_{k+1}(-)$

The insertion of the final $r_k \geq 0$ keys starting immediately after time $i + R_{k-1}$ did not result in the insertion of any links from $\mathrm{HT}_{h(k_i)}$ to cell HT_{j0}.

$\varepsilon_{k+1}(-)$ requires $h(k_{i+R_{k-1}+j}) \notin \{h(k_i), h(k_{j1}), \cdots, h(k_{jk-2})\}$ for $0 \leq j < r_k$, an event of probability

$$\left(1-\frac{k-1}{n}\right)^{r_k}$$

The constraint imposed on non-negative integers

$$\underline{r} \equiv (r_2, r_3, \cdots, r_k)$$

$$\underline{R} \equiv (R_2, R_3, \cdots, R_k) \quad R_j \equiv \begin{cases} r_2 + 1 & \text{if } j = 2 \\ R_{j-1} + r_j + 1 & \text{if } 3 < r < k \end{cases}$$

is $R_k = m - i$.

Because the events $\mathcal{E}_3(-)$, $\mathcal{E}_3(+)$, $\mathcal{E}_4(-)$, $\mathcal{E}_4(+)$ $\varepsilon_k(-)$, $\varepsilon_k(+)$, and $\varepsilon_{k+1}(-)$ conditional on $\mathcal{E}_2(+)$ are independent, the probability that $T_{1,m}(i) = k$ is obtained as

$$Pr\{\varepsilon_3(-)\times\varepsilon_3(+)\times\varepsilon_4(-)\times\varepsilon_4(+)\times\varepsilon_k(-)\times\varepsilon_k(+)\times\varepsilon_{k+1}(-)\} \tag{10.19}$$
$$= \sum_{\substack{r_2, r_3, \cdots, r_{k-1} \\ 0 \leq r_s \ (2 \leq s < k) \\ m-i-k = r_2 + r_3 + \cdots + r_{k-1}}} \prod_{s=1}^{k} \left(\frac{s}{n}\right)^{r_{s+1}} \left(1-\frac{s+1}{n}\right)^{r_{s+1}}$$

This multiple summation can be evaluated as follows, starting from the double and triple summations

$$\sum_{r=0}^{N}\left(1-\frac{a}{n}\right)^r\left(1-\frac{b}{n}\right)^{N-r} = \frac{n}{b-a}\left(1-\frac{a}{n}\right)^{N+1} + \frac{n}{a-b}\left(1-\frac{b}{n}\right)^{N+1} \tag{10.20a}$$

$$\sum_{\substack{r,s \\ r,s \geq 0 \\ r+s=N}}\left(1-\frac{a}{n}\right)^r\left(1-\frac{b}{n}\right)^s\left(1-\frac{c}{n}\right)^{N-(r+s)} = \frac{n^2}{(b-a)(c-a)}\left(1-\frac{a}{n}\right)^{N+2}$$
$$+\frac{n^2}{(a-b)(c-b)}\left(1-\frac{b}{n}\right)^{N+2} \frac{n^2}{(a-c)(b-c)}\left(1-\frac{c}{n}\right)^{N+2} \tag{10.20b}$$

an induction argument gives

$$\text{k}-\text{SUM}(a_1, a_2, \cdots, a_{k-1}) = \sum_{\substack{i_1, i_2, \cdots, i_{k-1} \\ 0 \geq i_j \ (1 \leq j < k) \\ i_1 + i_2 + \cdots + i_{k-1} = N}} \prod_{j=1}^{k-1} \left(1 - \frac{a_j}{n}\right)^{i_j} = \sum_{j=1}^{k-1} \frac{n^{k-1}}{\prod_{\substack{i=1 \\ i \neq j}}^{k-1} (a_i - a_j)} \left(1 - \frac{a_j}{n}\right)^{N+k-1} \quad (10.20c)$$

Applying equation (10.20*c*) with $(i_1, i_2, \cdots, i_{k-1}) = (1, 2, \cdots, k-1)$ completes the proof of (10.17*b*). ■

10.6 TO SEPARATE OR TO COALESCE; AND WHICH VERSION? THAT IS THE QUESTION

Knuth pointed out the drawback to separate chaining; although the chains are relatively short;

$$E\{L_C\} \sim \frac{\mu}{1 - e^{-\mu}}$$

which even for $\mu = 1.0$ is only 1.572, most of the hash-table cells are unoccupied.

Coalesced hashing begins with a sound idea—to use all of the hash-table cells to store keys and their address pointers. Theorem 10.3 showed that coalesced chaining has a smaller (expected) time for SEARCH(k) for a key already inserted in the table; Knuth refers to thus as a *successful search*. Both LICH and EICH insert keys that collide with an already inserted key in possibly different hash-table cells, although the costs of successful CC-SEARCH(k) will be the same.

Knott and Vitter both argue that if a key is known not to be in the table, then

- INSERT(k) with LICH has to wait until the end-of-chain cell is reached before find an unoccupied cell and inserting k, whereas
- INSERT(k) with EICH has to just determine the following:
  - If $HT_{h(k)}$ is an end-of-chain cell.
  - If $HT_{h(k)}$ is not an end-of-chain cell, say $HT_{h(k)} \xrightarrow{L} HT_j$, whether or not $k \in HT_j$.

Thus, as Benjamin Franklin noted in *Poor Richard's Almanac* (October 1935) ··· earliness has its virtues.[6]

Separate and coalesced chaining experience *unavoidable* interference when two keys k, k^* have the same hashing value. In separate chaining, this interference

[6]It is claimed that Benjamin might have done a bit of a plagiarizing. The proverb "Early to bed and early to rise, makes a man healthy wealthy and wise" is cited in John Clarke's *Paroemiologia Anglo-Latina* [Proverbs English, and Latine] published in 1639 almost 100 years before Franklin used it.

TABLE 10.4. Sample Memory Sizes, Costs, and Access Times During 1985 to 2007

Year	Size	Cost	Access Time
1985	512 KB	$210	200 ns
1989	4 MB	$753	120 ns
1993	4 MB	$110	80 ns
1997	16 MB	$58	70 ns
2001	128 MB [DIMM, PC133]	$39	
2003	512 MB [DIMM, PC133]	$39	
2007	1 GB [DIMM, DDR2-400]	$84	

requires a search of the cells in the chain (aka the nodes in the linked list) for the key k. Coalesced chaining with a true cellar in which all colliding keys are inserted behaves the same way; links in the cellar provides a interference-free path for CC-SEARCH(k).

The variants of coalesced chaining in which the entries of some colliding keys are inserted outside of the cellar does not limit interference to collisions in the hashing value. For example,

- Two keys k, k^* with different hashing values collide.
- k^* inserted before k and because of a real collision with another key k^{**} with $h(k^*) = h(k^{**})$.
- k^* is inserted into the $\text{HT}_{h(k)}$.

Voila! The key k suffers from identity theft and must be moved to a cell other than $\text{HT}_{h(k)}$. To be sure, Theorem 10.3 assures us that the effect of these unavoidable interferences is not serious.

Table 10.4 excerpted from http://www.jcmit.com/memoryprice.htm gives the random-access memory (RAM) costs from 1985 to 2003.

It may be presumptuous for a mathematician, but it seems that the emphasis on saving space may be misplaced. Hashing begins with the motto "Keep it simple, stupid", which refers to the cost of a hash-function evaluation. In 1953 when hashing was born and even in 1973 Knuth's seminal work was published, memory was much more costly than today. A truly accurate memory comparison of separate and coalesced hashing is hard to make and somewhat unimportant in 2008. What is relevant in a comparison of hashing protocols is the following:

- The total access time to locate a key
- The simplicity and correctness of the software program realizing SEARCH

In Chapter 13 we propose an extension of coalesced hashing to lessen the effect of this secondary interference.

REFERENCES

M. Abramowitz and I. A. Stegun (Editors), *Handbook of Mathematical Functions with Formulas, Graphs, and Mathematical Tables*, Dover Publications (New York), 1972.

R. Chelluri, L. B. Richmond, and N. Temme, "Asymptotic Estimates for Generalized Stirling Numbers", *Analysis*, **20**, p. 113, 2000.

W. C. Chen and J. S. Vitter, "Deletion Algorithms for Coalesced Hashing", *The Computer Journal*, **29**, #2, pp. 436–450, 1986.

D. Knuth, *The Art of Computer Programming: Volume 3/Sorting and Searching*, Addison-Wesley (Boston, MA), 1973.

B. Pittel, "On Probabilistic Analysis of a Coalesced Hashing Algorithm", *The Annals of Probability*, **15**, #3, pp. 1180–1202, 1987.

B. Pittel and J.-H Yu, "On Search Times for Early-Insertion Coalesced Hashing", *SIAM Journal of Computing*, **17**, #2, pp. 492–503, 1988.

J. S. Vitter, "Implementations for Coalesced Hashing", *Communications of the ACM*, **25**, #12, pp. 911–926, 1982.

J. S. Vitter, "Analysis of the Search Performance of Coalesced Hashing", *Communications of the ACM*, **30**, #2, pp. 231–258, 1983.

F. A. Williams, "Handling Identifiers as Internal Symbols and Language Processors", *Communications of the ACM*, **2**, #6, pp. 21–24, 1959.

CHAPTER 11

Perfect Hashing

11.1 OVERVIEW

In a perfect world, there would be no crime, deans, or collisions—hashing or other types. Even though the absence of the first two may be unattainable, the third may be possible to achieve. A hashing function h with a domain containing m keys, say $\mathcal{K} = \mathcal{Z}_m$ and range $\mathcal{Z}_n$, is a perfect hash function if $h(k_1) = h(k_2)$ if and only if $k_1 = k_2$. Perfection requires $m \leq n$. If h is perfect and its range is a set of m consecutive values, then h is a minimal perfect hash function. Knuth [Knuth 1973, p. 506] observes the following:

- The fraction of hashing functions from $\mathcal{Z}_m$ to $\mathcal{Z}_n$ that are perfect is

$$\frac{n(n-1)\cdots,(n-(m-1))}{n^m(n-m)!}$$

- The fraction of hashing functions from $\mathcal{Z}_n$ to $\mathcal{Z}_n$ that are minimal and perfect is $\frac{n!}{n^n} \approx \sqrt{2\pi n}e^{-n}$.

Both fractions are small for $n >> 10$.

Using a perfect hashing function only makes sense when the input set of keys is static, for example, as in the symbol table constructed by a compiler.

An early example of a minimal perfect hashing function is given in [Cichelli 1980]. A second example is Pearson's hashing [Pearson 1990], a third (from Jenkins) and fourth (GPERF) are described at www.burtleburtle.net and www.gnu.org/software/gperf/.

11.2 CICHELLI'S CONSTRUCTION

Cichelli [Cichelli 1980] described a clever technique for constructing perfect hashing functions. His method and improvements to it are described in the extensive study published by Czech et al. [Czech, Havas and Majewski 1997].

Suppose a perfect hashing function is sought for the words `SUNDAY`, `MONDAY`, · · ·, `SATURDAY`. Cichelli assigns a tentative hashing value equal to the number of

Hashing in Computer Science: Fifty Years of Slicing and Dicing, by Alan G. Konheim

TABLE 11.1. Initial Hashing Value Assignment

k	$h^{(0)}(k)$
SUN	3
MON	3
TUES	4
WEDNES	6
THURS	5
FRI	3
SATUR	5

TABLE 11.2. Letter Counts

S	N	T	M	W	F	I	R
5	2	2	1	1	1	1	1

TABLE 11.3. Cichelli's Initial Hashing Value Assignment

k	L	$f_1(k_0)$	$f_2(k_{L-1})$	$h^{(C)}(k)$
SUN	3	5	2	10
MON	3	1	2	6
TUES	4	2	5	11
WEDNES	6	1	5	12
THURS	5	2	5	12
FRI	3	1	1	5
SATUR	5	5	1	11

letters in each word. In Table 11.1, we suppress the common suffix DAY and next list the (modified) words and their lengths.

Cichelli proposed that the hashing value of the word $k = (k_0, k_1, \cdots, k_{L-1})$ composed of L characters be defined by

$$h(k) = L + f_1(k_0) + f_2(k_{L-1}) \tag{11.1}$$

where f_1 and f_2 are non-negative integer-valued functions on the set of characters $\{c\}$. Initially, $f_1 = f_2$ might be the counts of the letters in the set of words SUNDAY, MONDAY, $\cdots$, SATURDAY shown in Table 11.2.

Table 11.3 modifies Table 11.1 in which $f_1(k_0) = f_2(k_{L-1}) \equiv 0$.

The hashing function $h^{(C)}(k)$ is defective because there are conflicts $h^{(C)}(\text{TUES}) = h^{(C)}(\text{SATUR}) = 11$ and $h^{(C)}(\text{WEDNES}) = h^{(C)}(\text{THURS}) = 12$.

Chicelli proposes modifying the functions f_1 and f_2, for example, beginning with the highest letter count and assigning it the value $f_1(\text{S}) = 0$. If this does not result in a perfect hashing function, then another value should be tried, say $f_1(\text{S}) = 1$. The Chicelli backtracking technique has a potentially large cost.

TABLE 11.4. $f_1 = f_2$ **Assignments**

S	N	T	M	W	F	I	R
2	0	0	1	0	0	0	2

TABLE 11.5. Final Hashing Value Assignment Based on Table 11.4

k	L	$f_1(k_0)$	$f_2(k_{L-1})$	$h^{(C^*)}(k)$
SUN	3	2	0	5
MON	3	1	0	4
TUES	4	0	2	6
WEDNES	6	0	2	8
THURS	5	0	2	7
FRI	3	0	0	3
SATUR	5	2	2	9

TABLE 11.6. Actual Chicelli Perfect Hashing Function for SUNDAY · · · SATURDAY

$h(k)$						
SUNDAY	MONDAY	TUESDAY	WEDNESDAY	THURSDAY	FRIDAY	SATURDAY
↓	↓	↓	↓	↓	↓	↓
2	1	3	5	4	0	6

In Table 11.4, the $f_1 = f_2$ assignments yield the final hashing values shown in Table 11.5.

The actual Chicelli perfect hashing function is $h(k) = h^{(C*)}(k) - 3$ for SUNDAY, MONDAY, · · ·, SATURDAY, which is shown in Table 11.6.

Exercise 11.1. Construct a perfect hash for SUNDAY, MONDAY, · · ·, SATURDAY dropping the suffix DAY and using the letters k_0 and k_1 of the key $\underline{k} = (k_0, k_1, \cdots, k_{L-1})$.

REFERENCES

R. J. Cichelli, "Minimal Perfect Hash Functions Made Simple", *Communications of the ACM*, **23**, #1, pp. 17–19, 1980.

Z. J. Czech, G. Havas, and B. S. Majewski, "Fundamental Study Perfect Hashing", *Theoretical Computer Science*, **182**, pp. 1–143, 1997.

D. Knuth, *The Art of Computer Programming: Volume 3/Sorting and Searching*, Addison-Wesley (Boston, MA), 1973.

P. K. Pearson, "Fast Hashing of Variable-Length Text Strings", *Communications of the ACM*, **33**, #6, p. 677, 1990.

CHAPTER 12

The Uniform Hashing Model

12.1 AN IDEALIZED HASHING MODEL

Suppose the m keys $k_0, k_1, \cdots, k_{m-1}$ are inserted into the hash table with a capacity for n keys. The descriptor uniform hashing (UH) is applied to a collision resolution procedure (CRP) that results in hash-table insertions in such a way that all $\binom{n}{m}$ subsets of occupied hash-table addresses are equally likely to occur. We evaluate the efficiency of SEARCH and INSERT assuming UH.

Although there is no direct realization of UH, Yao [Yao 1985] proved that **double hashing**, which is defined in Chapter 14, behaves asymptotically (as $n, m \to \infty$) like uniform hashing in the sense that both have the same expected average SEARCH cost.

Here, we use Yao's lovely idea—the basis of this asymptotic equivalence—to identify the hashing function value $h(k)$ with a permutation $\underline{\pi} \equiv (\pi_0, \pi_1, \cdots, \pi_{n-1})$ of the hash-table addresses $0, 1, \cdots, n - 1$. The insertion of the key k proceeds by testing sequentially the hash-table addresses $\pi_0, \pi_1, \cdots, \pi_{n-1}$ until either an unoccupied address or an address containing the key k is encountered. Uniform hashing means that all permutations $\underline{\pi}$ are equally likely to occur; that is, $Pr\{h(k) = \underline{\pi}\} = \frac{1}{n!}$.

That Yao's UH-CRP results in the same occupancy as plain vanilla UH is because of the following Lemma:

Lemma 12.1. In the Yao UH-CRP, the probability $Pr\{\Gamma\}$ of some specific set Γ of m of the n cells being occupied depends only on m and is

$$Pr\{\Gamma\} = \frac{1}{\binom{n}{m}} \tag{12.1}$$

Proof. We use mathematical induction on m; the base case $m = 1$ is immediate. Assume that when $m - 1$ keys are inserted, each of the

Hashing in Computer Science: Fifty Years of Slicing and Dicing, by Alan G. Konheim

$$\binom{n}{m-1}$$

sets of hash tables addresses are equally likely to be occupied.

Suppose $k_0, k_1, \cdots, k_{m-2}$ are first inserted and thereafter an mth key k_{m-1} resulting in the occupancy of the following set of hash-table addresses

$$\Gamma = \{x_0, x_1, \ldots, x_{m-1}\} \tag{12.2a}$$

If x_j denotes the location at which the key k_{m-1} resides with $0 \le j < m$, the define

$$\Gamma_j = \Gamma - \{x_j\} \quad 0 \le j < m \tag{12.2b}$$

We will show that

$$Pr\{\Gamma\} = \frac{1}{\binom{n}{m}}$$

using the induction hypothesis

$$Pr\{\Gamma_j\} = \frac{1}{\binom{n}{m-1}} \quad 0 \le j < m$$

If $r \ge 1$ probes are required to insert k_{m-1}, the probe sequence $h(k_{m-1}) = (\pi_0, \pi_1, \cdots, \pi_{n-1})$ must meet the following three conditions:

1. The rth probe π_{r-1} must have tested hash-table address x_j.
2. The first $r-1$ probes must have tested only the hash-table addresses in Γ_j; there are

   $$\binom{m-1}{r-1}(r-1)!$$

   possible hash subsequences $(\pi_0, \pi_1, \cdots, \pi_{r-2})$ because the hash-table address colliding with $x_j = \pi_{r-1}$ can be chosen from Γ_j in

   $$\binom{m-1}{r-1}$$

   ways and then ordered in $(r-1)!$ ways.
3. The remaining $n-r$ probes $(\pi_r, \pi_{r+1}, \cdots, \pi_{n-1})$ must have tested the hash-table addresses in $\{0, 1, \cdots, n-1\} - \Gamma$ so that there are $(n-r)!$ possible hash subsequences $(\pi_r, \pi_{r+1}, \cdots, \pi_{n-1})$.

The probability of the previous three requirements *conditional* on Γ_j is

$$Pr\{\Gamma/\Gamma_j\} = \frac{(m-1)!(n-r)!}{(m-r)!n!}$$

leading to

$$\begin{aligned} Pr\{\Gamma\} &= \sum_{j=0}^{m-1}\sum_{r=1}^{m} Pr\{\Gamma/\Gamma_j\}Pr\{\Gamma_j\} \\ &= \sum_{j=0}^{m-1}\sum_{r=1}^{m} \frac{(m-1)!(n-r)!}{(m-r)!n!}\frac{1}{\binom{n}{m-1}} \\ &= \sum_{r=1}^{m}\frac{\binom{m}{r}}{\binom{n}{r}}\frac{1}{\binom{n}{m-1}} \end{aligned}$$

Using the formula **E10** in Appendix 1,

$$\sum_{r=1}^{m}\frac{\binom{m}{r}}{\binom{n}{r}} = \frac{m}{n-m+1}$$

yields

$$Pr\{\Gamma\} = \frac{1}{\binom{n}{m}} \qquad (12.3)$$

completing the proof. ■

Lemma 12.2 provides the correct preparation to evaluate the cost of SEARCH and INSERT in UH.

Theorem 12.2. The costs with uniform hashing are as follows:

a) If the $m-1$ keys $k_0, k_1, \cdots, k_{m-2}$ have been inserted, the probability $P_r(m, n)$ that r probes are required to INSERT(k_{m-1}) is[1]

$$P_r(m,n) = \frac{\binom{n-r}{m-r}}{\binom{n}{m-1}} = \frac{n-m+1}{n}\begin{cases} 1 & \text{if } r=1 \\ \prod_{j=1}^{r-1}\frac{m-j}{n-j} & \text{if } 1<r\le m+1 \end{cases} \qquad 0<m\le n-1 \quad (12.4)$$

[1]By convention, an empty product is assigned the value 1 so that $P_r(0, n) = 1$.

b) If the m keys $k_0, k_1, \cdots, k_{m-1}$ have been inserted, $P_r(m, n)$ is also the probability that r probes are required for SEARCH(k_{m-1}).

c) The expected number of probes $E\{N(m, n)\}$ required to INSERT(k_{m-1}) or SEARCH(k_{m-1}) is

$$E\{N(m,n)\} \equiv \sum_{r=1}^{n} rP_r(m,n) = \frac{n+1}{n-m+2} \tag{12.5}$$

d) The expected number of probes $E\{\bar{N}(m, n)\}$ required to INSERT(k_j) or SEARCH(k_j) _averaged_ over the m keys $k_0, k_1, \cdots, k_{m-1}$ is

$$E\{\bar{N}(m,n)\} \equiv \frac{1}{m}\sum_{j=1}^{m} E\{N(j,n)\} = \frac{n+1}{m}\sum_{j=1}^{m}\frac{1}{n-j+2} = \frac{n+1}{m}[H_{n+1} - H_{n-m+1}] \tag{12.6a}$$

where H_r is the rth Harmonic number

$$H_r = 1 + \frac{1}{2} + \cdots + \frac{1}{r} \tag{12.6b}$$

Proof. If Γ is the set of occupied hash-table addresses resulting from the insertion of $k_0, k_1, \cdots, k_{m-2}$ and $h(k_{m-1}) = (\pi_0, \pi_1, \cdots, \pi_{n-1})$, then

1. The $r-1$ probes $(\pi_0, \pi_1, \cdots, \pi_{r-2})$ must have tested only the hash-table addresses in Γ; there are $\binom{m-1}{r-1}(r-1)!$ possible hash subsequences $(\pi_0, \pi_1, \cdots, \pi_{r-2})$,
2. The remaining $n - r + 1$ probes $(\pi_{r-1}, \pi_r, \cdots, \pi_{n-1})$ must have tested the hash-table addresses in $\{0, 1, \cdots, n-1\} - \Gamma$; there are $(n - r + 1)!$ possible hash subsequences $(\pi_{r-1}, \pi_r, \cdots, \pi_{n-1})$.

Multiplying

$$\frac{(m-1)!}{(m-r)!} \text{ by } (n-r+1)!$$

and simplifying proves Theorem 12.2a. The number of comparisons required to SEARCH for k_{m-1} is clearly the same, which proves Theorem 12.2b.

Note that equation (12.4) implies the identity

$$\sum_{r=1}^{m}\binom{n-r}{m-r} = \binom{n}{m-1} \tag{12.7a}$$

Equation (12.7*a*) can be proved directly using Theorem 1.1 (Pascal's triangle) in Chapter 1.

To prove Theorem 6.2c, we write

$$\sum_{r=1}^{m} rP_r(m,n) = (n+1) - \sum_{r=1}^{m}(n+1-r)P_r(n,m)$$

$$= (n+1) - \sum_{r=1}^{m}(n+1-r)\frac{\binom{n-r}{m-r}}{\binom{n}{m-1}}$$

Next, replace $(n+1-r)\binom{n-r}{m-r}$ by $(n-m+1)\binom{n-r+1}{m-r}$ obtaining

$$\sum_{r=1}^{m} rP_r(m,n) = (n+1) - \sum_{r=1}^{m}(n-m+1)\frac{\binom{n-r+1}{m-r}}{\binom{n}{m-1}}$$

Finally, apply equation (12.7*a*) to obtain

$$\sum_{r=1}^{m} rP_r(m,n) = (n+1) - (n-m+1)\frac{\binom{n+1}{n-m+2}}{\binom{n}{m-1}} = \frac{n+1}{n-m+2} \qquad (12.7b)$$

Exercise 12.1. Prove that insertion of a the key k in a hash table of capacity n by uniform hashing can be described by the following *nonconstructive* algorithm:

procedure `U-INSERT`

1. `Success` := 0.
2. **Repeat while** {`Success` = 0};

Choose a random cell A with the uniform distribution $Pr\{A = a\} = \frac{1}{n}$.
If empty, then set `Success` := 1 and insert k into hash-table address a.
end

12.2 THE ASYMPTOTICS OF UNIFORM HASHING

The (table) occupancy of a hash table of size n containing entries for m keys $k_0, k_1, \cdots, k_{m-1}$ is the ratio $\mu = \frac{m}{n}$.

Theorem 12.3. Suppose keys $k_0, k_1, \cdots, k_{m-1}$ are inserted into a hash table of capacity n with uniform hashing. If $m, n \to \infty$ such that $m/n \to \mu$ with $\mu \in (0, 1)$, then

a) The asymptotic probability $P_r(\mu)$ that r probes are required to insert or search for the mth key k_{m-1} has the geometric distribution

$$\lim_{\substack{m,n\to\infty \\ m/n\to\mu}} P_r(m,n) \equiv P_r(\mu) = (1-\mu)\mu^{r-1} \quad 1 \le r < \infty \qquad (12.8a)$$

b) The asymptotic expected number of probes $E\{N(\mu)\}$ to insert or search for the mth key k_{m-1} is

$$\lim_{\substack{m,n\to\infty \\ m/n\to\mu}} E\{N(m,n)\} \equiv N(\mu) = \frac{1}{1-\mu} \qquad (12.8b)$$

c) The asymptotic expected number of probes $E\{\bar{N}(\mu)\}$ to insert or search averaged over the m keys $k_0, k_1, \cdots, k_{m-1}$ is

$$\lim_{\substack{m,n\to\infty \\ m/n\to\mu}} E\{\bar{N}(m,n)\} \equiv \bar{N}(\mu) = \frac{1}{\mu}\log\frac{1}{1-\mu} \tag{12.8c}$$

Proof. For each fixed j

$$\lim_{\substack{m,n\to\infty \\ m/n\to\mu}} \frac{m-j}{n-j-1} \to \mu$$

so that the right-hand side of equation (12.4) has the asymptotic value $(1-\mu)\mu^{r-1}$ proving Theorem 12.3a. The geometric distribution in equation (12.8*a*) implies that

$$\lim_{\substack{m,n\to\infty \\ m/n\to\mu}} E\{N(m,n)\} = \lim_{\substack{m,n\to\infty \\ m/n\to\mu}} \sum_{r=1}^{m} rP_r(m,n) = \sum_{r=1}^{\infty} r(1-\mu)\mu^{r-1} = \frac{1}{1-\mu}$$

proving equation (12.3*b* and 12.3*c*).

Finally, to prove Theorem 12.3c, we start with equation (12.6)

$$E\{\bar{N}(m,n)\} = \frac{1}{m}\sum_{j=0}^{m-1}\frac{1}{1-\frac{j}{n+1}}$$

and recognize that

$$\frac{1}{n+1}\sum_{j=0}^{m-1}\frac{1}{1-\frac{j}{n+1}} \approx \int_0^{\mu}\frac{1}{1-x}dx = \log(1-\mu)^{-1} \quad n\to\infty$$

completing the proof. ■

12.3 COLLISION-FREE HASHING

The insertion into a hash table of size n of the parameters of a key k is collision free if it requires $r = 1$ probes. As a hash table is filled, collision-free entries become less likely.

Exercise 12.2. Suppose keys $k_0, k_1, \cdots, k_{m-1}$ are inserted into a hash table of capacity n with uniform hashing.

a) Calculate the probability of the event
$E(m, n)$: The insertion of the key k_{m-1} is collision free.
b) Let $\text{CF}(m, n)$ be the total number of collision-free insertions when the m keys are inserted into a hash table. Calculate the expected value $E\{\text{CF}(m, n)\}$.

12.2c) Find a recursion for $\rho_r(m, n) \equiv Pr\{\text{CF}(m, n) = r\}$ for $1 \le r \le m \le n$.

12.2d) Calculate the expectation of $\text{CF}(m, n)$.

12.2e) Calculate the generating function $\text{CF}(m, n, z) \equiv \sum_{r=1}^{m} \rho_r(m, n) z^r$ of the sequence $\{\rho_r(m, n)\}$.

12.2f) Tabulate the values of $\rho_r(m, n)$ for $n = 1(1)10$, $m = 0(1)n$, $r = 0(1)m$.

12.2g) Browse through the entries in the table just constructed and make some reasonable conjectures, for example, monotonicity in the parameters m, n, r. Even better, prove some interesting properties of $\{\rho_r(m, n)\}$.

Exercise 12.3. Suppose keys $k_0, k_1, \cdots, k_{m-1}$ are inserted into a hash table of capacity n with uniform hashing. Calculate the following asymptotic values:

a) Calculate $\lim_{\substack{n \to \infty \\ m \to n\mu}} Pr\{E(m, n)\}$;

b) Calculate $\lim_{\substack{n \to \infty \\ m \to n\mu}} E\{\text{CF}(m, n)\}$.

REFERENCE

A. Yao, "Uniform Hashing is Optimal", *Journal of the ACM*, **32**, pp. 687–693, 1985.

CHAPTER 13

Hashing with Linear Probing

13.1 FORMULATION AND PRELIMINARIES

W. Wesley Peterson (1924–2009) evaluated the collision-resolution performance using a simulation of linear open addressing, which is also referred to as scatter storage and linear probing (LP).

We use the statistical model defined in Chapter 9, §9.1 to analyze linear probing; the cells of the m keys $\underline{k} = (k_0, k_1, \cdots, k_{m-1})$ are inserted by LP in a hash table (HT) of size n cells using a hash function h. The model postulates the independence and uniform distribution of hashing sequences; that is, the pair $(h, \underline{k})$ produces n^m equally likely hash sequences $h(\underline{k}) = (h(k_0), h(k_1), \cdots, h(k_{m-1}))$. Although most performance analysts question the assumption of a uniform distribution, there is little prognosis for success in modeling performance of LP if the assumption were replaced by another specific distribution, whose appropriateness would certainly be questioned. Furthermore, the results of McKenzie et al. [Mckenzie, Harries and Bell 1990] described in Chapter 9, §9.6 suggests the uniform distribution assumption is not unreasonable.

Hash-table cell HT_j contains a flag (F = 0/1); if F = 1, a key k^* and a pointer ADD[k^*] are stored at HT_j. LP involves a procedure by which the *calculated* hash-table cell $HT_{h(k_i)}$ of key k_i is translated into the *actual* hash-table cell HT_{a_i}. For example, the steps in the LP-SEARCH(k) are as follows:

0. Set $j = 0$ and compute $h_j(k) \equiv h(k)$.
1. If F = 0 at $HT_{h_j(k)}$, this is an unoccupied cell; no record with key k has yet been stored. When SEARCH(k) is the preparatory step in INSERT(k), the parameters (k, ADD[k])[1] may then be inserted at $HT_{h_j(k)}$ and the flag reset (F = 1).
2. If F = 1, the keys stored at $HT_{h_j(k)}$, say k^*, is compared with k.
   - 2.a If $k = k^*$, then LP-SEARCH uses the pointer ADD[k] to retrieve the record with k and SEARCH(k) is concluded.
   - 2.b If $k \neq k^*$, $h_j(k)$ is updated, $h_j(k) \to h_{j+1}(k) \equiv (h_j(k) + 1) \pmod n$[2]; SEARCH($k$) returns to step 1 using the updated $h_{j+1}(k)$ value.

[1]As in other hashing schema, ADD[k] is determined by the operating system.
[2]Hash tables are circular memories; hash-table cell HT_0 is the successor of HT_{n-1}.

The first analysis of hashing with probing is by Schay and Spruth [1962], they modify LP consistent with the following principles:

0. Keys are stored in the hash table with the order of their calculated hash-table cells; that is, consistently. If $h(k)$ cyclically precedes $h(k^*)$ $(h(k) \prec h(k^*))$, then the actual hash-table cell at which k is inserted cyclically precedes that of k^*.
1. The key k is inserted in the nearest successor hash-table cell to $\mathrm{HT}_h(k)$ consistent with step 0.

The steps in the Schay-Spruth LP_{SS}-`INSERT`(k) analogous to the LP-`SEARCH` steps 0 through 2 are:

0. Test whether $\mathrm{HT}_{h_0(k)}$ with $h_0(k) \equiv h(k)$ is unoccupied (F = 0); if so, store the parameters $(k, \mathrm{ADD}[k])$ there, concluding `INSERT`(k).
1. If F = 1 at $\mathrm{HT}_{h_0(k)}$, then sequentially search the successor cells

$$\mathrm{HT}_{h_1(k)}, \mathrm{HT}_{h_2(k)}, \cdots, \mathrm{HT}_{h_r(k)} \quad h_j(k) = h_{j+1}(k) \equiv (h_j(k)+1) (\mathrm{mod}\ n)$$

   determining the *largest* integer r for which the key currently stored at cell $\mathrm{HT}_{h_j(k)}$ cyclically precedes $h(k)$; that is, if k_j^* is the key stored at $\mathrm{HT}_{h_j(k)}$, then $h(k_j^*) \prec h(k)$ with $1 \le j \le r$.
   a) If the next hash-table cell $\mathrm{HT}_{h_{r+1}(k)}$ is unoccupied (F = 0), then store $(k, \mathrm{ADD}[k])$ in this cell, concluding `INSERT`(k).
   b) If the next hash-table cell $\mathrm{HT}_{h_{r+1}(k)}$ is occupied (F = 1), then do the following:
      - Find the nearest successor empty cell $\mathrm{HT}_{h_s(k)}$.
      - Right shift one cell the contents of each of the cells $\mathrm{HT}_{h_{r+1}(k)}$, $\mathrm{HT}_{h_{r+2}(k)}$, $\cdots$, $\mathrm{HT}_{h_{s-1}(k)}$.
      - Store $(k, \mathrm{ADD}[k])$ at cell $\mathrm{HT}_{h_{r+1}(k)}$, concluding `INSERT`(k).

In either instance, the actual hash-table address of k is $\mathrm{HT}_{h_{r+1}(k)}$. If step 2b of the Schay-Spruth LP_{SS}-`INSERT` analysis concludes `INSERT`(k), blocks of hash-table cells must be moved[3] to accommodate the key k so that the cells preserve the cyclical order of hash values $\{h(k_i)\}$.

Figure 13.1 depicts the Schay-Spruth modified LP-`INSERT`(k) in which

- The calculated addresses of the keys j_1, j_2, j_3 stored at cells $\mathrm{HT}_{h(k)}$, $\mathrm{HT}_{h(k)+1}$, $\mathrm{HT}_{h(k)+2}$ satisfy $h(j_i) \prec h(k)$ for $i = 1, 2, 3$.
- The calculated addresses of the keys j_4, j_5 stored at cells $\mathrm{HT}_{h(k)+3}$, $\mathrm{HT}_{h(k)+4}$ satisfy $h(k) \prec h(j_i)$ for $i = 4, 5$.
- Hash-table cell $\mathrm{HT}_{h(k)+5}$ is unoccupied.

[3]No hash-table cells would need to be moved if links are included in the hash-table as in Chapter 10.

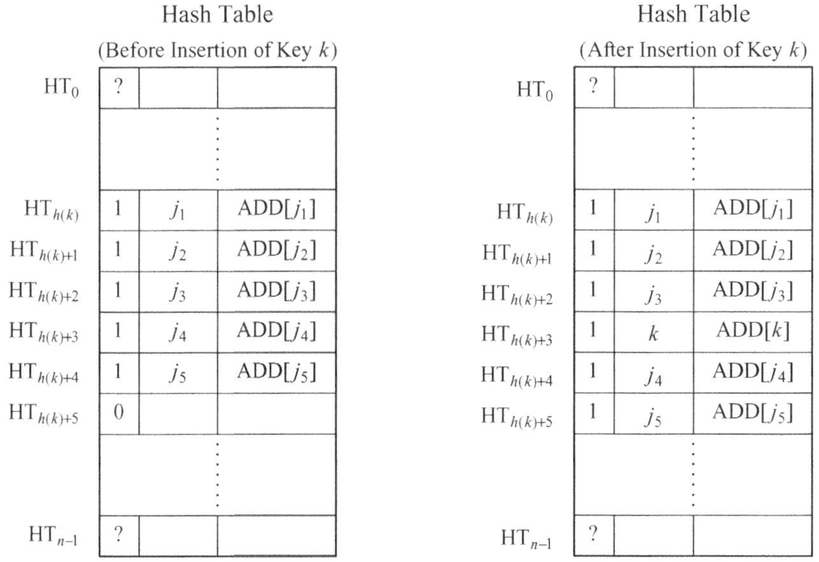

Hash Table
(Before Insertion of Key k)

HT_0	?		
$HT_{h(k)}$	1	j_1	ADD[j_1]
$HT_{h(k)+1}$	1	j_2	ADD[j_2]
$HT_{h(k)+2}$	1	j_3	ADD[j_3]
$HT_{h(k)+3}$	1	j_4	ADD[j_4]
$HT_{h(k)+4}$	1	j_5	ADD[j_5]
$HT_{h(k)+5}$	0		
HT_{n-1}	?		

Hash Table
(After Insertion of Key k)

HT_0	?		
$HT_{h(k)}$	1	j_1	ADD[j_1]
$HT_{h(k)+1}$	1	j_2	ADD[j_2]
$HT_{h(k)+2}$	1	j_3	ADD[j_3]
$HT_{h(k)+3}$	1	k	ADD[k]
$HT_{h(k)+4}$	1	j_4	ADD[j_4]
$HT_{h(k)+5}$	1	j_5	ADD[j_5]
HT_{n-1}	?		

$h(j_1) \leq HT_{h(k)}, h(j_2) \leq h(k), h(j_3) \leq h(k), h(j_1) \leq HT_{h(k)}, h(j_5) > h(k)$

Figure 13.1. Schay-Spruth INSERT(k).

Although the analysis of linear probing is essentially combinatorial, Schay and Spruth offer a different and simpler approach. First, they approximate the number of keys that hash to a particular cell by the Poisson distribution ([Feller 1957, pp. 135–145]). Next, they construct a Markov chain ([Feller 1957, Chapter IV] or Chapter 4, §4.9) associated with the insertion process and obtain the probability distribution of the displacement of a hash-table entry, for example, three cells in Figure 13.1.

[Konheim and Weiss 1966] contains the first *published* analytical evaluation of LP; other papers providing the analysis on hashing with linear probing and related topics have been published by Gaston H. Gonnet and Ian Munro ([Gonnet and Munro 1977, 1979, 1984]).

13.2 PERFORMANCE MEASURES FOR LP HASHING

Performance measures for LP include the distribution and expected number of key-comparisons needed for

INSERT(k), to insert a new key into the hash table.

SEARCH(k), to locate a key that may or may not be currently in the hash table.

DELETE(k), to delete the entries for a key currently in the hash table.

Notation. LP-INSERT computes the following:

i) The hash-sequence $\underline{h} = (h(k_0), h(k_1), \cdots, h(k_{m-1}))$ for the keys $k = (k_0, k_1, \cdots, k_{m-1})$.

ii) From $\underline{h}$ determines the actual hash-table address HT_{a_0}, HT_{a_1}, $\cdots$, $HT_{a_{m-1}}$ where subject to the following rules:
   - $HT_{a_0} = HT_{h(k_0)}$ is the actual hash-table cell for the record with key k_0.
   - The actual hash-table cell HT_{a_i} for the key k_i is $a_i = (h(k_i) + \delta_i) \pmod{n}$ where δ_i being the smallest non-negative integer such that $a_i \notin \{a_0, a_1, \cdots, a_{i-1}\}$.

Two performance measure for LP-SEARCH are as follows:

- The (random) number of comparisons (probes) $N_{m,n}$ needed to locate the cell containing the mth key k_{m-1} assuming all the keys $k_0, k_1, \cdots , k_{m-1}$ have *already* been stored in the hash table.
- The (random) average number of comparisons (probes)

$$\hat{N}_{m,n} = \frac{1}{m} \sum_{j=0}^{m-1} N_{j,n}$$

The cost of LP-SEARCH(k) will include the following:

1. The hash function evaluation $h(k)$.
2. $s \geq 1$ of comparisons of k with the keys at the occupied hash-table cells $HT_{h(k)}$, $HT_{h(k)+1}, \cdots , HT_{h(k)+s-1}$ are made with the following results:
   a) If $s > 1$, none of the keys stored at the s- cells $HT_{h(k])}, HT_{h(k)+1}, \cdots , HT_{h(k)+s-2}$ is equal to k.
   b) If $s \geq 1$, the key at this hash-table cell $HT_{h(k)+s-1}$ is equal to k.

When the key k has *not* previously been inserted in the hash table, LP-INSERT differs only in that step 2 is replaced by the following:

2′. s of comparisons of k with the keys at the largest contiguous block of occupied hash-table cells beginning with cell $HT_{h(k)}$; with the following results:
   a′. If $s \geq 1$, none of the keys stored at the s cells $HT_{h(k])}, HT_{h(k)+1}, \cdots , HT_{h(k)+s-1}$ is equal to k.
   b′. The flag is cleared (F = 0) at hash-table cell $HT_{h(k)+s}$.

We begin with an invariance principle first observed by Peterson [Peterson 1957, p. 137] and cited in [Knuth 1973; Theorem P, p. 530].

Theorem 13.1. When m keys are inserted, $\hat{N}_{(m,n)}$ is the same for the insertion orders $\underline{k} = (k_0, k_1, \cdots , k_{m-1})$ and $\underline{k}_\pi = (k_{\pi_0}, k_{\pi_1}, \cdots , k_{pi_{m-1}})$ where $\underline{\pi} = (\pi_0, \pi_1, \cdots , \pi_{m-1})$ is a permutation. $\hat{N}_{(m,n)}$ depends only on the number of keys hashed to each cell.

Proof. Every permutation π may be expressed as a product of two-element transpositions [Lederman 1953, p. 75]. Such transpositions can be constructed by combining repeated exchanges of *adjacent* terms. Thus, it suffices to prove the result when π is a transposition; that is,

$$h(\underline{k}) = (h(k_0), \cdots, h(k_{i-1}), h(k_i), h(k_{i+1}), \cdots, h(k_{m-1})$$
$$h(\underline{k}_\pi) = (h(k_0), \cdots, h(k_{i-1}), h(k_{i+1}), h(k_i), \cdots, h(k_{m-1})$$

produce the same value for $\hat{N}_{(m,n)}$. If the permutated order changes nothing, we are done; if $\underline{\pi}$ merely switches the cells into which k_i and k_{i+1} are inserted, the increase (*resp.* decrease) in the number of comparisons for k_{i+1} is balanced by a corresponding decrease (*resp.* increase) in the number of comparisons for k_i. ■

13.3 ALL CELLS OTHER THAN HT_{n-1} IN THE HASH-TABLE OF n CELLS ARE OCCUPIED

We begin by determining the number $T_{n-1,n}$ of hashing sequences

$$\underline{h} = (h_0, h_1, \cdots, h_{m-1}) \equiv (h(k_0), h(k_1), \cdots, h(k_{m-1}))$$

that insert the $m = n - 1$ keys $(k_0, k_1, \cdots, k_{m-1})$ into a hash table of n cells resulting in the hash-table cell HT_{n-1} being unoccupied.

Theorem 13.2. $T_{n-1,n} = n^{n-2}$ for $1 \le n < \infty$.

Proof. The set of hashing sequences $\mathcal{H}_{(m,n)} \equiv \{\underline{h} = (h_0, h_1, \cdots, h_{m-1})\}$ for $m = n - 1$ keys leaving cell HT_{n-1} unoccupied can be partitioned $\mathcal{H}_{(m,n)} = \bigcup_{s=1}^{n-1} \mathcal{H}_{(m,n)}(s)$ according to the actual hash-table address at which k_{m-1} is stored. Figure 13.2 shows the instance in which k_{m-1} is stored in cell HT_{n-s+1}. where $1 \le s \le n - 1$. Such an cell occupancy pattern is determined by

1a. The set $\Gamma_1 \subseteq \{k_0, k_1, \cdots, k_{m-2}\}$ of $n - s - 1$ keys that are inserted into the $n - s - 1$ hash-table cells $HT_0, HT_1, \cdots, HT_{n-s-2}$.
1b. The set $\Gamma_2 \subseteq \{k_0, k_1, \cdots, k_{m-2}\}$ of $s - 1$ keys with $\Gamma_1 \cap \Gamma_2 = \emptyset$ that are inserted into the $s - 1$ hash-table cells $HT_{n-s}, HT_{n-s+1}, \cdots, HT_{n-1}$.
2a. The hashing sequences $h(\Gamma_2)$ that result in the $s - 1$ keys in Γ_2 being inserted into the s cells $HT_{n-s}, HT_{n-s+1}, \cdots, HT_{n-1}$ leaving cell HT_{n-1} unoccupied.
2b. The hashing sequences $h(\Gamma_1)$ that result in the $n - s - 1$ keys in Γ_1 being inserted into the $n - s$ cells $HT_0, HT_1, \cdots, HT_{n-s-2}$ leaving cell HT_{n-s-1} unoccupied.
3. The hashing value $h(k_{m-1})$ that inserts key k_{m-1} into cell HT_{n-s-1}.

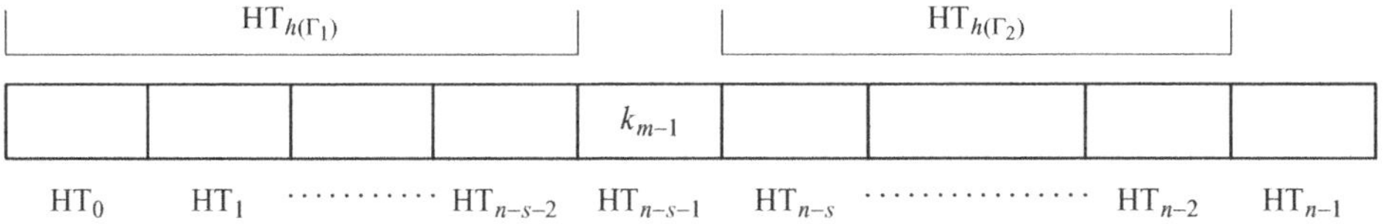

Figure 13.2. All Cells Other than HT_{n-1} in the Hash Table of n Cells are Occupied.[4]

[4]For simplicity, we display hash tables horizontally and only indicate the keys in the cell.

Note that

1a–b. There are $\binom{n-1}{s-1}$ ways of choosing the $s-1$ keys in Γ_1 and Γ_2.

2a. There are $T_{s-1,s}$ hashing sequences for the $s-1$ keys Γ_2 that result in these keys being stored at the hash-table cells $HT_{n-s}, HT_{n-s+1}, \cdots, HT_{n-1}$ leaving cell HT_{n-1} unoccupied.

2b. There are $T_{n-(s+1),n-s}$ hashing sequences for the remaining $n-s-1$ keys in Γ_1 that result in these keys being stored at the hash-table cells $HT_0, HT_1, \cdots, HT_{n-s-1}$ leaving cell HT_{n-s-1} unoccupied.

3. There are $n-s$ choices for calculated hash-table cells $h_{m-1} = h(k_{m-1}) \in \{0,1, \cdots, n-s-1\}$, which will cause k_{m-1} to be inserted in HT_{n-s-1}.

This leads to the recurrence

$$T_{n-1,n} = \sum_{s=1}^{n-1} \binom{n-1}{s-1} (n-s) T_{s-1,s} T_{n-(s+1),n-s} \tag{13.1a}$$

Introducing the (exponential) generating function $T(z)$ of the sequence $\{T_{n-1,n}\}$

$$T(z) = \sum_{n=1}^{\infty} \frac{T_{n-1,n}}{(n-1)!} z^{n-1} \tag{13.1b}$$

Equation (13.1a) leads to the differential equation for the generating function $T(z)$

$$z\frac{d}{dz}T(z) = zT(z)\frac{d}{dz}(zT(z)) \Leftrightarrow \frac{\frac{d}{dz}T(z)}{T(z)} = \frac{d}{dz}(zT(z)) \tag{13.2a}$$

with solution

$$T(z) = e^{zT(z)} \tag{13.2b}$$

It is shown in [Wright 1959] that the only solution of equation (13.2a) analytic at $z = 0$ is

$$T(z) = \sum_{n=1}^{\infty} \frac{n^{n-2}}{(n-1)!} z^{n-1} = \sum_{n=0}^{\infty} \frac{(n+1)^{n-1}}{n!} z^n \quad |z| < e^{-1} \tag{13.2c}$$

Because $0 < \mu e^{-\mu} < e^{-1}$ for $0 \le \mu < 1$, it follows that

$$T(\mu e^{-\mu}) = e^{\mu} \quad 0 \le \mu < 1 \tag{13.2d}$$

completing the proof of Theorem 13.2. ■

Hash-Table Decomposition Principle. A hash table with $m \ge 1$ occupied cells may be decomposed into one or more blocks each of which consists of some $s \ge 1$ con-

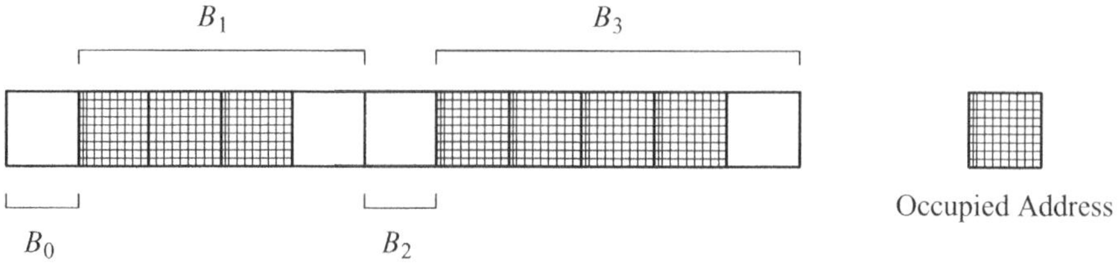

Figure 13.3. A Hash Table with Four Blocks.

tiguous occupied cells concatenated on the right by a single unoccupied cell. A trivial block is one with $s = 0$ occupied cells.

Figure 13.3 shows one decomposition of a hash table of size $n = 11$ consisting of four blocks; *i*) a trivial block, *ii*) a block of length 3, *iii*) a second trivial block, and finally *iv*) a block of length 4. The hash sequence $\underline{h} = (h(k_0), h(k_1), \cdots, h(k_6)) = (1, 2, 1, 7, 8, 6, 6)$ for the 7 keys $\underline{k} = (k_0, k_1, \cdots, k_6)$ results in this occupancy configuration.

Assume

1. m keys $k_0, k_1, \cdots, k_{m-1}$ are inserted into a hash table of n cells.
2. Hashing produces a configuration of r blocks $\mathcal{B} = (B_0, B_1, \cdots, B_{r-1})$ of lengths $b_0, b_1, \cdots, b_{r-1}$ with $b_i \geq 1$ for $0 \leq i < r$ and $b_0 + b_1 + \cdots + b_{r-1} = m + r = n$.

After selecting the keys Γ_i appearing in block B_i for $0 \leq i < r$ in

$$\binom{m}{b_0 - 1\; b_1 - 1 \cdots b_{r-1} - 1}$$

ways the blocks $\{B_i\}$ are *independent* because the number of ways $N\mathcal{B}$ of producing the configuration $\mathcal{B}$ is the *product* of the number of hashing sequences $\prod_{i=0}^{r-1} T_{b_i-1,b_i}$, which will result in the occupancy of the r-blocks $\mathcal{B}$ where $T_{b_i-1,b_i} = |h(\Gamma_i)|$ is the number of hashing subsequences for the keys in Γ_i. These keys result in filling block B_i, leaving the rightmost address of this block vacant for $0 \leq i < r$.

Remark 13.1. Because the hash table is a circular memory, there is nothing special about cell HT_{n-1}. Any result that contains the phrase "$\cdots$ which leave cell HT_{n-1} unoccupied is $\cdots$" may replace cell HT_{n-1} by cell HT_j. For example, see Corollary 13.4.

Corollary 13.4. For each fixed hash-table address i with $0 \leq i < n$, $T_{n-1,n}$ counts the number of hash sequences $(h(k_0), h(k_1), \cdots, h(k_{m-1}))$ that hash the $m = n - 1$ keys $\{k_0, k_1, \cdots, k_{m-1}\}$ into the hash table, which leaves cell HT_i unoccupied.

13.4 *m*-KEYS HASHED INTO A HASH TABLE OF *n* CELLS LEAVING CELL HT_{n-1} UNOCCUPIED

Next, we generalized Theorem 13.2 as follows: For each fixed i with $0 \leq i < n$, we determine the number $T_{m,n}$ of hashing sequences $\underline{h} = (h_0, h_1, \cdots, h_{m-1}) = (h(k_0), h(k_1),$

$\cdots, h(k_{m-1}))$, which hash the $m(1 \le m < n)$ keys $\{k_0, k_1, \cdots, k_{m-1}\}$ into a hash-table of n cells. This strategy results in cell HT_i being unoccupied.

Theorem 13.4 [Konheim and Weiss 1966].

a) $T_{m,n}$ satisfies the recurrence

$$T_{m,n} = \begin{cases} \sum_{j=1}^{m} \binom{m}{j-1} T_{j-1,j} T_{m-(j-1),n-j} & \text{if } 1 \le m < n \\ 0 & \text{if } 0 < n \text{ and } m = n \\ 1 & \text{if } 0 \le n \text{ and } m = 0 \end{cases} \tag{13.3}$$

b) with solution

$$T_{m,n} = n^{m-1}(n-m) \quad 0 \le m < n \tag{13.4}$$

Proof. If m keys are hashed to n cells, there are $n - m$ unoccupied cells; therefore, the hash-table contains $n - m$ blocks $B_0, B_1, \cdots, B_{n-(m+1)}$. We may assume that $i = n - 1$, by *Remark 13.1*.

The Hash-Table Decomposition Principle implies the set of hashing sequences $\mathcal{H}_{(m,n)} = \{\underline{h}\}$ counted by $T_{m,n}$ are partitioned into the m subsets

$$\mathcal{H}_{(m,n)} = \bigcup_{j=0}^{m-1} \mathcal{H}_{(m,n)}(j)$$

where $\mathcal{H}_{(m,n)}(j)$ consists of those hashing sequences for which the j contiguous hash-table cells $HT_{n-j}, HT_{n-j+1}, \cdots, HT_{n-1}$ constitute the right-most block $B_{n-(m+1)}$. The cardinality of $\mathcal{H}_{(m,n)}(j)$ is determined by the following:

1. The $\binom{m}{j-1}$ ways of choosing the set $\Gamma_{n-(m+1)}$ consisting of the $j - 1$ keys $\{k_{i_0}, k_{i_1}, \cdots, k_{i_{j-2}}\}$ that form the rightmost block $B_{n-(m+1)}$ of size j

2a. The number $T_{j-1,j}$ of hash subsequences $h(\Gamma_{n-(m+1)})$ that result in the actual occupied cells of the $j - 1$ keys in $\Gamma_{n-(m+1)}$ being $HT_{n-j}, HT_{n-j+1}, \cdots, HT_{n-2}$

2b. The number $T_{m-(j-1),n-j}$ of hash sequences for the remaining $m - (j - 1)$ keys $(k_0, k_1, \cdots, k_{m-1}) - \Gamma_{n-(m+1)}$ into the hash table consisting of the $n - j$ cells HT_0, $HT_1, \cdots, HT_{n-j-1}$, which leaves cell HT_{n-j-1} unoccupied.

This decomposition proves equation (13.3), which may be written as

$$\frac{T_{m,n}}{m!} = \sum_{\substack{j \ge 1 \\ j \le m+1}} \frac{T_{j-1,j}}{(j-1)!} \frac{T_{m-(j-1),n-j}}{(m-(j-1))!} \tag{13.5a}$$

Applying the same argument to $\bigcup_{i=0}^{n-(m+2)} B_i$ generalizes equation (13.5a) to

$$\frac{T_{m,n}}{m!} = \sum_{\substack{j_0, j_1, \cdots, j_{n-(m+1)} \geq 1 \\ j_0 + j_1 + \cdots + j_{n-(m+1)} = m}} \prod_{\ell=0}^{n-(m+1)} \frac{T_{i_\ell - 1, i_\ell}}{(i_\ell - 1)!} \tag{13.5b}$$

Equation (13.5*b*) implies that $T_{m,n}$ is the constant terms in the mth derivative of $T^{n-m}(z)$

$$T_{m,n} = [z^0]\left(\frac{d^m}{dz^m} T^{n-m}\right)(z) \tag{13.5c}$$

Using equation (13.2*b*)

$$T^{n-m}(z) = e^{(n-m)T(z)} = \sum_{k=0}^{\infty} (n-m)^k T^k(z) \frac{z^k}{k!} = \sum_{r=0}^{\infty} z^r \sum_{k=0}^{r} \frac{(n-r)^k}{k!} \frac{T_{m-k,m}}{(m-k)!} \tag{13.5d}$$

which leads directly to the recurrence

$$T_{m,n} = \sum_{k=0}^{m} \binom{m}{k} (n-m)^k T_{m-k,m} \tag{13.5e}$$

An induction argument and the binomial theorem completes the proof of equation (13.4) and Theorem 13.4. ■

13.5 THE PROBABILITY DISTRIBUTION FOR THE LENGTH OF A SEARCH

Let $N_{(m,n)}$ be the number of comparisons needed for

SEARCH(k_{m-1}): to locate the cell containing the key k_{m-1} when the hash table currently contains entries for the m keys $\{k_0, k_1, \cdots, k_{m-1}\}$, or equivalently

INSERT(k_{m-1}): to insert the key k_{m-1}, when the hash table currently contains entries for the $m-1$ keys $\{k_0, k_1, \cdots, k_{m-2}\}$

in the hash table with n cells. The probability of the event $\{N(m,n)) = r\}$ is the sum over all cell configurations (i, b) $(0 \leq i < b < m)$ defined by

1. Some $r + i + 1$ of the $m - 1$ keys have already been inserted into the cells HT_{b-i}, $HT_{b-i+1}, \cdots, HT_{b+r-1}$.
2. Cells HT_{b-i-1} and HT_{b+r} are unoccupied.
3. $h(k_{m-1}) = b$.

Figure 13.4 depicts an (i, b) configuration yielding $Pr\{N(m, n) = r\}$.

Theorem 13.5. The distribution and length for a search are

$$Pr\{N_{m,n} = r\} = \frac{1}{n^{m-1}} \sum_{i=r}^{m-1} \binom{m-1}{i} (i+1)^{i-1} (n-m)(n-i-1)^{m-i-2} \quad 0 \leq r < m \tag{13.6a}$$

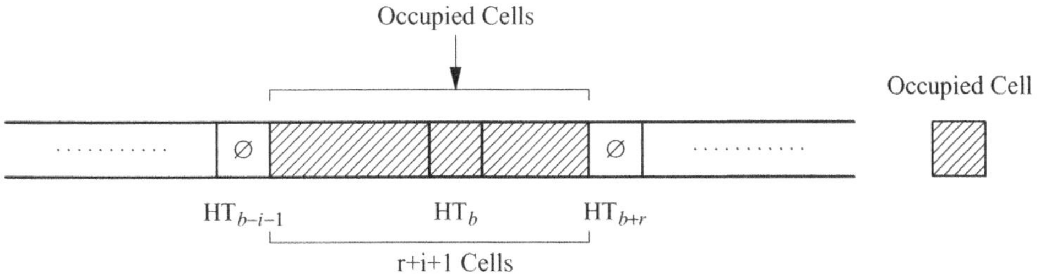

Figure 13.4. Hash-Table Cell Configuration Yielding $Pr\{N_{m,n} = r\}$.

$$E\{N_{m,n}\} = \frac{1}{2}\frac{n-m}{n^{m-1}}\sum_{i=1}^{m-1}\binom{m-1}{i}(i+1)^i\, i(n-i-1)^{m-i-2} \tag{13.6b}$$

Proof

1a. There are

$$\binom{m-1}{r+i}$$

ways of choosing the $i + r$ keys $\Gamma \subset \{k_{0,1}, \cdots, k_{m-1}\}$ that have been inserted into the $r + i + 1$ hash-table cells $\mathrm{HT}_{b-i}, \cdots, \mathrm{HT}_{b-1}, \mathrm{HT}_b, \mathrm{HT}_{b+1}, \cdots, \mathrm{HT}_{b+r-1}$.

1b. By Theorem 13.2, there are $T_{r+i,r+i+1} = (r + i + 1)^{r+i-1}$ hash subsequences $\underline{h}(\Gamma)$ that result in the keys in Γ being inserted into the hash-table cells $\mathrm{HT}_{b-i}, \cdots, \mathrm{HT}_{b-1}, \mathrm{HT}_b, \mathrm{HT}_{b+1}, \cdots, \mathrm{HT}_{b+r-1}$, which leaves cell HT_{b+r} unoccupied

2a. The remaining $m - (r + i + 1)$ keys $\Gamma_1 \equiv \{k_{0,1}, \cdots, k_{m-1}\} - \Gamma$ must have been inserted into the remaining $n - (r + i + 1)$ cells $\mathrm{HT}_{b+r+1}, \cdots, HT_{n-1}, HT_0, HT_1, \cdots, \mathrm{HT}_{b-i-1}$, which leaves the right-most cell HT_{b-i-1} unoccupied.

2b. By Theorem 13.2, there are $T_{m-(r+i+1),n-(r+i+1)}$ hash sequences $\underline{h}(\Gamma_1)$, which result in these cell assignments. This gives

$$\begin{aligned} Pr\{N_{m,n} = r\} &= \frac{1}{n^{m-1}}\sum_{i=0}^{m-(i+1)}\binom{m-1}{r+i}T_{r+i,r+i+1}T_{m-(r+i+1),n-(r+i+1)} \\ &= \frac{1}{n^{m-1}}\sum_{i=r}^{m-1}\binom{m-1}{i}T_{i,i+1}T_{m-(i+1),n-(i+1)} \\ &= \frac{1}{n^{m-1}}\sum_{i=r}^{m-1}\binom{m-1}{i}(i+1)^{i-1}(n-m)(n-(i+1))^{m-(i+2)} \end{aligned}$$

proving Theorem 13.5a. Next,

$$E\{N_{m,n}\} = \sum_{r=1}^{m-1} rPr\{N_{m,n} = r\} = \sum_{r=1}^{m-1} r\frac{1}{n^{m-1}}\sum_{i=r}^{m-1}\binom{m-1}{i}(i+1)^{i-1}(n-m)(n-(i+1))^{m-(i+2)}$$

which yields Theorem 13.5*b* by interchanging the order of the summation. ■

Exercise 13.1 (Collision). $N_{m,n}$ continues to denote the number of comparisons needed to locate the mth record when keys $\{k_0, k_1, \cdots, k_{m-1}\}$ are inserted into the hash table.

a) Calculate the probability that the insertion of k_{m-1} is collision free; that is, $Pr\{N_{m,n} = 0\}$.
b) Calculate the expected number $\text{CF}_{m,n}$ of collision-free insertions experienced by the insertions of the m keys.
c) Calculate the asymptotic behavior ($n \to \infty, m = \mu n($ of $E\{\text{CF}_{m,n}\}$.
d) Write a recursion for $\rho_{r,m,n} = Pr\{\text{CF}_{m,n} = r\}$.
e) Calculate the generating function $\text{CF}_{m,n}(z) = \sum_{r=1}^{m} \rho_{r,m,n} z^r$.

Exercise 13.2. If m keys $K = \{k_0, k_1, \cdots, k_{m-1}\}$ have been inserted into the hash table of n cells, calculate the probability of the event

$\mathcal{O}_{[a,a+b)}$: the contiguous block of b cells $\text{HT}_a, \cdots, \text{HT}_{b-1}$ are occupied where $a < b$ and $1 \le b \le m$.

Exercise 13.3. If m keys $K = \{k_0, k_1, \cdots, k_{m-1}\}$ have been inserted into the hash table of n cells, calculate the probability of the event

$\mathcal{U}_{[a,a+b)}$: the contiguous block $\text{HT}_a, \cdots, \text{HT}_{b-1}$ of exactly b cells are unoccupied with $a < b$ and $1 \le b \le n - m$.

Exercise 13.4. How does $Pr\{\mathcal{U}_{[a,a+b)}$ change if $\mathcal{U}_{[a,a+b)}$ is replaced by

$\bar{\mathcal{U}}_{[a,a+b)}$: the contiguous block of at least b cells including $\text{HT}_a, \cdots, \text{HT}_{b-1}$ are unoccupied with $a < b$ and $1 \le b \le n - m$?

13.6 ASYMPTOTICS

Determining the asymptotic behavior ($\mu n \equiv \lfloor n\mu \rfloor$)

$$E\{N_{\mu n}\} \equiv \lim_{\substack{m,n\to\infty \\ m=\mu n}} E\{N_{m,n}\}$$

$$Pr\{N_{\mu n} = j\} \equiv \lim_{\substack{m,n\to\infty \\ m=\mu n}} Pr\{N_{m,n} = j\} \text{ and}$$

with $0 < \mu < 1$ is formally straightforward, but the technical details are more delicate. We will prove Theorem 13.6.

Theorem 13.6. If the table occupancy is μ with $0 < \mu < 1$, then

$$\lim_{n\to\infty} E\{N_{n\mu}\} = \frac{1}{2}\mu \frac{2-\mu}{(1-\mu)^2} \tag{13.7a}$$

$$\lim_{n\to\infty} Pr\{N_{n\mu} = j\} = (1-\mu)e^{-\mu} \sum_{i=j}^{\infty} \frac{1}{i!}(i+1)^{i-1}(\mu e^{-\mu})^i \quad 0 \le j < \infty \tag{13.7b}$$

Some additional asymptotic analysis shows that the average cost to search for a record is

$$\lim_{n\to\infty}\frac{1}{\mu n}\sum_{k=1}^{n\mu}[1+E\{N_k\}]=\frac{1}{2}\left[1+\frac{1}{1-\mu}\right] \tag{13.7c}$$

We start with equation (13.6*a*); replacing *m* by $\mu n \equiv \lfloor \mu n \rfloor$ gives

$$E\{N_{\mu n}\}=\sum_{i=1}^{\infty}\alpha_{i,\mu,n} \qquad \alpha_i \equiv \alpha_{i,\mu,n} \tag{13.8a}$$

where

$$\alpha_i=\frac{1-\mu}{2n^{\mu n-2}}\binom{\mu n-1}{i}(i+1)^i\, i(n-i-1)^{\mu n-i-2} \tag{13.8b}$$

with

$$\binom{\mu n-1}{i}\equiv\frac{(\mu n-1)(\mu n-2)\cdots(\mu n-i)}{i!} \tag{13.8c}$$

Equations (13.8*b* and 13.8*c*) show that for fixed *i* and $n \to \infty$

$$\alpha_i \to \frac{(1-\mu)e^{-\mu}}{2}\mu^i(i+1)^i\frac{i}{i!}e^{-\mu i} \tag{13.8d}$$

Although these formal manipulations are correct, there is a technical complication involving the interchange of the order of summation over *i*, and the limit as $n \to \infty$; $E\{N_{\mu n}\}$ is a summation over *i* with $i \to \infty$. One way to justify the interchange is to split the *i* summation into two parts

$$T_1(n)=\sum_{i=1}^{i_0(n)-1}\alpha_i \qquad T_2(n)=\sum_{i=i_0(n)}^{\mu n-1}\alpha_i \tag{13.8e}$$

and to prove that an appropriate choice of $i_0(n)$:

- The first summation converges $\lim_{n\to\infty} T_1(n)<\infty$.
- The second summation (the tail) is negligible $\lim_{n\to\infty} T_2(n)=0$.

We begin by using Stirling's formula, giving an estimate for $\frac{1}{i!}$ [see Chapter 3, equation (3.25*a*)]

$$\frac{1}{i!}=\frac{1}{\sqrt{2\pi}}i^{-\left(i+\frac{1}{2}\right)}e^{i-\frac{1}{12i}}=\frac{1}{\sqrt{2\pi}}i^{-\left(i+\frac{1}{2}\right)}e^{i}\left(1+o(i^{-1})\right) \tag{13.9a}$$

Next, because

$$e^{-x} = 1 - x + \frac{x^2}{2!} - \frac{x^3}{3!} + \cdots = 1 - x + \frac{x^2}{2!}\left(1 - \frac{x}{3}\right) + \frac{x^4}{4!}\left(1 - \frac{x}{5}\right) + \cdots \qquad (13.9b)$$

we obtain the bound $(1 - x)^m \leq e^{-mx}$ for $0 < x < 1$, $m > 0$.

Applying this bound to the two of the terms in equation (13.8a), we obtain

$$(\mu n - 1)(\mu n - 2)\cdots(\mu n - i) = (n\mu)^i \prod_{j=1}^{i}\left(1 - \frac{j}{n\mu}\right) \leq (n\mu)^i e^{-\frac{i(i+1)}{2n\mu}} \qquad (13.10a)$$

$$(n - i - 1)^{\mu n - i - 2} = n^{\mu n - i - 2}\left(1 - \frac{i+1}{n}\right)^{\mu n - i - 2} \leq n^{\mu n - i - 2} e^{-i\mu} e^{-\mu} e^{\frac{-(i+1)(i+2)}{n}} \qquad (13.10b)$$

If $i \leq i_0(n) \equiv \lfloor \sqrt[4]{n} \rfloor$, the expressions

$$e^{-\frac{i(i+1)}{2n\mu}}$$

[equation (13.10a)] and

$$e^{-i\mu} e^{-\mu} e^{\frac{-(i+1)(i+2)}{n}}$$

[equation (13.10b)] are equal to 1 with a negligible error $o\left(\sqrt[-2]{n}\right)$ as $n \to \infty$.

$$\begin{cases} e^{-\frac{i(i+1)}{2n\mu}} & \leq e^{-C_1 \frac{1}{\sqrt[2]{n}}} \\ e^{\frac{-(i+1)(i+2)}{n}} & \leq e^{-C_2 \frac{1}{\sqrt[2]{n}}} \end{cases} = 1 + o\left(\sqrt[-2]{n}\right) \qquad (13.10c)$$

where $C_1 > 0$ and $C_2 > 0$ are constants.

Equations (13.8b and 13.8c) and (13.10a through 13.10c) show that for $i \leq i_0(n)$

$$\alpha_i \leq C_3 \sqrt{i} e^{i(1-\mu+\log\mu)} \quad 1 \leq i \leq i_0(n) \qquad (13.11a)$$

where $C_3 > 0$ is another constant. Because $(1 - \mu + \log\mu) < 0$ for $0 < \mu < 1$, we conclude that

$$\alpha_i \leq C_3 \sqrt{i} e^{-C_4 \sqrt[4]{i}} \qquad (13.11b)$$

where $C_4 > 0$ is still another constant proving that $\lim_{n\to\infty} T_1(n)$ exists.

To prove the tail $T_2(n)$ is negligible, choose $\varepsilon > 0$ and evaluate the ratio of consecutive terms

$$\alpha_i / \alpha_{i-1} \leq \frac{n\mu - 1}{i} \frac{(i+1)^i}{i^{i-1}} \frac{i}{i-1} \frac{(n-i-1)^{n\mu-i-2}}{(n-i)^{n\mu-i-1}} \qquad (13.12a)$$

Repeated use of the two bounds $(1-x)^m \le e^{-mx}$ and $\frac{(i+1)^i}{i^i} \le e$ gives

$$(n\mu - i)\frac{(n-i-1)^{n\mu-i-2}}{(n-i)^{n\mu-i-1}} \le \frac{n\mu - i}{n-i}\left(1-\frac{1}{n-i}\right)^{n\mu-i}\left(1-\frac{1}{n-i}\right)^{-2}$$
$$\le \frac{n\mu-i}{n-i}\exp\left\{-\frac{n\mu-i}{n-i}+\frac{2}{n-i}\right\}$$

so that equation (13.12*a*) simplifies to

$$\alpha_i/\alpha_{i-1} \le \frac{i}{i-1}\exp\left\{\frac{2}{n-i}\right\}\exp\left\{1+\log\frac{n\mu-i}{n-i}-\frac{n\mu-i}{n-i}\right\} \tag{13.12b}$$

The function

$$1+\log\frac{n\mu-i}{n-i}-\frac{n\mu-i}{n-i}$$

increases on the interval $i_0(n) \le i < \mu n$ with maximum value $1-\mu+\log\mu < 0$ for $0 < \mu < 1$ so that the bound in (equation 13.12*b*) yields

$$\alpha_i/\alpha_{i-1} \le \frac{i}{i-1}\exp\left\{\frac{2}{n-i}\right\} \le 1-\varepsilon \quad \text{for } n \ge N \tag{13.12c}$$

implying

$$\alpha_{i_0(n)+j} \le \alpha_{i_0(n)}(1-\varepsilon)^j \tag{13.12d}$$

We finally conclude

$$\lim_{n\to\infty} E\{N_{\mu n}\} = \frac{1}{2}(1-\mu)e^{-\mu}\sum_{i=1}^{\infty}\mu^i(i+1)^i\frac{i}{i!}e^{-i\mu} = \frac{(1-\mu)e^{-\mu}}{2}\mu e^{-\mu}\left.\frac{d^2U(z)}{dz}\right|_{z=\mu e^{-\mu}} \tag{13.13a}$$

with

$$U(z) = zT(z) \tag{13.13b}$$

The functional equation (13.2*b*) $T(z) = e^{zT(z)}$ and equation (13.2*d*) yields

$$U(z) = ze^{U(z)} \tag{13.13c}$$
$$U(\mu e^{-\mu}) = \mu \tag{13.13d}$$

which gives

$$z\frac{d^2U(z)}{dz} = \frac{U(z)}{1-U(z)}\frac{1}{z}\left\{\frac{1}{(1-U(z))^2}-1\right\} \tag{13.13e}$$

$$\left.\frac{d^2U(z)}{dz}\right|_{z=\mu e^{-\mu}} = \mu e^{\mu}\frac{2-\mu}{(1-\mu)^3} \tag{13.13f}$$

which with equation (13.13*a*) completes the proof of equation (13.7*a*). ■

Starting with equation (13.6*b*) with $m = n\mu \approx \lfloor \mu n \rfloor$

$$Pr\{N_{n\mu,n} = j\} = \frac{1}{n^{\mu-1}} \sum_{i=j}^{n\mu-1} \binom{n\mu-1}{i} (i+1)^{i-1} \left(n - \mu(n-i-1)\right)^{n\mu-i-2} \quad (13.14a)$$

If $n \to \infty$, the formal limit $Pr\{N_{n\mu} = j\} = \lim_{n\to\infty} Pr\{N_{n\mu,n} = j\}$ is evaluated as

$$Pr\{N_{n\mu} = j\} = (1-\mu)e^{-\mu} \sum_{i=j}^{\infty} \frac{1}{i!}(i+1)^{i-1}\left(\mu e^{-\mu}\right)^i \quad (13.14b)$$

It is also necessary to justify the interchange of the limit $n \to \infty$ and the summation form $i = j$ to ∞. The steps are the same as in the proof of (13.6*a*). ■

Minor modifications of this process yields the cost of SEARCH(k) *averaged* over the existing $m = n\mu$ inserted keys as $n \to \infty$, which yields

$$\lim_{n\to\infty} \frac{1}{\mu n} \sum_{k=1}^{m=n\mu} [1 + E\{N_{k,n}\}] = \frac{1}{2}\left[1 + \frac{1}{1-\mu}\right] \quad (13.15)$$

Table 13.1 compares the values of $\pi_{\mathrm{LP}}(\mu, j) = \lim_{n\to\infty} Pr\{N_{n\mu} = j\}$ and $\pi_{\mathrm{U}}(\mu, j) = \lim_{n\to\infty} Pr\{N_{n\mu} = j\}$ for μ = 0.2(0.2)0.8 and j = 0(1)20 where U denotes uniform.

TABLE 13.1. The Probability of a Search Length *r* With Uniform and Linear Probing

j	$\pi_{LP}(0.2, j)$	$\pi_U(0.2, j)$	$\pi_{LP}(0.4, j)$	$\pi_U(0.4, j)$	$\pi_{LP}(0.6, j)$	$\pi_U(0.6, j)$	$\pi_{LP}(0.8, j)$	$\pi_U(0.8, j)$
0	0.800000	0.800000	0.600000	0.600000	0.400000	0.400000	0.200000	0.200000
1	0.145015	0.160000	0.197808	0.240000	0.180475	0.240000	0.110134	0.160000
2	0.037764	0.032000	0.089969	0.096000	0.108189	0.144000	0.077831	0.128000
3	0.011421	0.006400	0.046597	0.038400	0.072484	0.086400	0.060413	0.102400
4	0.003753	0.001280	0.025923	0.015360	0.051583	0.051840	0.049282	0.081920
5	0.001300	0.000256	0.015096	0.006144	0.038140	0.031104	0.041467	0.065536
6	0.000467	0.000051	0.009076	0.002458	0.028962	0.018662	0.035643	0.052429
7	0.000173	0.000010	0.005588	0.000983	0.022429	0.011197	0.031117	0.041943
8	0.000065	0.000002	0.003504	0.000393	0.017636	0.006718	0.027492	0.033554
9	0.000025	0.000000	0.002229	0.000157	0.014036	0.004031	0.024520	0.026844
10	0.000010	0.000000	0.001435	0.000063	0.011283	0.002419	0.022039	0.021475
11	0.000004	0.000000	0.000933	0.000025	0.009145	0.001451	0.019936	0.017180
12	0.000002	0.000000	0.000612	0.000010	0.007464	0.000871	0.018131	0.013744
13	0.000001	0.000000	0.000404	0.000004	0.006129	0.000522	0.016566	0.010995
14	0.000000	0.000000	0.000269	0.000002	0.005060	0.000313	0.015197	0.008796
15	0.000000	0.000000	0.000180	0.000001	0.004196	0.000188	0.013990	0.007037
16	0.000000	0.000000	0.000121	0.000000	0.003494	0.000113	0.012920	0.005629
17	0.000000	0.000000	0.000081	0.000000	0.002920	0.000068	0.011964	0.004504
18	0.000000	0.000000	0.000055	0.000000	0.002449	0.000041	0.011107	0.003603
19	0.000000	0.000000	0.000037	0.000000	0.002060	0.000024	0.010335	0.002882
20	0.000000	0.000000	0.000026	0.000000	0.001737	0.000015	0.009636	0.002306
	1.000000	1.000000	0.999944	1.000000	0.989871	0.999978	0.819720	0.990777

13.7 HASHING WITH LINEAR OPEN ADDRESSING: CODA

LP-SEARCH(k) probes the cells in the hash table to the right; that is, it examines the contents of $HT_{h(k)}$, $HT_{h(k)+1}$, $HT_{h(k)+2}$, $\cdots$. It should be clear that if LP-SEARCH(k) is modified probing the cells are to-the-left $HT_{h(k)}$, $HT_{h(k)-1}$, $HT_{h(k)-2}$, $\cdots$, the same conclusions relating to costs result.

More generally, let $c \geq 1$ be relatively prime to the table size n and require SEARCH(k) to test whether k has been inserted in the cells $HT_{h(k)}$, $HT_{h(k)+c}$, $HT_{h(k)+2c}$, $\cdots$ until either of the following occur:

- A cell containing the key k is found.
- An unoccupied cell is encountered.

When c is relatively prime to the table size n, the sequence $i, i + c, i + 2c, \cdots$ (*modulo* n) traverses all the cells in the hash table.

LP_cSEARCH(k) modifies LP-SEARCH(k) by testing whether k is in a cell in the order $HT_{h(k)}$, $HT_{h(k)+c}$, $HT_{h(k)+2c}$, $\cdots$.

Exercise 13.5. Prove that the distribution of the number of probes for LP_cSEARCH(k) is the same for each $c \geq 1$, which is relatively prime to the table length n.

13.8 A POSSIBLE IMPROVEMENT TO LINEAR PROBING

We study an environment in which a hash table remains relatively static, meaning infrequent insertions occur after the initial LP insertion of m keys $(k_0, k_1, \cdots, k_{m-1})$ into an n-cell hash table. The examples we have in mind include the following:

- An airline reservation system, keys derived from the flight numbers
- The student records at a university, keys derived from student IDs.

In both instances, there will be additional entries; however, keep in mind the following:

- New airline flights might be introduced infrequently compared with the time scale at which SEARCH is carried out.
- New student records might only be created infrequently; for example, only once a quarter or semester.

If there are only infrequent insertions, the primary cost is for SEARCH(k), which depends on the length of the chain of cells $HT_{h(k)}$, $HT_{h(k)+1}$, $\cdots$ to be examined. We shall now investigate a modification LP→LP* intended to reduce the cost of SEARCH. LP* requires the following three amendments to LP:

1. Additional fields in a hash-table cell
2. Altering the steps in LP-SEARCH
3. The potential for some rearrangement of the hash-table entries

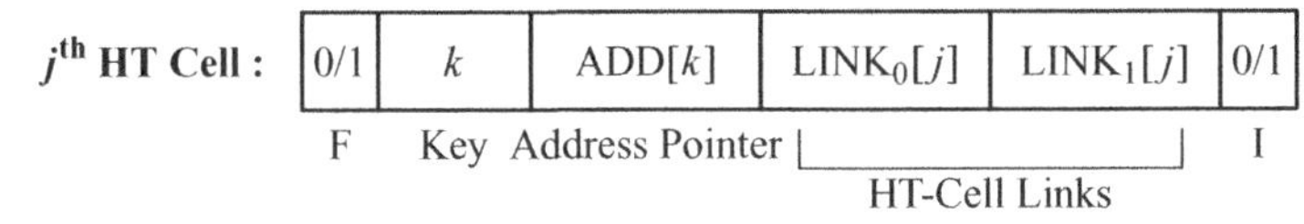

Figure 13.5. Modified Hash-Table Cell.

TABLE 13.2. Possible Values of (F_j, LINK$_0$[j], LINK$_1$[j], I_j)

F_j	LINK$_0$[j]	LINK$_1$[j]	I_j
1	j_0'		0
1	j_1'	j_1''	1
0	0		0

Figure 13.5 illustrates an LP* cell; in addition to the current flag (F), key (k), and the address pointer (ADD[k]), the cell contains an indicator flag (I) and two link fields LINK$_0$[j] and LINK$_1$[j], which is similar to those in coalesced hashing.

The only values for (F_j, LINK$_0$[j], LINK$_1$[j], I_j) are given in Table 13.2.

The indicator and links (LINK$_0$[j], LINK$_1$[j]) are assigned as the keys $K = \{k_0, k_1, \cdots, k_{m-1}\}$ are inserted. The meanings given to the LP* four-tuple (F_j, LINK$_0$[j], LINK$_1$[j], I_j) in Figure 13.5 are as follows:

(1,0): if (F_j,I_j) = (1,0), only LINK$_0$[j] has an assigned value.

- L_0 Because $I_j = 0$, the key $k \in HT_j$ satisfies $h(k) = h(j)$.

  If $LINK_0[j] = j_0' = 0$, HT_j is the only cell in the hash table that contains a key k' satisfying

  $h(k') = j = h(k)$ and we write $HT_j \xrightarrow{L_0} \emptyset$.[5]

  If $LINK_0[j] = j_0' > 0$, this link points to the next successor cell $HT_{j_0'}$ containing a key k' satisfying $h(k') = j = h(k)$, and we write $HT_j \xrightarrow{L_0} HT_{LINK_0[j]}$.

(1,1): If (F_j, I_j) = (1, 1), then both LINK$_0$[j] and LINK$_1$[j] have an assigned values.

- L_0 Because $I_j = 1$, the key $k \in HT_j$ satisfies $h(k) \neq h(j)$.

  If $LINK_0[j] = j_1' = 0$, no cell with hashing value $h(k)$ where $k \in HT_j$ has been stored, and we write $HT_j \xrightarrow{L_0} \emptyset$;

  If $LINK_0[j] = j_1' > 0$, some successor cell contains a key k' with hashing value $h(k' = h(k)$ where $k \in HT_j$ has been stored; $LINK_0[j] = j_1'$ points to the first such successor cell $HT_{j_1'}$, and we write $HT_j \xrightarrow{L_0} HT_{LINK_0[j]}$.

- L_1 If $LINK_1[j] = j_1'' = 0$, then HT_j is the only cell in the hash table that contains a key k' satisfying $h(k') = h(k)$ where $k \in HT_j$, and we write $HT_j \xrightarrow{L_1} \emptyset$

  If $LINK_1[j] = j_1'' > 0$, some successor cell contains a key k' satisfying $h(k') = h(k)$ where $k \in HT_j$; $LINK_0[j] = j_1''$ points to the first such successor cell $HT_{j_1''}$, and we write $HT_j \xrightarrow{L_1} HT_{LINK_1[j]}$.

(0,0): If (F_j, I_j) = (0, 0), then neither LINK$_0$[j] and LINK$_1$[j] are defined.

[5]Because 0-index origins have been used for cells, one coding for $\emptyset$ might be LINK$_?$[j] = j.

The Role of I_j. If $F_j = 0$, HT_j contains key parameters and the indicator I_j signals the comparison of $h(k)$ ($k \in HT_j$) with $h(j)$; that is,

$$\begin{cases} I_j = 0 & \text{if and only if } h(k) = j \\ I_j = 1 & \text{if and only if } h(k)/j \end{cases}$$

The Chains at HT_j. If $F_j = 1$, then LP* creates one chain $\mathcal{C}_{0,j}$ when $I_j = 0$ and two chains ($\mathcal{C}_{0,j}$, $\mathcal{C}_{1,j}$) when $I_j = 1$.

$\mathcal{C}_{0,j}$ —

0. If $I_j = 0$, the key $k \in HT_j$ satisfies $h(k) = h(j)$.
   The chain $\mathcal{C}_{0,j}$ of length ℓ_j links all hash-table cells that contain currently inserted keys k that satisfy $h(k) = h(j)$.

$$\mathcal{C}_{0,j} : HT_j \xrightarrow{L_0} HT_{LINK_0[j]} \xrightarrow{L_0} HT_{LINK_0^2[j]} \xrightarrow{L_0} \cdots \xrightarrow{L_0} HT_{LINK_0^{\ell_j}[j]} \xrightarrow{L_0} \emptyset$$

1. if $I_j = 1$, the key $k \in HT_j$ satisfies $h(k) \neq h(j)$.
   $LINK_0[j] \in \mathcal{C}_{0,j}$ points to the first hash-table cell containing a currently inserted key k that satisfies $h(k) = h(j)$.

$$\mathcal{C}_{0,j} : HT_j \xrightarrow{L_0} HT_{LINK_0[j]}$$

$\mathcal{C}_{1,j}$ —

0. If $I_j = 0$, the key $k \in HT_j$ satisfies $h(k) = h(j)$.
   $LINK_1$ is not assigned a value.
1. If $I_j = 1$, the key $k \in HT_j$ satisfies $h(k) \neq h(j)$.
   Either $LINK_1[j] = \emptyset$ if k has not been inserted or $LINK_1[j] \in \mathcal{C}_{0,j}$ points to the first hash-table cell containing a currently inserted key k, which satisfies $h(k) = h(j)$.

SEARCH*(k): k in the hash table

$I_{h(k)} = 0$ The key k' stored at $HT_{h(k)}$ has the same hashing value as k; the link $LINK_0[h(k)]$ points to the next cell number with a key having the same hashing value. Following the yellow brick road always leads to success because we have assumed k was previously inserted.

If we have bad luck ($I_{h(k)} = 1$) (and who does not once in a while), all the successors cells $LINK_0^s[h(k)]$ for $s = 1, 2, \cdots$ contain keys with the same hashing-value; if $I_j = 0$, then we may still be unlucky if k is at the end of the long chain $\mathcal{C}_{0,h(k)}$.

$I_{h(k)} = 1$ The key k' stored at $HT_{h(k)}$ does not have the same hashing value as k; nevertheless, the link $LINK_1[h(k)]$ provides the next cell number containing a key with the hashing value $h(k)$. Again, the yellow brick road always leads to success because we have assumed k was previously inserted.

If k has not been inserted, we learn this in one of the following three ways:

1. $I_{h(k)} = 0$ and cell $HT_{LINK_0^{\ell_j}[h(k)]}$ is reached without finding the key k, or
2. $I_{h(k)} = 1$ and
   a. $LINK_1[h(k)] = \emptyset$, or

TABLE 13.3. Hashing Values for Example 13.1.

k:	Alan	Beth	Carol	David	Dawn	Jay	Joshua	Keith	Madelyn	Pikun	Seth
***h*(*k*)**:	0	2	0	2	3	3	3	1	3	1	3

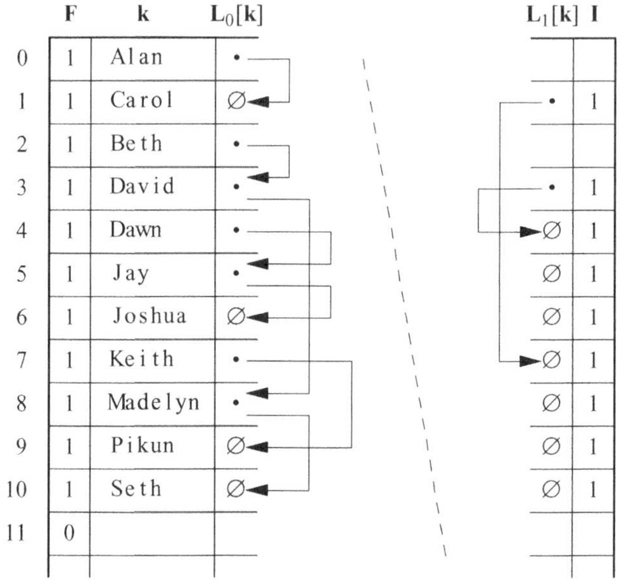

Figure 13.6. Example of LP* Links.

b. $\mathrm{LINK}_1[h(k)]$ points to a cell containing the *first* inserted key with hash value $h(k)$ that subsequently reveals as in the first item 1 that k has not been inserted.

Example 13.1. The 11 names of my immediate family are the keys; they are inserted in alphabetic order using the hashing values in Table 13.3.

Figure 13.6 shows the hash-table with the following characteristics:

- The address field omitted and
- The table split into left and right segments to better display the L_0 and L_1 links.

The hash table in Figure 13.6 contains only two nontrivial L_1 links, as follows:

- The key `Keith` was displaced from HT_1 by his mother `(Carol)` and was inserted at $HT_{\mathrm{LINK}_1^1[1]}$.
- The key `Dawn` was displaced from HT_3 by her nephew `(David)` and was inserted at $HT_{\mathrm{LINK}_1^1[3]}$.

We have hopes to extend the hash table; if we had prematurely tried `SEARCH(Malikon-suay)`[6] using the hashing value 1, the key comparisons visiting the $C_{0,1}$-chain of cells

[6]Thai for "most beautiful flower."

$$\mathcal{C}_{0,1}: \mathrm{HT}_1 \xrightarrow{\mathrm{L}_1} \mathrm{HT}_7 \xrightarrow{\mathrm{L}_0} \mathrm{HT}_9 \xrightarrow{\mathrm{L}_0} \cdots \xrightarrow{\mathrm{L}_0} \emptyset$$

would be unsuccessful.

The cost of SEARCH*(k) depends slightly on the value of $\mathrm{I}_{h(k)}$, as follows:

- if $\mathrm{I}_{h(k)} = 0$, the C-SEARCH*(k) is the average length $\frac{1}{2}(L_{h(k)}+1)$ of the path $\mathrm{HT}_{0,h(k)} \xrightarrow{\mathrm{L}_0} \cdots \xrightarrow{\mathrm{L}_0} \mathrm{HT}_j$ to the cell containing k where $L_{h(k)}$ is the length of $\mathcal{C}_{0,h(k)}$;
- if $\mathrm{I}_{h(k)} = 1$, the cost of C-SEARCH*(k) is the same provided we neglect the cost of detecting if the I flag is set.

Theorem 10.1 (Chapter 10) gave formulas for in separate chaining for the following:

- The expected number $E\{N_{\mathcal{C}}\} \sim n(1 - e^{-\mu})$
- The expected length $E\{L_{\mathcal{C}}\} \sim \dfrac{\mu}{1-e^{-\mu}}$

of chains when $m = \mu n$ keys are inserted as $n \to \infty$. For $\mu = 0.8, 0.9, 1.0$, we have $E\{L_{\mathcal{C}}\} \sim 1.453, 1.517$ and 1.572, so that the cost of C-SEARCH*(k) differs markedly from the cost of LP-SEARCH(k).

Insert*(k) for the new key, UA #7837. "New, nonstop service from Santa Barbara to the South Pole, effective ··· " may be carried out exactly as with LP-INSERT; it requires finding the first unoccupied cell HT_j after $\mathrm{HT}_{h(k)}$. If deletions (like insertions) are infrequent, then we may assume that no deletions have occurred. In this case, the steps in LP-INSERT might even be improved on, because the sought after cell HT_j must be a successor of $h(k) < j$. The search can begins at $\mathrm{HT}_{h(k)}$ but can be sped up; the various outcomes are as follows:

- If $\mathrm{F}_{h(k)} = 0$, truly, the God(s) have smiled on us; we would have found an occupied cell.
- If $\mathrm{F}_{h(k)} = 1$ and $\mathrm{I}_{h(k)} = 0$, we can follow the chain $\mathcal{C}_{0,hk}$ of length $\ell_{h(k)} > 0$ until the end, knowing that the first unoccupied slot will be found right after the end of the rainbow.
- If $\mathrm{F}_{h(k)} = 1$, $\mathrm{I}_{h(k)} = 1$ and $\mathrm{LINK}_1[h(k)] \neq \emptyset$, we can first use $\mathrm{LINK}_0[h(k)]$ to the HT_{j^*} with $j^* = \mathrm{LINK}_0[h(k)]$ containing a key k^* with $h(k^*) = h(k)$, and then we can follow the chain $\mathcal{C}_{0,j^*}$ until the end, knowing that the first unoccupied slot will be found right after the end of the rainbow. If we are really unlucky, the next option will be true.
- If $\mathrm{F}_{h(k)} = \mathrm{I}_{h(k)} = 1$ and $\mathrm{LINK}_0[h(k)] = \emptyset$, it is confirmed that k has not yet been inserted in the table.

Table 13.1 shows that when $\mu = 0.8$, LP-SEARCH(k) requires ≥4 probes with probability ≥0.448. However, the search for an unoccupied cell as described above may reduce the cost of an insertion C-INSERT*(k) to $\sim 1 + 1.453$.

But the additional link fields in LP*'s hash-table cell provides an opportunity to modify the hash table on the basis of how often keys are accessed to shorten the number of probes in SEARCH(k) dependent on their frequency of usage. If $K_a \equiv \{k_0, k_1, \cdots, k_{\ell_a-1}\}$ are all the keys with $h(k_i) = a$) $(i = 0, 1, \cdots, \ell_a - 1)$, the average cost of SEARCH*(k) with $k \in K_a$ is

$$\text{C-SEARCH}*(K_a) = \frac{1}{\ell}\sum_i (i+1) \tag{13.16a}$$

C-SEARCH*(K_a) ignores an additive term equal to 1 if the first comparison of SEARCH* starts at the cell HT_a for which $I_a = 1$. Even though this cost function is consistent with the uniform hashing assumption, it neglects the frequency with which keys are accessed. We are tempted to replace equation (13.16a) by

$$\text{C-SEARCH}*(K_a) = \frac{1}{\ell}\sum_i (i+1)q_i \tag{13.16b}$$

where $\{q_i\}$ are the probabilities that SEARCH*(k) will be for $k = k_i$.

The plan is for LP* to reorder the keys in a chain to minimize C-SEARCH*(K_a). The optimal order for the keys in a chain of length N is easy to describe.

Lemma 13.7. If $c_0, c_1, \cdots, c_{N-1}$ are strictly increasing integers, $\underline{q} = (q_0, q_1, \cdots, q_{r-1})$ is a probability mass function and $v = (v_0, v_1, \cdots, v_{N-1})$ if a permutation of $0, 1, \cdots, N-1$, then $\sum_{j=0}^{r-1} c_j q_{v_j}$ is minimized by any permutation v that orders the $\{q_{v_j}\}$ in decreasing order; that is, $q_{v_0} \ge q_{v_1} \ge \cdots \ge q_{v_{N-1}}$.

Applying the case $c_j = j + 1 (0 \le j < N)$, it seems appropriate to push the most frequently accessed keys to the front of the chain just as in the case of memory stacks. LP* is complicated by the following two factors:

1. The absence of information about $\underline{q}$
2. The overhead of reordering

We cannot get around the first factor, to lessen the overhead of the second factor, we carry out reordering in an incremental manner.

Suppose SEARCH*(k) reveals the chain of $s + 1$ cells

$$HT_{h(k_0)} \xrightarrow{L_0} HT_{h(k_1)} \xrightarrow{L_0} \cdots \xrightarrow{L_0} HT_{h(k_s)} \xrightarrow{L_0} HT_{h(k)}$$

of equal hashing values $k_0, k_1, \cdots, k_s$ with $\begin{cases} k \ne k_i \\ h(k_i) = h(k) \end{cases}$ $(i = 0, 1, \cdots, s)$. The rules for modification of the links depending on s and illustrated in Figures 13.7A–B are as follows;

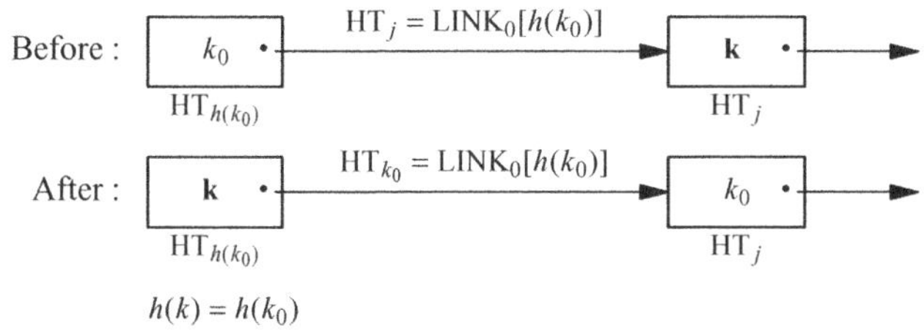

Figure 13.7A. Relinking Hash-Table Cells.

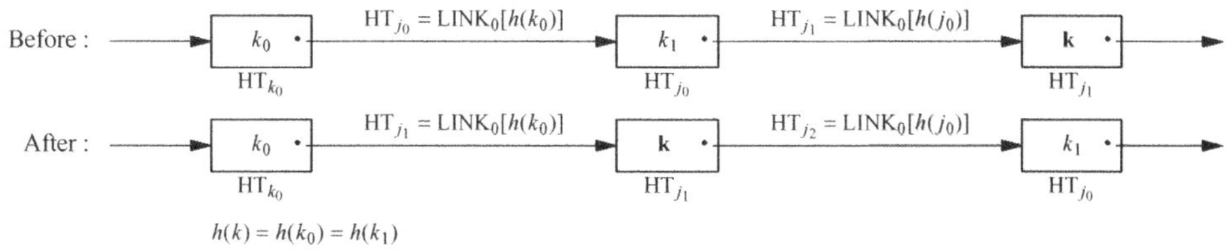

Figure 13.7B. Relinking Hash-Table Cells.

Case 1. $(s = 0)$

k has exactly one predecessor in the chain $\mathcal{C}_{0,h(k_0)}$, say $h(k_0) = h(k)$ and is stored in HT_j.

Case 2. $(s > 0)$

k has two or more predecessors in the chain $\mathcal{C}_{0,h(k_0)}$, say $h(k_0) = h(k_1) = h(k)$. k_1 is stored in HT_{j_0} and k is stored in HT_{j_1}.

In *Case 1*, the contents (key k and its associated parameters) of cells $HT_j = HT_{\text{LINK}[h(k_0)]}$ and $HT_{h(k_0)}$ are exchanged; in *Case 2*, the key, F-flag and address parameters are not moved, only I and the LINK_0s are changed.

Is LP* a real improvement? The answer is difficult to give, but we strive to sketch a somewhat flawed solution in the next and final exercise.

Exercise 13.6 (Technical preparation). Suppose $K = \{k_0, k_1, \cdots, k_{N-1}\}$ is a set of keys with equal hashing value; that is, $Pr\{h(K_i) = j\}$ is independent of i. The repeated use of SEARCH(k) with $k \in K$ may determine a chain of cells (of length ≥ 4) where we have suppressed the subscript 0 on L_0.

$$\mathcal{C}_{h(k)} : HT_{h(k)} \xrightarrow{\ L\ } HT_{\text{LINK}[h(k)]} \xrightarrow{\ L\ } HT_{\text{LINK}^2[h(k)]} \xrightarrow{\ L\ } HT_{\text{LINK}^3[h(k)]}$$

For example with $N = 4$, if the keys (k_0, k_1, k_2, k_3) have the same hashing value $h(k_0) = h(k_1) = h(k_2) = h(k_3) = h(k)$ (say) and are inserted in the order $\underline{x} = (3, 1, 0, 2)$, then they produce a single chain with

k_3 is inserted in $HT_{h(k)}$	k_1 is inserted in $HT_{\text{LINK}[h(k)]}$
k_0 is inserted in $HT_{\text{LINK}^2[h(k)]}$	k_2 is inserted in $HT_{\text{LINK}^3[h(k)]}$

The position of a particular k^* in $\mathcal{C}_{h(k)}$ may change as a result of relinking. We model these changes by a discrete-time stationary Markov chain $\mathcal{M}_N = \{\underline{X}_n\colon n = 0, 1, \cdots\}$ where $\underline{X}_n = (x_0, x_1, \cdots, x_{N-1})$, which is a permutation of the integers $0, 1, \cdots, N - 1$.

Let $\mathcal{M}_N = \{X_n\colon n = 0, 1, \cdots\}$. We imagine $\mathcal{M}_N$ to be a synchronous time process whose states are observed at times $0, \tau, 2\tau, \cdots$, at which point SEARCH* is carried out; let q_j be the probability that SEARCH*(k_j) is requested at time $n\tau$.

The rules governing changes in the table state $\underline{x}$ are as follows:

· If at time n, SEARCH*(k_{x_0}) is requested, $\underline{x}$ remains unchanged, with probability q_{x_0}

$$Pr\{\underline{X}_n = \underline{x} / \underline{X}_{n-1} = \underline{x}\} = q_{x_0}$$

- If at time n, SEARCH*(k_{x_j}) for $0 < j < N$ is requested, the state makes the transition

$$Pr\{\underline{X}_n = \underline{x}_{[j-1\leftrightarrow j]} / \underline{X}_{n-1} = \underline{x}\} = q_{x_j}$$

$$\underline{x} = (x_0, x_1, \cdots, x_{j-1}, x_j, \cdots, x_{N-1}) \rightarrow \underline{x}_{[j-1\leftrightarrow j]} = (x_0, x_1, \cdots, x_j, x_{j-1}, \cdots, x_{N-1})$$

where $\underline{x}_{[j-1\leftrightarrow j]}$: $(\cdots, x_j, x_{j-1}, \cdots) \rightarrow (\cdots, x_{j-1}, x_j, \cdots)$ indicates that the $(j-1)^{\text{st}}$ and j^{th} components of $\underline{x}$ are interchanged for $1 \le j < N$.

The probabilities $\{q_j\}$ determine the simplex

$$\mathcal{S}_N : q_j \ge 0 \quad 1 = q_0 + q_1 + \cdots + q_{N-1}$$

a) Construct the state transition diagrams of $\mathcal{M}_N$ for $N = 2, 3$ and conclude that $\mathcal{M}_N$ is irreducible for $N = 2, 3$.
b) Prove that $\mathcal{M}_N$ is irreducible for general $N \ge 2$.
c) Write the backward equations for $\mathcal{M}_N$ for $N = 2, 3$.
d) Write the backward equations for $\mathcal{M}_N$ with general N.
e) Show that if $Q_{\underline{x}} \equiv \prod_{j=0}^{N-1} q_{x_j}^{N-j-1}$, then the probability $\pi_{x_0,x_1,\cdots,x_{N-1}} \equiv \dfrac{Q_{\underline{x}}}{\sum_{\underline{y}\in P_N} Q_{\underline{y}}}$ satisfies the recurrence

$$\frac{\pi(x_0, x_1, \cdots, x_{j-2}, x_{j-1}, x_j, x_{j+1}, \cdots, x_{N-1})}{\pi(x_0, x_1, \cdots, x_{j-2}, x_j, x_{j-1}, x_{j+1}, \cdots, x_{N-1})} = \frac{q_{x_{j-1}}}{q_{x_j}} \quad 1 \le j < N$$

f) Use the results in item e) to show that $\pi_{x_0,x_1,\cdots,x_{N-1}}(N)$ given in the equation above is the solution of the backward equation—the equilibrium probabilities for $\mathcal{M}_N$.

Have we been spinning our wheels? Does LP* achieve an improvement over plain vanilla LP?

$$\text{C-SEARCH}*(\underline{v}) = \frac{1}{N} \sum_i (i+1) q_{v_i} \tag{13.17a}$$

is the average cost of SEARCH*(K_a) when the N keys in K_a that hashed to cell HT_a:

- Are initially inserted into the hash table in the order determined by to the permutation v
- There is no reordering of the keys in the hash table whereas

$$\text{C-SEARCH}*(\underline{\pi}_x) = \frac{1}{N} \sum_{\underline{x}\in\mathcal{M}_N} \left\{ \sum_{i=0}^{N-1} (i+1) q_{x_i} \right\} \underline{\pi}_{\underline{x}} \tag{13.17b}$$

is the average cost of SEARCH*(K_a) when the N keys in K_a that hashed to cell HT_a

- Are initially inserted into the hash table in the order determined by the probabilities $\underline{q}$.
- There is reordering of the keys in the hash-table according to LP*.

To compare the costs in equations (13.17*a* and 13.17*b*) we need to determine the permutation v if the keys are inserted into the table in the order $\underline{k_{\prec}} : k_{v0} \prec k_{v1} \prec \cdots \prec k_{v_{N-1}}$. Note that the order $\underline{v}$ requires

0. k_{v_0} be first inserted followed by j_0 requests SEARCH* k_{v_0}.
1. k_{v_1} be next inserted followed by j_1 requests for either SEARCH* k_{v_0} or SEARCH* k_{v_0}.
2. k_{v_2} be next inserted followed by j_2 requests for either SEARCH*(k) for one of the keys in $\{k_{v_0}, k_{v_1}, k_{v_2}\}$ and so forth. The probability of the order $\underline{v}$ is

$$Pr\{\underline{v}\} = \left[\prod_{i=0}^{N-1} \frac{q_{v_i}}{1 - \sum_{j=0}^{i} q_{v_j}} \right\} q_{v_{N-1}} \tag{13.18a}$$

so that

$$q_{v_i} = \sum_{j=0}^{N-1} q_j \sum_{\substack{v \\ v_i = j}} Pr\{\underline{v}\} \tag{13.18b}$$

Example 13.2. Table 13.4 gives the values of $\underline{q}_v$ for three examples of $\underline{q} = (q_0, q_1, q_2, q_3, q_4)$ chosen by specifying the ratios $\frac{q_j}{q_4}$ for $0 \le j < 5$.

- Row #0: $\underline{q}$ ratio 16:8:4:2:1 $\rightarrow \underline{q}_v$ ratio 6.1:1.8:3.3:2.0:1.0.
- Row #1: $\underline{q}$ ratio 5:4:3:2:1 $\rightarrow \underline{q}_v$ ratio 2.0:1.9:1.7:1.5:1.0.

We will prove Theorem 13.8 next.

Theorem 13.8. The equilibrium C-SEARCH* satisfies $\sum_{\underline{x} \in \mathcal{M}_N} \left\{ \sum_{j=0}^{N-1} (j+1) q_{x_j} \right\} \pi_{\underline{x}} \le \frac{N+1}{2}$ with equality if and only if $q_j = \frac{1}{N}$ for $0 \le j < N$.

Solution. Define

$$Q_{\underline{x}} = \sum_{j=0}^{n-1} q_{x_j}^{n-j-1} \qquad \pi_{\underline{x}} = \frac{Q_{\underline{x}}}{\sum_{\underline{y} \in \mathcal{M}_N} Q_{\underline{y}}} \tag{13.19a}$$

$$\pi_{\underline{x}} \ge 0 \, (\underline{x} \in \mathcal{M}_N), \quad 1 = \sum_{\underline{x}} \pi_{\underline{x}} \tag{13.19b}$$

TABLE 13.4. The Transformation $\underline{q} - \underline{q}_v$

$\underline{q}$					$\underline{q}_v$				
0.5161	0.2581	0.1290	0.0645	0.0323	0.3548	0.2807	0.1930	0.1133	0.0581
0.3333	0.2667	0.2000	0.1333	0.0667	0.2444	0.2328	0.2150	0.1840	0.1238
0.2000	0.2000	0.2000	0.2000	0.2000	0.2000	0.2000	0.2000	0.2000	0.2000

$$F\left(\underline{q}\right)=\sum_{\underline{x}\in\mathcal{M}_N}\pi_{\underline{x}}\sum_{\ell=0}^{N-1}(1+\ell)q_{x_\ell} \tag{13.20}$$

where

- $\underline{x}=(x_0, x_1, \cdots, x_{N-1}) \in \mathcal{M}_N$ be a permutation of the integers $\{0, 1, \cdots, N-1\}$.
- $\underline{q}=(q_0, q_1, \cdots, q_{N-1})$ be a discrete probability distribution on the integers, $\{0, 1, \cdots, N-1\}$.

$F(\underline{q})$ is a symmetric function of the variables $\{q_j\}$ on the simplex

$$\mathcal{S}_N:\left\{\underline{q}:q_j\geq 0, 0\leq i<N, 1=q_0+q_1+\cdots+q_{N-1}\right\}$$

and

a) $F(\underline{q})=1$ when $\underline{q}=(0_j, 1, 0_{N-j-1})$ for $0\leq j<N$.

b) $F\left(\underline{q}\right)=\dfrac{N+1}{2}$ when $\underline{q}=\underline{o}\equiv\left(\dfrac{1}{N},\dfrac{1}{N},\cdots,\dfrac{1}{N}\right)$.

c) $F\left(\underline{q}\right)<\dfrac{N+1}{2}$ if $\underline{q}\in\mathcal{S}_N-\{\underline{o}\}$.

Properties a) and b) are immediate; the main issue is property c). Suppose $\underline{q}$ maximizes $F(\underline{q})$. We may assume that $q_r>0$ for all r; otherwise, $F(\underline{q})$ is really defined on a subsimplex and we may argue the correctness of the result using mathematical induction. Write

$$F\left(\underline{q}\right)=\frac{\text{NUM}}{\text{DENOM}} \tag{13.21}$$

$$\text{NUM}=\sum_{\underline{x}\in\mathcal{M}_N}Q_{\underline{x}}\sum_{\ell=0}^{N-1}(1+\ell)q_{x_\ell} \tag{13.22a}$$

$$\text{DENOM}=\sum_{\underline{y}\in\mathcal{M}_N}Q_{\underline{y}} \tag{13.22b}$$

and observe that

$$\frac{\partial}{\partial q_r}F\left(\underline{q}\right)=\frac{d_r\text{NUM}}{(\text{DENOM})^2} \tag{13.23a}$$

with

$$d_r\text{NUM}=\text{DENOM}\left(\frac{\partial}{\partial q_r}\text{NUM}\right)-\text{NUM}\left(\frac{\partial}{\partial q_r}\text{DENOM}\right) \tag{13.23b}$$

We need to evaluate the derivatives $\dfrac{\partial}{\partial q_r}$ DENOM and $\dfrac{\partial}{\partial q_r}$ NUM, and we begin first by computing the derivative $\dfrac{\partial}{\partial q_r}$ DENOM. For this purpose, we write

$$\text{DENOM}=\sum_{k=0}^{N-1}\text{DENOM}_k \tag{13.24a}$$

$$\text{DENOM}_k = \sum_{\substack{\underline{x} \in \mathcal{M}_N \\ x_k = r}} Q_{\underline{x}} \tag{13.24b}$$

The factor $q_r^{N-(k+1)}$ appears in each term of DENOM_k so that we have

$$\frac{\partial}{\partial q_r} \text{DENOM}_k = \frac{N-(k+1)}{q_r} \sum_{\substack{\underline{x} \in \mathcal{M}_N \\ x_k = r}} Q_{\underline{x}} \tag{13.24c}$$

and

$$\frac{\partial}{\partial q_r} \text{DENOM} = \frac{1}{q_r} \sum_{k=0}^{N-1} (N-(k+1)) \sum_{\substack{\underline{x} \in \mathcal{M}_N \\ x_k = r}} Q_{\underline{x}} \tag{13.24d}$$

The computation of the derivative $\frac{\partial}{\partial q_r}$ NUM is more complicated.

$$\text{NUM} = \sum_{j=0}^{N-1} \sum_{\substack{\underline{y} \in \mathcal{M}_N \\ y_j = r}} Q_{\underline{y}} \sum_{\ell=0}^{N-1} (1+\ell) q_{y_\ell} \tag{13.25a}$$

It will convenient to split

$$\sum_{\ell=0}^{N-1} (1+\ell) q_{y_\ell}$$

for $\underline{y}$ with $y_j = q_r$ into two terms; the first with $\ell \neq j$ and the second with $\ell = j$ writing

$$\text{NUM} = \sum_{j=0}^{N-1} [\text{NUM}_j(1) + \text{NUM}_j(2)] \tag{13.25b}$$

$$\text{NUM}_j(1) \equiv \sum_{j=0}^{N-1} \sum_{\substack{\underline{y} \in \mathcal{M}_N \\ y_j = r}} Q_{\underline{y}} \sum_{\substack{\ell=0 \\ \ell \neq j}}^{N-1} (1+\ell) q_{y_\ell} \tag{13.25c}$$

$$\text{NUM}_j(2) \equiv \sum_{j=0}^{N-1} \sum_{\substack{\underline{y} \in \mathcal{M}_N \\ y_j = r}} Q_{\underline{y}} (1+j) q_r \tag{13.25d}$$

In a term in the summation for $\text{NUM}_j(1)$ with $y_j = r$, the following is true:

- The factor $q_r^{N-(j+1)}$ appears in $Q_{\underline{y}}$, whereas
- The q variable q_r does not occur in $\sum_{\substack{\ell=0 \\ \ell \neq j}}^{N-1} (1+\ell) q_{y_\ell}$. Therefore, we have

$$\frac{\partial}{\partial q_r} \text{NUM}_j(1) = \frac{1}{q_r} \sum_{j=0}^{N-1} (N-(j+1)) \sum_{\substack{\underline{y} \in \mathcal{M}_N \\ y_j = r}} Q_{\underline{y}} \sum_{\substack{\ell=0 \\ \ell \neq j}}^{N-1} (1+\ell) q_{y_\ell} \tag{13.26a}$$

Next, if $y_j = r$, the factor q_r^{N-j} appears in NUM_j (2), being the product of the following:

- The factor $q_r^{N-(j+1)}$ from the $Q\underline{y}$-term.
- The factor q_r from the term $(1 + j)q_r$.

Therefore, we have

$$\frac{\partial}{\partial q_r}\mathrm{NUM}_j(2)=\sum_{j=0}^{N-1}(N-j)(j+1)\sum_{\substack{\underline{y}\in\mathcal{M}_N\\ y_j=r}}Q_{\underline{x}} \tag{13.26b}$$

We want to add equations (13.27a) and (13.27b), but first we modify these terms by

Δ_1 correcting $\frac{\partial}{\partial q_r}\mathrm{NUM}_j(1)\to\frac{\partial}{\partial q_r}\mathrm{NUM}_j^*(1)$ by restoring the contribution of term $(1+j)q_r$ in the truncated summation $\sum_{\substack{\ell=0\\ \ell\neq j}}^{N-1}(1+\ell)q_{y_\ell}$ to $\frac{\partial}{\partial q_r}\mathrm{NUM}_j(1)$, which had previously been deleted

$$\frac{\partial}{\partial q_r}\mathrm{NUM}_1(j)\to\frac{\partial}{\partial q_r}\mathrm{NUM}_1(j)+\frac{1}{q_r}\sum_{j=0}^{N-1}(N-(j+1))\sum_{\substack{\underline{y}\in\mathcal{M}_N\\ y_j=r}}Q_{\underline{y}}(1+j)q_{y_j}\equiv\frac{\partial}{\partial q_r}\mathrm{NUM}_1^*(j)$$

and

Δ_2 correcting $\frac{\partial}{\partial q_r}\mathrm{NUM}_j(2)\to\frac{\partial}{\partial q_r}\mathrm{NUM}_j^*(2)$ by subtracting the term just added in Δ_1 above

$$\frac{\partial}{\partial q_r}\mathrm{NUM}_2(j)\to\frac{\partial}{\partial q_r}\mathrm{NUM}_2(j)-\frac{1}{q_r}\sum_{j=0}^{N-1}(N-(j+1))\sum_{\substack{\underline{y}\in\mathcal{M}_N\\ y_j=r}}Q_{\underline{y}}(1+j)q_{y_j}\equiv\frac{\partial}{\partial q_r}\mathrm{NUM}_2^*(j)$$

Note that

$$\frac{\partial}{\partial q_r}\mathrm{NUM}=\sum_{j=0}^{N-1}\left(\frac{\partial}{\partial q_r}\mathrm{NUM}_j(1)+\frac{\partial}{\partial q_r}\mathrm{NUM}_j(2)\right)=\sum_{j=0}^{N-1}\left(\frac{\partial}{\partial q_r}\mathrm{NUM}_j^*(1)+\frac{\partial}{\partial q_r}\mathrm{NUM}_j^*(2)\right) \tag{13.27}$$

To avoid the additional variables with the superscript *, we suppress the asterisk * in Δ_1 and Δ_2 and abuse the notation by writing

$$\frac{\partial}{\partial q_r}\mathrm{NUM}_j(i)(\text{ for }\frac{\partial}{\partial q_r}\mathrm{NUM}_j^*(i)$$

for $i=1,2$ and $0\le j<N$.

Now, we use equations (13.23d, 13.25a, and 13.25b) to evaluate d_rNUM [equation (13.22e)].

$$d_r\mathrm{NUM}=\sum_{k=0}^{N-1}\sum_{\substack{\underline{x}\in\mathcal{M}_N\\ x_k=r}}Q_{\underline{x}}\left\{\left[\frac{1}{q_r}\sum_{j=0}^{N-1}(N-(j+1))\sum_{\substack{\underline{y}\in\mathcal{M}_N\\ y_j=r}}Q_{\underline{y}}\sum_{\ell=0}^{N-1}(1+\ell)q_{y_\ell}\right]+\left[\sum_{j=0}^{N-1}(j+1)\sum_{\substack{\underline{y}\in\mathcal{M}_N\\ y_j=r}}Q_{\underline{y}}\right]\right\}$$
$$-\left[\frac{1}{q_r}\sum_{k=0}^{N-1}(N-(k+1))\sum_{\substack{\underline{x}\in\mathcal{M}_N\\ x_k=r}}Q_{\underline{x}}\right]\left[\sum_{j=0}^{N-1}\sum_{\substack{\underline{y}\in\mathcal{M}_N\\ y_j=r}}Q_{\underline{y}}\sum_{\ell=0}^{N-1}(1+\ell)q_{y_\ell}\right] \tag{13.27}$$

We next rearrange the terms in equation (13.27); the process is carried out in two steps: first,

$$d_r\text{NUM} = \frac{1}{q_r}\sum_{j=0}^{N-1}\sum_{k=0}^{N-1}[(N-(j+1))-(N-(k+1))]\sum_{\substack{\underline{x}\in\mathcal{M}_N\\ x_k=r}}\sum_{\substack{\underline{y}\in\mathcal{M}_N\\ y_j=r}} Q_{\underline{x}}Q_{\underline{y}}\sum_{\ell=0}^{N-1}(1+\ell)q_{y_\ell}$$
$$+\sum_{j=0}^{N-1}(j+1)\sum_{k=0}^{N-1}\sum_{\substack{\underline{x}\in\mathcal{M}_N\\ x_k=r}}\sum_{\substack{\underline{y}\in\mathcal{M}_N\\ y_j=r}} Q_{\underline{x}}Q_{\underline{y}}$$

and second, simplify the first line on the right in the previous equation by combining the two terms obtaining

$$d_r\text{NUM} = \frac{1}{q_r}\sum_{j=0}^{N-1}\sum_{k=0}^{N-1}(k-j)\sum_{\substack{\underline{x}\in\mathcal{M}_N\\ x_k=r}}\sum_{\substack{\underline{y}\in\mathcal{M}_N\\ y_j=r}} Q_{\underline{x}}Q_{\underline{y}}\sum_{\ell=0}^{N-1}(1+\ell)q_{y_\ell}$$
$$+\sum_{j=0}^{N-1}(j+1)\sum_{k=0}^{N-1}\sum_{\substack{\underline{x}\in\mathcal{M}_N\\ x_k=r}}\sum_{\substack{\underline{y}\in\mathcal{M}_N\\ y_j=r}} Q_{\underline{x}}Q_{\underline{y}}$$

which we will write as

$$d_r\text{NUM} = \left(\frac{1}{q_r}d_r\text{NUM}_1\right) + d_r\text{NUM}_2 \tag{13.28a}$$

$$d_r\text{NUM}_1 = \sum_{j=0}^{N-1}\sum_{k=0}^{N-1}(k-j)\sum_{\substack{\underline{x}\in\mathcal{M}_N\\ x_k=r}}\sum_{\substack{\underline{y}\in\mathcal{M}_N\\ y_j=r}} Q_{\underline{x}}Q_{\underline{y}}\sum_{\ell=0}^{N-1}(1+\ell)q_{y_\ell} \tag{13.28b}$$

$$d_r\text{NUM}_2 = \sum_{j=0}^{N-1}(j+1)\sum_{k=0}^{N-1}\sum_{\substack{\underline{x}\in\mathcal{M}_N\\ x_k=r}}\sum_{\substack{\underline{y}\in\mathcal{M}_N\\ y_j=r}} Q_{\underline{x}}Q_{\underline{y}} \tag{13.28c}$$

Fix the indices r, s with $0 \le r, s < N$ and $r \ne s$ and define the mapping $\lambda \equiv \lambda_{r,s}$ on $\mathcal{M}_N$, which interchanges the positions in which the integers r, s appear in a state $\underline{x}$ (and $\underline{y}$), leaving the remaining values unchanged.

$$\lambda : \underline{x} \to \underline{x}' \equiv \lambda(\underline{x})$$
$$\text{if } (x_\alpha, x_\beta) = (r, s) \text{ then } \begin{cases}(x'_\alpha, x'_\beta) = (s, r) \\ x'_\gamma = x_\gamma \quad \text{if} \quad \gamma \ne \alpha, \beta\end{cases} \qquad \begin{cases}x'_\alpha \equiv \lambda(\underline{x})_\alpha \\ x'_\beta \equiv \lambda(\underline{x})_\beta\end{cases}$$

For example, when $N = 6$ and $(r, s) = (4, 1)$,

$$\lambda : (5, 1, 3, 0, 2, 4) \to (5, 4, 3, 0, 2, 1)$$

λ is clearly a 1-1 transformation $\mathcal{M}_N$. Using λ as a change of variables gives

$$\sum_{\substack{\underline{x}\in\mathcal{M}_N\\ x_k=r}} \to \sum_{\substack{\lambda(\underline{x})\in\mathcal{M}_N\\ \lambda(\underline{x})_k=s}} \qquad \sum_{\substack{\underline{y}'\in\mathcal{M}_N\\ y_j=r}} \to \sum_{\substack{\lambda(\underline{y})\in\mathcal{M}_N\\ \lambda(\underline{y})_j=s}}$$

Conclusion. $d_r\text{NUM}_1$ and $d_r\text{NUM}_2$ have the same value for all r which implies that

$$\frac{\partial}{\partial q_r} F(\underline{q}) = \left(\frac{1}{q_r} C_1\right) + C_2 \tag{13.29}$$

where $C_1 \equiv C_1(\underline{q})$ and $C_2 \equiv C_2(\underline{q})$ do not depend on r.

By the method of Lagrange multipliers, a necessary condition for $\underline{q}$ to be an extreme point of $F(\underline{q})$ subject to the constraint $q_0 + q_1 + \cdots + q_{N-1} = 1$ is

$$0 = \frac{\partial}{\partial q_r}\left[F(\underline{q}) - \lambda(q_0 + q_1 + \cdots + q_{N-1})\right] \tag{13.30}$$

By virtue of equation (13.29)

$$0 = \left(\frac{1}{q_r} C_1\right) + C_2 - \lambda \tag{13.31}$$

which requires

$$0 = \left(\frac{1}{q_r} C_1\right) - \left(\frac{1}{q_0} C_1\right)$$

which implies

$$q_0 = q_1 = \cdots = q_{N-1}$$

concluding the proof. ■

An anonymous hasher has the more interesting idea of rearranging the hash-table entries so as to minimize the cost. Within each chain of records with the same hash-value, the optimal permutation of the records has already been given in **Lemma 7.7.** If the following occurs:

- $c_0, c_1, \cdots, c_{N-1}$ are strictly increasing integers,
- $\underline{q} = (q_0, q_1, \cdots, q_{N-1})$ is a probability mass function, and
- $\underline{v} = (v_0, v_1, \cdots, v_{N-1})$ if a permutation of $0, 1, \cdots, N-1$

then $\sum_{j=0}^{N-1} c_j q_{v_j}$ is minimized by any permutation $\underline{v}$ which orders the $\{q_{v_j}\}$ in decreasing order; that is $q_{v_0} \geq q_{v_1} \geq \cdots \geq q_{v_{N-1}}$.

Unfortunately, the hasher's procedure does not arrange the entries in the hash table to minimize the access cost.

Example 13.3. Table 13.5 below lists the probabilities

$Pr\{X_i = j\}$ ($\underline{X} = (X_0, X_1, X_2, X_3, X_4)$) for each state of $\mathcal{M}_5$ for two distributions $\underline{q} = (q_0, q_1, q_2, q_3, q_4)$.

The largest j entry in each row printed in boldface.

In these two small examples, the records are ordered correctly (in-probability). Alas, the *hasher*'s reorganization strategy while simple to implement is not optimal.

While it is certainly not true that "there's nothing new in hashing under the sun,"

TABLE 13.5. Values for Example 13.3

q	0.0500	0.1000	0.1500	0.2000	0.5000
$i\downarrow$	$Pr\{X_i = j\}$	$j\rightarrow$			
0	0.0026494	0.0210935	0.0839398	0.2564506	**0.6358666**
1	0.0303304	0.1213702	0.2500670	**0.3707544**	0.2274780
2	0.0940365	0.2561753	**0.3314249**	0.2254878	0.0928755
3	0.1796947	**0.3728908**	0.2711008	0.1341977	0.0421160
4	**0.6932890**	0.2284702	0.0634675	0.0131095	0.0016639
q	0.2000	0.0100	0.5000	0.190	0.1000
$i\downarrow$	$Pr\{X_i = j\}$	$j\rightarrow$			
0	0.1628260	**0.3314097**	0.3166756	0.1794210	0.0096678
1	0.0000049	0.0001914	0.0041544	0.0726055	**0.9230439**
2	**0.6615148**	0.2400221	0.0793719	0.0186890	0.0004022
3	0.1463621	0.3131096	**0.3312043**	0.1980931	0.0112309
4	0.0292922	0.1152673	0.2685939	**0.5311913**	0.0556552

the hasher should have visited the library more often; the idea of improving the performance of search schemes by reordering has been considered by others including [Rivest 1978], [Hester and Hirschberg 1985], [Hofri and Shachnai 1992], and [Flajolet, Gardy and Thimonier 1992].

The last three papers deal with optimizing linear lists in which the hasher's protocol is referred to as the transpose method. Rivest's paper is about hash-table performance and includes the formula in equation (13.19*a*).

It is not good to be discouraged in research. You have to kiss a lot of frags before finding your prince or princess! As an anonymous *Scottish Tune* advises,

It's a lesson all should heed,
Try, try, try again;
If at first you don't succeed,
Try, try, try again;
Then your courage will appear,
If you only persevere,
You will conquer, never fear!
Try, try, try again.

REFERENCES

W. Feller, *An Introduction to Probability Theory and Its Applications*, Volume 1 (Second Edition), John Wiley & Sons (New York), 1957; (Third Edition), John Wiley & Sons (New York), 1967.

P. Flajolet, D. Gardy, and L. Thimonier, "Birthday Paradox, Coupon Collectors, Cache Algorithms, and Self-Organizing Linear Search", *Discrete Applied Mathematics*, **39**, pp. 207–229, 1992.

G. H. Gonnet and J. I. Munro, "The Analysis of an Improved Hashing Technique", *ACM Symposium on Theory of Computing (STOC)*, pp. 113–121, 1977.

G. H. Gonnet and J. I. Munro, "Efficient Ordering of Hash Tables", *SIAM Journal of Computing*, **8**, #3, pp. 463–478, 1979.

G. H. Gonnet and J. I. Munro, "The Analysis of Linear Probing Sort by the Use of a New Mathematical Transform", *Journal of Algorithms*, **5**, #4, pp. 451–470, 1984.

J. H. Hester and D. S. Hirschberg, "Self Organizing Linear Search", *Computing Surveys*, **17**, #3, pp. 295–311, 1985.

M. Hofri and H. Shachnai, "The Application of Restricted Counter Schemes to Three Models of Linear Search", *Probability in Engineering Information Science*, **6**, pp. 371–389, 1992.

A. G. Konheim and B. Weiss, "An Occupancy Discipline and Applications", *SIAM Journal of Applied Mathematics*, **14**, #6, pp. 1266–1274, 1966.

D. Knuth, *The Art of Computer Programming: Volume 3/Sorting and Searching*, Addison Wesley (Boston, MA), 1973.

Walter Lederman, *Introduction to the Theory of Finite Groups*, Oliver and Boyd (London), 1953.

B. J. Mckenzie, R. Harries, and T. Bell, "Selecting a Hashing Algorithm", *Software Practice and Experience*, **20**, #2, pp. 209–224, 1990.

W. W. Peterson, "Addressing for Random Access Storage", *IBM Journal*, **1**, #2, pp. 130–146, 1957.

R. L. Rivest, "Optimal Arrangement of Keys in a Hash Table", *Journal of the ACM*, **25**, #2, pp. 200–209, 1978.

G. Schay Jr. and W. G. Spruth, "Analysis of a File Addressing Method", *Communications of the ACM*, **5** #8, pp. 459–462, 1962.

E. M. Wright, "Solution of the Equation $ze^z = a$", *Bulletin of the American Mathematical Society*, **65**, pp. 89–93, 1959.

CHAPTER 14

Double Hashing

14.1 FORMULATION OF DOUBLE HASHING[1]

Double hashing (DH) is an alternative to linear probing and chaining; it replaces the single integer-valued hashing function h from keys $k \in \mathcal{K}$ to a cell in a hash table with n cells

$$h : \mathcal{K} \to \mathcal{Z}_n \equiv \{0, 1, \cdots, n-1\}; \quad h(k) = i$$

with a pair $h = (h_1, h_2)$ of integer-valued hashing functions

$$\begin{cases} h_1 : \mathcal{K} \to \mathcal{Z}_n \\ h_2 : \mathcal{K} \to \mathcal{Z}_n - \{0\} \end{cases}; \quad h(k) = (h_1(k), h_2(k)) = (i, c)$$

Bell and Kaman [Bell and Kaman 1970] implemented double hashing $h = (h_1, h_2)$ with their linear quotient code defined by the formulas

- $h_1(k) = i$, the remainder, and
- $h_2(k) = c$, the (modified) quotient

resulting from the modular-like division $\frac{k}{n}$ of the key k by the size n of the hash table. Limited simulation results cited in [Bell and Kaman 1970] suggest that the search cost for double hashing might be those of uniform hashing. Maurer [Maurer 1968] and Radke [Radke 1970] also constructed pairs $h = (h_1, h_2)$ with other arithmetic functions.

DH-SEARCH (k) with $h(k) = (i, c)$ generates the probe sequence of hash-table cells in decreasing order HT_i, HT_{i-c}, HT_{i-2c}, $\cdots$

- If cell HT_i contains k, DH-SEARCH (k) has been successful and ends.
- Otherwise, each key in the cells HT_{i-c}, HT_{i-2c}, $\cdots$ is compared sequentially with k until either a match is found or an empty cell is encountered.

DH-INSERT (k) probes the same sequence of cells and requires j probes to insert k into cell HT_{i-jc} if the following is true:

[1] "Two heads are better than one." Polish proverb: "Co dwie glowy to nie jedna."

- Cells HT_i, HT_{i-c}, HT_{i-2c}, $\cdots$, $HT_{i-(j-1)c}$ are all occupied, and
- Cell HT_{i-jc} is unoccupied.

The cost of DH-INSERT (k) is the number of probes required to find the first unoccupied cell.

The first analysis of DH-INSERT in [Guibas and Szemeredi 1978] is rooted in Guibas' 1976 dissertation at Stanford University. The performance of a hashing scheme is degraded by clustering, which results when several keys follow the same probe sequence, thus interfering with one another. This occurs, for example in linear probing, when two keys have the same hashing value; double hashing reduces the probability of the confluence of hashing values for two keys from

$$\frac{1}{n} \text{ to } \frac{1}{n(n-1)} 5$$

potentially resulting in improved performance. [Guibas and Szemeredi 1978, p. 233] note that their analysis of double hashing will "use techniques that have a different flavor from those previously employed in algorithmic analysis." Instead of deriving recurrence relations and generating functions for inserting the next key into an existing hash table, [Guibas and Szemeredi 1978] prove that as $n \to \infty$, the configurations of occupied cells in hash tables of size n and occupancy μ for double and uniform hashing have the same probability. Building on the techniques implicit in [Guibas and Szemeredi 1978], Lueker and Molodowitch [Lueker and Molodowitch 1993] simplified and extended their results.

The insertion cost for tables created by double hashing are evaluated within the same statistical model by both [Guibas and Szemeredi 1978, p. 230] and [Lueker and Molodowitch 1993, p. 84]. It assumes the following:

i) The hashing values $(h_1(k), h_2(k)) = (i, c)$ for each key k are the outcomes of a chance experiment ε

$$Pr\{h_1(k)=i, h_2(k)=c\} = \frac{1}{n(n-1)} \qquad 0 \le i < n, 1 \le c < n \tag{14.1a}$$

ii) If $m \le n$ keys $\underline{k} = (k_0, k_1, \cdots, k_{m-1})$ are inserted into a hash-table of n cells, their hashing-values

$$(h, k_0) \to (i_0, c_0) \quad (h, k_0) \to (i_0, c_0) \quad \cdots \quad (h, k_{m-1}) \to (i_{m-1}, c_{m-1}) \tag{14.1b}$$

result as if generated by independent and identically distributed trials of the chance experiment ε.

Yao's optimality theorem (Chapter 15) proves the asymptotic expected cost of a successful search for uniform hashing (UH)

$$\lim_{n\to\infty} C_S^{(UH)}(m, n) = \frac{1}{1-\mu} \log \frac{1}{1-\mu}$$

is a lower bound among all open-addressing hashing schema. In conclusion, Yao asks whether the optimality of UH remains true for the asymptotic expected insertion cost [Yao 1985, p. 693]. Guibas and Szemeridi prove [Guibas and Szemeredi 1978] the equality of the asymptotic expected insertion costs $Y_{m,n}^{(\mathrm{DH})}$ and $Y_{m,n}^{(\mathrm{UH})}$ for double and uniform hashing when $\mu \le \mu_0 \cong 0.31. \cdots$

$$\lim_{\substack{n,m\to\infty \\ m=\mu n}} C_1^{(\mathrm{DH})}(m,n) = \lim_{\substack{n,m\to\infty \\ m=\mu n}} C_1^{(\mathrm{DH})}(m,n) = \frac{1}{1-\mu}$$

We provide an exposition of the analysis of DH-INSERT contained in [Lueker and Molodowitch 1993], which both simplifies the arguments used in [Guibas and Szemeredi 1978] and at the same time also removes the restriction $\mu \le \mu_0$. They prove tight bounds relating the probabilities $P_{m,n}^{(\mathrm{DH})} = Pr\{Y_{m,n}^{(\mathrm{DH})} \ge r\}$ and $P_{m,n}^{(\mathrm{UH})} = Pr\{Y_{m,n}^{(\mathrm{UH})} \ge r\}$ as $n \to \infty$ with $m \approx \mu n$ and $0 < \mu < 1$. Sections 14.2 through 14.4 develop the apparatus to prove these results; an outline of the essential ideas in [Lueker and Molodowitch 1993] is deferred until §14.5; the proof is contained in the sections that follow.

Technical Assumption. The size n of the hash table is fixed prime in both [Guibas and Szemeredi 1978] and [Lueker and Molodowitch 1993]; §14.3 will explain why this assumption is made. Examples will use the value $n = 31$.

14.2 PROGRESSIONS AND STRIDES

The state of a hash table with n cells is specified by the Boolean vector $\underline{x} = (x_0, x_1, \cdots, x_{n-1})$ with $x_i = 1$, if cell HT_i is occupied and 0 otherwise.

[Guibas and Szemeredi 1978, p. 234] and [Lueker and Molodowitch 1993, p. 86] define an (arithmetic) progression of length ℓ and stride c leading to the cell HT_i as the chain of cells $\mathcal{C}_i(\ell, c): \mathrm{HT}_{i+\ell c} \to \mathrm{HT}_{i+(\ell-1)c} \to \cdots \to \mathrm{HT}_{i+c}$.

Example 14.1. The progressions $\mathcal{C}_4\,(5, c_j)$ $j = 1, 2, 3$ length $\ell = 5$ with strides $(c_1, c_2, c_3) = (1, 7, 10)$ are

$$\mathcal{C}_4(5,1): \mathrm{HT}_9 \to \mathrm{HT}_8 \to \mathrm{HT}_7 \to \mathrm{HT}_6 \to \mathrm{HT}_5$$
$$\mathcal{C}_4(5,7): \mathrm{HT}_8 \to \mathrm{HT}_1 \to \mathrm{HT}_{25} \to \mathrm{HT}_{18} \to \mathrm{HT}_{11}$$
$$\mathcal{C}_4(5,10): \mathrm{HT}_{23} \to \mathrm{HT}_{13} \to \mathrm{HT}_3 \to \mathrm{HT}_{24} \to \mathrm{HT}_{14}$$

Double hashing inserts the key k into the unoccupied cell HT_i in table state $\underline{x}$ if and only if for some (ℓ, c)

1. The hashing value of k is $h(k) = (i + \ell c, c)$.
2. All the ℓ cells in the progression $\mathcal{C}_i(\ell, c)$: $\mathrm{HT}_{i+\ell c}, \mathrm{HT}_{i+(\ell-1)c}, \cdots, \mathrm{HT}_{i+c}$ are occupied.

When double hashing is used to create a hash table, the event

$$E_i(\ell, c): \text{"All cells in } \mathcal{C}_i(\ell, c): \mathrm{HT}_{i+\ell c} \to \mathrm{HT}_{i+(\ell-1)c} \to \cdots \to \mathrm{HT}_{i+c} \text{ are occupied"}$$

is the intersection of the dependent events "Cell HT_{i+jc} is Occupied" for $j = 1, 2, \cdots, \ell$. The complex combinatorics to enumerate double-hashing insertion sequences that lead to table state $\underline{x}$ associated with the event $\bigcup_{i=0}^{n-1} \bigcup_{\ell \geq r-1} E_i(\ell, c)$ $E_i(\ell, c)$ is described in [Guibas and Szemeredi 1978].

14.3 THE NUMBER OF PROGRESSIONS WHICH FILL A HASH-TABLE CELL

In this section, we accomplish the following:

i) Assume m keys have been inserted into a hash table of n cells by uniform hashing creating $\underline{x}$.
ii) Calculate upper and lower bounds for the probability P_i that double hashing would insert the next key k into an unoccupied cell HT_i as $n \to \infty$.

When the hash table is filled by uniform hashing, independently inserting a key into each cell with probability p, then $Pr\{E_i(\ell, c)\} = p^\ell$ for $0 \leq i, \ell < n$ and $1 \leq c < n$.

A partial explanation for this perhaps curious mixture of *i*) creating a table using uniform hashing and then *ii*) inserting a key using double hashing is provided in the next exercise.

Exercise 14.1. Suppose m keys are inserted into a hash table of n cells by uniform hashing, n is a prime, and (i, c) satisfy $0 \leq i < n, 1 \leq c < n$.

a) Prove if $r < m$, then the probability $p_{m,n}(r)$ of the event
   - Uniform hashing inserts keys into all the cells $HT_i, HT_{i-c}, \cdots, HT_{i-(r-2)c}$.
   - Uniform hashing does not insert a key into the cell $HT_{i-(r-1)c}$ is

$$p_{m,n}(r) = \frac{m}{n}\frac{m-1}{n-1}\cdots\frac{m-(r-2)}{n-(r-2)}\left(1-\frac{m-(r-1)}{n-(r-1)}\right)$$

b) Prove $p_{m,n}(r) \to \mu^{r-1}(1-\mu)$ if $n, m \to \infty$ with $m = \mu n$ and $0 < \mu < 1$.
c) Conclude that the number of probes using double or uniform hashing to insert a key into the hash table created by uniform hashing have the same probability distribution as $n \to \infty$.

Define the following counting functions:

- $N_{i,\ell}$: the number of progressions $C_i(\ell, c)$ of cells leading to HT_i having a fixed length ℓ but any stride c, assigning the value $N_{i,\ell} = n - 1$ when $\ell = 0$
- N_i: the number of progressions $C_i(\ell, c)$ of cells leading to cell HT_i of any length ℓ and any stride c.

N_i and $N_i(\ell)$ are related by

$$N_i = \sum_{\ell=0}^{n-1} N_{i,\ell} \tag{14.2a}$$

$$N_{i,\ell} = \sum_{c=1}^{n-1} \chi\{E_i(\ell, c)\} \tag{14.2b}$$

Because the table is created by uniform hashing, the probability P_i that the next key would inserted by DH into an unoccupied cell HT_i is

$$P \equiv P_i = \frac{1}{n(n-1)} \sum_{\ell=0}^{n-1} N_{i,\ell} \tag{14.2c}$$

[Lueker and Molodowitch 1993] write that two progressions $\mathcal{C}_i(\ell, c_1)$ and $\mathcal{C}_i(\ell, c_2)$ with $c_1 \neq c_2$

$$\mathcal{C}_i(\ell, c_1): HT_{i+\ell c_1} \to HT_{i+(\ell-1)c_1} \to HT_{i+(\ell-2)c_1} \to \cdots \to HT_{i+c_1}$$
$$\mathcal{C}_i(\ell, c_2): HT_{i+\ell c_2} \to HT_{i+(\ell-1)c_2} \to HT_{i+(\ell-2)c_2} \to \cdots \to HT_{i+c_2}$$

- Interact if $\mathcal{C}_i(\ell, c_1) \cap \mathcal{C}_i(\ell, c_2) \neq \emptyset$.
- Are independent if $\mathcal{C}_i(\ell, c_1) \cap \mathcal{C}_i(\ell, c_2) = \emptyset$.

For fixed i and ℓ, equation (14.2*b*) expresses

- $N_{i,\ell} = \sum_{c=1}^{n-1} \chi_{\{E_i(\ell,c)\}}$ as the sum of (0,1)-valued (or Bernoulli) random variables $\{\chi_{\{E_i(\ell,c)\}} : 1 \leq c < n\}$.
- The $\{\mathcal{C}_i(\ell, c) : 1 \leq c < n\}$ are independent progressions if and only if the random variables $\{\chi_{\{E_i(\ell,c)\}} : 1 \leq c < n\}$ are independent.

If $\{\mathcal{C}_i(\ell, c) : 1 \leq c < n\}$ are independent, then $N_{i,\ell} = \sum_{c=1}^{n-1} \chi_{\{E_i(\ell,c)\}}$ has the binomial distribution BIN $(n-1, p^\ell)$.

Example 14.1. shows that the $\{\mathcal{C}_i(\ell, c) : 1 \leq c < n\}$ does not generally consist of independent progressions, because of the following reasons:

i) $\mathcal{C}_4(5, 1)$ and $\mathcal{C}_4(5, 7)$ are *not* independent progressions because $\mathcal{C}_4(5, 1) \cap \mathcal{C}_4(5, 7) = HT_8$, whereas
ii) The progressions in $\mathcal{C}_4(5, 1)$ and $\mathcal{C}_4(5, 10)$ are independent because $\mathcal{C}_4(5, 1) \cap \mathcal{C}_4(5, 10) = \emptyset$, and
iii) The progressions in $\mathcal{C}_4(5, 7)$ and $\mathcal{C}_4(5, 10)$ are independent because $\mathcal{C}_4(5, 7) \cap \mathcal{C}_4(5, 10) = \emptyset$.

14.3.1 Progression Graphs

A graph $\mathcal{G} = (V, E)$ consists of the following:

- A (finite) set V of vertices.
- A set E of edges, two-element subsets $e = \{v, v'\}$ whose elements v, v' are adjacent vertices.

We write $v \leftrightarrow v'$ if $e = \{v, v'\}$ is an edge and $v \not\leftrightarrow v'$ otherwise.

We begin with a few basic concepts as follows from graph theory needed in the sequel ([Welsh 1976], [Even 1979], and [Tutte 1984]):

- The valence of a vertex v is the number of vertices that are adjacent to it.
- An r coloring of the graph $\mathcal{G} = (V, E)$ is an assignment of one of r colors to each vertex v with different colors assigned to adjacent vertices v and v'.
- An r coloring of the graph $\mathcal{G}$ partitions the set of vertices V into sets $S[0], S[1], \cdots, S[r-1]$ consisting of vertices assigned the same color.
- Theorem IX.33 [Tutte 1984, p. 233]: An r coloring of $\mathcal{G}$ exists if the maximum valence of each vertex is $r - 1$.

For each (n, i, ℓ), the progression graph $\mathcal{G}(n, i, \ell) = (V, E)$ is defined as follows:

- Its vertices correspond to the strides $V = \{c : 1 \le c < n\}$.
- Two strides c and c' are adjacent if and only if $\mathcal{C}_i(\ell, c)$ and $\mathcal{C}_i(\ell, c')$ are adjacent; that is, $\mathcal{C}_i(\ell, c) \cap \mathcal{C}_i(\ell, c') \ne \emptyset$.
- An r coloring of $\mathcal{G}(n, i, \ell)$ partitions $S[0], S[1], \cdots, S[r-1]$ the set of chains $\{\mathcal{C}_i(\ell, c) : 1 \le c < n\}$ such that $\mathcal{C}_i(\ell, c), \mathcal{C}_i(\ell, c') \in S[j]$ if and only if $\mathcal{C}_i(\ell, c), \mathcal{C}_i(\ell, c')$ are independent.

For a fixed c, the following is true:

- The stride $c' \ne c$ satisfies $\mathcal{C}_i(\ell, c) \cap \mathcal{C}_i(\ell, c') \ne \emptyset$, provided
- A pair (j, j') with $1 \le j, j' \le \ell$ exists such that $i + jc = i + j'c'$.

Pippenger[2] observed that when n is a prime, the solution of the equation $i + jc = i + j'c'$ is either

- $c = c'$ and $j = j'$.
- If $c \ne c'$ then j determines a unique j'.

It follows that the valence of each vertex in $\mathcal{G}(n, i, \ell)$ is $\ell^2 - 1$ so that $\mathcal{G}(n, i, \ell)$ has an ℓ^2-coloring; the set of $n - 1$ strides c may be partitioned into ℓ^2 subsets

$$S[0], S[1], \cdots S[\ell^2 - 1].$$

The partition of $\{\mathcal{C}_i(\ell, c): 1 \le c < n\}$ allows us to write $N_i(\ell)$ as the sum of r independent random variables

$$N_i(\ell) = \sum_{0 \le j < \ell^2} N_i(\ell)[j] = \sum_{c \in S[j]} \chi_{\{E_i(\ell, c)\}}.$$

Since the progressions in $S[j]$ are independent, $N_i(\ell)[j]$ has the binomial distribution $\text{BIN}(s[j], p^{\ell})$ with $s[j] = \|S[j]\|$.

[2]In a private communication cited in [Lueker and Molodowitch 1993].

TABLE 14.1. The Progressions $\mathcal{C}_4(5, c)$ in $\mathcal{G}(31, 4, 5)$

c		c	
1	$HT_9 \to HT_8 \to HT_7 \to HT_6 \to HT_5$	16	$HT_{22} \to HT_6 \to HT_{21} \to HT_5 \to HT_{20}$
2	$HT_{14} \to HT_{12} \to HT_{10} \to HT_8 \to HT_6$	17	$HT_{27} \to HT_{10} \to HT_{24} \to HT_7 \to HT_{21}$
3	$HT_{19} \to HT_{16} \to HT_{13} \to HT_{10} \to HT_7$	18	$HT_1 \to HT_{14} \to HT_{27} \to HT_9 \to HT_{22}$
4	$HT_{24} \to HT_{20} \to HT_{16} \to HT_{12} \to HT_8$	19	$HT_6 \to HT_{18} \to HT_{30} \to HT_{11} \to HT_{23}$
5	$HT_{29} \to HT_{24} \to HT_{19} \to HT_{14} \to HT_9$	20	$HT_{11} \to HT_{22} \to HT_2 \to HT_{13} \to HT_{24}$
6	$HT_3 \to HT_{28} \to HT_{22} \to HT_{16} \to HT_{10}$	21	$HT_{16} \to HT_{26} \to HT_5 \to HT_{15} \to HT_{25}$
7	$HT_8 \to HT_1 \to HT_{25} \to HT_{18} \to HT_{11}$	22	$HT_{21} \to HT_{30} \to HT_8 \to HT_{17} \to HT_{26}$
8	$HT_{13} \to HT_5 \to HT_{28} \to HT_{20} \to HT_{12}$	23	$HT_{26} \to HT_3 \to HT_{11} \to HT_{19} \to HT_{27}$
9	$HT_{18} \to HT_9 \to HT_0 \to HT_{22} \to HT_8$	24	$HT_0 \to HT_7 \to HT_{14} \to HT_{21} \to HT_{28}$
10	$HT_{23} \to HT_{13} \to HT_3 \to HT_{24} \to HT_{14}$	25	$HT_5 \to HT_{11} \to HT_{17} \to HT_{23} \to HT_{29}$
11	$HT_{28} \to HT_{17} \to HT_6 \to HT_{26} \to HT_{15}$	26	$HT_{10} \to HT_{15} \to HT_{20} \to HT_{25} \to HT_{30}$
12	$HT_2 \to HT_{21} \to HT_9 \to HT_{28} \to HT_{16}$	27	$HT_{15} \to HT_{19} \to HT_{23} \to HT_{27} \to HT_0$
13	$HT_7 \to HT_{25} \to HT_{12} \to HT_{30} \to HT_{17}$	28	$HT_{20} \to HT_{23} \to HT_{26} \to HT_{29} \to HT_1$
14	$HT_{12} \to HT_{29} \to HT_{15} \to HT_1 \to HT_{18}$	29	$HT_{25} \to HT_{27} \to HT_{29} \to HT_0 \to HT_2$
15	$HT_{17} \to HT_2 \to HT_{18} \to HT_3 \to HT_{19}$	30	$HT_{30} \to HT_0 \to HT_1 \to HT_2 \to HT_3$

TABLE 14.2. The Progressions Adjacent to $\mathcal{C}_4(5, 1)$ in $\mathcal{G}(31, 4, 5)$

c		c	
1	$HT_9 \to HT_8 \to HT_7 \to HT_6 \to HT_5$	2	$HT_{14} \to HT_{12} \to HT_{10} \to HT_8 \to HT_6$
3	$HT_{19} \to HT_{16} \to HT_{13} \to HT_{10} \to HT_7$	4	$HT_{24} \to HT_{20} \to HT_{16} \to HT_{12} \to HT_8$
5	$HT_{29} \to HT_{24} \to HT_{19} \to HT_{14} \to HT_9$	7	$HT_8 \to HT_1 \to HT_{25} \to HT_{18} \to HT_{11}$
8	$HT_{13} \to HT_5 \to HT_{28} \to HT_{20} \to HT_{12}$	9	$HT_{18} \to HT_9 \to HT_0 \to HT_{22} \to HT_8$
11	$HT_{28} \to HT_{17} \to HT_6 \to HT_{26} \to HT_{15}$	12	$HT_2 \to HT_{21} \to HT_9 \to HT_{28} \to HT_{16}$
13	$HT_7 \to HT_{25} \to HT_{12} \to HT_{30} \to HT_{17}$	16	$HT_{22} \to HT_6 \to HT_{21} \to HT_5 \to HT_{20}$
17	$HT_{27} \to HT_{10} \to HT_{24} \to HT_7 \to HT_{21}$	18	$HT_1 \to HT_{14} \to HT_{27} \to HT_9 \to HT_{22}$
19	$HT_6 \to HT_{18} \to HT_{30} \to HT_{11} \to HT_{23}$	21	$HT_{16} \to HT_{26} \to HT_5 \to HT_{15} \to HT_{25}$
22	$HT_{21} \to HT_{30} \to HT_8 \to HT_{17} \to HT_{26}$	24	$HT_0 \to HT_7 \to HT_{14} \to HT_{21} \to HT_{28}$
25	$HT_5 \to HT_{11} \to HT_{17} \to HT_{23} \to HT_{29}$		

We will show how the sum $N_i(\ell) = \sum_j N_i(\ell)[j]$ for large n may be accurately bounded.

Example 14.2. The progressions $\mathcal{C}_4(5, c)$ in $\mathcal{G}(31, 4, 5)$ are shown in Table 14.1.

Tables 14.2 and 14.3 list the progressions and strides in $\mathcal{G}(31, 4, 5)$, which are adjacent and independent to $\mathcal{C}_4(5, 1)$.

Exercise 14.2. Prove that $\sim$ defined on the progressions $\mathcal{C}_i(\ell, c)$ in $\mathcal{G}(i, n, \ell)$ by $c_1 \sim c_2$ if and only if $\mathcal{C}_i(\ell, c_1) \cap \mathcal{C}_i(\ell, c_2) = \emptyset$. is an equivalence relation?

Table 14.4 is a 12 coloring of the progression graph $\mathcal{G} \equiv \mathcal{G}(31, 4, 5)$; $S[j]$ denotes the set of independent progressions assigned the jth color with $0 \le j < 12$.

Pippenger's observation leads to the following tight bounds on N_i.

TABLE 14.3. The Progressions Independent of $\mathcal{C}_4(5, 1)$ in $\mathcal{G}(31, 4, 5)$

c		c	
6	$HT_3 \to HT_{28} \to HT_{22} \to HT_{16} \to HT_{10}$	10	$HT_{23} \to HT_{13} \to HT_3 \to HT_{24} \to HT_{14}$
14	$HT_{12} \to HT_{29} \to HT_{15} \to HT_1 \to HT_{18}$	15	$HT_{17} \to HT_2 \to HT_{18} \to HT_3 \to HT_{19}$
20	$HT_{11} \to HT_{22} \to HT_2 \to HT_{13} \to HT_{24}$	23	$HT_{26} \to HT_3 \to HT_{11} \to HT_{19} \to HT_{27}$
26	$HT_{10} \to HT_{15} \to HT_{20} \to HT_{25} \to HT_{30}$	27	$HT_{15} \to HT_{19} \to HT_{23} \to HT_{27} \to HT_0$
28	$HT_{20} \to HT_{23} \to HT_{26} \to HT_{29} \to HT_1$	29	$HT_{25} \to HT_{27} \to HT_{29} \to HT_0 \to HT_2$
30	$HT_{30} \to HT_0 \to HT_1 \to HT_2 \to HT_3$		

Lemma 14.1. For a fixed table occupancy μ with $0 < \mu < 1$, fill an initially empty hash table of n cells by uniform a hashing and independently inserting a key into each cell with probability p where

$$0 \le p \le \frac{2+\mu}{3}$$

For some C, the number of progressions of occupied cells leading to any cell HT_i of any length ℓ and stride c satisfies

$$\frac{n}{1-p}\left(1-\frac{(C\log n)^{5/2}}{\sqrt{n}}\right) \overset{P}{\le} N_i \overset{P}{\le} \frac{n}{1-p}\left(1+\frac{(C\log n)^{5/2}}{\sqrt{n}}\right) \tag{14.3}$$

where the notation $\overset{P}{\le}$ indicates $\le$ is valid except for a polynomial small correction.[3]

Proof. As the sum of independent Bernoulli random variables, $N_i(\ell) = \sum_{j=0}^{\ell^2-1} N_i(\ell)[j]$ has the binomial distribution BIN$(s[j], p^\ell)$. Tight bounds for $N_i(\ell)$ as $n \to \infty$ are obtained using the following method.

Hoeffding's Bound [Hoeffding 1963, p. 15]. If the random variable X has the binomial distribution BIN(t, q) and $\beta \ge 0$, then

$$Pr\{X \le (q-\beta)t\} \le e^{-2t\beta^2} \quad Pr\{X \ge (q+\beta)t\} \le e^{-2t\beta^2} \tag{14.4}$$

Equations (14.5 through 14.7) are derived from Hoeffding's bounds in which C will be some large enough constant, $\ell \ge 2$ and

$$\beta_j = \sqrt{\frac{C\log n}{s(j)}}$$

First, equation (14.4) implies the following inequalities hold for $N_{i,\ell,j}$

[3]If X is a non-negative random variable, an equality $X = B(n)$ (or a bound, say $X \le B(n)$) holds except for a polynomial small correction (in n), provided $Pr\{X > B(n)\}$ (or $Pr\{X \ne B(n)\}$) is polynomially small (in n); that is, $n^r Pr\{X > B(n)\}$) or $n^r Pr\{X \ne B(n)\}$) is bounded as $n \to \infty$ for all positive r.

TABLE 14.4. A 12 Coloring of the Graph $\mathcal{G} \equiv \mathcal{G}(31, 4, 5)$

	c			c	
S[0]	1	$HT_{9} \to HT_{8} \to HT_{7} \to HT_{6} \to HT_{5}$	**S[1]**	6	$HT_{3} \to HT_{28} \to HT_{22} \to HT_{16} \to HT_{10}$
				5	$HT_{29} \to HT_{24} \to HT_{19} \to HT_{14} \to HT_{9}$
				19	$HT_{6} \to HT_{18} \to HT_{30} \to HT_{11} \to HT_{23}$
S[2]	10	$HT_{23} \to HT_{13} \to HT_{3} \to HT_{24} \to HT_{14}$	**S[3]**	14	$HT_{12} \to HT_{29} \to HT_{15} \to HT_{1} \to HT_{18}$
	7	$HT_{8} \to HT_{1} \to HT_{25} \to HT_{18} \to HT_{11}$		3	$HT_{19} \to HT_{16} \to HT_{13} \to HT_{10} \to HT_{7}$
	16	$HT_{22} \to HT_{6} \to HT_{21} \to HT_{5} \to HT_{20}$			
S[4]	15	$HT_{17} \to HT_{2} \to HT_{18} \to HT_{3} \to HT_{19}$	**S[5]**	20	$HT_{11} \to HT_{22} \to HT_{2} \to HT_{13} \to HT_{24}$
	2	$HT_{14} \to HT_{12} \to HT_{10} \to HT_{8} \to HT_{6}$		11	$HT_{28} \to HT_{17} \to HT_{6} \to HT_{26} \to HT_{15}$
	21	$HT_{16} \to HT_{26} \to HT_{5} \to HT_{15} \to HT_{25}$			
S[6]	23	$HT_{26} \to HT_{3} \to HT_{11} \to HT_{19} \to HT_{27}$	**S[7]**	26	$HT_{10} \to HT_{15} \to HT_{20} \to HT_{25} \to HT_{30}$
	4	$HT_{24} \to HT_{20} \to HT_{16} \to HT_{12} \to HT_{8}$		8	$HT_{13} \to HT_{5} \to HT_{28} \to HT_{20} \to HT_{12}$
	24	$HT_{0} \to HT_{7} \to HT_{14} \to HT_{21} \to HT_{28}$		18	$HT_{1} \to HT_{14} \to HT_{27} \to HT_{9} \to HT_{22}$
S[8]	27	$HT_{15} \to HT_{19} \to HT_{23} \to HT_{27} \to HT_{0}$	**S[9]**	28	$HT_{20} \to HT_{23} \to HT_{26} \to HT_{29} \to HT_{1}$
	12	$HT_{2} \to HT_{21} \to HT_{9} \to HT_{28} \to HT_{16}$		9	$HT_{18} \to HT_{9} \to HT_{0} \to HT_{22} \to HT_{8}$
	13	$HT_{7} \to HT_{25} \to HT_{12} \to HT_{30} \to HT_{17}$		17	$HT_{27} \to HT_{10} \to HT_{24} \to HT_{7} \to HT_{21}$
S[10]	29	$HT_{25} \to HT_{27} \to HT_{29} \to HT_{0} \to HT_{2}$	**S[11]**	30	$HT_{30} \to HT_{0} \to HT_{1} \to HT_{2} \to HT_{3}$
	22	$HT_{21} \to HT_{30} \to HT_{8} \to HT_{17} \to HT_{26}$		25	$HT_{5} \to HT_{11} \to HT_{17} \to HT_{23} \to HT_{29}$

$$1-e^{-2Cn\log n} \leq Pr\{N_i(\ell)[j] \leq (p^\ell + \beta_j)s(j)\} \leq 1 \tag{14.5a}$$

$$1-e^{-2Cn\log n} \leq Pr\{N_i(\ell)[j] \geq (p^\ell - \beta_j)s(j)\} \leq 1 \tag{14.5b}$$

Because $\beta_j \to \infty$ as $n \to \infty$, equations (14.5*a* and 14.5*b*) are equivalent to the statements

$$N_i(\ell)[j] \overset{P}{\leq} (p^\ell + \beta_j)s[j] \quad N_i(\ell)[j] \overset{P}{\geq} (p^\ell - \beta_j)s[j] \tag{14.6}$$

Summing equations (14.6) over j and noting $n = \sum_{j=0}^{\ell^2-1} s[j]$ gives the upper bound for $N_i(\ell)$

$$N_i(\ell) = \sum_{j=0}^{\ell^2-1} N_i(\ell)[j] \overset{P}{\leq} \sum_{j=0}^{\ell^2-1} (p^\ell + \beta_j)s[j] \overset{P}{\leq} p^\ell n + \sqrt{C\log n}\sum_{j=0}^{\ell^2-1}\sqrt{s(j)} \tag{14.7a}$$

and the lower bound for $N_i(\ell)$

$$N_i(\ell) = \sum_{j=0}^{\ell^2-1} N_i(\ell)[j] \overset{P}{\geq} \sum_{j=0}^{\ell^2-1} (p^\ell - \beta_j)s[j] \overset{P}{\geq} p^\ell n - \sqrt{C\log n}\sum_{j=0}^{\ell^2-1}\sqrt{s(j)} \tag{14.7b}$$

The term involving $\sqrt{\ldots}$ in the summation in equations (14.7*a* and 14.7*b*) may both be bounded in the correct needed direction using the concavity[4] of the $\sqrt{x}$ function

$$\sum_{j=0}^{\ell^2-1}\sqrt{s[j]} = \ell^2 \sum_{j=0}^{\ell^2-1}\frac{\sqrt{s[j]}}{\ell^2} \leq \ell^2\sqrt{\sum_{j=0}^{\ell^2-1}\frac{s[j]}{\ell^2}} \leq \ell\sqrt{n}$$

This replaces (14.7*a* and 14.7*b*) by the upper and lower bounds

$$N_i(\ell) \overset{P}{\leq} p^\ell n + \ell\sqrt{Cn\log n} \quad N_i(\ell) \overset{P}{\geq} p^\ell n - \ell\sqrt{Cn\log n} \tag{14.8}$$

The condition $\ell \geq 2$ is unnecessary; First, the initial condition $N_{i,0} = n - 1$ show that the bounds in equation (14.8) certainly remain true if $\ell = 0$. Next, $N_i(1)$ is the number of filled cells other than HT_i so that Hoeffding's bound also validates the bounds in equations (14.8) for $\ell = 1$.

The condition $\ell \leq \kappa = C\log n$ is also unnecessary; if $\ell > \kappa = C\log n$, then $p^\ell < p^\kappa < e^{C\log n \log p} = n^{C\, n\log p}$. Because $0 \leq p \leq \frac{2+\mu}{3}$ and $\mu < 1$, then $C\log p < 0$, which implies $p < 1$ so that p^ℓ is polynomial small and $N_i(\ell) \overset{P}{=} 0$ for $\ell > \kappa$.

This allows us to write

$$N_i \overset{P}{=} \sum_{\ell=0}^{\kappa} N_i(\ell) \overset{P}{\leq} \sum_{\ell=0}^{\kappa}\left(p^\ell n + \ell\sqrt{Cn\log n}\right) \leq \frac{n}{1-p} + \frac{\kappa(\kappa+1)}{2}\sqrt{Cn\log n}$$

$$\overset{P}{\leq} \frac{n}{1-p} + (C'\log n)^{5/2}\sqrt{n} \overset{P}{\leq} \frac{n}{1-p}\left(1 + \frac{(C'\log n)^{5/2}}{\sqrt{n}}\right) \tag{14.9a}$$

[4] $f(x)$ is concave if $\alpha f(x) + (1-\alpha) f(y) \geq f(\alpha x + (1-\alpha)y)$ for $0 \leq \alpha \leq 1$. If the components of the vector $\underline{\alpha} = (\alpha_0, \alpha_1 \cdots)$ are non-negative with $1 = \sum_j \alpha_j$, then $\sum_j \alpha_j f(x_j) \geq f\left(\sum_j \alpha_j x_j\right)$ for concave function. A function f with a continuous second derivative is concave if and only if $f'' \leq 0$.

where C' is another constant. Similarly, the lower bound

$$N_i = \sum_{\ell=0}^{\kappa} N_i(\ell) \overset{P}{\ge} \sum_{\ell=0}^{\kappa} \left(p^\ell n - \ell\sqrt{Cn\log n}\right) \overset{P}{\ge} \frac{n}{1-p} - \frac{\kappa(\kappa+1)}{2}\sqrt{Cn\log n}$$

$$\overset{P}{\ge} \frac{n}{1-p} - (C''\log n)^{5/2}\sqrt{n} \overset{P}{\ge} \frac{n}{1-p}\left(1 - \frac{(C''\log n)^{5/2}}{\sqrt{n}}\right) \qquad (14.9b)$$

where C'' is still another constant, from the inexhaustible supply of constants made available to computer scientists in their O and o-arguments. This completes the proof of Lemma 14.1. ■

14.4 DOMINANCE

The idea of hash-table dominance is used in [Lueker and Molodowitch 1993] to compare the probability of the number of probes to insert a key by double hashing in two table states; they write $\underline{x}_1 \succ \underline{x}_2$, read $\underline{x}_1$ dominates $\underline{x}_2$ if $\underline{x}_1 \ge \underline{x}_2$ point wise as vectors.

For example, $\underline{x}_1 = (1, 1, 1, 0, 0, 0, 0, 1) \ge \underline{x}_2 = (0, 1, 1, 0, 0, 0, 0, 1)$.

Dominance is preserved under DH insertion; if $\underline{x}_1 \succ \underline{x}_2$ and a new key k is inserted into each of the two tables changing their states $\underline{x}_1 \xrightarrow{\text{DH}} \underline{x}_1'$ and $\underline{x}_2 \xrightarrow{\text{DH}} \underline{x}_2'$, then dominance is preserved $\underline{x}_1' \succ \underline{x}_2'$.

In particular, if we start with a hash table with state x_2 and insert in some cell a fictitious or Ghost key[5] k obtaining the state $\text{INSERT}_{\text{Ghost}}: \underline{x}_2 \to \underline{x}_1$, then the following is true:

- $\underline{x}_1 \succ \underline{x}_2$, and
- The DH insertion of a real key k into both the hash table $\underline{x}_2 \xrightarrow{\text{DH}} \underline{x}_2'$ and into the Ghost hash table $\underline{x}_1 \xrightarrow{\text{DH}} \underline{x}_1'$ preserves dominance $\underline{x}_1' \succ \underline{x}_2'$, although k might be inserted into a different cell of $\underline{x}_1'$ because of the Ghost-key in $\underline{x}_1$.

As a consequence of dominance,

- If keys fictitiously inserted into $\underline{x}_1$ creating the state $\underline{x}_1 \equiv \text{INSERT}_{\text{Ghost}}(\underline{x}_2)$ so that $\underline{x}_1 \succ \underline{x}_2$, and
- Corresponding additional real insertions are made to both the fictitious $\underline{x}_1$ and to $\underline{x}_2$

then the fictitious insertions will never cause some cell in $\underline{x}_1$ to remain **un**occupied if it would have been otherwise filled in $\underline{x}_2$.

Exercise 14.3. (Monotonicity of $\succ$). Let $N_i^{(j)}$ be the number of chains of any length or stride leading to cell HT_i in the hash table with state $\underline{x}^{(j)}$. Prove that $\underline{x}^{(1)} \succ \underline{x}^{(2)}$ implies $N_i^{(1)} \ge N_i^{(2)}$.

[5]The ghost of Christmas past appears in Charles Dickens' *A Christmas Carol*. Central to the comparison between uniform and double hashing in [Lueker and Molodowitch 1993] is the procedure `UsuallyDoubleHash`, which refers to the keys inserted into the hash table it creates as either red or green keys.

The monotonicity of $\succ$ is central to the comparison of uniform and double hashing and we begin with the following Lemma.

Lemma 14.2. For a fixed hash-table occupancy $\mu \in (0,1)$, create the hash-table state $\underline{x}$ by inserting $m = \nu n$ keys into an initially empty hash table of n cells by choosing their m cells independently using the uniform distribution with probability ν satisfying

$$0 \leq \nu \leq \frac{1+\mu}{2}$$

The number of progressions of occupied cells leading to cell HT_i of any length ℓ and any stride c satisfies

$$\left(1-\frac{(C\log n)^{5/2}}{\sqrt{n}}\right) \stackrel{P}{\leq} N_i \stackrel{P}{\leq} \frac{n}{1-\nu}\left(1+\frac{(C\log n)^{5/2}}{\sqrt{n}}\right) \tag{14.11}$$

Proof. If

$$p = \nu\left(1+\frac{\log n}{\sqrt{n}}\right),$$

$$\text{then } 0 \leq \nu \leq \frac{1+\mu}{2} \text{ implies}$$

$$0 \leq p \leq \frac{2+\mu}{3}$$

for sufficiently large n. Therefore, if an initially empty hash table of n cells is filled by uniform hashing with probability

$$p = \nu\left(1+\frac{\log n}{\sqrt{n}}\right)$$

the random number of cells filled will be $pn = \nu n\left(1+\frac{\log n}{\sqrt{n}}\right) > \nu n$ [c–polysmall]. The monotonicity of $\succ$ and Lemma 14.1 completes the proof. ■

Dominance that provides a way to compare double hashing and uniform hashing if the *Ghost* occur insertions infrequently will be explored in the following sections.

14.5 INSERTION-COST BOUNDS RELATING UNIFORM AND DOUBLE HASHING

Notations

1. μn with $\mu \in (0,1)$ is the integer defined by $\mu = \lfloor \mu n \rfloor / n$.
2. $Y_{m,n}^{(\mathrm{UH})}$ is the number of probes to insert the *next* key into the hash table of n cells created by the insertion of m keys by uniform hashing; $P_{m,n}^{(\mathrm{UH})}(r) = Pr\{Y_{m,n}^{(\mathrm{UH})} \geq r\}$.

3. $Y_{m,n}^{(DH)}$ is the number of probes to insert the *next* key into the hash table of n cells created by the insertion of m keys by double hashing; $P_{m,n}^{(DH)}(r) = Pr\{Y_{m,n}^{(DH)} \geq r\}$.

The equality of the asymptotic expected insertion cost for double hashing and uniform hashing is a consequence of the following theorem:

Theorem 14.3. [Lueker and Molodowitch 1993]. For a hash-table occupancy $\mu \in (0, 1)$

$$P_{\mu n,n}^{(DH)}(r) \leq P_{(1+\delta)\mu n,n}^{(UH)}(r) + O(n^{-b}) \quad b \geq 1, n \to \infty \tag{14.12a}$$

$$P_{\mu n,n}^{(DH)}(r) \leq P_{(1-\delta)\mu n,n}^{(UH)}(r) - O(n^{-b}) \quad b \geq 1, n \to \infty \tag{14.12b}$$

$$\delta = Cn^{-1/2}(\log n)^{5/2} \tag{14.13}$$

for a sufficiently large constant C depending on b.

Equations (14.12 and 14.3) yield

Corollary 14.4. [Lueker and Molodowitch 1993]. For double hashing with table occupancy $\mu \in (0, 1)$

$$C_{\mu n,n}^{(DH)} = \frac{1}{1-\mu} + O\left(n^{-1/2}(\log n)^{5/2}\right) \quad n \to \infty \tag{14.14}$$

14.5.1 Outline of the Proof of Equation (14.12*a*)

Write $\texttt{INSERT} \xrightarrow{A} \underline{x}$ to mean hashing procedure A has created the hash-table; examples of A include the following:

- Uniform hashing (UH)
- Double hashing (DH)
- `UsuallyDoubleHash` (UDH) (§14.6), and
- `UsuallyDoubleHash′` (UDH′) (§14.9).

1. The procedure `UsuallyDoubleHash` (UDH) with parameters (m, n, δ) creates the hash table of n cells with state $\texttt{INSERT} \xrightarrow{UDH} \underline{x}^{(UDH)}$ by attempting to insert $m = \mu n$ original keys.

   For large enough n, UDH
   - Very frequently (with probability ≈ 1) inserts an original key (colored green) using double hashing,
   - Very infrequently (with probability ≈ 0) inserts a Ghost key (colored red).
   - UDH succeeds, if all m original keys are inserted;
   - UDH fails ($\mathcal{F}$), if UDH does *not* insert the m original keys.
2. Lemma 14.5: The state $\underline{x}^{(UDH)}$ is created as if UDH filled the n cells by inserting keys by uniform-hashing.
3. Lemma 14.6: If n is large enough, $P_{(1+2\delta)\mu n,n}^{(UH)}(r) = P_{\mu n,n}^{(DH)}(r)$.
4. Lemma 14.7: If $\mathcal{F}$ does *not* occur, then UDH inserts $m = \mu n$ *green* keys and therefore $P_{\mu n,n}^{(DH)}(r) \geq P_{\mu n,n}^{(UDH)}(r)$.

5. Lemma 14.8: $Pr\{\mathcal{F}\}$ is c-polysmall.
   Steps 1 through 5 give equation (14.12a). ■

Outline of the Proof of Equation (14.12*b*)

1′. Using the procedure `UsuallyDoubleHash′` (UDH′) with parameters (m, n, δ), creates a hash table of n cells with state $\texttt{INSERT}\xrightarrow{\text{UDH}'}\underline{x}^{(\text{UDH}')}$ by inserting $\leq m = \mu n$ original keys.
   For large n, UDH′
   - Always inserts a key using double hashing, but
   - Very infrequently (with probability $p_{\text{RJ}} \sim 0$), rejects the insertion and discards the key.

2′. Lemma 14.9: The event that "$\geq 2\delta\mu n$ rejections occur in UDH′" is [c-polysmall].

3′. Lemma 14.10: The state $\underline{x}^{(\text{UDH}')}$ results as if UDH′ filled the n cells by inserting keys using uniform hashing.

4′. Lemma 14.11: $P^{(\text{UH})}_{(1-2\delta)\mu n,n}(r) \overset{P}{\leq} P^{(\text{DH})}_{\mu n,n}(r)$.
   Steps 1′ through 4′ gives equation (14.12b). ■

14.6 USUALLYDOUBLEHASH

`UsuallyDoubleHash` (UDH) is a randomized mixture of uniform and double hashing; it creates the state $\underline{x}^{(\text{UDH})}$ in a hash table of n cells by inserting the following:

- $g \leq m \approx \mu n$ of the m original keys $k_0, k_1, \cdots, k_{m-1}$, colored green, and
- r Ghost keys, colored red.

procedure `UsuallyDoubleHash`

Parameters: m, the number of green keys to be inserted;
n, the number of hash-table cells;
$\delta = \dfrac{(C \log n)^{5/3}}{\sqrt{n}}$, with C a constant to be specified

UDH state variables: Current table state $\underline{x}$;
g, the current number of green keys inserted in table state $\underline{x}$;
f, the current number of table slots filled in table state x.

Initialization: $g = f = 0$.

begin /*UDH*/
  while $0 \leq g < m$ and $0 \leq f < \lfloor(1 + 2\delta)m\rfloor$) **do**
    begin /***while**-loop*/▶
      if $P_j(\underline{x}) > \dfrac{1+\delta}{n-f}$ for at least one unoccupied cell HT_j, **then**
        exit this **while**-loop
      else

$$\mathit{UsualCase}.\ P_j(\underline{x})\begin{cases} = 0 & \text{if } \mathrm{HT}_j \text{ is occupied} \\ \le \dfrac{1+\delta}{n-f} & \text{if } \mathrm{HT}_j \text{ is } \underline{\text{un}}\text{occupied} \end{cases}$$

Randomize [FLIP]. Toss a coin; probability of heads $\frac{1}{1+\delta}$;

Outcome. *heads*;

- Update $g \to g + 1$, and
- Insert the original key k_g using double hashing coloring it green into the unoccupied cell HT_i selected with probability $P_i(\underline{x})$ where
$$\begin{cases} 0 \le P_i(\underline{x}) & (0 \le j < n) \\ 1 = \sum_{0 \le j < n} \chi_{\{x_j=0\}} P_i(\underline{x}) \end{cases}$$

Outcome: *tails*;

- Insert a key coloring it red into the unoccupied cell HT_i selected with probability $Q_i(\underline{x}) \equiv \delta^{-1}\left(\frac{1+\delta}{n-f} - P_i(\underline{x})\right)$ where
$$\begin{cases} 0 \le Q_j(\underline{x}) & (0 \le j < n) \\ 1 = \sum_{0 \le j < n} \chi_{\{x_j=0\}} Q_i(\underline{x}) \end{cases}$$

update $f : f \to f + 1$

end/***while**-loop*/ ▶

VeryUnlikelyCase. $P_j(\underline{x}) > \frac{1+\delta}{n-f}$ for at least one unoccupied cells $\{\mathrm{HT}_j\}$.

while $f < \lfloor (1 + 2\delta)m \rfloor$ **do**

insert a key according to uniform hashing coloring it *red*.

end; /*UDH*/

$P_i(x)$ in UDH is defined as in equation (14.2*c*); x indicates the presumed dependence in UDH on the current table state.

$$P_i(\underline{x}) = \frac{1}{n(n-1)} \sum_{\ell=0}^{n-1} N_{i,\ell}(\underline{x})$$

$$N_{i,\ell}(\underline{x}) = \sum_{c=1}^{n-1} \chi_{\{E_i(\ell,c)\}}(\underline{x})$$

14.7 THE UDH CHANCE EXPERIMENT AND THE COST TO INSERT THE NEXT KEY BY DOUBLE HASHING

- $Y_{m,n}(\underline{x}^{(\mathrm{UDH})})$ denotes the number of probes needed to insert the next key k by double hashing into the hash table of n cells whose state $\underline{x}^{(\mathrm{UDH})}$ was created by the insertion of keys in `UsuallyDoubleHash` (UDH); $P_{m,n}^{(\mathrm{UDH})}(r) = Pr\{Y_{m,n}^{(\mathrm{UDH})}(\underline{x}^{(\mathrm{UDH})} \ge r\}$.
- $Y_{m,n}(\underline{x}^{(\mathrm{UDH}')})$ denotes the number of probes needed to insert the next key k by double hashing into the hash table of n cells whose state $\underline{x}^{(\mathrm{UDH}')}$ was created by the insertion of keys in `UsuallyDoubleHash'` (UDH′) and $P_{m,n}^{(\mathrm{UDH}')}(r) = Pr\{Y_{m,n}^{(\mathrm{UDH}')}(\underline{x}^{(\mathrm{UDH}')} \ge r\}$.

Lueker and Molodowitch use $P_{m,n}^{(\mathrm{UDH})}(r)$ and $P_{m,n}^{(\mathrm{UDH}')}(r)$—computed for the hash table states $\underline{x}^{(\mathrm{UDH})}$, $\underline{x}^{(\mathrm{UDH}')}$ are used to bound $P_{m,n}^{(\mathrm{UH})}(r)$ and $P_{m,n}^{(\mathrm{DH})}(r)$.

The output of UDH is the result of a chance experiment $\mathcal{E}^{(\mathrm{UDH})}$ with state variables f, g initialized by $f = g = 0$; during the fth step, a key is inserted and $f \to f + 1$.

- If the color of the key inserted during the fth step is green, then $g \to g + 1$.
- If the color of the key inserted during the fth step is red, then g is unchanged.

During the evolution of UDH, either

(A) UDH begins, remains and ends *entirely* in the
- *UsualCase*
  * The state variable f increases through the sequence of values $f = 0, 1, \cdots, f^* - 1$.
  * The table state $\underline{x}$ changes as a result of the insertion of F keys
    - The color of a key inserted is either *green* or *red*.
    - The cell into which the key is inserted depends on its color and table state $\underline{x}$

    Both of the previous items determined by randomization using probabilities $P_i(\underline{x})$ and $Q_i(\underline{x})$.

(B) UDH begins in the following situations:
- *UsualCase*
  * The state variable f increases through the sequence of values $f = 0, 1, \cdots, F - 1$.
  * The table state $\underline{x}$ changes as a result of the insertion of F keys in item (**A**), and thereafter, UDH enters and remains until its termination in the next case.
- *VeryUnlikelyCase*
  * The state variable f increases through the sequence of values $f = F, F + 1, \cdots, f^* - 1$.
  * The table state $\underline{x}$ changes as a result of the insertion of an additional $f^* - F$ keys:
    - The colors of these $f^* - F$ inserted keys are *red*.
    - The cells into which they are inserted are determined by uniform hashing.

We define the chance experiment $\mathcal{E}^{(\mathrm{UDH})}$ by explaining the random variables involved in its outcome.

Flip: $(T_0, T_1, \cdots, T_{f^*-1})$ tosses of a coin; $T_f = \begin{cases} 1 & \text{Flip outcome is heads} \\ 0 & \text{Flip outcome is tails} \end{cases}$
$\{T_f\}$ are i.i.d. with

$$Pr\{T_j = 1\} = \frac{1}{1+\delta}$$

G-INSERT: $H_0, H_1, \cdots, H_{f^*-1}$

The random double-hashing values for the insertion of *green* keys
$\{H_f\}$ are independent and identically distributed random variables [equation (14.1)].
R-INSERT: $U_0, U_1, \cdots, U_{f^*-1}$
Selecting cells for the insertion of red keys using the uniform distribution occurs as follows:

a) If $T_f = 1$ and $g < m$, then H_f is used to insert the green key by double hashing.
b) If $T_f = 0$, then use U_f to insert the key by uniform hashing and color it red.

UDH terminates in one of three ways:
$\mathcal{N}$: $g = m, f < \lfloor 2(1 + \delta)m \rfloor$.
$\mathcal{F}_1$: $g < m$, $f \leq \lfloor 2(1 + \delta)m \rfloor$ and *VeryUnlikelyCase is* entered before termination.
$\mathcal{F}_2$: $g < m$, $f \leq \lfloor 2(1 + \delta)m \rfloor$ and *VeryUnlikelyCase* is not entered before termination.

14.8 PROOF OF EQUATION (14.12*a*)

Figure 14.1 is a graphical representation of the possible outcomes of `UsuallyDoubleHash` in the *Usual Case.*

Lemma 14.5. In the *UsualCase*, the probability that a key (green or red) is inserted into cell HT_i in state $\underline{x}$ is $P_i^{(\text{UDH})}(\underline{x}) = \chi_{\{x_i=0\}} \dfrac{1}{n-f}$; equivalently, if the color of a key inserted is ignored, UDH is filling cells in the *UsualCase* using uniform-hashing.

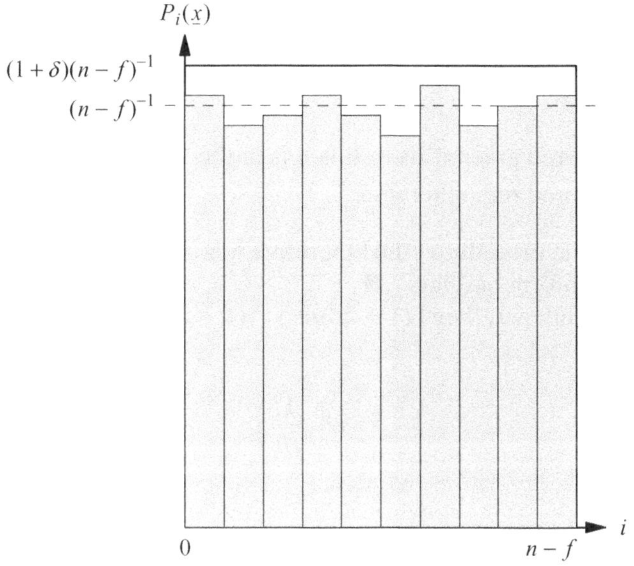

Figure 14.1. Outcomes of `UsuallyDoubleHash`.

Proof. The probability $P_i^{(\text{UDH})}(\underline{x})$ that a key (green or red) is inserted in cell HT_i in the *UsualCase* in state $\underline{x}$ is the sum of the following two terms.

- The probability that the outcome of the coin toss is heads

$$\frac{1}{1+\delta}$$

multiplied by the probability $P_i(\underline{x})$ that HT_i is selected as the cell into which the green key is inserted

- The probability that the outcome of the coin toss is tails

$$\frac{\delta}{1+\delta}$$

multiplied by the probability $Q_i(\underline{x})$ that HT_i i is selected as the cell into which the red key is inserted.

$$P_i^{(\text{UDH})}(\underline{x}) = \underbrace{\chi_{\{x_i=0\}} \frac{1}{1+\delta} P_i(\underline{x})}_{\text{green}} + \underbrace{\chi_{\{x_i=0\}} \frac{\delta}{1+\delta} Q_i(\underline{x})}_{\text{red}}$$

$$= \chi_{\{x_i=0\}} \left(\frac{1}{1+\delta} P_i(\underline{x}) + \frac{\delta}{1+\delta} \delta^{-1} \left\{ \frac{1+\delta}{n-f} - P_i(\underline{x}) \right\} \right)$$

yielding

$$P_i^{(\text{UDH})}(\underline{x}) = \chi_{\{x_i=0\}} \frac{1}{n-f} \tag{14.15}$$

Combining Figure 14.1 and equation (14.15) we conclude that the insertion of a key in the *UsualCase* in state $\underline{x}$ with variable (n, f) is equivalent to choosing a point with $1 \le i \le n - f$ in Figure 14.1

- The key is colored green if its ordinate is in the shaded region.
- The key is colored red, otherwise.

In the *UsualCase*, the procedure UDH therefore fills cells in which keys are inserted as if performing uniform hashing. ■

If $m = \mu n$ is an integer, then $\lfloor (1 + 2\delta)\mu n \rfloor = (1 + 2\delta)\mu n$ is also an integer for n because

$$\delta = O\left(\frac{1}{\sqrt{n}}\right)$$

Lemma 14.6. If $m = \mu n$ is an integer, then for a large enough n, UDH inserts $(1 + 2\delta)\mu n$ keys by uniform hashing into a hash table of n cells creating the table state $\underline{x}^{(\text{UH})}$ and

$$P_{\mu n,n}^{(\mathrm{UDH})}(r) = P_{(1+2\delta)\mu n,n}^{(\mathrm{UH})}(r) \tag{14.16}$$

where $Y_{(1+2\delta)\mu n,n}^{(\mathrm{UH})}(\underline{x}^{(\mathrm{UH})})$ is the cost to insert the next key in the hash table with state $\underline{x}^{(\mathrm{UH})}$ and

$$P_{(1+2\delta)\mu n,n}^{(\mathrm{UH})}(r) = Pr\{Y_{(1+2\delta)\mu n,n}^{(\mathrm{UH})}(\underline{x}) \geq r\}$$

Proof. Lemma 14.5 implies the *UsualCase* in UDH may be replaced by *UsualCaseVariant*: In state (f, g)

i) Choose the next hash-table cell HT_i to be filled according to the uniform probability $\frac{1}{n-f}$.

ii) With probability

$$\frac{\frac{P_i(\underline{x})}{1+\delta}}{n-f}$$

color the inserted key green updating $g \to g + 1$.

iii) Update $f \to f + 1$.

If UDH is modified in this way, the colors of the keys are essentially ignored, and when $m = \mu n$ is an integer, UDH implements uniform hashing inserting $(1 + 2\delta)m$ keys into a hash table of n cells. *Exercise 14.1* yields equation (14.16). ■

Lemma 14.7. If UDH does not fail, then all the $m = \mu n$ green (or original) keys $k_0, k_1, \cdots, k_{m-1}$ are inserted, along with some number of additional (red)keys.
Dominance implies $Y_{\mu n,n}^{(\mathrm{DH})}(\underline{x}^{(\mathrm{UDH})}) \leq Y_{\mu n,n}^{(\mathrm{UDH})}(\underline{x}^{(\mathrm{UDH})})$. ■

Lemma 14.8. The event $\{\mathcal{F}\}$ is polynomial small.

Proof. If *VeryUnlikelyCase* is entered when exiting UDH, then for some cell HT_j we have

$$P_j(\underline{x}) > \frac{1+\delta}{n-f}$$

which implies

$$N_j(\underline{x}) > (1+\delta)\frac{n-1}{1-\nu_f}$$

with $f = n\nu_f$ which is a small polynomial.

$\mathcal{F}_2$: If the *VeryUnlikelyCase* was *not* entered when exiting `UusuallyDoubleHash`, then

- The state counters (f, g) after the exit to satisfy $g \leq m - 1$ and $f = \lfloor(1 + 2\delta)m\rfloor$ so that
- The FLIP outcome heads occurred fewer than m times in $(1 + 2\delta)m$ Bernoulli trials
- The probability of heads is $\frac{1}{1+\delta}$.

The previous event is also a small polynomial by Hoeffding's bound. ■

14.9 USUALLYDOUBLEHASH′

`UsuallyDoubleHash′` (UDH′) is a procedure with parameters (m, n, δ) with $m \approx \mu n$ which creates the state $\underline{x}^{(\mathrm{UDH}')}$ of a hash-table of n cells as follows:
for $g = 0, 1, \cdots, m - 1$

- Inserts an original key using double hashing.
- Using a state-dependent rejection probability, rejects the insertion and deletes the just inserted key.

procedure `UsuallyDoubleHash′`
begin
Parameters: m, the number of green keys to be inserted;
n, the number of hash-table cells;
$\delta = \frac{(C\log n)^{5/3}}{\sqrt{n}}$, with C a constant to be specifieds
State Counters: $\underline{x}$, current table state;
g, the number of green keys inserted in table state $\underline{x}$;
f, number of table slots filled in table state $\underline{x}$.
Initialization: $g = f = 0$.
while $0 \leq g < m$ **do**
begin /***while**-loop*/
if an empty cell HT_j exists for which $\left|(n-f)P_j^{(\mathrm{UDH}')}(\underline{x})-1\right| > \delta$
then *VeryUnlikelyCase′*: report failure and **return**.
else *UsualCase′*
begin /* *UsualCase′**/
1. Insert a key k according to double hashing in cell HT_i with probability $P_i(\underline{x})$.
2. *Randomize* [Flip]: Toss a coin [probability of heads $p_{RJ}(i) = 1 - \frac{1-\delta}{n-f}\frac{1}{P_i(\underline{x})}$
a) **If** heads, **then** discard k making cell HT_i empty.
b) **If** tails, **then** update: $f \to f + 1$.
3. update: $g \to g + 1$;

end /* *UsualCase′**/
end /***while**-loop*/

$P_i(\underline{x})$ is defined as in equation (14.2*c*) where $\underline{x}$ is the current state (f, g) in UDH′.

$$P_i(\underline{x}) = \frac{1}{n(n-1)} \sum_{\ell=0}^{n-1} N_{i,\ell}(\underline{x})$$

$$N_{i,\ell}(\underline{x}) = \sum_{c=1}^{n-1} \chi_{\{E_i(\ell,c)\}}(\underline{x})$$

14.10 PROOF OF EQUATION (14.12*b*)

Lemma 14.9.

a) $\sum_{i=0}^{n-1} \chi_{\{x_i=0\}} P_i(\underline{x}) = 1$

b) The *Usual Case′* is entered whenever for *at least* one empty cell HT_j

$$|(n-f)P_j(\underline{x})-1| > \delta \Leftrightarrow 0 \le P_j(\underline{x}) < \frac{1-\delta}{n-f} \text{ or } \frac{1+\delta}{n-f} < P_j(\underline{x}) \le 1$$

$$|(n-f)P_j(\underline{x})-1| \le \delta \Leftrightarrow \frac{1-\delta}{n-f} \le P_j(\underline{x}) \le \frac{1+\delta}{n-f}.$$

c) $p_{RJ}(j)$ satisfies $0 \le p_{RJ}(j) \le \frac{2\delta}{1-\delta}$

d) The random number $N_{\mu n,n}^{(\text{UDH}')}$ of keys inserted in the hash table with state $\underline{x}^{(\text{UDH}')}$ satisfies $N_{\mu n,n}^{(\text{UDH}')} \ge (1-2\delta)\mu n$ [*c*–polysmall].

Lemma 14.10.

a) If the counter values in UDH′ are (f, g), then the probability that cell HT_i
   - is selected in Step 1 and
   - the key is rejected in Step 3

   is

$$\chi_{\{x_i=0\}} \frac{1-\delta}{n-f} = \underbrace{\chi_{\{x_i=0\}} P_i(\underline{x})}_{\text{Select HT}_j} \times \underbrace{(1-p_{RJ}(i))}_{\text{No Rejection}} \tag{14.17a}$$

b) If the counter values are (f, g), the probability that cell HT_i is unoccupied is

$$P_i(\underline{x}) = \chi_{\{x_i=0\}} \frac{1-\delta}{n-f} \tag{14.17b}$$

c) Equivalently, the state $\texttt{INSERT} \xrightarrow{\text{UDH}'} \underline{x}^{(\text{UDH}')}$ is as if UDH′ inserted keys by uniform hashing.

Lemma 14.11. If state of the hash-table created by UDH′ contains $(1-2\delta)\mu n$ keys and $Y^{(\text{DH})}(\underline{x}^{(\text{UDH}')})$ denotes the cost to insert the next key using double hashing into the hash-table with state $\underline{x}^{(\text{UDH}')}$, then

$$P_{\mu n,n}^{(\mathrm{DH})}(r) \geq P^{(\mathrm{DH})}(r) \tag{14.18a}$$

$$P_{(1-2\delta)\mu n,n}^{(\mathrm{UH})}(r) \overset{P}{=} P^{(\mathrm{DH})}(r) \tag{14.18b}$$

where $P^{(\mathrm{DH})}(r) = Pr\{Y^{(\mathrm{DH})}(\underline{x}^{(\mathrm{UDH}')}) \geq r\}$.

Proof. The table state $\underline{x}^{(\mathrm{UDH}')}$ contains the $(1 - 2\delta)\mu n$ original keys that were not deleted; equation (14.18*a*) follows from dominance and equation (14.18*b*) is a consequence of *Exercise 14.1* and Lemma 14.9(d). ■

REFERENCES

J. R. Bell, and C. H. Kaman, "The Linear Quotient Hash Code", *Communications of the ACM*, **13**, #11, pp. 675–677, 1970.

Shimon Even, *Graph Algorithms*, Computer Science Press (Maryland), 1979.

L. J. Guibas and E. Szemeredi, "The Analysis of Double Hashing", *Journal of Computer and System Sciences*, **16**, pp. 226–274, 1978.

Wassily Hoeffding, "Probability Inequalities for Sums of Bounded Random Variables", *Journal of the American Statistical Association*, **58**, #301. (March, 1963), pp. 13–30.

G. S. Lueker and M. Molodowitch, "More Analysis of Double Hashing", *Combinatorica*, **13**, #1, pp. 83–96, 1993.

W. D. Maurer, "An Improved Hash Code for Scatter Storage", *Communications of the ACM*, **11**, #1, pp. 35–38, 1968.

C. E. Radke, "The Use of Quadratic Residue Research", *Communications of the ACM*, **13**, #3, pp. 103–105, 1970.

W. T. Tutte, *Graph Theory*, Addison-Wesley (California), 1984.

D. J. A. Welsh, *Matroid Theory*, Academic Press (New York), 1976.

A. C. Yao, "Uniform Hashing is Optimal", *Communications of the ACM*, **32**, #3, pp. 687–693, 1985.

CHAPTER 15

Optimum Hashing

15.1 THE ULLMAN–YAO FRAMEWORK[1]

Ullman [Ullman 1972] conjectured that uniform hashing minimizes the expected SEARCH-cost in the class of open-addressing schema. A proof of his conjecture appeared 13 years later in Yao's paper [Yao 1985], an exposition of which is the subject of this chapter.

15.1.1 The Ullman–Yao Hashing Functions

The hashing functions h in Chapters 10 to 13 were modeled as integer-valued mappings $h: \mathcal{K} \to \mathcal{Z}_n$ from the set of keys $\mathcal{K}$ to hash-table (HT) cells HT_i with $i \in \mathcal{Z}_n$. Double hashing (DH) (Chapter 14) extended the definition of a hashing function $h \to (h_1, h_2)$; the domain remained unchanged, but the range $h(k)$ specified a probe sequence, which is a sequence of cells generated by an arithmetic progression. If $(h_1(k), h_2(k)) = (i, c)$, then

$$\text{HT}_i, \text{HT}_{i-c}, \text{HT}_{i-2c}, \cdots$$

- The choice of the initial cell HT_i into which an attempt to insert the k would be made.
- In the event of a collision, the order that the cells should be probed thereafter.

Ullman made another extension in [Ullman 1972] by defining h as a mapping h: $\mathcal{K} \to \mathcal{H}_n$ from keys $\mathcal{K}$ to probe sequences $\mathcal{H}_n$; accordingly, $h = (h_0, h_1, \cdots, h_{n-1})$ is a permutation the integers in $\mathcal{Z}_n$ and $\mathcal{H}_n$ is some subset of $\mathcal{P}_n$, the set of all permutations.

- The probe sequence for $k \in \mathcal{K}$ generated by h is the n-tuple of hash-table cells $h \equiv h(k) \equiv (h_0(k), h_1(k), \cdots, h_{n-1}(k)) \in_{(i)} \mathcal{P} \subseteq \mathcal{H}_n$.
- The probe sequences for m keys $\underline{k} = (k_0, k_1, \cdots, k_{m-1})$ generated by $h \in \mathcal{H}_n$ is an m-tuple of permutations $h(k_0, k_1, \cdots, k_{m-1}) = (h(k_0), h(k_1), \cdots, h(k_{m-1}))$ where $h(k_i) \in \mathcal{H}_n$ for $0 \le i < m$.

[1] "Tis a lesson you should heed, Try, try again." Thomas H. Palmer (1782–1861) Teacher's Manual [Palmer 1840] and Frederick Maryat's novel *The Children of the New Forest* [Maryat 1847].

15.1.2 Ullman–Yao INSERT(k) and SEARCH(k)

SEARCH(k) using $h \in \mathcal{H}_n$ probed the sequence of cells $HT_{h_0(k)}, HT_{h_1(k)}, \cdots, HT_{h_{n-1}(k)}$ whose keys are compared in sequence with k until either of the following occur:

- An empty cell is encountered indicating that *no* record with key k has been stored.
- A cell containing the key k is found.

The cost of UY-SEARCH(k) is the number of key comparisons.

INSERT(k) using $h \in \mathcal{H}_n$, probes the cells $HT_{h_0(k)}, HT_{h_1(k)}, \cdots, HT_{h_{n-1}(k)}$ until an empty cell is encountered into which k may be inserted. The cost of UY-INSERT(k) is the number of cells probed.

As in the case of (separate and coalesced) chaining, linear probing (LP), and double hashing (DH), UY-SEARCH(k) might be the precursor of UY-SEARCH(k).

15.1.3 The Ullman–Yao Statistical Model

The statistical models for hashing used in Chapters 10 to 13 assumed the following:

i) The system specified a hashing function $h \in \mathcal{H}_n$.
ii) The unknown keys $\underline{k} = (k_0, k_1, \cdots, k_{m-1})$ were generated by independent and identically distributed trials of a chance experiment $\mathcal{E}$ of choosing a key.
iii) Each m-tuple of integers from $\mathcal{H}_n$ was assigned probability n^{-m}.

[Yao 1985] is not evaluating the performance of SEARCH/INSERT in a system with a specific hashing function but in the ensemble of all systems and hashing functions. He replaces the chance experiment

$\mathcal{E}$: Randomly chooses *keys* $\underline{k} = (k_0, k_1, \cdots, k_{m-1})$ by
$\mathcal{E}^{(Y)}$: Randomly choose *hashing functions* $\underline{h} = (h_0, h_1, \cdots, h_{m-1})$

Yao's performance model begins with a probability distribution $P(h)$ defined on a set $\mathcal{H}_n$ of hashing functions and also assumes the following:

i) The insertion of a k corresponds to the selection of the hashing function h with probability $P(h)$ by an oracle[2].
ii) The insertion of m keys $\underline{k} = (k_0, k_1, \cdots, k_{m-1})$ consists of m independent and identically distributed trials of $\mathcal{E}^{(Y)}$ with outcomes $\underline{h} = (h_0, h_1, \cdots, h_{m-1})$.
iii) Each m-tuple outcome $\underline{h} = (h_0, h_1, \cdots, h_{m-1})$ of $\mathcal{E}^{(Y)}$ occurs with probability $\prod_{i=0}^{m-1} P(h_i)$.

[2] An oracle is a person providing wise counsel and/or prophetic opinion—an infallible authority, often spiritual in nature. The major temple to Phoebus Apollo in Delphi (Greece) was the site of a famous prehistoric oracle. Oracles exist in computer science, which are slightly less divine but more prescient.

Although keys $\underline{k} = (k_0, k_1, \cdots, k_{m-1})$ are really secondary in Yao's model, we still will use the phrase "the insertion of key", hopefully to clarify.

Yao says that the $\mathcal{E}^{(Y)}$ insertion of m keys $\underline{k}$ in a hash table of n cells produces an (m, n)-scenario,[3] a hash-table configuration $\underline{h} \equiv h(\underline{k}) \equiv (h(k_0), h(k_1), \cdots, h(k_{m-1}))$.

We simplify the notation by omitting the argument $\underline{k}$ and writing $\underline{h}$.

Yao calls $h \in \mathcal{H}_n$ regular if $P(h) > 0$ for every h in $\mathcal{H}$. The following examples are instances of regular hashing functions.

Example 15.1 (Uniform hashing)
$\mathcal{H}_n$ contains all $n!$ permutations in ${}_{(i)}\mathcal{P}$ and $P(h) = \frac{1}{n!}$ for each $h \in \mathcal{H}_n$.

Example 15.2 (Linear probing)
$\mathcal{H}_n$ consists of the n permutations of the form $h = (i, i + 1, \cdots) \pmod n$ with $0 \le i < n$ and $P(h) = \frac{1}{n}$ for each $h \in \mathcal{H}_n$.

Example 15.3 (Double hashing)
$\mathcal{H}_n$ consists of the $n\phi(n)$ permutations of the form $h = (i, i - c, i - 2c, \cdots) \pmod n$

$$\begin{cases} 0 \le i < n \\ 1 \le c < n \end{cases}$$

where $\phi(n)$ is the Euler totient function [Chapter 5, §5.2, $gcd\{c, n\} = 1$ and $P(h) = \frac{1}{n\phi(n)}$ for each $h \in \mathcal{H}_n$.

The average cost $C_{\underline{h}}(n)$ of UY-SEARCH for m hashing functions $\underline{h} = (h_0, h_1, \cdots, h_{m-1})$ in a hash table of n cells is defined as follows:

- $C_{\underline{h}}(n) = \frac{1}{m} \sum_{i=0}^{m-1} C_{h_i}(n)$ is the UY-SEARCH-cost $C_{h_i}(n)$ averaged over the hashing functions using $\{h_i : 0 \le i < m\} \subset \mathcal{H}_n$.

For uniform hashing (Theorem 6.2 in Chapter 12)

- The UH-SEARCH cost is the (expected) search cost averaged over the insertion of m keys in a hash table of n cells

$$C_{\mathrm{UH}}(m, n) = \frac{n+1}{m}[H_{n+1} - H_{n-m+1}]$$

where $H_r = 1 + \frac{1}{2} + \cdots + \frac{1}{r}$ is the rth Harmonic number.

- The asymptotic[4] formula $H_r \sim \log r + \gamma$ gives $\lim_{\substack{n,m\to\infty \\ m=\mu n}} [H_{n+1} - H_{n-m+1}] \sim \log \frac{1}{1-\mu}$ so that

$$\lim_{\substack{n,m\to\infty \\ m=\mu n}} C_{\mathrm{UH}}(m, n) = \frac{1}{\mu} \log \frac{1}{1-\mu}$$

$$C_{\mathrm{UH}}(n, n) = \frac{n+1}{n}[H_{n+1} - 1] \sim \log n + O\left(\frac{1}{n}\right) \text{ as } n \to \infty.$$

[3]We use the undersore $\underline{h} = (h_0, h_1, \cdots, h_{m-1})$ to distinguish a sequence of hash functions $\{h_j\}$ defining an (m, n)-scenario from individual hashing functions.

[4]$\gamma \approx 0.5772156649 \cdots$ is the Euler-Mascheroni constant.

Yao's main result is

Theorem 15.1. For any $\in > 0$, there exists a constant $C_1 \equiv C_1(\in)$ such that for all integers $m, n > 1$

$$\mu = \frac{m}{n} \quad \varepsilon < \mu < 1 - \varepsilon$$

and any $\underline{h} = (h_0, h_1, \cdots, h_{m-1})$ with $h_i \in \mathcal{H}_n$ for $0 \le i < m$

$$C_{\underline{h}}(n) \ge \frac{1}{1-\mu} \log \frac{1}{1-\mu} - C_1 \frac{\log n}{n} \tag{15.1}$$

Moreover, there exists an abosolute constant C_2 such that for $n > 1$ and any $\underline{h} = (h_0, h_1, \cdots, h_{n-1})$ with $h_i \in \mathcal{H}_n$ for $0 \le i < n$

$$C_{\underline{h}}(n) \ge \log n - \log \log n - C_2 \tag{15.2}$$

The main ideas in Yao's proof of the optimality of uniform hashing are as follows:

- To use RANDOM-SCENARIO, to generate (m, n)-scenarios randomly.
- To show the cost $C_{\underline{h}}(n)$ for <u>every</u> (m, n)-scenario $\underline{h}$ averaged over the m keys
  1. First, the RANDOM-SCENARIO expected value of the number of times δ_i any specific cell HT_i is probed averaged over the cells $\{\text{HT}_i\}$ [Lemma 15.2a],
  2. Next, to show δ_i is bounded from below by a carefully chosen function $f(v_i)$ of the RANDOM-SCENARIO probability v_i that cell HT_i is occupied after most of the m keys have been inserted [Lemma 15.2b and Lemma 15.6]
  3. Finally, to show that a suitable choice of *most* $f(v_i)$ is bounded from below by the <u>*average expected*</u> uniform hashing SEARCH-cost [§15.6].

The proof of Theorem 15.1 here and in [Yao 1985] requires that several preliminary results be established, a task to which we now turn.

> In order to limit unnecessary subscripting, we omit all references to the size n of the hash table. For example: $\mathcal{P}$ replaces $\mathcal{P}_n$, $\mathcal{H}$ replaces $\mathcal{H}_n$, and m-scenario replaces (m, n)-scenario.

15.2 THE RATES AT WHICH A CELL IS PROBED AND OCCUPIED

Let $\underline{h}$ be the m-scenario that results when m keys are inserted into a table of n cells. Let d be an integer satisfying $1 \le d < m$ and define the following:

- The average number of times δ_i that h makes a key comparison involving cell HT_i during the insertion of the m keys

- The probability v_i that cell HT_i is occupied after $m - d$ of the keys $\underline{k}$ have been inserted.

Lemma 15.2. Then

a) $C_{\underline{h}} = \frac{1}{m}\sum_{i=0}^{n-1}\delta_i$, and

b) $m-d=\sum_{i=0}^{n-1} v_i$.

Proof. Define the following indicator functions:

- The indicator function

$$\chi_i^{(\mathrm{P})}(h_j) \text{ with values} = \begin{cases} 1 & \text{if the insertion by } h_j \text{ probes cell } \mathrm{HT}_i \\ 0 & \text{otherwise} \end{cases}$$

- The indicator function

$$\chi_i^{(\mathrm{I})}(h_j) \text{ with values} = \begin{cases} 1 & \text{if } h_j \text{ causes cell } \mathrm{HT}_i \text{ to be occupied} \\ 0 & \text{otherwise} \end{cases}$$

If $N(h_j) \equiv \sum_{i=0}^{n-1}\chi_i^{(\mathrm{P})}(h_j)$ is the number of probes (comparisons) used by h_j, then

$$N(h_j) = \sum_{i=0}^{n-1}\chi_i^{(\mathrm{P})}(h_j)$$

which gives

$$C_{\underline{h}} = \frac{1}{m}\sum_{j=0}^{m-1} N(h_j) = \frac{1}{m}\sum_{j=0}^{m-1}\sum_{i=0}^{n-1}\chi_i^{(\mathrm{P})}(h_j)$$

But $\delta_i = \sum_{j=0}^{m-1}\chi_i^{(\mathrm{P})}(h_j)$, so that interchanging the order of summations gives

$$C_{\underline{h}} = \frac{1}{m}\sum_{i=0}^{n-1}\delta_i$$

proving Lemma 15.2a.
Next

$$1 = \sum_{i=0}^{n-1}\chi_i^{(\mathrm{I})}(h_j) \quad n-d = \sum_{i=0}^{n-1}\chi_i^{(\mathrm{I})}(h_j)$$

which gives

$$v_i = \sum_{j=0}^{m-d+1}\sum_{i=0}^{n-1}\chi_i^{(\mathrm{I})}(h_j)$$

The proof of Lemma 15.2b is completed by interchanging the order of summation

$$m-d = \sum_{i=0}^{n-1} v_i \quad \blacksquare$$

15.3 PARTITIONS OF ${}_{(i)}$SCENARIOS, ${}_{(i)}$SUBSCENARIOS, AND THEIR SKELETONS

Fix some particular cell HT_i in the hash table of n cells; if the insertion m keys $\underline{h} = (h_0, h_1, \cdots, h_{m-1})$ with $0 \leq m < n$, creating the m-scenario leaves cell HT_i unoccupied, then we will call $\underline{h}$ an ${}_{(i)}m$-scenario. Denote the set of occupied cells by $B_{\underline{h}}$.

The ${}_{(i)}m$-scenario partitions $\mathcal{P}$ into two sets $({}_{(i)}\mathcal{P}[\underline{h}], {}_{(i)}\mathcal{P}'[\underline{h}])$ as follows:

- If $i \in B_{\underline{h}}$, then[5] ${}_{(i)}\mathcal{P}[\underline{h}] = \emptyset$.
- If $i \notin B_{\underline{h}}$, then h^* is in ${}_{(i)}\mathcal{P}[\underline{h}]$ if and only if the hashing sequence h^* would insert a key $k \notin \{k_0, k_1, \cdots, k_{m-1}\}$ into HT_i.

Note that HT_i will be occupied after the insertion of an additional key into the ${}_{(i)}m$-scenario $\underline{h}$ if and only if the hashing function h^* is chosen from ${}_{(i)}\mathcal{P}[\underline{h}]$.

Example 15.4. A ${}_{(5)}4$-scenario $\underline{h}$ results when four keys are inserted into a hash-table of $n = 8$ cells using the hashing functions in Table 15.1.

The ${}_{(5)}4$-scenario $\underline{h}$ occupies the cells $B_{\underline{h}} = \{0, 2, 3, 4\}$ and produces the hash-table configuration shown in Figure 15.1

Example 15.5. The ${}_{(5)}4$-scenario $\underline{h}$ is that described in *Example 15.4.* For the insertion of an additional key k, the oracle has chosen one of the following two hashing functions:

Case 1. $h^* = (3, 1, 4, 2, 5, 7, 6, 0)$; the hash-table changes the state to that shown in Figure 15.2A.

h^* is in ${}_{(5)}\mathcal{P}'[\underline{h}]$.

TABLE 15.1. Hashing Functions in Example 15.4

$\underline{h}$	
$h_0 = (4, 1, 6, 7, 5, 0, 3, 2)$	k_0 inserted into HT_4
$h_1 = (2, 5, 0, 1, 3, 4, 7, 6)$	k_1 inserted into HT_2
$h_2 = (2, 3, 5, 6, 1, 4, 7, 0)$	k_2 inserted into HT_3
$h_3 = (0, 4, 5, 6, 7, 3, 1, 2)$	k_3 inserted into HT_3

k_3	$\emptyset$	k_1	k_2	k_0	$\emptyset$	$\emptyset$	$\emptyset$
0	1	2	3	4	5	6	7

Figure 15.1. Hash-Table State in Example 15.4.

k_3	k	k_1	k_2	k_0	$\emptyset$	$\emptyset$	$\emptyset$
0	1	2	3	4	5	6	7

Figure 15.2A. Hash-Table State in Case 1, Example 15.5.

[5] If h is a $(0, n)$ scenario so that $B_h = \emptyset$, then $\underline{h}^* \in \begin{cases} {}_{(i)}\mathcal{P}[\underline{h}] & \text{if } h_0^* = i \\ {}_{(i)}\mathcal{P}'[\underline{h}] & \text{if } h_0^* \neq i \end{cases}$.

k_3	∅	k_1	k_2	k_0	k	∅	∅
0	1	2	3	4	5	6	7

Figure 15.2B. Hash-Table State in Case 2, Example 15.5.

Case 2. $h^* = (3, 4, 2, 5, 1, 7, 6, 0)$; the hash-table state changes to that shown in Figure 15.2B.

h^* is in ${}_{(5)}\mathcal{P}[\underline{h}]$.

Example 15.6. What hashing functions h^* are in ${}_{(i)}\mathcal{P}[\underline{h}]$ for different values of i with $0 \le i < 8$ for the 4-scenario in *Example 15.4*.

Solution. For each h^* in $\mathcal{P}$, cell HT_i and $B_{\underline{h}} = \{0, 2, 3, 4\}$, define $s \equiv s_i$ by $h_s^* = i \ (0 \le s < n)$. Note the following:

i) If $i \in B_{\underline{h}}$, then ${}_{(i)}\mathcal{P}[\underline{h}] = \emptyset$.

ii) If $i \notin B_{\underline{h}}$, then h^* is in ${}_{(i)}\mathcal{P}[\underline{h}]$ if and only if $h_j^* = \begin{cases} \in B_{\underline{h}} & \text{for } every\ j \text{ with } 0 \le j < s \\ = i & \text{if } j = s \end{cases}$;

that is, h^* is in ${}_{(i)}\mathcal{P}[\underline{h}]$ if and only if the first cell probed by h^* which is *not* in $B_{\underline{h}}$ is HT_i. The number of such hashing functions is $M_s = (4)_s(7 - s)!$ because

a $(4)_s$ is the number of sequences $h_0^*, h_1^*, \cdots, h_{s-1}^*$ without repetition with $h_j^* \in B_{\underline{h}}$ for every j with $0 \le j < s$, and

b $(7 - s)!$ is the number of sequences $h_{s+1}^*, \cdots, h_{n-1}^*$ without repetition with who entries are the remaining elements in $\mathcal{Z}_8 - \{i\}$ other than those selected in the previous item;

iii) If $i \notin B_{\underline{h}}$, then $h^* \in \mathcal{P}'[\underline{h}]$ if and only if $h_j^* = \begin{cases} \notin B_{\underline{h}} & \text{for some } j \text{ with } 0 \le j < s \\ = i & \text{if } j = s \end{cases}$;

that is, h^* is in ${}_{(i)}\mathcal{P}'[\underline{h}]$ if and only if some cell not in $B_{\underline{h}}$ is probed by h^* before probing HT_i. The number of such hashing functions is $M_s' = [(7)_s - !(4)_s](7 - s)! = 7! - M_s$ because

a $(7)_s$ is the number of sequences $h_0^*, h_1^*, \cdots, h_{s-1}^*$ without repetition with $h_j^* \in \mathcal{Z}_8 - \{i\}$ for $0 \le j < s$.

b $(4)_s$ is the number of sequences $h_0^*, h_1^*, \cdots, h_{s-1}^*$ without repetition with $h_j^* \in B_{\underline{h}}$ for $0 \le j < s$.

c $(7)_s - (4)_s$ is the number of sequences $h_0^*, h_1^*, \cdots, h_{s-1}^*$ without repetition with $h_j^* \notin B_{\underline{h}}$ for *some* $0 \le j < s$.

d $(7 - s)!$ is the number of sequences $h_{s+1}^*, \cdots, h_{n-1}^*$ without repetition with who entries are the remaining elements in $\mathcal{Z}_8 - \{i\}$ other than those selected in the previous item.

Table 15.2 lists the following:

- The conditions required for a hashing function h^* be in either ${}_{(i)}\mathcal{P}[\underline{h}]$ or ${}_{(i)}\mathcal{P}'[\underline{h}]$.
- The sizes of these partitions ${}_{(i)}\mathcal{P}[\underline{h}]$ and ${}_{(i)}\mathcal{P}'[\underline{h}]$.

for each cell HT_i $(0 \le i < 8)$ in the hash table.

TABLE 15.2. Conditions on h^* for Membership in $_{(i)}\mathcal{P}[\underline{h}]$ and $_{(i)}\mathcal{P}'[\underline{h}]$ For Cell HT_i, $(0 \le i < 8)$

i	When is h^* in $_{(i)}\mathcal{P}[\underline{h}]$?	$\|_{(i)}\mathcal{P}[\underline{h}]\|$	When is h^* in $_{(i)}\mathcal{P}'[\underline{h}]$?	$\|_{(i)}\mathcal{P}'[\underline{h}]\|$
0		0		7!
1	$h_t^* \in B_{\underline{h}}$, for all t, $0 \le t \le s$	M_1	$h_t^* \notin B_{\underline{h}}$, for some t, $0 \le t \le 1$	M_1'
2		0		7!
3		0		7!
4		0		7!
5	$h_t^* \in B_{\underline{h}}$, for all t, $0 \le t \le s$	M_4	$h_t^* \notin B_{\underline{h}}$, for some t, $0 \le t \le 1$	M_4'
6	$h_t^* \in B_{\underline{h}}$, for all t, $0 \le t \le s$	M_4	$h_t^* \notin B_{\underline{h}}$, for some t, $0 \le t \le 1$	M_4'
7	$h_t^* \in B_{\underline{h}}$, for all t, $0 \le t \le s$	M_4	$h_t^* \notin B_{\underline{h}}$, for some t, $0 \le t \le 1$	M_4'

15.3.1 $_{(i)}$Subscenarios

Let $\underline{h} = (h_0, h_1, \cdots, h_{m-1})$ be an $_{(i)}m$ scenario. I write $\underline{h}_{(j)} = (h_0, h_1, \cdots, h_j)$ for what Yao refers to as a $_{(i)}j$ subscenario.

The $_{(i)}j$ subscenario $\underline{h}_{(j)}$ determines a partition $(_i\mathcal{P}[\underline{h}_{(j)}], {}_i\mathcal{P}'[\underline{h}_{(j)}])$ of $_{(i)}\mathcal{P}$ using the same rules as for scenarios.

15.3.2 Skeletons

If

- $\underline{h} = (h_0, h_1, \cdots, h_{m-1})$ produces an m scenario, and
- Cell HT_i is not occupied after the insertion of the first $m - d$ of the m keys,

then the resulting $_{(i)}(m - d)$ subscenario is a skeleton.

Exercise 15.1. Prove that for a fixed cell HT_i, an $(m - d, n)$ scenario $\underline{h} = (h_0, h_1, \cdots, h_{m-d-1})$ is a skeleton if and only if $h_j \in \mathcal{P}'[\underline{h}_{(0,j)}]$ for $0 \le j < m - d$.

15.4 RANDOMLY GENERATED *m*-SCENARIOS

If $\mathcal{V} \subseteq \mathcal{P}$ is not empty, $P_\mathcal{V}(h)$ denotes the restriction of P to $\mathcal{V}$

$$P_\mathcal{V}(h) \equiv \begin{cases} \dfrac{P(h)}{P(\mathcal{V})} & \text{if } h \in \mathcal{V} \\ 0 & \text{otherwise} \end{cases} \qquad P(\mathcal{V}) \equiv \sum_{h \in \mathcal{V}} P(h)$$

The output of `RANDOM-SCENARIO` is an m scenario $\underline{h}$ generated by hashing functions h in $\mathcal{H}$.

Procedure `RANDOM-SCENARIO`

<u>Parameters:</u> m, d with $1 \le d < m < n$ and a hash-table cell HT_i

<u>Output:</u> ω, an m scenario composed of hashing functions h in $\mathcal{H}$

S1. Generate a random skeleton $\underline{h} = (h_0, h_1, \cdots, h_{m-d-1})$ recursively, as follows:

- Randomly choose the hashing function $h_0 \in \mathcal{V}_0$ with probability $P_{\mathcal{V}_0}(h_0)$, $\mathcal{V}_0$ consists of those hashing functions $\{h^*\}$ for which $h_0^* \neq i$.
- The $_{(i)}1$ subscenario $\underline{h}_{(0)} \equiv (h_0)$ determines a partition $({}_{(i)}\mathcal{P}[\underline{h}_{(1)}], {}_{(i)}\mathcal{P}'[\underline{h}_{(1)}])$ of $_{(i)}\mathcal{P}$.

Next, for each integer $1 \leq j < m - d$

- Randomly choose a hashing function $h_j \in \mathcal{V}_j$ with probability $P_{\mathcal{V}_j}(h_j)$ where $\mathcal{V}_j \equiv {}_{(i)}\mathcal{P}'[\underline{h}_{(j-1)}]$.
- The $_{(i)}j$ subscenario $\underline{h}_{[0,j)} = (h_0, h_1, \cdots, h_j)$ is formed by concatenating $\underline{h}_{[0,j-1)}$ on the right by h_j; $\underline{h}_{(0,j)}$ determines a partition $({}_{(i)}\mathcal{P}[\underline{h}_{[0,j)}], {}_{(i)}\mathcal{P}'[\underline{h}_{[0,j)}])$ of $_{(i)}\mathcal{P}$.

> $\underline{h}$ is a skeleton because cell HT_i is not occupied in the $(m - d)$ scenario $\underline{h} = (h_0, h_1, \cdots, h_{m-d-1})$.

S2. For each integer $0 \leq j < m - d$ start with the $_{(i)}(j - 1)$ subscenario $\underline{h}_{(j-1)} = (h_0, h_1, \cdots, h_{j-1})$ composed of the first $j - 1$ of the hashing functions generated in step S1.
Generate a scenario $\underline{h}^{(j)}$ with a randomly determined number of hashing functions as follows:

- $u_j \equiv P_{\mathcal{W}j}$ with $\mathcal{W}_j \equiv {}_{(i)}\mathcal{P}[\underline{h}_{(j-1)}]$ and R_j is a geometrically distributed random variable with parameter u_j; that is, $Pr\{R_j = \ell\} = (1 - u_j)u_j^\ell$, $(0 \leq l < \infty)$.
- If $R_j = r_j$, then generate the r_j scenario $\underline{h}^{(j)} = (h_{j,0}, h_{j,1}, \cdots, h_{j,r_j-1})$ where for $0 \leq t < r_j$, $h_{j,t}$ is chosen independently of all previous hashing functions with probability Pw_j.

> If $r_j > 0$, each of the hashing functions in the r_j scenario $\underline{h}^{(j)} = (h_{j,0}, h_{j,1}, \cdots, h_{j,r_j-1})$ probes HT_i, and one of the keys inserted with $\underline{h}^{(j)}$ may this occupy this cell.

S3a. If $r = \sum_{j=1}^{m-d} r_j$ satisfies $r = d$, then $m - d + r = r_0 + 1 + r_1 + 1 + \cdots + r_{m-d-1} + 1 = m$.
The interleaving of the following will occur:

- The $m - d$ hashing functions $(h_0, h_1, \cdots, h_{m-d-1})$ generated in S1.
- The $m - d$ scenarios generated in S2.

$\underline{h}^{(0)} = (h_{0,0}, h_{0,1}, \cdots, h_{0,r_0-1})$, which is the r_0 scenario generated form $\underline{h}_{(-1)}$.
$\underline{h}^{(1)} = (h_{1,0}, h_{1,1}, \cdots, h_{1,r_1-1})$, which is the r_1 scenario generated from $\underline{h}_{(0)}$.
$\underline{h}^{(m-d-1)} = (h_{m-d-1,0}, h_{m-d-1,1}, \cdots, h_{m-d-1,r_{m-d-1}-1})$; the r_{m-d} scenario generated from $\underline{h}_{(m-d-2)}$ produces a m scenario

$$\omega = \left(\underbrace{\overbrace{h_{0,0}, h_{0,1}, \cdots, h_{0,R_0-1}}^{r_0 \text{ scenario } \underline{h}^{(0)}}, h_0}_{0^{\text{th}} \text{ term}}, \underbrace{\overbrace{h_{1,0}, h_{1,1}, \cdots, h_{1,r_1-1}}^{r_1 \text{ scenario } \underline{h}^{(1)}}, h_1}_{1^{\text{st}}\text{-term}}, \cdots \underbrace{\overbrace{h_{m-d-1,0}, h_{m-d-1,1}, \cdots, h_{m-d-1,r_{m-d-1}-1}}^{r_{m-d-1} \text{ scenario } \underline{h}^{(m-d-1)}}, h_{m-d-1}}_{(m-d-1)^{\text{st}} \text{ term}} \right) \tag{15.3}$$

S3a-END `RANDOM-SEARCH`.

S3b. If $r = \sum_{j=1}^{m-d} r_j$ satisfies $r = d$, then $m - d + r = r_0 + 1 + r_1 + 1 + \cdots + r_{m-d-1} + 1 > m$; the interleaving of the following:

- The $m - d$ hashing functions $(h_0, h_1, \cdots, h_{m-d-1})$ generated in S1
- The $m - d$ subscenarios generated in S2

generates more hashing functions than required, $m^* > m$. In this instance, ω is defined to be the leftmost m hashing functions of equation (15.3)-ω; the rightmost $m^* - m$ hashing functions in equation (15.3) are deleted.

S3b-END `RANDOM-SEARCH`.

S3c. If $r = \sum_{j=1}^{m-d} r_j$ satisfies $r < d$, then $m - d + r = r_0 + 1 + r_1 + 1 + \cdots + r_{m-d-1} + 1 < m$; the interleaving of the following:

- The $m - d$ hashing functions $(h_0, h_1, \cdots, h_{m-d-1})$ generated in S1
- The $m - d$ scenarios generated in S2

produces fewer hashing functions than needed, $m^* < m$. In S3c, ω is the equation (15.3). ω is suffixed by a $(d - r)$ scenario $\omega' = (\omega_0, \omega_1, \cdots, \omega_{d-r-1})$ consisting of an additional $d - r$ hashing functions in $\mathcal{H}$, generated independently with probability P.

S3c-END `RANDOM-SEARCH`.

Remark 15.2. The construction of the r_j scenario in step S2 is properly defined;

- The skeleton $\underline{h} = (h_0, h_1, \cdots, h_{m-d-1})$ generated in step S1 requires h_j to be chosen with probability $P_{\mathcal{V}_j}(h_j)$ where $\mathcal{V}_j \equiv {}_{(i)}\mathcal{P}'[\underline{h}_{(j-1)}]$).
  It follows that cell HT_i is not in any of the subscenarios $\underline{h}_{[0,j)]}$.
- Because neither $\mathcal{W}_j \equiv {}_{(i)}\mathcal{P}[\underline{h}_{(j-1)}]$ nor ${}_{(i)}\mathcal{P}[\underline{h}_{(j-1)}]$ are vacuous and $P(h) > 0$ for $h \in \mathcal{H}$, it follows that $u_j = P(\mathcal{W}_j)$ is in the open interval (0,1).

Therefore, each of the $\{R_j: 0 \le j < m - d\}$ random variables has a nondegenerate distribution.

Lemma 15.3. The output ω of `RANDOM-SCENARIO` generates an m scenario.

Proof. It is required to show that the random outcome $\omega = (\omega_0, \omega_1, \cdots, \omega_{m-1})$ of RANDOM-SCENARIO satisfies the following:

- The component hashing functions $\{\omega_i: 0 \le i < m\}$ are independent.
- The probability that $\underline{\underline{\omega}}$ occurs is $\prod_{i=0}^{m-1} P(\omega_i)$.

RANDOM-SCENARIO generates a m scenario ω, which is the interleaving of the following:

- Hashing functions $\{h_j\}$, which leave HT_i empty,
- Scenarios $\{\underline{h}^{(j)}\}$, which probe and may occupy HT_i, and possibly
- An s scenario $\underline{\rho} = (\rho_0, \rho_1, \cdots, \rho_{s-1})$ consisting of additional s hashing functions in $\mathcal{H}_n$, which are generated independently with probability P.

The three terminations of RAMDOM-SCENARIO correspond to the value of

$$\sum_{j=1}^{M-d} r_j - d; \text{ if } \sum_{j=1}^{M-d} r_j = r > d.$$

- At least one of the final hashing functions in the skeleton $(h_0, h_1, \cdots, h_{M-d-1})$
- Some of the hashing functions in the final segment of scenarios $(\underline{h}^{(0)}, \underline{h}^{(1)}, \cdots, \underline{h}^{(M-d-1)})$

are omitted from the scenario ω in equation (15.3) to produce RANDOM-SCENARIO's output m scenario

$$\underline{\omega} = \left(\underline{h}^{(0)}, h_0, \underline{h}^{(1)}, h_1, \cdots, \underline{h}^{(M-d-1)}, h_{M-d-1}, \underline{h}^{(M-d)}\right) \quad M < m \tag{15.4}$$

The m scenario generated by RANDOM-SCENARIO consists of the following:

- The leftmost $M - d$ hashing functions $(h_0, h_1, \cdots, h_{M-d-1})$ from the S1-generated skeleton of $m - d$ hashing functions, interleaved with
- The leftmost $M - d + 1$ scenarios $(\underline{h}^{(0)}, \underline{h}^{(1)}, \cdots, \underline{h}^{(M-d-1)}, \underline{h}^{(M-d)})$ from the sequence of S2-generated scenarios $(\underline{h}^{(0)}, \underline{h}^{(1)}, \cdots, \underline{h}^{(m-d-1)})$. This segment contains $(r_0, r_1, \cdots, r_{M-d-1}, r_{M-d})$ hashing functions whose their lengths satisfy

$$m > M - d + r_0 + r_1 + \cdots + r_{M-d-1}$$

$$m = M - d + r_0 + r_1 + \cdots + r_{M-d-1} + r_{M-d}$$

$$r_j \begin{cases} \ge 0 & \text{if } 0 \le j < M - d \\ \ge d - r' & \text{if } j = M - d \end{cases} \quad \text{where } r' \equiv M + \sum_{0 \le \ell < M-d} r_\ell$$

$$M - d + r_0 + r_1 + \cdots + r_{M-d-1} < m = M - d + r_0 + r_1 + \cdots + r_{M-d-1} + r_{M-d}$$

We need to prove that the probability of ω in equation (15.4) is the product of the P probabilities of the hashing functions contained in ω. For this purpose, define following the events:

$E_{1,j}$: Step S1 of `RANDOM-SCENARIO` generated the hashing function h_j for $0 \le j < M - d$.

$E_{2,j}$: Step S2 of `RANDOM-SCENARIO` generated the following:

- First, the value of the random variable $R_j = r_j$
- Second, the r_j scenario $\underline{h}^{(j)} = (h_{j,0}, h_{j,1}, \cdots, h_{j,r_j-1})$

for each j with $0 \le j \le M - d$.

E_3: Step S3 of `RANDOM-SCENARIO` generated the following:

- First, the value of the random variable $R_{M-d} \ge r' - d$.
- Second, the $(r' - d)$ scenario $\underline{h}^{(M-d)} = (h_{M-d,0}, h_{M-d,1}, \cdots, h_{M-d,r'-d-1})$.

We begin by expressing the event "the output is ω" in terms of the events $\{E_{1,j}, E_{2,j}: 0 \le j < M - d\}$ and E_3.

$$\underline{\omega} = (\underline{h}^{(0)}, h_0, \underline{h}^{(1)}, h_1, \cdots, \underline{h}^{(M-d-1)}, h_{M-d-1}, \underline{h}^{(M-d)}) \Leftrightarrow E \equiv \left\{ \bigcap_{j=0}^{M-d-1} [E_{1,j} \cap E_{2,j}] \right\} \cap E_3 \qquad (15.5a)$$

$$Pr\{E\} = \left\{ \prod_{j=0}^{M-d-1} Pr\{E_{1,j}\} Pr\{E_{2,j} / E_{1,j}\} \right\} Pr\{E_3 / E_{1,M-d}\} \qquad (15.5b)$$

Equation (15.5*b*) uses the following statements:

- $E_{1,j}$ is independent of $E'_{1,j}$ for $j' > j$ conditional on $E''_{1,j}$ for $j'' < j$.
- $Pr\{E_{1,j} \cap E_{2,j}\} = Pr\{E_{1,j}\} Pr\{E_{2,j} / E_{1,j}\}$.

The probabilities in the equation (15.5*b*) product are evaluated as follows:

i) Each hashing function h_j in the skeleton $(h_0, h_1, \cdots, h_{M-d-1})$ is chosen conditional on the $(j - 1)$ subscenario $\underline{h}_{(j-1)} = (h_0, h_1, \cdots, h_{j-1})$; the value of h_j is in the set $\mathcal{V}_j \equiv {}_{(i)}\mathcal{P}'[\underline{h}_{(j-1)}]$. Because $Pr\{\mathcal{V}_j\} = 1 - u_j$, this gives

$$\prod_{j=0}^{M-d-1} Pr\{E_{1,j}\} = \prod_{j=0}^{M-d-1} \frac{P(h_j)}{1 - u_j} \qquad (15.6a)$$

ii) For each j with $0 \le j < M - d$, the random variables $\{R_j\}$ are independent (in j), each with the geometric distribution $Pr\{R_j = \ell\} = (1 - u_j) u_j^{\ell} \quad (0 \le \ell < \infty)$. Having determined $R_j = r_j$, each hashing function $h_{j,t}$ in the r_j scenario $\underline{h}^{(j)} = (h_{j,0}, h_{j,1}, \cdots, h_{j,r_j-1})$ is chosen independently (in j and t), conditional on the value of the jth skeleton $\underline{h}_{(j-1)} = (h_0, h_1, \cdots, h_{(j-1)})$ generated in S1. $h_{j,t}$ has a value in the set ${}_{(i)}\mathcal{P}[\underline{h}_{(j-1)}]$ with probability $\frac{P(h_{j,t})}{u_j}$, where $u_j = Pr\{{}_{(i)}\mathcal{P}[\underline{h}_{(j-1)}]\})$. This yields the following formula:

$$\prod_{j=0}^{M-d-1} Pr\{E_{2,j} / E_{1,j}\} = \left\{ \prod_{j=0}^{M-d-1} (1 - u_j) u_j^{r_j} \left(\prod_{t=0}^{r_j-1} \frac{P(h_{j,t})}{u_j} \right) \right\} = \left\{ \prod_{j=0}^{M-d-1} (1 - u_j) \prod_{t=0}^{r_j-1} P(h_{j,t}) \right\} \qquad (15.6b)$$

iii) The random variable R_{M-d} is independent of the previous variables $\{R_j\}$ and has the geometric distribution $Pr\{R_{M-d} = \ell\} = (1-u_{M-d})u_{M-d}^{\ell} \quad (0 \le \ell < \infty)$.
The event E_j requires $R_{M-d} \ge r' - d$, which is satisfied if $R_{M-d} = t$ for some $t \ge r - d$, an event of probability $(1-u_{M-d})u_{M-d}^{t}$.
If $R_{M-d} = t$, each hashing function $h_{M-d,t}$ in the r_{M-d}-scenario $\underline{h}^{(M-d)} = (h_{M-d,0}, h_{M-d,1}, \cdots, h_{M-d,r'-d-1})$ is chosen independently (in l) conditional on the value of the $(M-d)^{st}$-skeleton $\underline{h}_{(M-d)} = (h_0, h_1, \cdots, h_{M-d})$ generated in S1.
$h_{M-d,l}$ has a value in the set ${}_{(i)}\mathcal{P}[\underline{h}_{(M-d)}]$ with probability

$$\frac{P(h_{M-d,t})}{u_{M-d}}$$

where $u_{(M-d)} = Pr\{{}_{(i)}\mathcal{P}[\underline{h}_{(M-d)}]\})$. This gives the formula

$$Pr\{E_3/E_{1,M-d}\} = (1-u_{M-d})\left\{\sum_{t \ge r-d} u_{M-d}^{t}\left(\prod_{t=0}^{r-d-1}\frac{P(h_{M-d,t})}{u_{M-d}}\right)\right\} \tag{15.6c}$$

multiplying the probabilities in equations (15.6*a* through 15.6*c*) proves Lemma 15.3, at least in the case $r = \sum_{j=1}^{m-d} r_j > d$. The other cases yield the same result, but we leave the reader the task of modifying the argument to show this.

Lemma 15.4. If S_i is the number of times HT_i is probed in the *m* scenario generated by `RANDOM-SCENARIO` (with input parameter HT_i and output ω) and *R* is the total (random) number of hashing functions in the output excluding those generated in step S1, then

a) $S_i \ge \min\{R, d\}$
b) $E\{S_i/\underline{\omega}\} = \sum_{r=0}^{d} Pr\{R \ge r/\underline{\omega}\}$

Proof. Lemma 15.2 proved $\delta_i = E\{S\}$. If ω is the output of `RANDOM-SCENARIO`, then

- HT_i is not probed by any of the hashing functions in the skeleton $(h_0, h_1, \cdots, h_{m-d-1})$ generated in step S1 by `RANDOM-SCENARIO`;
- HT_i is probed by each of the hashing functions in the all of the $m - d$ scenarios $\{\underline{h}^{(j)}: 0 \le j < m - d\}$ generated in step S2 by `RANDOM-SCENARIO`.

S_i is therefore the number *R* of hashing functions in all of the $m - d$ scenarios $\{\underline{h}^{(j)}: 0 \le j < m - d\}$ generated in step S2 by `RANDOM-SCENARIO`. Examining the three possible ways in which `RANDOM-SCENARIO` terminates, we see that $S_i = \min\{R, d\}$. Using the first Lemma provided previously, we have

$$E\{S_i/\underline{\omega}\} = E\{\min\{R,d\}/\underline{\omega}\} = \sum_{r=0}^{\infty}\min\{r,d\}Pr\{R = r/\underline{\omega}\}$$
$$= \sum_{r=0}^{d} rPr\{R = r/\underline{\omega}\} = \sum_{r=0}^{d} Pr\{R \ge r/\underline{\omega}\} \tag{15.7}$$

completing the proof of Lemma 15.4. ■

15.5 BOUNDS ON RANDOM SUMS

Exercise 15.2. Prove that if X_0, X_1 are independent random variables with the Poisson distributions

$$Pr\{X_i \le s\} = e^{-\lambda_i} \sum_{j=0}^{s} \frac{\lambda_i^j}{j!} \quad i = 0, 1, 0 \le s < \infty$$

then

$$Pr\{X_0 + X_1 \le s\} = e^{-(\lambda_0+\lambda_1)} \sum_{j=0}^{s} \frac{(\lambda_0 + \lambda_1)^j}{j!} \quad 0 \le s < \infty$$

Exercise 15.3. Prove that when $|z| \le 1$, $z \ne -1$

$$\sum_{j=0}^{r} z^j \le \sum_{j=0}^{r} \frac{\left(\log \frac{1}{1-z}\right)^j}{j!} \quad 0 \le r < \infty \tag{15.8}$$

Lemma 15.5. If $Y_0, Y_1, \cdots, Y_{r-1}$ are r independent and identically distributed random variables each with the geometric distribution $Pr\{Y_i = s\} = (1-u_i)u_i^s$, $0 \le s < \infty$, $0 \le i < r$ and $Y = \sum_{i=0}^{r-1} Y_i$, then

$$Pr\{Y \ge s\} \ge e^{-\lambda} \sum_{j=s}^{\infty} \frac{\lambda^j}{j!} \quad 0 \le s < \infty \tag{15.9}$$

with $\lambda = -\log(1 - U)$ and $U = \prod_{j=0}^{r-1}(1-u_i)$.

Proof. We prove by induction on the number of summands r that

$$Pr\{Y \le s\} \le U \sum_{j=0}^{s} \frac{\left(\log \frac{1}{1-U}\right)^j}{j!} \quad 0 \le s < \infty \tag{15.10}$$

First, for $r = 1$, $Y = Y_0$, we use Exercise 15.3

$$Pr\{Y_0 \le s\} = (1-u_0) \sum_{j=0}^{s} u_0^j \le \sum_{j=0}^{s} \frac{\left(\log \frac{1}{1-u_0}\right)^j}{j!} \quad 0 \le s < \infty$$

If

$$Y = Y_{[0,r-2]} + Y_{r-1}, \quad Y_{[0,r-2]} \equiv \sum_{i=0}^{r-2} Y_i, \quad \lambda_{[0,r-2]} = -\log\left(\prod_{i=0}^{r-2}(1-u_i)\right)$$

the induction hypothesis gives

$$Pr\{Y^{(r-2)} \geq s\} \geq e^{-\lambda_{[0,r-2]}} \sum_{j=s}^{\infty} \frac{\lambda_{[0,r-2]}^j}{j!} \quad Pr\{Y^{(r-2)} \leq s\} \leq e^{-\lambda_{[0,r-2]}} \sum_{j=0}^{s} \frac{\lambda_{[0,r-2]}^j}{j!}$$

But

$$Pr\{Y \leq s\} = \sum_{j=0}^{s} Pr\{Y_{r-1} = j, Y_{[0,r-2]} \leq j-s\} = \sum_{j=0}^{s} (1-u_{r-1}) u_{r-1}^j Pr\{Y_{[0,r-2]} \leq j-s\}$$

Next, we use the same argument as in *Exercise 15.3*; namely, that the discrete distribution $(1-u_{r-1})u_{r-1}^j$ bounds from above the discrete distribution $(1-u_{r-1})\frac{\left(\log\frac{1}{1-u_{r-1}}\right)^j}{j!}$ to conclude

$$\sum_{j=0}^{s}(1-u_{r-1})u_{r-1}^j Pr\{Y_{[0,r-2]} \leq j-s\} \leq \sum_{j=0}^{s}(1-u_{r-1})\frac{\left(\log\frac{1}{1-u_{r-1}}\right)}{j!} Pr\{Y_{[0,r-2]} \leq j-s\}$$

$$\leq \sum_{j=0}^{i}(1-u_{v-1})\frac{\left(\log\frac{1}{1-u_{r-1}}\right)^j}{j!} e^{-\lambda_{[0,r-2]}} \sum_{j=0}^{s} \frac{\lambda_{[0,r-2]}^j}{j!} \tag{15.11}$$

The right-hand side of equation (15.11) is the convolution of two Poisson distributions; the first with parameter $\log\frac{1}{1-u_{r-1}}$, the second with parameter $\lambda_{[0,(r-2]}$; therefore, by *Exercise 15.2* it is Poisson with parameter $\log\frac{1}{1-u_{r-1}} + \lambda_{[0,v-2]}$. ■

Lemma 15.6. If $f_d(x) = \lambda - e^{-\lambda}\sum_{t>d}(t-d)\frac{\lambda^t}{t!}$ with $\lambda = -\log(1-x)$, then

$$\delta_i \geq f_d(v_i) \quad 0 \leq i < n \tag{15.12}$$

Proof. We first prove that $f_d(x)$ is convex; that is, $f_d\left(\sum_t r_t x_t\right) \leq \sum_t r_t f_d(x_t)$ if the components of $\underline{r} = (r_0, r_1, \cdots)$ are non-negative with $1 = r_0 + r_1 + \cdots$. For a twice-differentiable function $f_d(x)$, to prove convexity it suffices [Courant 1957, pp. 323–26] to show

$$\frac{d^2}{dx^2} f_d(x) \geq 0$$

$$\frac{d}{dx} f_d(x) = \frac{1}{1-x}\left(\frac{d}{d\lambda} f_d(x)\right)$$

$$= \left(1 + e^{-\lambda}\sum_{t\geq d}(t-d)\frac{\lambda^t}{t!} - e^{-\lambda}\sum_{t\geq d-1}(t+1-d)\frac{\lambda^t}{t!}\right)\frac{1}{1-x}$$

$$= \left(1 - e^{-\lambda}\sum_{t>d}\frac{\lambda^t}{t!}\right)\frac{1}{1-x}$$

$$\frac{d^2}{dx^2}f_d(x)=\frac{1}{1-x}\left(\frac{d^2}{d\lambda^2}f_d(x)\right)+\frac{1}{(1-x)^2}\left(\frac{d}{d\lambda}f_d(x)\right)$$
$$=\left(1-e^{-\lambda}\sum_{t>d}\frac{\lambda^d}{t!}\right)\frac{1}{(1-x)^2}+\frac{e^{-\lambda}}{1-x}\frac{\lambda^d}{d!}>0$$

Because $e^{-\lambda}\sum_{t>d}\frac{\lambda^t}{t!}<1$, we have proved $f_d(x)$ is convex.

Let ω be the output of the m scenario generated by RANDOM-SCENARIO with input m, d, n and fixed cell HT_i and

- S be the number of times HT_i is probed in ω
- R be the total (random) number of hashing function in the output ω, excluding those generated in step S1.

R is the sum of independent geometrically distribution random variables so that by Lemma 15.5

$$Pr\{R\geq r/\underline{\omega}\}=e^{-\lambda}\sum_{t=r}^{\infty}\frac{\lambda^t}{t!}$$

where $\lambda\equiv\lambda[\omega]$. Applying the bound [equation (15.9)] to the inequality [equation (15.7)] from Lemma 15.4 gives

$$E\{S/\underline{\omega}\}=\sum_{r=0}^{d}Pr\{R\geq r/\underline{\omega}\}=\sum_{r=0}^{d}e^{-\lambda}\sum_{t=r}^{\infty}\frac{\lambda^t}{t!}=\lambda-e^{-\lambda}\sum_{t>d}^{\infty}(t-d)\frac{\lambda^t}{t!}\tag{15.13}$$

where the final equality presented previously follows from an interchange of the t and r summations.

Because the following is true:

- HT_i is not occupied in the skeleton $\underline{h}=(h_0, h_1, \cdots, h_{m-d-1})$ generated in step S1 of RANDOM-SCENARIO,
- all of the R_i scenarios generated in RANDOM-SCENARIO, excluding those generated in step S1 probe HT_i

it follows that HT_i remains unoccupied in the m scenario output ω_i of RANDOM-SCENARIO if and only if $R=0$; but

$$Pr\{R=0\}=\prod_{j=0}^{N-d-1}(1-u_j)$$

where u_j is the geometric distribution parameter of the r_j-scenario $\underline{h}^{(j)}$ generated in step 2 of RANDOM-SCENARIO. Equation (15.13) gives

$$\delta_i=E\{S_i\}\geq f(Pr\{R=0\})\tag{15.14}$$

Using equations (15.12) and the convexity of f_d, we obtain

$$\delta_i = \sum_{\underline{\omega}} Pr\{\underline{\omega}\} E\{S_i/\underline{\omega}\}$$
$$\geq f_d\left(\sum_{\underline{\omega}} Pr\{\underline{\omega}\} Pr\{R>0\}\right)$$

completing the proof of Lemma 15.5. ■

The first few terms of the sequence of functions $\{E_d(x)\}$ defined by

$$E_d(x) = \sum_{t=d}^{\infty} \frac{\lambda^t}{t!} \quad 0 \leq d, x < \infty, \quad \lambda = \log(1-x) \tag{15.15a}$$

are

$$E_0(x) = e^{\lambda} \qquad E_1(x) = e^{\lambda} - 1$$
$$E_2(x) = e^{\lambda} - 1 - \lambda \quad E_3(x) = e^{\lambda} - 1 - \lambda - \frac{\lambda^2}{2!}$$

This $\{E_d(x)\}$ sequence is clearly related to the sequence of $\{f_d(x)\}$ defined in Lemma 15.6

$$f_d(x) = \lambda - e^{-\lambda} \sum_{t>d} (t-d) \frac{\lambda^t}{t!} \quad 0 \leq d, x < \infty, \quad \lambda = \log(1-x) \tag{15.15b}$$

whose corresponding initial terms are

$$f_0(z) = 0 \qquad f_1(x) = 1 - e^{-\lambda}$$
$$f_2(x) = 2 - e^{-\lambda}(2+\lambda) \quad f_3(z) = 3 - e^{-\lambda}\left(3 + 2\lambda + \frac{1}{2}\lambda^2\right)$$

Exercise 15.4. Prove the sequences $\{E_d(x)\}$ and $\{f_d(x)\}$ defined in equations (15.15*a*,*b*) are related by

$$f_d(x) = \lambda - e^{-\lambda}[\lambda E_d(x) - dE_{d+1}(x)] \quad d = 0, 1, \cdots \lambda = -\log(1-x) \tag{15.15c}$$

Table 15.3 gives the leading terms in the Taylor series in λ about for $\{f_d(x)\}$ and suggests their behavior with d as $\lambda \downarrow 0$.

Exercise 15.5. Prove that $f_d(x) = \lambda - \dfrac{\lambda^{d+1}}{(d+1)!} + o(\lambda^{d+2})$ for $0 \leq d < \infty$.

TABLE 15.3. Initial Terms in the Taylor Series of $f_d(x)$, $0 \leq d < 4$

$f_0(z) = 0$	$f_1(x) = \lambda - \frac{1}{2!}\lambda^2 + o(\lambda^2)$
$f_2(x) = \lambda - \frac{1}{3!}\lambda^3 + o(\lambda^3)$	$f_3(z) = \lambda - \frac{1}{4!}\lambda^4 + o(\lambda^4)$

15.6 COMPLETING THE PROOF OF THEOREM 15.1

We now have all the preliminary results needed to complete the proof of Theorem 15.1.

We start with equation (15.1) writing

$$C_{\underline{h}}(n) = \frac{1}{m} \sum_{i=0}^{n-1} \delta_i$$

Lemma 15.5 gives a lower bound

$$C_{\underline{h}}(n) \geq \frac{1}{m} \sum_{i=0}^{n-1} f_d(v_i)$$

The convexity of f_d (in λ) yields

$$C_{\underline{h}}(n) \geq \frac{n}{m} f_d\left(\frac{\sum_{i=0}^{n-1} v_i}{n}\right) = \frac{n}{m} f_d\left(\frac{m-d}{n}\right) \quad (15.16a)$$

Now let n, m and $d \to \infty$ with $m \approx \mu n$, write $\frac{d}{n} \approx d^*(n)$ and assume $d^*(n) \to 0$.

In the limit as $n \to \infty$, the bound in equation (15.16a) is replaced by

$$C_{\underline{h}}(n) \geq \frac{1}{\mu} f_d(\mu - d^*(n)) \quad (15.16b)$$

Because $d^*(n) \to 0$ as $n \to \infty$ and

$$\lambda = -\log(1-x) = -\log[1-\mu+d^*(n)] = -\log(1-\mu) - \log\left[1 + \frac{d^*(n)}{1-\mu}\right]$$

$$= -\log(1-\mu) - \frac{d^*(n)}{1-\mu} + o(d^*(n))$$

it follows that

$$0 = \lim_{d \to \infty} \frac{\lambda^{d+1}}{(d+1)!}$$

Exercise 15.5 gives

$$f_d(x) \approx \lambda - \frac{\lambda^{d+1}}{(d+1)!} + o(\lambda^{d+2}) \quad q \geq 1$$

so that

$$f(\mu - d^*(n)) = \frac{1}{1-\mu} - \frac{d^*(n)}{1-\mu}$$

and

$$C_{\underline{h}}(n) \geq \frac{1}{\mu}\log\frac{1}{1-\mu} - \frac{d^*(n)}{1-\mu} \tag{15.16c}$$

If we take $d \approx \frac{C_1}{1-\mu}\log n$, then

$$d^*(n) \approx C_1 \frac{\log n}{n} \tag{15.16d}$$

which substituted into equation (15.16*c*) gives

$$C_{\underline{h}}(n) \geq \frac{1}{\mu}\frac{1}{1-\mu} - C_1 \frac{\log n}{n} \tag{15.16e}$$

completing proof of equation (15.1).

The bound in equation (15.2) is proved by a similar argument; we begin equation (15.16*a*) setting $m = n$ to obtain

$$C_{\underline{h}}(n) \geq f_d\left(1 - \frac{d}{n}\right) \tag{15.17a}$$

If we take $d \approx \frac{C_2^*}{n}\log n$, then

$$x = 1 - \frac{d}{n} \Rightarrow \lambda = -\log(1-x) = -\log\left(\frac{C_2^*}{n}\log n\right) = \log n - \log\log n - \log C_2^* \tag{15.17b}$$

Because

$$0 = \lim_{n\to\infty} \log\left(\frac{C_2^*}{n}\log n\right)$$

equations (15.17*a* and 15.17*b*) implies the conclusion in equation (15.2), completing the proof of Theorem 15.1 and this chapter! ■

REFERENCES

R. Courant, *Differential and Integral Calculus*, Volume 2, Wiley-Interscience (New York), 1957.

F. Maryat, *The Children of the New Forest*, Wordsworth Books (Little Rock, AK), 1847.

T. H. Palmer, *The Teachers Manual*, General Books LLC (Seattle, WA), 1840.

J. D. Ullman, "A Note on the Efficiency of Hashing Functions", *Journal of the ACM*, **19**, #3, pp. 569–575, 1972.

A. C. Yao, "Uniform Hashing is Optimal", *Communications of the ACM*, **32**, #3, pp. 687–693, 1985.

PART III

SOME NOVEL APPLICATIONS OF HASHING

CHAPTER 16

Karp-Rabin String Searching

16.1 OVERVIEW

Let $\underline{y} = (y_0, y_1, \cdots, y_{n-1})$ be string of characters of length $|\underline{y}$ equal to n from an alphabet $\mathcal{A}$.

A basic string search problem **P** is

Given: a string $\underline{x} = (x_0, x_1, \cdots, x_{m-1})$ of m characters from the alphabet $\mathcal{A}$

Determine: whether $\underline{x}$ is a substring of $\underline{y} = (y_0, y_1, \cdots, y_{n-1})$.

When $\underline{x}$ is (*resp.* is not) a substring of $\underline{y}$, we write $\underline{x} \subseteq \underline{y}$ (*resp.* $\underline{x} \not\subseteq \underline{y}$). In the first case, extensions of the search problem **P** include the following:

P1. Find the first/last occurrence of $\underline{x}$ in $\underline{y}$; the first/last index a such that $x_i = y_{i+a}$ for $0 \leq i < m$, or

P2. The set of all occurrences of $\underline{x}$ in $\underline{y}$.

Algorithm #1 below solves **P** by making m bit-comparisons of $\underline{x}$ in each of the $n - m$ possible substrings $\underline{y}_{[i,i+m)} \equiv (y_i, y_{i+1}, \cdots, y_{i+m-1})$ for $i = 0, 1, \cdots, n - m$.

Algorithm #1: **P**
for $i = 0$ to $n - m$ do
Set $\text{IND}_i = 1$
for $j = 0$ to $m - 1$ do
$\text{IND}_i = \begin{cases} 1 & \text{if } y_{i+j} = x_i \\ 0 & \text{otherwise} \end{cases}$

An extensive literature including the paper by Knuth et al. [Knuth, Morris and Pratt 1977]. The performance issues include the solution's running time (number of comparisons and arithmetic operations) and memory required (the number of the registers required).

We describe a novel application of hashing in the paper by R. Karp and M. Rabin [Karp and Rabin 1987] to obtain a solution to the search problem **P** and a multi-dimensional extension. Hashing as used in [Karp and Rabin 1987] is an example of

Hashing in Computer Science: Fifty Years of Slicing and Dicing, by Alan G. Konheim

a digital fingerprint—a quantity derived from a mathematical algorithm—that might be used to identify a data object. The fingerprint algorithm F maps objects in a large domain to their fingerprints in a much smaller domain. The equality of the fingerprints $F(x) \stackrel{?}{=} F(y)$ is used to compare x and y. The advantage, if any, of comparing $F(x)$ and $F(y)$ instead of x and y derives from the simplicity of the computation of and the comparison of the fingerprints. Examples of fingerprints include checksum, hash functions, and randomly generated poly-nomials as described in earlier paper of Rabin [Rabin 1981].

16.2 THE BASIC KARP-RABIN HASH-FINGERPRINT ALGORITHM

We begin with a simple version of **P** assuming the alphabet $\mathcal{A} = \{0, 1\}$. The fingerprints compared in [Karp and Rabin 1987] are the hashing values

$$H(\underline{x}) \equiv x_0 2^{m-1} + x_1 2^{m-2} + \cdots + x_{m-1}$$

and

$$H\left(\underline{y}_{[i,i+m)}\right) \equiv y_i 2^{m-1} + y_{i+1} 2^{m-2} + \cdots + y_{i+m-1} \quad 0 \le i \le n-m$$

Of course, $\underline{x} \subseteq \underline{y}$ if and only if $H(x) = H(\underline{y}_{[i,i+m)})$ for some index i with $0 \le i \le n - m$.

Of course, comparing the fingerprints $H(\underline{y}_{[i-1,i-1+m)})$ and $H(\underline{y}_{[i,i+m)})$ may not necessarily be an improvement over comparing $\underline{x}$ and $\underline{y}_{[i,i+m)}$ if the calculation of $H(\underline{y}_{[i,i+m)})$ for different values of i must be made afresh, independent from one another. The Karp-Rabin fingerprint algorithm performs a shifting hash that recursively calculates the hash value $H(\underline{y}_{[i,i+m)})$, reducing the fingerprint overhead. To test if $\underline{x} \subseteq \underline{y}$, define

$$X \equiv H(\underline{x}) = x_0 2^{m-1} + x_1 2^{m-2} + \cdots + x_{m-1} \tag{16.1a}$$

Note that X may be computed using the recursion

$$X_i = \begin{cases} x_0 & \text{if } i = 0 \\ 2X_{i-1} + x_i & \text{if } 1 \le i < m \end{cases} \tag{16.1b}$$

with

$$H(\underline{x}) = X = X_{m-1} \tag{16.1c}$$

Next, define the sequence $Y_0, Y_1, \cdots, Y_{n-m-1}$ by

$$Y_i = H\left(\underline{y}_{[i,i+m]}\right) = y_i 2^{m-1} + y_{i+1} 2^{m-2} + \cdots + y_{i+m-1} \quad 0 \le i \le n-m \tag{16.2a}$$

The $\{Y_i\}$ satisfy the recursion

$$Y_{i+1} = 2\left(Y_i - y_i 2^{m-1}\right) + y_{i+m} \quad 0 \le i < n-m \tag{16.2b}$$

The computation of (the integer) Y_{i+1} from (the integer) Y_i requires an integer subtraction, an integer addition, and a multiplication (shift). Unfortunately, this new hashing-based implementation of *Algorithm #1* may require as many as $m(n-m+1)$ bit operations. Karp-Rabin offer several versions of their fingerprint algorithm, which improves this by reducing the average number of times the comparison test $\underline{x} \stackrel{?}{=} \underline{y}_{[i,i+m)}$ is made from $n - m + 1$ to a much smaller value.

16.3 THE PLAIN VANILLA KARP-RABIN FINGERPRINT ALGORITHM

Randomly choose a prime p with $p < n^2m$; for integer b, write ${}_pb = b(mod\ p)$.

The plain vanilla Karp-Rabin algorithm **P*** begins by replacing equations (16.1*a* through 16.1*c*) by

$$ {}_pX \equiv {}_pX_{n-1} \tag{16.3b}$$

$$ {}_pX_i = \begin{cases} x_0 \text{ (mod } p) & \text{if } i=0 \\ 2({}_pX_{i-1}+x_{i-1}) \text{(mod } p) & \text{if } 1 \le i < m \end{cases} \tag{16.3a}$$

The recursive computation of ${}_pX_{i+1}$ from ${}_pX_i$ requires one addition and one shift.

Next, replace the sequence $Y_0, Y_1, \cdots, Y_{n-m-1}$ by ${}_pY_0, {}_pY_1, \cdots, {}_pY_{n-m-1}$ defined recursively by

$$ {}_pY_i = \begin{cases} Y_0 \text{ (mod } p) & \text{if } i=0 \\ 2({}_pY_{i-1}-2^{m-1}y_{i-1}+y_{i+m}) \text{(mod } p) & \text{if } 1 \le i < n-m \end{cases} \tag{16.4}$$

The recursive computation of (the integer) ${}_pY_{i+1}$ from (the integer) ${}_pY_i$ requires one (modulo p integer) subtraction, one (modulo p integer) addition, and one shift (modulo p multiplication).

If $\underline{x} \subseteq \underline{y}$, then

$$\underline{x} \subseteq \underline{y} \Leftrightarrow X = Y_i \quad \text{for some } i\ (0 < i < n-m) \tag{16.5a}$$

$$\Rightarrow {}_pX = {}_pY_i \quad \text{for some } i\ (0 < i < n-m) \quad \text{and every } p \tag{16.5b}$$

The reverse implication ${}_pX = {}_pY_i$ for some $i(0 < i < n - m)$ and some $p \Rightarrow \underline{x} \subseteq \underline{y}$ is not true.

Example 16.1. If $\underline{x} = (1, 1)$ and $\underline{y} = (0, 1, 0, 1, 0)$, then $m = 2, n = 5, X = 2^4 + 2^3 = 24$ and

a) $Y_i = 16y_i + 8y_{i+1} = 8$ or 16 so that $X \neq Y_i$ for *any* i with $0 \le i < 3$, whereas
b) If $p = 2$, then ${}_pX = {}_pY_i = 0$.

Although little has been gained by replacing the comparison (16.5*a*) by (16.5*b*) if the comparisons have to be made for every i and p, a big improvement is achieved if different randomly chosen primes are used for ${}_pX \stackrel{?}{=} {}_pY_i$.

Algorithm **P***
0. Randomly choose p (prime) in the interval $1 < p < n^2m$; set $i = 0$.
1. Although $i \le n - m$, compute ${}_pX$, ${}_pY_i$ and compare ${}_pX \stackrel{?}{=} {}_pY_i$.
2a. If ${}_pX \ne {}_pY_i$, increment $i \to i + 1$ and return to **1**.
2b. If ${}_pX = {}_pY_i$, test if $\underline{x} = \underline{y}_{[i,i+m)}$;
* if yes, END (Success);
* if no, $\begin{cases} \text{chose another } p \text{ and GO TO 1} & \text{if } i < n-m \\ \text{END (Failure)} & \text{if } i = n-m \end{cases}$

Remarks

1. The computation of each bit of ${}_pY_{i+1}$ from ${}_pY_i$ requires a (mod p integer) subtraction, a (mod p integer) addition, and a (mod p integer) multiplication (shift).
2. The comparison of the bits in $\underline{x} \ne \underline{y}_{[i,i+m)}$ takes place only when a match occurs.
3. A false match occurs if ${}_pX = {}_pY_i$ but $\underline{x} \ne \underline{y}_{[i,i+m)}$.
4. There are two types of costs in Algorithm **P***;
   a) The updating ${}_pY_i \to {}_pY_{i+1}$ of $O(\log_2 p)$ bit operations, and
   b) The bit-comparison test $\underline{x} \ne \underline{y}_{[i,i+m)}$ performed when ${}_pX \stackrel{?}{=} {}_pY_i$. The costly part of the plain vanilla Rabin-Karp is the m-bit comparisons in the previous step.
5. If ${}_pX = {}_pY_i$ and the test $\underline{x} \stackrel{?}{=} \underline{y}_{[i,i+m)}$ is successful, Algorithm **P*** has a true match, and the search ends with $i = i^*$ and the current value of p.
6. If either $\underline{x} \subseteq \underline{y}$ or $\underline{x} \not\subseteq \underline{y}$, there may be multiple false matches.
7. If $\underline{x} \subseteq \underline{y}$, the execution of Algorithm **P*** may be described as follows:
   a) i^* is the first index of i for which there is a match; that is, $\underline{x} = \underline{y}_{[i^*,i^*+m)}$. Furthermore,
   b) There are r false matches with indices $i_0, i_1, \cdots, i_{r-1}$ with $0 \le i_0 < i_1 \cdots < i_{r-1} < i_r = i^*$ and primes $p_0, p_1, \cdots, p_{r-1}$.
   That is,
   - ${}_{p_0}X \ne {}_{p_0}Y_j$ for $0 \le j < i_0$
   - $\begin{cases} {}_{p_k}X = {}_{p_k}Y_j \text{ and } \underline{x} \ne \underline{y}_{[i_k,i_k+m)} & \text{for } j = i_k \\ {}_{p_{k+1}}X \ne {}_{p_{k+1}}Y_j & \text{for } i_k < j < i_{k+1} \end{cases} \quad \text{for } 0 \le k \le r-1$
   - ${}_{p_r}X = {}_{p_r}Y_{i^*}$ and $\underline{x} = \underline{y}_{[i^*,i^*+m)}$
8. If $\underline{x} \not\subseteq \underline{y}$, the execution of Algorithm **P*** may be described as follows:
   a) There are no true matches.
   b) There are r false matches with indices $i_0, i_1, \cdots, i_{r-1} < i_r = n - m$ with $0 \le i_0 < i_1 \cdots < i_{r-1}$ and primes $p_0, p_1, \cdots, p_{r-1}, p_r$.
   That is,
   - ${}_{p_0}X \ne {}_{p_0}Y_j$ for $0 \le j < i_0$
   - $\begin{cases} {}_{p_k}X = {}_{p_k}Y_j \text{ and } \underline{x} \ne \underline{y}_{[i_k,i_k+m)} & \text{for } j = i_k \\ {}_{p_{k+1}}X \ne {}_{p_{k+1}}Y_j & \text{for } i_k < j < i_{k+1} \end{cases} \quad \text{for } 0 \le k \le r-1$

9. If the match in step **2b** of Algorithm **P*** only occurs when $X = Y_i$, the cost of plain vanilla Rabin-Karp would consist of $O(n - m + 1)$—the cost of step 1—plus the cost a single m-bit comparison.
10. The performance improvement in the plain vanilla Rabin-Karp fingerprint algorithm occurs because they prove many false matches are unlikely. In fact, if the primes are chosen from the interval $[1, M]$ with M large enough, it is unlikely to have even a single false match.

16.4 SOME ESTIMATES ON PRIME NUMBERS

Plain vanilla requires the random selection of a prime number and so we begin with some refinements of the material in Chapter 5. A randomly chosen prime (in some interval) can be obtained by either of the following steps:

1. Randomly (uniformly) selecting an integer N in the interval $[1, M]$
2. Checking whether N is a prime

Number 2 in the previous list might be implemented as described in Chapter 5 (§5.7) using either the Miller-Rabin algorithm [Rabin 1981] or the earlier Solovay-Strassen algorithm [Solovay and Strassen 1977].

We need some preliminary estimates relating to the distribution of prime numbers. Let $\pi(n)$ denote the number of prime numbers $\leq n$; we use the Prime Number Theorem (Chapter 5, Theorem 5.1) and a refinement of it (Chapter 5, Lemma 5.2) from [Rosser and Schoenfeld 1962] where $\mathcal{P}_n$ denotes the set of primes $\leq n$.

Exercise 16.1. Use Lemma 5.2a (Chapter 16) to prove $\prod_{p \in \mathcal{P}_n} p > 2^n$ if $n \geq 49$.

Exercise 16.2. Tabulate $n, \pi(n), 2^n$ and $Q_n \equiv \prod_{p \in \mathcal{P}_n} p$ for $n = 29(1)48$ to prove $\prod_{p \in \mathcal{P}_n} p > 2^n$ if $29 \leq n \leq 48$.

Exercise 16.3. Prove that if $n \geq 29$ and $\eta \leq 2^n$, then η has fewer that $\pi(n)$ different prime divisors.

16.5 THE COST OF FALSE MATCHES IN THE PLAIN VANILLA KARP-RABIN FINGERPRINT ALGORITHM

Choose a large integer M and assume primes are chosen from the interval $[1, M]$. The performance analysis in [Karp and Rabin 1987] for fixed X and sequence Y_i begins by observing that a false match occurs for a pair in the set $\{({}_pX, {}_pY_i) : 0 \leq i \leq n - m\}$, for the prime $p \in [1, M]$ is equivalent to any of the following statements:

1. For some i, p is a divisor of ${}_pX - {}_pY_i$ and $\underline{x} \neq \underline{y}_{[i,i+m)}$.
2. For some i such that $\underline{x} \neq \underline{y}_{[i,i+m)}$, we have $p/\{H(\underline{x}) - H(\underline{y}_{[i,i+m)})\}$.
3. $p \Big/ \prod_{i, \underline{x} \neq \underline{y}_{[i,i+m)}} \{H(\underline{x}) - H(\underline{y}_{[i,i+m)})\}$.

Theorem 16.1. If n, m satisfy $m(n - m + 1) \geq 29$ and the prime p in the plain vanilla Rabin-Karp algorithm (§16.2) is chosen uniformly from the set of integers $\mathcal{P}_M$, then the probability of a false match for the pair $({}_pX, {}_pY_i)$ is bounded above by

$$\frac{\pi(m(n-m+1))}{\pi(M)}$$

Proof. For each i with $0 \leq i \leq n - m + 1$, we have

$$\left|H(\underline{x}) - H\left(\underline{y}_{[i,i+m)}\right)\right| < 2^m$$

and therefore

$$\prod_{i, \underline{x} \neq \underline{y}_{[i,i+m)}} \left|H(\underline{x}) - H\left(\underline{y}_{[i,i+m)}\right)\right| < 2^{(n-m)m}$$

There are $\pi(M)$ primes p for the set $[1, M]$, and only $\pi(m(n - m + 1))$ will result in a false match so that Theorem 16.1 follows from *Exercise 16.3*. ■

Corollary 16.2

a) The probability of at least one false match $Pr\{\mathcal{F}_{n,m,M}\}$ with $M \geq m(n - m + 1)^2$ and $m(n - m + 1) \geq 29$ satisfies $Pr\{\mathcal{F}_{n,m,M}\} \leq \frac{2.511012}{n-m+1}$.

b) The probability of r or more false match $Pr\{\mathcal{F}^{(r)}_{n,m,M}\}$ with $M \geq m(n - m + 1)^2$ and $m(n - m + 1) \geq 29$ satisfies $Pr\{\mathcal{F}^{(r)}_{n,m,M}\} \leq \left(\frac{2.511012}{n-m+1}\right)^r$.

Proof. We use both of the bounds on $\pi(n)$ given in Chapter 5, Lemma 5.2b valid when $n \geq 17$. Thus

$$\frac{n}{\log n} \leq \pi(n) \leq 1.255506 \frac{n}{\log n}$$

so that by Theorem 3.1

$$Pr\{\mathcal{F}_{n,m,M}\} \leq 1.255506 \frac{n-m+1}{\pi(M)\log(n-m+1)}$$

and if $M \geq m((n - m + 1)^2$, then

$$\pi(M) \geq \frac{m(n-m+1)^2}{\log\left(m(n-m+1)^2\right)} = \frac{m(n-m+1)^2}{2\log(m(n-m+1))}$$

which when combined proves Corollary 16.2a.

Since the primes are chosen independently, the matching renews itself and Corollary 1.2b follows. ■

16.6 VARIATIONS ON THE PLAIN VANILLA KARP-RABIN FINGERPRINT ALGORITHM

Karp and Rabin describes two variations on (what I refer to as) the plain vanilla search algorithm.

Algorithm KR1

---

0. Randomly choose k primes $p_1, p_2, \cdots, p_k$ in the interval $1 < p < M$; set $i = 0$.
1. While $i \le n - m$, compute ${}_{p_j}X$, ${}_{p_j}Y_i$ and compare ${}_{p_j}X \stackrel{?}{=} {}_{p_j}Y_i$ for all $j = 1, 2, \cdots, k$.

2a. If ${}_{p_j}X \neq {}_{p_j}Y_i$, for some j, then increment $i \to i + 1$ and return to **1**.

2b. If ${}_{p_j}X = {}_{p_j}Y_i$ for all $j = 1, 2, \cdots, k$, then test if $\underline{x} = \underline{y}_{[i,i+m)}$;
  * if yes, END (Success);
  * if no, then increment $i \to i + 1$ and return to **1**.

END (Failure) if $i = n - m$

Remarks. The same argument used in §16.5 shows that if $m(n - m + 1) \ge 29$, then a false match at position i

$${}_{p_j}X \stackrel{?}{=} {}_{p_j}Y_i \quad \text{for } j = 1, 2, \cdots, k$$

occurs with probability $\left(\dfrac{\pi(m(n-m+1))}{\pi(M)}\right)^k$.

The Karp-Rabin algorithm KR2 uses only one prime, that is, $k = 1$.

16.7 A NONHASHING KARP-RABIN FINGERPRINT

We conclude with a second Karp-Rabin example of a fingerprint, one that does not involve hashing, which leaves most of the details to the reader.

We need some ideas described in Chapter 5; the Fibonacci sequence $\{F_n : 0 \le n < \infty\}$ was defined recursively in §2.1 of Chapter 2 by

$$F_n = \begin{cases} 1 & \text{if } n = 0, 1 \\ F_{n-2} + F_{n-1} & \text{if } 2 \le n < \infty \end{cases} \qquad (16.6a)$$

The solution was shown in *Example 2.4* of in Chapter 2 to be

$$F_n = \frac{1}{\sqrt{5}}\left(\frac{1+\sqrt{5}}{2}\right)^{n+1} - \frac{1}{\sqrt{5}}\left(\frac{1-\sqrt{5}}{2}\right)^{n+1} \quad (0 \le n < \infty) \qquad (16.6b)$$

from which we deduce

$$F_n \approx \frac{1}{\sqrt{5}}\left(\frac{1+\sqrt{5}}{2}\right)^{n+1} \approx 0.72360679775(1.6180339888)^n \quad \text{as } n \to \infty \qquad (16.6c)$$

The second Karp-Rabin fingerprint function is matrix valued; let $\mathbf{I} = \begin{pmatrix} 1 & 0 \\ 0 & 1 \end{pmatrix}$ and

$$\mathbf{K}_0 = \begin{pmatrix} 1 & 0 \\ 1 & 1 \end{pmatrix} \quad \mathbf{K}_0^{-1} = \begin{pmatrix} 1 & 0 \\ -1 & 1 \end{pmatrix} \tag{16.7a}$$

$$\mathbf{K}_1 = \begin{pmatrix} 1 & 1 \\ 0 & 1 \end{pmatrix} \quad \mathbf{K}_1^{-1} = \begin{pmatrix} 1 & -1 \\ 0 & 1 \end{pmatrix} \tag{16.7b}$$

The mapping **K** on strings $\underline{y} = (y_0, y_1, \cdots, y_{n-1})$ is defined by

$$\mathbf{K}_{\underline{y}} = \begin{cases} \mathbf{I} & \text{if } n = 0 \\ \mathbf{K}_{y_0} \times \mathbf{K}_{y_1} \times \cdots \times \mathbf{K}_{y_{n-1}} & \text{if } 1 \le n < \infty \end{cases} \tag{16.7c}$$

Table 16.1 gives the values of $\mathbf{K}\,\underline{y}$ for all strings $\underline{y}$ of length $|\underline{y}| = 5$.

Exercise 16.4. Prove that if $\underline{y}$ and $\underline{z}$ are (0,1)-strings of the same length, then $\mathbf{K}_{\underline{y}} = \mathbf{K}_{\underline{z}}$ if and only if $\underline{y} = \underline{z}$.

Query: Did we need to add the quantifier "··· of the same length"?

Define

$$\underline{u}_0 = (1, 0) \quad \underline{u}_1 = (0, 1)$$

TABLE 16.1. $\mathbf{K}_{\underline{y}}$ for Strings $|\underline{y}| = 5$

$\underline{y}$	$K_{\underline{y}}$	$\underline{y}$	$K_{\underline{y}}$	$\underline{y}$	$K_{\underline{y}}$	$\underline{y}$	$K_{\underline{y}}$
00000	$\begin{pmatrix} 1 & 0 \\ 5 & 1 \end{pmatrix}$	00001	$\begin{pmatrix} 1 & 1 \\ 4 & 5 \end{pmatrix}$	10000	$\begin{pmatrix} 5 & 1 \\ 4 & 1 \end{pmatrix}$	10001	$\begin{pmatrix} 4 & 5 \\ 3 & 4 \end{pmatrix}$
00010	$\begin{pmatrix} 2 & 1 \\ 7 & 4 \end{pmatrix}$	00011	$\begin{pmatrix} 1 & 2 \\ 3 & 7 \end{pmatrix}$	10010	$\begin{pmatrix} 7 & 4 \\ 5 & 3 \end{pmatrix}$	10011	$\begin{pmatrix} 3 & 7 \\ 2 & 5 \end{pmatrix}$
00100	$\begin{pmatrix} 3 & 1 \\ 8 & 3 \end{pmatrix}$	00101	$\begin{pmatrix} 2 & 3 \\ 5 & 8 \end{pmatrix}$	10100	$\begin{pmatrix} 8 & 3 \\ 5 & 2 \end{pmatrix}$	10101	$\begin{pmatrix} 5 & 8 \\ 3 & 5 \end{pmatrix}$
00110	$\begin{pmatrix} 3 & 2 \\ 7 & 5 \end{pmatrix}$	00111	$\begin{pmatrix} 1 & 3 \\ 2 & 7 \end{pmatrix}$	10110	$\begin{pmatrix} 7 & 5 \\ 4 & 3 \end{pmatrix}$	10111	$\begin{pmatrix} 2 & 7 \\ 1 & 4 \end{pmatrix}$
01000	$\begin{pmatrix} 4 & 1 \\ 7 & 2 \end{pmatrix}$	01001	$\begin{pmatrix} 3 & 4 \\ 5 & 7 \end{pmatrix}$	11000	$\begin{pmatrix} 7 & 2 \\ 3 & 1 \end{pmatrix}$	11001	$\begin{pmatrix} 5 & 7 \\ 2 & 3 \end{pmatrix}$
01010	$\begin{pmatrix} 5 & 3 \\ 8 & 5 \end{pmatrix}$	01011	$\begin{pmatrix} 2 & 5 \\ 3 & 8 \end{pmatrix}$	11010	$\begin{pmatrix} 8 & 5 \\ 3 & 2 \end{pmatrix}$	11011	$\begin{pmatrix} 3 & 8 \\ 1 & 3 \end{pmatrix}$
01100	$\begin{pmatrix} 5 & 2 \\ 7 & 3 \end{pmatrix}$	01101	$\begin{pmatrix} 3 & 5 \\ 4 & 7 \end{pmatrix}$	11100	$\begin{pmatrix} 7 & 3 \\ 2 & 1 \end{pmatrix}$	11101	$\begin{pmatrix} 4 & 7 \\ 1 & 2 \end{pmatrix}$
01110	$\begin{pmatrix} 4 & 3 \\ 5 & 4 \end{pmatrix}$	01111	$\begin{pmatrix} 1 & 4 \\ 1 & 5 \end{pmatrix}$	11110	$\begin{pmatrix} 5 & 4 \\ 1 & 1 \end{pmatrix}$	11111	$\begin{pmatrix} 1 & 5 \\ 0 & 1 \end{pmatrix}$

and note that

$$\mathbf{K}_0\underline{u}_0 = \underline{u}_0 + \underline{u}_1 \quad \mathbf{K}_0\underline{u}_1 = \underline{u}_1 \tag{16.8a}$$

$$\mathbf{K}_1\underline{u}_0 = \underline{u}_0 \quad \mathbf{K}_1\underline{u}_1 = \underline{u}_1 + \underline{u}_1 \tag{16.8b}$$

so that

$$\mathbf{K}_{\underline{y}}\underline{u}_0 = A_{\underline{y}}\underline{u}_0 + B_{\underline{y}}\underline{u}_1 \tag{16.9a}$$

$$\mathbf{K}_{\underline{y}}\underline{u}_1 = C_{\underline{y}}\underline{u}_0 + D_{\underline{y}}\underline{u}_1 \tag{16.9b}$$

Table 16.1 lists the values of the right-hand sides of equations (16.8*a* and 16.8*b*) for all strings of length 5. A modicum of concentration enables one to recognize the pattern.

Exercise 16.5. Prove that if $|\underline{y}| = n$, then each element in $\mathbf{K}_{\underline{y}}$ is bounded above by F_n.

Karp and Rabin define $K_{\underline{x}}$ to be the fingerprint of the sequence $\underline{x}$. Note that *Exercise 16.4* asserts that the fingerprint of $K_{\underline{x}}$ unambiguously identifies $\underline{x}$.

We use the fingerprint function ${}_pK$ with $p \le M$ to test if $\underline{x} \in \{\underline{y}_t : 1 \le t \le T\}$. A false match occurs if

$$\underline{x} \notin \{\underline{y}_t : 1 \le t \le T\} \quad \text{and} \quad {}_pK\underline{x} = {}_pK\underline{y}_t \quad \text{for } p \in \mathcal{P}_M, 1 \le t \le T \tag{16.10}$$

Theorem 16.2. If algorithm KR1 and the fingerprint function ${}_pK$ is used to compare $\underline{x}$ with $\{\underline{y}_t : 1 \le t \le T\}$, then

16.2a. if $p \in \mathcal{P}_M$, the probability of a false match satisfies

$$Pr\{\mathcal{F}\} \le \frac{\pi(\lceil 4T\log_2 F_n \rceil)}{\pi(M)} \quad \text{if} \quad \lceil 4T\log_2 F_n \rceil \ge 29 \tag{16 .11a}$$

16.2b. if $p \in \mathcal{P}nT^2$, the probability of a false match satisfies

$$Pr\{\mathcal{F}\} \le \frac{6.971}{T} \tag{16.11b}$$

Proof. Suppose

$$K_{\underline{x}} \ne K_{\underline{y}_t} \quad 1 \le t \le T \tag{16.12a}$$

$${}_pK_{\underline{x}} = {}_pK_{\underline{y}_t} \quad 1 \le t \le T, p \in \mathcal{P}_M \tag{16.12b}$$

It follows from equation (16.12*b*) that p divides every nonzero element in $K_{\underline{x}} - K_{\underline{y}_t}$. But every element in $K_{\underline{z}}$ is bounded by F_n so that the product of the nonzero elements in $K_{\underline{x}} - K_{\underline{y}_t}$ is bounded by $F_n^{4T} \le 2^{\lceil 4T\log_2 F_n \rceil}$. By *Exercise 16.3*, the number of such primes that divide this product is $\le \pi(\lceil 4T\log_2 F_n \rceil)$, which completes the proof of Theorem 16.2a.

The second assert follows from the asymptotic growth in equation (16.6*e*).

REFERENCES

R. M. Karp and M. O. Rabin, "Efficient Randomized Pattern-Matching Algorithms", *IBM Journal of Research and Development*, **31** #2, pp. 249–260, 1987.

D. E. Knuth, J. H. Morris, and V. R. Pratt, "Fast Pattern Matching in Strings", *SIAM Journal of Computing*, **6**, pp. 323–350, 1977.

M. O. Rabin, "Probabilistic Algorithm for Testing Primality", *Journal of Number Theory*, **12**, pp. 128–138, 1980.

M. O. Rabin, "Fingerprinting by Random Functions", Report TR-15-81, Center for Research in Computing Technology, Harvard University, 1981.

J. B. Rosser and L. Schoenfeld, "Approximate Formulas for Some Functions of Prime Numbers", *Illinois Journal of Mathematics*, **6**, pp. 64–94, 1962.

R. Solovay and V. Strassen, "A Fast Monte-Carlo Test for Primality", *SIAM Journal of Computing*, **6**, pp. 84–85, 1977.

CHAPTER 17

Hashing Rock and Roll

17.1 OVERVIEW OF AUDIO FINGERPRINTING

You are a contestant on Triple Jeopardy and Alex Trebek asks "For $10,000, what performance group popularized this song?"

Is this Glenn Miller, Mozart, or U2?

Even though the reader might not actually be faced with such a question, a more practical application is provided by the requirements of The Music Performance Trust Fund (MPTF). Their web page explains that the MPTF is "··· the world's largest sponsor of live admission-free, professional musical programs." Founded in 1948, the MPTF sponsors the public performance of free live music. An agreement between the recording industry and the American Federation of Musicians (AFM) sets forth the terms under which royalties on the sale and performance of recorded music are paid into the MPTF by recording companies. **Audio fingerprinting** is a technique to identify an audio signal; it makes it possible the following:

Hashing in Computer Science: Fifty Years of Slicing and Dicing, by Alan G. Konheim

- To generate efficiently playlists for automatic royalty collection
- To verify automatically that a commercial for some useless, and perhaps even harmful, drug product has been broadcast according to the advertiser's contact.
- To enhance people metering; big brother/sister not only can watch you but also can check your musical tastes. Are you listening to Osama's favorite jihad nasheed, "I'm Going to Kill, Kill, Kill ⋯"?

17.2 THE BASICS OF FINGERPRINTING MUSIC

An (analog) audio signal $s(t)$ is a (real-valued) function representing the waveform (shape) of sound as recorded by a microphone and converted to an electrical signal.

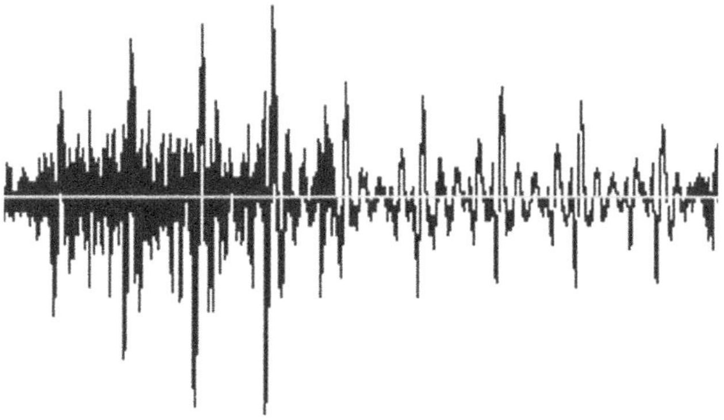

The fingerprinting technique described in Chapter 16 mapped a string with values in a large domain into a numerical fingerprint in a much smaller domain. Fingerprinting has to be adjusted for audio applications because the human ear perceives audio signals in different ways. Audio compression systems, for example MPEG-1 Audio Layer 3 (MP3) or Windows Media Audio (WMA), result in different binary encodings of the signal $s(t)$ of a song.

An audio signal $s(t)$ is usually modeled as a periodic signal; the derived fingerprint must take into account the perceptual characteristics of $s(t)$, including the follwing:

1. The average zero crossing rate—the rate at which the voltage of the signal attains the value 0;
2. The tempo of the signal; *andante* (slow) and *presto* (fast), which is often measured in units of bits per minute (BPM);
3. Characteristics of the spectrum, the Fourier-transform of the time-signal $s(t) \rightarrow \hat{s}(\omega) \equiv \int s(t)e^{-2\pi i t}dt$;
4. The bandwidth of the signal.

So far, no hashing! [Özer, Sankurm and Memom 1984] observe the following:

- Hashing for data storage which is described in Chapters 7 through 15
- Cryptography, which is described in Chapters 16 through 18

is intended to be both

- *Un*predictable (random) and
- Fragile in the sense that hash values are sensitive to input data; a small change in the input should be detected in the hashed-value output.

Fingerprinting, as applied in audio identification, it is intended on the contrary to be robust, largely insensitive to variability in signal-processing operations (processing schema, filtering, and compression). In fact, various papers in audio fingerprinting refer to perceptual hashing.[1]

The two components to audio identification are as follows:

i) The fingerprinting process
ii) The retrieval process

If the song database contains 10000 songs and many fingerprints are created from a single audio sample, we are faced with searching for a match in a file of ten billion entries. Not only will we need to find *Song X*, but perhaps we will need to identify the performing group (artist), such as *Amazing Joy Buzzards* or *Baldwin and the Whiffles* or *The Barbusters* or *The Barn Burners* or *The Bedbugs* ··· This is truly a daunting challenge, but audio fingerprinting is up to the task!

The hashing aspect of audio identification is related to searching large dimensional databases. This has been studied extensively; for example, our own National Security Agency (NSA) has been generous in support of such endeavors [Bramford 2008]. One approach is described in [Gionis, Indyk and Motwani 1998].

We follow the Waveprint, which is an audio fingerprinting system described by Baluja and Covell in [2007] and improved on in [Baluja, Covell and Ioffe 2008]. Their work has much in common with earlier work by Haitsma and Kalker [2002].

[Baluja and Covell 2007] derive a fingerprint by combining many spectral images obtained from frequency information derived by sampling an audio signal $s(t)$ on overlapping time intervals (windows). Sampling even a pure signal with one frequency component would introduce extraneous frequencies, but changing the shape of the time-domain window reduces this. [Baluja and Covell 2007] derive use a 370-ms Hanning window [Harris 1978] with a short-time Fourier transform (STFT),[2] discarding the phase component of the STFT and combining with triangular weighting the 32 nonoverlapping logarithmically spaced frequency bands from 318 Hz to 2000 Hz.

[1] *Perceptual hash* n. a fingerprint of an audio, video, or image file that is mathematically based on the audio or visual content contained within. Unlike cryptographic hash functions that rely on the avalanche effect of small changes in input leading to drastic changes in the output, perceptual hashes are "close" to one another if the inputs are visually or auditorially similar.

[2] $s(t) \to \hat{s}[f] \equiv \int s(t+\tau)w(\tau)e^{-2\pi i f \tau}d\tau$

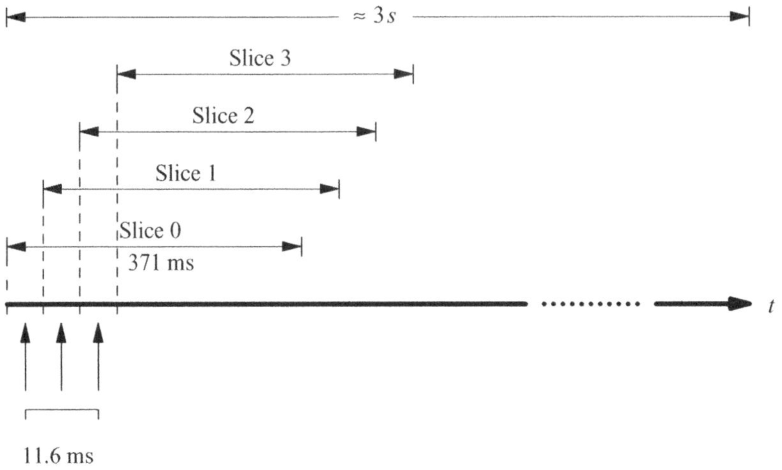

Figure 17.1. Baluja-Covell audio slices.

The audio signal $s(t)$ is sampled over 256 overlapping time slices each of length 371 ms taken every 11.6 ms over an interval of average duration $256 \times 11.6 \approx 3$ s as portrayed in Figure 17.1.

Each (time-slice, frequency-band) pair gives a 32-bit floating point number that is largely insensitive to the high-frequency content in music (e.g., cymbals) and to across-frequency phase changes (e.g., echo- or amplifier-induced effects).

The steps in the Baljua-Covell audio fingerprinting process are as follows:

1. Compute the Haar-wavelets on the spectral images
2. Extract the top-T wavelets, measured by magnitude
3. Create a binary representation of the top-T Haar-wavelets

We need to make two digressions to explain the previous retrieval processes.

17.3 HAAR WAVELET CODING

The Haar wavelets are a special sequence of functions; the Haar wavelet transformation is used to code audio information. The scaling function $\phi(t)$ is defined by

$$\phi(t) = \begin{cases} 1 & \text{if } 0 \le t < 1 \\ 0 & \text{otherwise} \end{cases} \tag{17.1a}$$

From $\phi(t)$ scaled and translated functions $\{\Phi_{i,j}(t) : 0 \le j \le 2^i - 1\}$ are defined for $i = 0, 1, \cdots$ by

$$\Phi_{i,j}(t) = \phi(2^i t - j) \quad 0 \le i < \infty, \quad 0 \le j < 2^i \tag{17.1b}$$

Figures 17.2 and 17.3 show these box functions for $i = 0, 1, 2$.

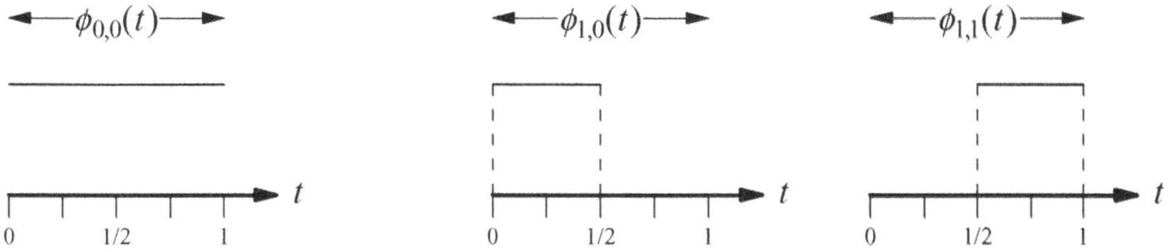

Figure 17.2. $\Phi_{0,j}$ for $j = 0$ and $\Phi_{1,j}$ (t) for $0 \le j < 2$.

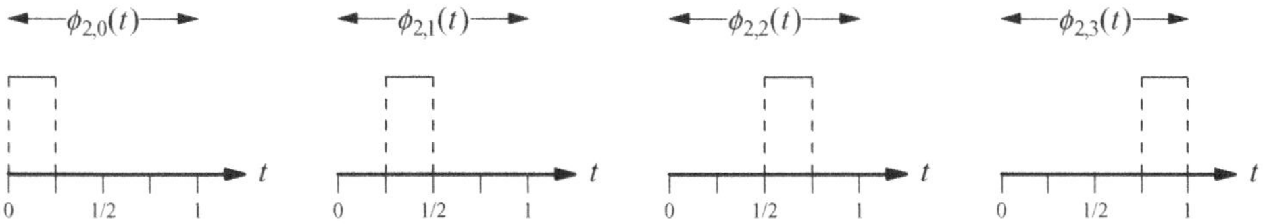

Figure 17.3. $\Phi_{2,j}(t)$ for $0 \le j \le 3$.

The Haar wavelet functions are defined by

$$\psi(t) = \begin{cases} 1 & \text{if } 0 \le t < \dfrac{1}{2} \\ -1 & \text{if } \dfrac{1}{2} < t \le 1 \\ 0 & \text{otherwise} \end{cases} \tag{17.2a}$$

and the family of Haar wavelets by

$$\Psi_{i,j} = \psi(2^i t - j) \quad 0 \le i < \infty, \quad 0 \le j < 2^i \tag{17.2b}$$

Figure 17.4 depicts $\Psi_{0,0}$ and $\Psi_{1,j}$ for $j = 0, 1$.

Figure 17.5 depicts $\Psi_{2,j}$ for $j = 0, 1, 2, 3$.

The functions $\{\Psi_{i,j}(t)\}$ first appeared in the paper [Haar 2010] by Alfréd Haar (1885, 1933). Defined in terms of the square wave $\psi(t)$, these functions are also related to the Walsh-Rademacher functions named for both Joseph Leonard Walsh (1895–1973) and Hans Adolph Rademacher (1892–1969). Wavelet coding arises in a natural way when images are represented in digital form. An image is identified with a positive integer-valued function f on a rectangle in the plane. Digital images are obtained when a grid divides the rectangle into pixels (or picture elements); the luminance in each pixel is assigned a value, normally from 0 to 255 (8 bits). The value of $f(x)$ at a pixel x is the luminance (or brightness) of the image at x. If a rectangle is divided into N^2 pixels, say $N = 512$, a canonical digital image may require $512 \times 512 \times 8 = 2\,097\,152 = 2^{21}$ bits. Wavelet coding is a method of image compression needed to reduce the storage. When wavelet coding is applied to audio fingerprinting, there are two dimensions; the first is time, and the second is power in a specific frequency band.

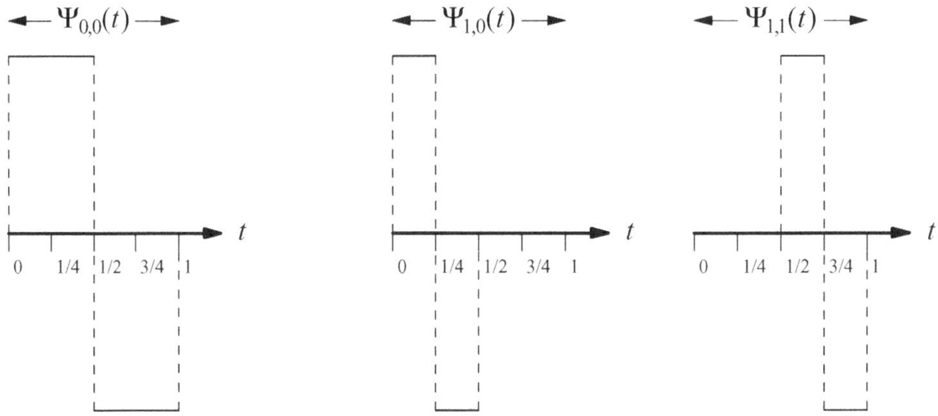

Figure 17.4. The Haar Wavelets $\Psi_{0,0}$ and $\Psi_{1,j}$ for $j = 0, 1$.

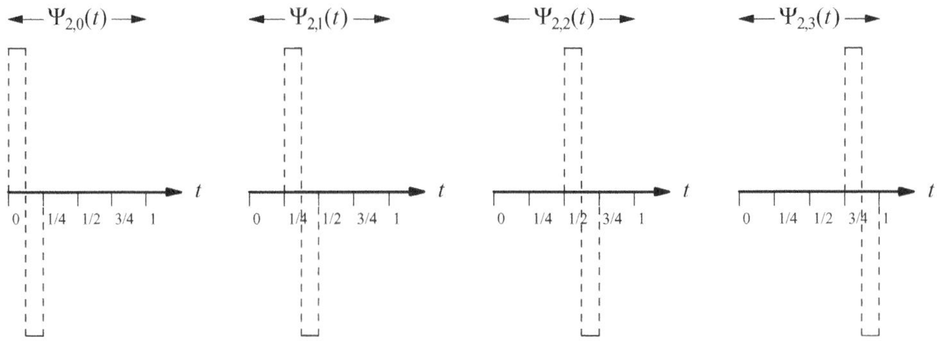

Figure 17.5. The Haar Wavelets $\Psi_{2,j}$ for $j = 0, 1, 2, 3$.

We begin by describing the one-dimensional wavelet coding, next the more natural two-dimensional setting and finally show is application to audio fingerprinting.

Let f be a (real-valued) function on the set of integers $\mathcal{Z}_{2^n} = \{0, 1, 2, \cdots, 2^n - 1\}$. The coefficient vector $\underline{f} = (f_0, f_1, \ldots, f_{2^n-1})$ may be viewed as representing the the function on the interval [0, 1] whose value in the subinterval

$$\left[\frac{i}{2^n}, \frac{i+1}{2^n}\right)$$

is f_i.

Example 17.1. If $n = 3$ and $\underline{f} = (6, 9, 8, 3, 1, 3, 7, 3)$, the function f is shown in Figure 17.6.

The vector space of the canonical representation of such functions can be coded into the box functions $\Phi_{n,j}(t)$ with $0 \le j < 2^n$.

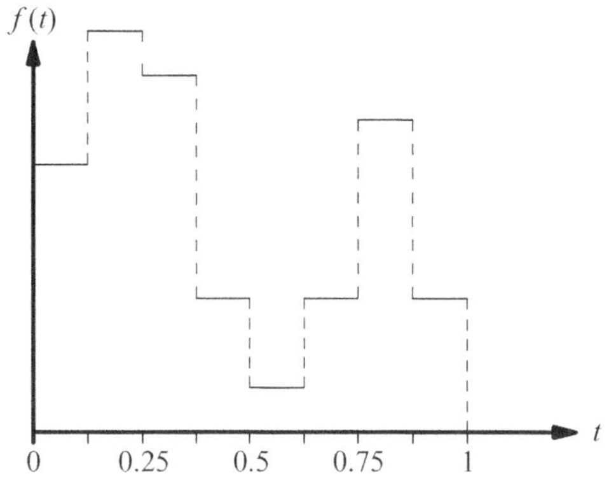

Figure 17.6. Graph of the *Example 17.1* function.

Example 17.1 (continued): If $\underline{f}$ = (6, 9, 8, 3, 1, 3, 7, 3), then

$$\underline{f} = 6\Phi_{3,0} + 9\Phi_{3,1} + 8\Phi_{3,2} + 3\Phi_{3,3} + \Phi_{3,4} + 3\Phi_{3,5} + 7\Phi_{3,6} + 3\Phi_{3,7} \qquad (17.3a)$$

However, it is possible to code $\underline{f}$ by replacing each pair (f_{2j}, f_{2j+1}) by (g_j, h_j), where $g_j = \frac{1}{2}[f_{2j} + f_{2j+1}]$ and the detail coefficient [Stollnitz, DeRose and Salesin 1995] is $h_j = f_{2j} - g_j$.

$$\begin{aligned} \underline{f} &\to 7.5\Phi_{2,0} - 1.5\Psi_{2,0} + 5.5\Phi_{2,1} + 2.5\Psi_{2,1} + 2\Phi_{2,2} - \Psi_{2,2} + 5\Phi_{2,3} + 2\Psi_{2,2} \\ &= 7.5\Phi_{2,0} + 5.5\Phi_{2,1} + 2\Phi_{2,2} + 5\Phi_{2,3} + \{-1.5\Psi_{2,0} + 2.5\Psi_{2,1} - \Psi_{2,2} + 2\Psi_{2,2}\} \qquad (17.3b) \end{aligned}$$

Now, we repeat the process, replacing the $\{\Phi_{2,\cdot}\}$ functions by $\{\Phi_{1,\cdot}\}$ and $\{\Psi_{1,\cdot}\}$ functions

$$\underline{f} \to 4.5\Phi_{1,0} + 3.5\Phi_{1,1} + \{3\Psi_{1,0} - 1.5\Psi_{1,1} - 1.5\Psi_{2,0} + 2.5\Psi_{2,1} - \Psi_{2,2} + 2\Psi_{2,2}\} \qquad (17.3c)$$

The final reduction yields

$$\underline{f} \to 4\Phi_{0,0} + \{0.5\Phi_{0,0} + 3\Psi_{1,0} - 1.5\Psi_{1,1} - 1.5\Psi_{2,0} + 2.5\Psi_{2,1} - \Psi_{2,2} + 2\Psi_{2,2}\} \qquad (17.3d)$$

The wavelet transformation is

$$\underline{f} \to \mathcal{H}[\underline{f}] = 4\Phi_{0,0} + \{0.5\Psi_{0,0} + 3\Psi_{1,0} - 1.5\Psi_{1,1} - 1.5\Psi_{2,0} + 2.5\Psi_{2,1} - \Psi_{2,2} + 2\Psi_{2,2}\} \qquad (17.3e)$$

To simplify the notation, we write

$$\mathcal{H}[\underline{f}] = 4, 0.5, 3, -1.5, -1.5, 2.5, -1, 2 \qquad (17.3f)$$

for the Haar wavelet transformation of data $\underline{f}$.

The important observation is that no information has been lost in the transformation $\underline{f} \to \mathcal{H}[\underline{f}]$; the original data $\underline{f}$ can be recovered from $\mathcal{H}[\underline{f}]$.

Example 17.1. (continued). From $\mathcal{H}[\underline{f}] = 4, 0.5, 3, -1.5, -1.5, 2.5, -1, 2$, we can obtain $\underline{f}$. First, $\mathcal{H}[\underline{f}]$ in equation (17.3*e*) recovers $\mathcal{H}[\underline{f}]$ in equation (17.3*d*) and so forth back to $\underline{f}$ in equation (17.3*a*).

Exercise 17.1. Prove that the Haar wavelet transformation $\underline{f} \to \mathcal{H}[\underline{f}]$ is invertible.

What is the advantage of Haar wavelet coding? Three examples will suggest the answer.

Exercise 17.2. Calculate $\mathcal{H}[\underline{f}]$ if $\underline{f} = (c)_8 \equiv \underbrace{(c, c, \cdots, c)}_{8 \text{ copies}}$.

Exercise 17.3. Calculate $\mathcal{H}[\underline{f}]$ if

a) $\underline{f} = (c)_4, (d)_4$
b) $\underline{f} = (c)_4, (d)_3$

Exercise 17.4. Prove that if $\underline{f} = (f_0, f_1, \ldots, f_{2^{n-1}})$, then $\mathcal{H}[\underline{f}] = \frac{1}{2^n}[f_0 + f_1 + \cdots + f_{2^{n-1}}], \cdots$.

For the sampled function in two dimensions, $\underline{f} = (f_{i,j})$ with $0 \leq i < 2^{n-1}$ and $0 \leq j < 2^{m-1}$, the wavelet transformation may be applied to each of the functions $\underline{f}_j = (f_{0,j}, f_{1,j}, \cdots, f_{2^{n-1},j})$.

In [Baluja and Covell 2007], $f_{i,j}$ corresponds to the power in the jth frequency band in the ith slice with $0 \leq i < 2^{n-1}$ and $0 \leq j < 2^{m-1}$ for $n = 8$, $m = 5$. Each Haar wavelet is a vector of 256 (32-bit) floating point values, the first component of which is the average $\frac{1}{256}\Sigma_i f_{i,j}$. The Haar wavelets $\{\mathcal{H}[\underline{f}_j]:j\}$ are sorted (in decreasing size of the first wavelet (32-bit floating point) component creating the wavelet signature

$$\mathcal{H}[\underline{f}_{\pi 0}], \mathcal{H}[\underline{f}_{\pi 1}], \cdots, \mathcal{H}[\underline{f}_{\pi 255}]$$

The wavelet *sub*signature is

$$\mathcal{H}[\underline{f}_{\pi 0}], \mathcal{H}[\underline{f}_{\pi 1}], \cdots, \mathcal{H}[\underline{f}_{\pi T-1}]$$

is formed from the top T values. It will be used to determine the similarity with songs in the Waveprint database. And now, we are ready to hash!

17.4 MIN-HASH

No matter what measurements are made, there will be too many renditions of Glenn Miller's *In The Mood* to allow all the fingerprints to be stored as a precursor to song recognition. Additionally, the measurements of a "song" are a (real) vector in a high-dimensional space. If recognition of song x is viewed as finding the nearest neighbor in the set of all fingerprints; the problem is, if not unmanageable, is certainly daunting.

The first step in finding a computationally feasible solution is to reduce the dimension of the fingerprints so as to reduce the complexity of the nearest nature search. This allows the computation of the hamming distance across the low-dimensional output space. The second step is to have some statistical measure of how likely that second distance will give the same distance ordering than the original distance.

Min hashing provides a viable solution; it clusters the fingerprint vectors so that the following occurs:

1. $\rho(f_1, f_2)$ is small for the fingerprints of two recordings regarded as the same
2. $\rho(f_1, f_2)$ is big for the fingerprints of two recordings regarded as distinct

where ρ is some distance measure.

Similarity measures for a collection of objects have been studied by several researchers. The Baluja-Covell [2007] formulation starts with an array $X = (x_{k,i})$ of dimension (n, N) consisting of (0,1)-vectors; the ith column $\underline{x}_i = (x_{0,i}, x_{1,i}, \cdots, x_{n-1,i})$ specifies whether the kth feature is present ($x_{k,j} = 1$) in the ith object.

One similarity measure of the ith- and j^{th}-objects is to compare the rows in sets C_i and C_j on which the kth-feature (*aka* row) is present; that is

$$C_i \equiv \{k : 0 \le k < n, x_{k,i} = 1\} \tag{17.4}$$

The Jaccard index [1901] defined by $\text{sim}(C_i, C_j) \equiv \frac{|C_i \cap C_j|}{|C_i \cup C_j|}$ where $|\cdots|$ denotes the cardinality of $\cdots$.

Fix i and j and classify the rows of X as one of the following four types:

Paul Jaccard (b. 1868, d. 1994) was a professor of botany and biology at the ETH Zurich. In addition to the Jaccard index of similarity, which he referred to as the **coeficient de communauté**, he also defined a species-to-genus ratio in biogeography.

Type	C_i	C_j
a	1	1
b	1	0
c	0	1
d	0	0

Let $_{(1,1)}n_{i,j}$, $_{(1,0)}n_{i,j}$, $_{(0,1)}n_{i,j}$ and $_{(0,0)}n_{i,j}$ denote the number of rows that are of type a (*resp. b, c, d*).

Exercise 17.5. Prove the formulas

Lemma 17.1.

$$\text{sim}(C_i, C_j) = \frac{_{(1,1)}n_{i,j}}{_{(1,1)}n_{i,j} + {}_{(1,0)}n_{i,j} + {}_{(0,1)}n_{i,j}} \tag{17.5a}$$

and show that

$$\text{sim}(C_i, C_j) = \frac{|C_i| + |C_j| - d_H(C_i, C_j)}{|C_i| + |C_j| + d_H(C_i, C_j)} \tag{17.5b}$$

where $d_H(C_i, C_j)$ is the Hamming distance[3] between columns C_i and C_j).

[3]The number of entries (rows) in which C_i and C_j differ.

To use the sim measure, an economical **signature** of $\mathcal{C}$, say sig($\mathcal{C}$) needs to be defined in such a way that $d(\text{sig}(\mathcal{C}_1), \text{sig}(\mathcal{C}_2))$ enjoys the properties in section 17.4.

The difficulty with sim($\mathcal{C}_i$, $\mathcal{C}_j$) as a similarity measure occurs when the columns of the array X are sparse so that most rows are of type d. One way out of this dilemma, using the concept of min-hashing, appears in Cohen et al. [2001], and in [Chum, Philbin and Zisserman 2008]. A very concise and clear exposition can be found in lecture notes by Professor Ullman.[4]

h_i is defined as the smallest value of k (row in X) for which the entry $x_{k,i}$ in (the ith column in X) is 1. The function h is referred to, most appropriately, as the min-hash function.

$$h_i \equiv \min_{x_{k,i}=1} k \tag{17.6}$$

Exercise 17.6. Prove

Lemma 17.2. If the rows of X are permuted randomly,[5] the random variables $H = (H_0, H_1, \cdots, H_{N-1})$ satisfy

$$Pr\{H_i = H_j\} = \text{sim}(\mathcal{C}_i, \mathcal{C}_j) \tag{17.7a}$$

The idea is to repeat the Lemma 17.2 experiment 100 times, deriving a signature

$$\text{sig}(\mathcal{C}) \equiv (H_{\underline{\pi}_0}(\mathcal{C}), H_{\underline{\pi}_1}(\mathcal{C}), \ldots, H_{\underline{\pi}_{99}}(\mathcal{C})) \tag{17.7b}$$

where $H_{\underline{\pi}_r}(\mathcal{C})$ is the hash of $\mathcal{C}$ using the rth permutation.

In this way, a signature for column $\mathcal{C}$ consisting of approximately 100 $\lceil \log_2 n \rceil$-bit integers is derived. Furthermore, if the sample size of 100 is large enough, then we expect

$$\text{sig}(\mathcal{C}_i) \approx \text{sig}(\mathcal{C}_j) \Leftrightarrow \text{sim}(\mathcal{C}_i, \mathcal{C}_j) \quad \text{is high} \tag{17.7c}$$

Consider the array $X = (x_{k,i})$ of dimension (n, N) in which the ith column is the (0,1) feature vector $\underline{x}_i = (x_{0,i}, x_{1,i}, \cdots, x_{n-1,i})$, specifying whether the kth feature is present $(x_{k,i} = 1)$ in the ith object. Let ${}_{(0)}n_i$ (*resp.* ${}_{(0)}n_i$) count the number of rows in which the kth feature is present $x_{k,i} = 1$ (*resp.* the kth feature is missing $x_{k,i} = 0$).

If $\underline{\pi} = (\pi_0, \pi_1, \cdots, \pi_{n-1})$ is a random permutation of the rows, then $H(\mathcal{C}_i)$ is a random variable with values in $0, 1, \cdots, n - {}_{(0)}n - 1$.

Exercise 17.7. Use the counting methods in Chapter 1 to give a proof of

Theorem 17.3.

$$Pr\{H(\mathcal{C}_i) = r\} = \frac{{}_{(1)}n_i}{n} \frac{\binom{n - {}_{(1)}n_i}{r}}{\binom{n-1}{r}} \qquad 0 \le r \le n - {}_{(1)}n_i \tag{17.8}$$

[4]At http://infolab.stanford.edu/ ullman/mining/minhash.pdf.

[5]Using the uniform distribution on the $n!$ permutation of rows.

One cannot look at equation (17.8) without wondering whether the identify $1=\sum_{r=0}^{n-{}_{(1)}n_i} Pr\{H(C_i)=r\}$ is a consequence of some obvious binomial coefficient identity. Although "obvious" is perhaps an exaggeration, the identify $1=\sum_{r=0}^{n-{}_{(1)}n_i} Pr\{H(C_i)=r\}$ is a special instance of formula (4.1) in [Gould 1972].

$$\sum_{\kappa=j}^{m} \frac{\binom{z}{\kappa}}{\binom{x}{\kappa}} = \frac{x+1}{x-z+1}\left\{\frac{\binom{z}{j}}{\binom{x+1}{j}} - \frac{\binom{z}{m+1}}{\binom{x+1}{m+1}}\right\} \tag{17.9}$$

Exercise 17.8. Prove equation (17.9) and indicate the range of integer parameters (j, m, x, z) for which it is valid.

The binomial series identity in equation (17.10) also provides a formula for $E\{H(C_i)\}$.

Exercise 17.9. Prove

Theorem 17.4.

$$E\{H(C_i)\} = \frac{n - {}_{(1)}n_i}{{}_{(i)}n_i + 1} \tag{17.10}$$

If the developers are judicious in their choice of features, we may assume n is large and

$$\alpha = \frac{{}_{(1)}n_i}{n}$$

Exercise 17.10. Use the methods in Chapter 3 to show

Theorem 17.5.

$$\lim_{\substack{n\to\infty \\ \frac{{}_{(1)}n_i}{n}=\alpha}} Pr\{H(C_i)=r\} = \alpha(1-\alpha)^r \tag{17.11a}$$

$$\lim_{\substack{n\to\infty \\ \frac{{}_{(1)}n_i}{n}=\alpha}} E\{H(C_i)\} = \frac{1-\alpha}{\alpha} \tag{17.11b}$$

The combinatorial analysis of a "random" $H(C)$ can be extended to evaluate sim(C_i, C_j). We start with equation (17.7a) and modify the idea used in the proof of Lemma 17.3. The event $E_{(i,j)} \equiv \{H(C_i) = H(C_j)\}$ is the union of the mutually exclusive events $E_{(i,j)}(r) \equiv \{H(C_i) = H(C_j) = r\}$ for $r = 0, 1,$ whose occurrence requires

$$(x_{\pi_k,i}, x_{\pi_k,j}) = \begin{cases} (0,0) & \text{if } 0 \le k < r-1 \\ (1,1) & \text{if } k = r \end{cases} \quad \text{with } 0 \le r \le {}_{(0,0)}n_{(i,j)}$$

This gives a "two-dimensional" version of equation (17.8).

$$Pr\{E_{(i,j)}(r)\} = \frac{\left({}_{(0,0)}n_{(i,j)}\right)_r \; {}_{(1,1)}n_{(i,j)} (n-r-1)!}{n!}$$

$$= \frac{{}_{(1,1)}n_{(i,j)}}{n} \frac{\binom{{}_{(0,0)}n_{(i,j)}}{r}}{\binom{n-1}{r}} \quad 0 \leq r \leq {}_{(0,0)}n_{(i,j)} \tag{17.12a}$$

$$Pr\{E_{(i,j)}\} = \frac{{}_{(1,1)}n_{(i,j)}}{n} \sum_{r=0}^{{}_{(0,0)}n_{(i,j)}} \frac{\binom{{}_{(0,0)}n_{(i,j)}}{r}}{\binom{n-1}{r}} \tag{17.12b}$$

Applying the summation formula equation (17.9) gives

Theorem 17.6

$$Pr\{H(C_i) = H(C_j)\} = \frac{{}_{(1,1)}n_{(i,j)}}{n - {}_{(0,0)}n_{(i,j)}} \tag{17.13a}$$

$$\lim_{\substack{n\to\infty \\ \frac{(1)^{n_i}}{n}=\alpha}} Pr\{H(C_i) = H(C_j)\} = \text{sim}(C_i, C_j) \tag{17.13b}$$

Unfortunately, a computational dilemma is that the number n of rows (features) is likely to be large and the process of generating $m \approx 100$, say, "random" permutations will be prohibitive. Instead, the min-hash function h is replaced by m randomly chosen uniformly distributed mappings of the row space $\{0, 1, \cdots, n-1\}$

$$H_s(i) : \{0, 1, \cdots, n-1\} \to \{0, 1, \cdots, n-1\} \quad s = 0, 1, \cdots, m-1 \tag{17.14a}$$

and the signature of C_i is defined by

$$\underline{H}(i) \equiv \text{sig}(C_i) = (H_0(i), H_1(i), \cdots, H_{m-1}(i)) \tag{17.14b}$$

Each $H_s(i)$ with $0 \leq s < m$ will be referred to as a min-hash value for C_i. The essential point of Theorem 17.6 is that if m is large enough, then $\underline{H}(i)$ and $\underline{H}(j)$ will be similar if and only if sim(C_i, C_j) is large.

We now consider the comparison regime described in [Baluja, Covell and Ioffe 2008]; a song database $\mathcal{D}$ is constructed to contain a sequence of min-hash values of some collection of songs. It might contain an entry for Tony Bennett's signature song, *I Left My Heart In San Francisco* recorded January 23, 1962, and, perhaps, a second entry for *Ol' Blue Eyes*'s August 27 1962 rendition by the Nelson Riddle orchestra of the Neal Hefti arrangement of *I Left My Heart.*

There still is the problem of how to search $\mathcal{D}$ to identify a performance that we have represented by a sequence $\underline{H}^*$ of min-hash values. [Baluja, Covell and Ioffe

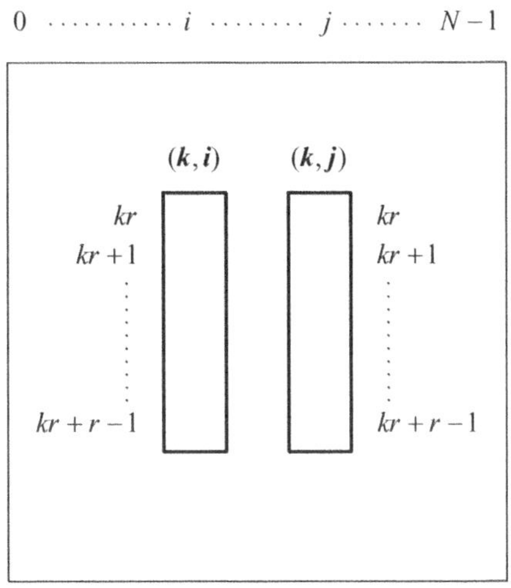

Figure 17.7. The kth band in the ith and j^{th} columns of H.

2008] mentions a $\mathcal{D}$ of 2×10^6 entries; if each entry contains m ($30 \le m \le 80$) 32-bit min-hash values, we have a lot of data to compare with $\underline{H}^*$.

[Cohen et al. 2001] describes **locally sensitive hashing** (LSH), which is an additional processing step originally described by Indyk and Motwani [Indyk and Motwani 1998]. LSH is introduced to produce a more economical representation; the m-by-N array of min-hash values $H = H(\underline{H}(0), \underline{H}(1), \cdots, \underline{H}(N-1))$ (aka signatures) is first partitioned into l bands of r rows ($m = \ell \times r$) as depicted in in Figure 17.7.

$$H = (\underline{H}(0), \underline{H}(1), \ldots \underline{H}(N-1))$$

The entries in the kth band in the ith column are interpreted as a key. To identify the song $\underline{H}^*$, its ℓ keys are compared with those in the song database $\mathcal{D}$. Why do we expect similarity? How are the LSH parameters (ℓ, r) chosen?

Because the permutations used to construct the array $\boldsymbol{H}$ are independent, the values

$$H[kr, (k+1)r) \equiv (\underline{H}(kr), \underline{H}(kr+1), \cdots, \underline{H}((k+1)r-1))$$

are independent and identically distributed. $P_{r,\ell}(C_i, C_j)$ is the probability that $H_s(C_i) = H_s(C_j)$ for at least one permutation in the kth band. The event $Pr\{H(C_i) = H(C_j)\}$ for a randomly chosen permutation π of the n rows of X only depends on the similarity sim(C_i, C_j). It follows that if s = sim(C_i, C_j), then for the min-hash value $H_s(i)$ determined by π_s, the following occur:

- The probability of the events $\{H_t(i) = H_t(j)\}$ agree for a fixed t with $0 \le t < m$ is s
- The probability of the events $\{H_{t_u}(i) = H_{t_u}(j)\}$ for the r values $\{t_u\}$ with $kr \le t_1 < t_2 < \cdots < t_r < k(r+1)$ is s^r
- The probability of at least one of the events $\{H_t(i) = H_t(j)\}$ for some t with $kr \le t < k(r+1)$ is $P_{r,l}(C_i, C_j) = 1 - (1 - s^r)^\ell$

leading to the following lemma.

Lemma 17.7. [Cohen et al. 2001, Lemma 2, p. 70]. If $s = \text{sim}(C_i, C_j)$, then given s^* and $0 < \delta, \varepsilon < 1$, parameters (ℓ, r) exist such that

a) For any $s \geq (1 + \delta)s^*$, we have $P_{r,l}(C_i, C_j) \geq 1 - \varepsilon$.
b) For any $s \leq (1 - \delta)s^*$, we have $P_{r,l}(C_i, C_j) \leq \varepsilon$.

A threshold s^* determines parameters (ℓ, r) such that similarity in columns C_i and C_j is unmistakable.

Example 17.2. We illustrate the use of min-hash with a synthetic example[6] in which the array $X = (x_{k,i})$ is of dimension $(n, N) = (400, 8)$. The number of ones in each column ($N_i = \Sigma_k x_{k,i}$) and the similarity matrix $\text{sim}(C_i, C_j)$ are shown as Tables 17.1A and 17.1B.

These synthetic data were derived as follows:

i) *Forcing* for each of the column pairs (C_{2j}, C_{2j+1}) $(0 \leq j < 4)$ 30 rows randomly chosen from the 400 rows to have the entry 1 in the array X
ii) For each of the 8 columns C_j $(0 \leq j < 8)$, "randomly" choosing 20 rows from the 400 rows to have the entry 1 in the array X.

TABLE 17.1A. Column counts in *X*

		$N_i \equiv \sum_{k=0}^{n-1} x_{k,i}$		$0 \leq i < 8$				
i→	**0**	**1**	**2**	**3**	**4**	**5**	**6**	**7**
	40	40	42	42	48	38	48	48

TABLE 17.1B. Similarity matrix for *X*

			$\text{sim}(C_i, C_j)$	$0 \leq i < j < 8$			
***i*↓*j*→**	1	2	3	4	5	6	7
0	0.4035	0.0250	0.0123	0.0244	0.0130	0.0353	0.0233
1	0.4035	0.0513	0.0380	0.0120	0.0263	0.0115	0.0233
2	0.0250	0.0513	0.4237	0.0488	0.0667	0.0345	0.0465
3	0.0123	0.0380	0.4237	0.0617	0.0811	0.0465	0.0345
4	0.0244	0.0120	0.0488	0.0617	0.4138	0.0455	0.0455
5	0.0130	0.0263	0.0667	0.0811	0.4138	0.0361	0.0238
6	0.0353	0.0115	0.0345	0.0465	0.0455	0.0361	0.4328
7	0.0233	0.0233	0.0465	0.0345	0.0455	0.0238	0.4328

[6]By synthetic, we mean the entries in the array $X = (x_{k,i})$ are randomly generated and do not correspond to actual audio measurements.

TABLE 17.1C. Min-Hash Values ($\underline{H}(0)$, $\underline{H}(1)$)

0:	0	27	27	27	27	38	39	47	47	52	55	55	56	56	56	56	59	62	63	71
	71	71	79	83	83	89	96	147	249	311	335	379								
1:	0	22	22	27	27	38	48	52	56	56	56	62	62	64	64	71	71	79	89	96
	143	143	187	187	192	203	241	295	295	351	351	351								

TABLE 17.1D. The Matrix $M = (m_{i,j})$

	j →							
i↓	**0**	**1**	**2**	**3**	**4**	**5**	**6**	**7**
0:		34	2	1	2	1	2	2
1:	34		5	5	2	1	1	7
2:	2	5		37	1	13	2	11
3:	1	5	37		4	17	3	9
4:	2	2	1	4		31	5	2
5:	1	1	13	17	31		2	1
6:	2	1	2	3	5	2		23
7:	2	7	11	9	2	1	23	

iii) The rows selected in Step *ii* are exclusive ORed with those selected in Step *i*.

Next, 32 randomly selected permutations are generated and the min-hash values are calculated. Table 17.1C lists the 32 min-hash values ($\underline{H}(0)$, $\underline{H}(1)$) for columns 0 and 1.

The matrix $M = (m_{i,j})$ shown in Table 17.1D with

$$m_{i,j} = \begin{cases} \sum_{k_1=0}^{31} \sum_{k_2=0}^{31} \chi_{\{h_{k1}(i)=h_{k1}(i)\}} & i \neq j \\ \text{undefined} & \text{If } i = j \end{cases}$$

measures the similarity of the columns of min-hash values and roughly corresponds to a single band of 32 rows.

17.5 SOME COMMERCIAL FINGERPRINTING PRODUCTS

Vendors include the following:

- Shazam (http://www.shazam.com/music/portal)
- Audio Magic (http://www.audiblemagic.com/products-services/custom/)
- MusicBrainz(http://musicbrainz.org/)

These vendors offer an audio identification service for a fee. For example, Shazam customers call an entry number, play music from CD through their speaker to the telephone, and have their selection compared to Shazam's database. Shazam computes a hash of the music that is used for the identification process. MusicBrainz hashes the speaker-to-telephone music and creates a fingerprint for the audio file and compares it with its community-driven database.

REFERENCES

S. Baluja and M. Covell, "Audio Fingerprinting: Combining Computer Vision and Data Stream Processing", in Proceedings of the International Conference on Acoustics, Speech, and Signal Processing, 2007.

S. Baluja and M. Covell, "Waveprint: Efficient Wavelet-Based Audio Fingerprinting", *Pattern Recognition*, **41**, pp. 3467–3480, 2008.

S. Baluja, M. Covell, and S. Ioffe, "Permutation Grouping: Intelligent Hash Function Design for Audio and Image Retrieval", in Proceedings of the International Conference on Acoustics, Speech, and Signal Processing, 2008.

J. Bramford, *The Shadow Factory*, Doubleday (New York), 2008.

O. Chum, J. Philbin, and A. Zisserman, "Near Duplicate Image Detection: *min-Hash and tf-idf* Weighting", in Proceedings of the 19th British Machine Vision Conference, 2008, pp. 1–10.

E. Cohen, M. Datar, S. Fujiwara, A. Gionis, P. Indyk, R. Motwani, J. D. Ullman, and C. Yang, "Finding Interesting Associations without Support Pruning", *IEEE Transactions on Knowledge and Data Engineering*, **13**, #1, pp. 64–77, 2001.

A. Gionis, P. Indyk, and R. Motwani, "Similarity Search in High Dimensions via Hashing", in Proceedings of the 25th-VLDB Conference (Edinburgh, Scotland), 1999.

H. W. Gould, *Combinatorial Identities*, Author (Morgantown, WV), 1972.

A. Haar, "Zur Theorie der Orthogonalen Funktionensysteme", *Mathematische Annalen*, **69**, pp. 331–371, 1910.

J. Haitsma and T. Kalker, "A Highly Robust Audio Fingerprinting System", in Proceedings of the International Symposium on Music Information Retrieval, 2002, pp. 107–115.

F. J. Harris, "On the Use of Windows for Harmonic Analysis with the Discrete Fourier Transform", *Proceedings of the IEEE*, **66**, pp. 51–83, 1978.

P. Indyk and R. Motwani, "Approximate Nearest Neighbor: Towards Removing the Curse of Dimensionality", in Proceedings of the 30th Annual ACM Symposium on Theory of Computing, 1998.

P. Jaccard, "Etude Comparative de la Distribution Florale dans une Portion des Alpes et des Jura", *Bulletin del la Societe Vaudoise des Sciences Naturelle*, **37**, pp. 547–579, 1901.

H. Özer, B. Sankurm, and N. Memom, "Robust Audio Hashing for Audio Identification", Paper 1091, 12th European Signal Processing Conference (EUSIPCO), 2004.

E. J. Stollnitz, T. D. DeRose, and D. H. Salesin, "Wavelets for Computer Graphics: A Primer, Part 1", *IEEE Computer Graphics and Applications*, **15**, #3, pp. 76–84, 1995.

CHAPTER 18

Hashing in E-Commerce

18.1 THE VARIED APPLICATIONS OF CRYPTOGRAPHY

The three principal roles of cryptography are *secrecy*, *access control*, and *authentication*. Secrecy is needed to deny information contained in text by disguising its form; for example, to do the following:

i) To prevent an eavesdropper from learning the content of the communication when two users communicate over an open or insecure network
ii) To hide information in a file stored on some system

Cryptographic techniques are used to limit access to other facilities in information processing systems, but now the purpose is to authenticate or verify the user's identity. For example, when a customer engages in an automated teller machine transaction (ATM), the customer:

i) Must be in possession of a valid ATM card on which the customers's personal account number (PAN) is recorded
ii) Must know the corresponding personal identification number (PIN).

The Web has provided the third and most widespread application; when two parties communicate over an open and possibly insecure network, the following is true:

- Each party needs to be confident of the identity of the other
- Secrecy or privacy of the information exchanged must be provided.

Webster's dictionary defines authentication as "a process by which each party to a communication verifies the identity of the other." There are significant risks if my credit card information is revealed when purchasing goods or services during a Web transaction. Clever but criminal hackers[1] often can replicate the logo of a merchant,

[1]A hacker is someone with technical expertise in computing systems and a delight in circumventing the controls placed in computing systems. Although not necessarily evil or criminal, they often engage in improper and illegal activities. The website http://catb.org/~esr/faqs/hacker-howto.html in fact offers free instruction in hacking.

e.g. www.amazon.com. During such a Web session, you reveal personal information think that you are dealing with the real secure website. Isn't wonderful that every new industry, spawns others?

The risks in e-commerce—transactions implemented through the Web—and in particular, the role played by hashing, is the subject of this chapter.

18.2 AUTHENTICATION

A new class of security problems in the twentieth century and the ubiquitous nature of public computer networks has given rise to e-commerce and in the process has enlarged the area in which cryptography is needed. Transactions over the Web have changed the scale and environment in which the problems of secrecy and authentication exist. The principal security issues connected with e-commerce are as follows:

1. *Privacy*—Users want their data transmitted on the Web to be hidden from any parties who monitor communications, for example, during the exchange of personal health data.
2. *Authentication of the user identity*—Because users communicating over a network are not in physical proximity (for example, they do not necessarily see or talk to one another), both need to be confident of the identity of the other party.
3. *Authentication: Message integrity*—When users communicate over a network, each wants to be certain that the other party has not maliciously or unintentionally modified the transmitted data. Although it is not possible to prevent transaction data from being altered, a scheme must be implemented that will be likely to detect changes.

A transaction between two users involves one or more exchanges of data. Each transmission of *transaction data* is suffixed by a message authentication code (MAC) or digital signature (SIG); the MAC/SIG authenticates both the (sender, receiver) pair and the content of the communication (Figure 18.1).

The MAC is a sequence of 0s and 1s functionally dependent on the transaction data and the identities of the corresponding parties.

- If privacy is required, the transaction data and MAC must be enciphered.
- The authenticity of participants in a transaction must be established.
- To ensure the integrity of the exchange of information, the MAC must depend on the transaction data in such a way that a secret element is involved in the

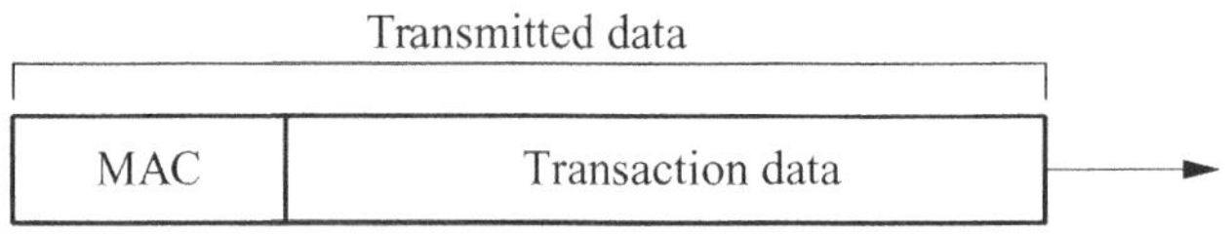

Figure 18.1. The MAC appended to transaction data.

construction of the MAC, no user can expect to construct a valid MAC for the transaction data without knowledge of the secret element, and any change in the transaction data will likely change the MAC.

Web-based electronic transactions require a framework in which the purchaser and seller can be confident of the integrity of their transactions.

Hashing's role in cryptography relates to its role in authentication.

18.3 THE NEED FOR CERTIFICATES

Thirty-five years ago, cryptography was used only in a limited community; military and diplomatic communications were needed to be secure between Washington, U.S. military bases in Europe, Japan, and the Mideast. Moreover, parties with the ability to monitor and decipher and/or alter communications in a timely manner were similarly limited.

All this has changed because of the Internet (Web); in 1990, there were more than 300000 hosts (mainframe machines). Vincent Cerf, who was a computer scientist often referred to as one of the "founding fathers of the Internet," claimed several years ago that there are over 60 million Internet users. In 2008, the estimates of 1574313184 Web users in a world population of 6710029070 is provided at www.internetworldstats.com/stats.htm. The number of potential user-to-user endpoints is staggering. Public key cryptography provided a mechanism to replace the $\binom{N}{2}$ key pairs required to protect all possible user-to-user communications with N users to one of complexity N. Nevertheless, User_ID[i] must make available the public key PuK(ID[i]) to any User_ID[j] with whom an enciphered communication is to take place. The thought of a server maintaining a file containing several million keys is absurd. Moreover, even if such a key-exchange server is implemented, there is the need to prevent spoofing attacks, wherein User_ID[i]s public key is temporarily replaced by that of the spoofer; for example, see [Konheim 2007, pp. 448–449].

A proposed solution using of certificates for user authentication was discussed in Chapter 6, §6.9. A certificate provides a link between User_ID[i]'s network identifier ID[i] and public key PuK(ID[i]). Various certificate authorities (CAs) would be authorized to issue certificates. To provide *trust* on the Web, the CA needs to be a trusted party, and the certificate needs to be computationally infeasible to forge.

If User_ID[i] wishes to enter into a transaction with User_ID[j], a User_ID[i]-certificate is delivered (or otherwise made available) to User_ID[j]. The data on the certificate bind the pair (ID[i], PuK(ID[i])). User_ID[j] verifies the binding by testing the certificate. Implicit is the assumption that only a valid CA could construct a certificate. Certificates use the same paradigm as public key cryptography; namely, the signature on the certificate is encipherment of certificate data using the (secret) private key of the CA. One possible scenario for the secure exchange of information on the Web involves the following steps:

- User_ID[i] signs DATA using User_ID[i]s public key appending the certificate.
- User_ID[j] first uses the certificate to verify that PuK(ID[i]) is the public key of User_ID[i].

- If the agreement is verified, User_ID[j] can then examine the DATA and decide on some action.

What has been gained? Only the CA's public key must be secured rather than securely storing all public keys. In contrast, if someone can learn the CA's private key or can otherwise forge a valid certificate, then all the CA-issued certificates become worthless for authentication.

Computer applications are global and connect entities with different political and scientific outlooks. International cooperation and agreements are necessary so that we can talk to one another.

Certificates involve a signature by the issuing authority of a compressed form (hashing) of the critical certificate data. Before proceeding further, we need to describe the cryptographic hash.

18.4 CRYPTOGRAPHIC HASH FUNCTIONS

As we have already noted in the preface, the verb *to hash* means to chop into small pieces. A hashing function h is a mapping from values x in some finite set $\mathcal{X}$ (of text) into a value y contained in another (larger) set $\mathcal{Y}$, which mixes up and hides the values x. Hashing is a synonym for a (uniformly distributed) random mapping in cryptography.

One operational inconvenience with the hashing Figure 18.2 is the dependence of the hash length on the text's hash length. A message digest (or cryptographic hash function) is a hash function that derives a fixed-length hash value for every message in some message domain. The processes of computing the message digest with a public key cryptosystem (PKC), and verifying the message digest is depicted next in Figures 18.3A and 18.3B.

A hash function h is as follows:

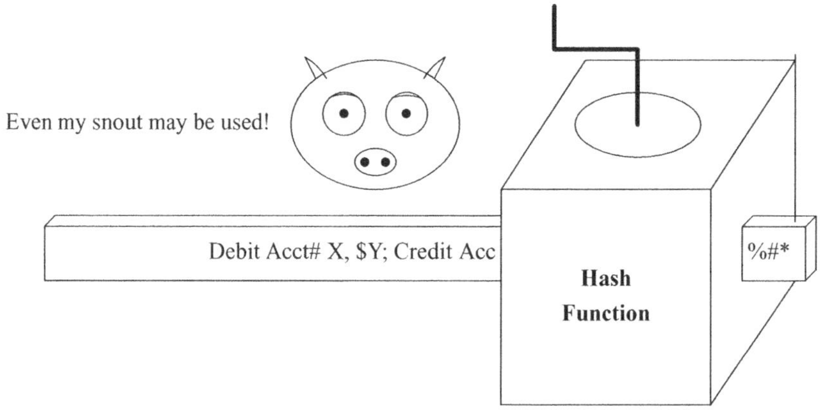

Figure 18.2. High-cholesterol hashing.

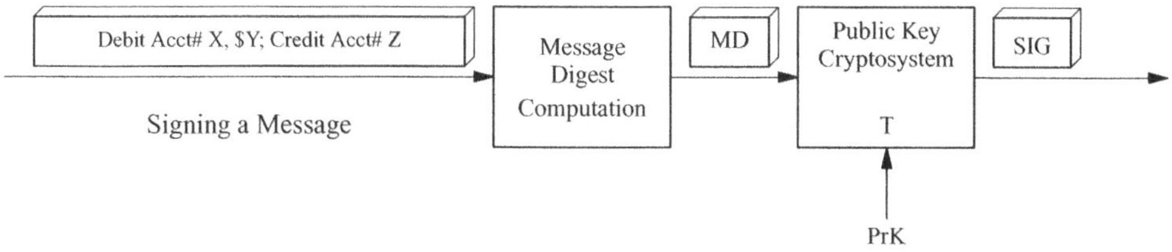

Figure 18.3A. Deriving a message digest.

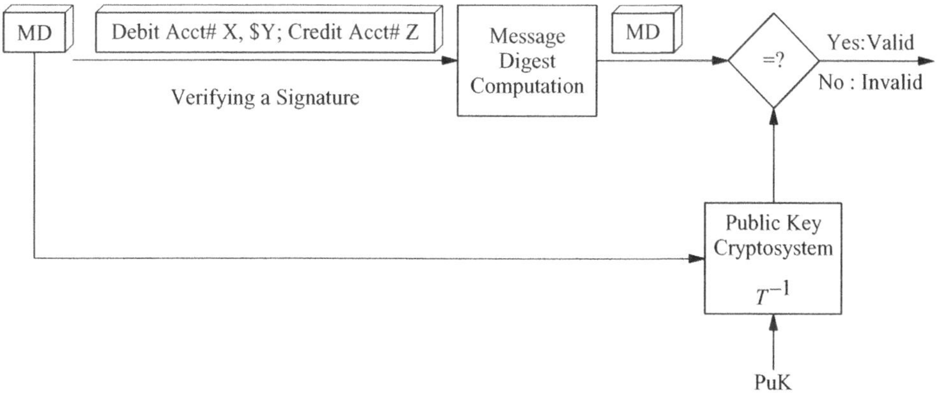

Figure 18.3B. Verifying the correctness of a message digest.

- A one-way hash function if given the hash of message $h(m)$, it is computationally infeasible to determine the message m.
- A collision-resistant hash function if given the hash of message $h(m)$, it is computationally infeasible to determine any other message m^* with the same hash value $h(m) = h(m^*)$.
- A second pre-image-resistant hash function if given the message m and hash of message $h(m)$, it is computationally infeasible to determine any other message m^* with the same hash value $h(m) = h(m^*)$.

Because a hash function maps a large, indeed potentially unbounded domain of text into fixed-length hash values, there are many, indeed a potentially unbounded number, of texts with the same hash value, which are referred to as a hash function collision. Thus, it is not too surprising that two messages were found in 2004 which have hash-function collisions for a specific hash function. Although not surprising, it is a daunting task to construct a pair of colliding messages. Their construction is ingenious and has been simplified considerably, and we take up these ideas in §18.9–10. The following remains to be shown:

1. Whether the exposure for certificate-based security will be completely compromised.
2. Whether their construction can be modified so that

   given a message m with a specific structure,

find a colliding message m' with the same structure can be constructed; for example,

m: "Deposit $1 000 000 to my bank account"

m': "Deposit $1 000 000 to your bank account"

3. What obvious improvements can be made to a specific hash-function to prevent the second item above?

18.5 X.509 CERTIFICATES AND CCIT STANDARDIZATION

The international Standards Organization (ISO) recognized the need in 1977 for the standardization of information-network architectures. For users on two systems to communicate, a common set of rules must be implemented. ISO's TC97 committee, which was responsible for information systems, created the Open Systems Interconnection (OSI) model in 1979. Influenced by the earlier network architectures of IBM [systems network architecture (SNA) in 1974], the Digital Equipment Corporation (DNA in 1975) and the TCP/IP suite of the ARPANET. The ISO Open System Interconnection model divided the services need to implement computer communication into layers.

Maintaining worldwide compatibility in communications is the charter of the International Telegraph Union (ITU), which is an agency of the United Nations. The Comité Consultatif Internationale de Télégraphique et Téléphonique (CCITT) was formed in 1956 by merging the CCIT (Télégraphique) and CCIF (Téléphonique). A reorganization in 1989 divided the ITU into the following three sectors:

- International Radiocommunications Sector (ITU-R)
- International Telecommunication Development Sector (ITU-D)
- International Telecommunication Standardization Sector (ITU-T)

[X.509] updated the original November 1988 recommendation providing additional functionality. The fields in a X.509 v3 certificate include the following:

Serial number: Unique identifier for the certificate.

Algorithm identifier: Specifies algorithm used to sign certificate by CA.

Issuer distinguished name: The name[2] of entity that issued the certificate.

Validity period: The starting and ending dates during which the certificate is valid.

Distinguished subject name: The name of entity to whom the certificate is used.

Public key information: The name of the public key cryptosystem and the public key.

Version 3 Extensions: Special optional features, most important of which are the following:

- CA = TRUE, if this is a CA certificate, and FALSE, is a user certificate;
- Path length field: maximum number of intermediate CAs between root-CA and user.

[2]The term distinguished name (DN) is used to describes the identifying information in a certificate and is part of the certificate itself and is not a reference to the superior character of the issuing entity.

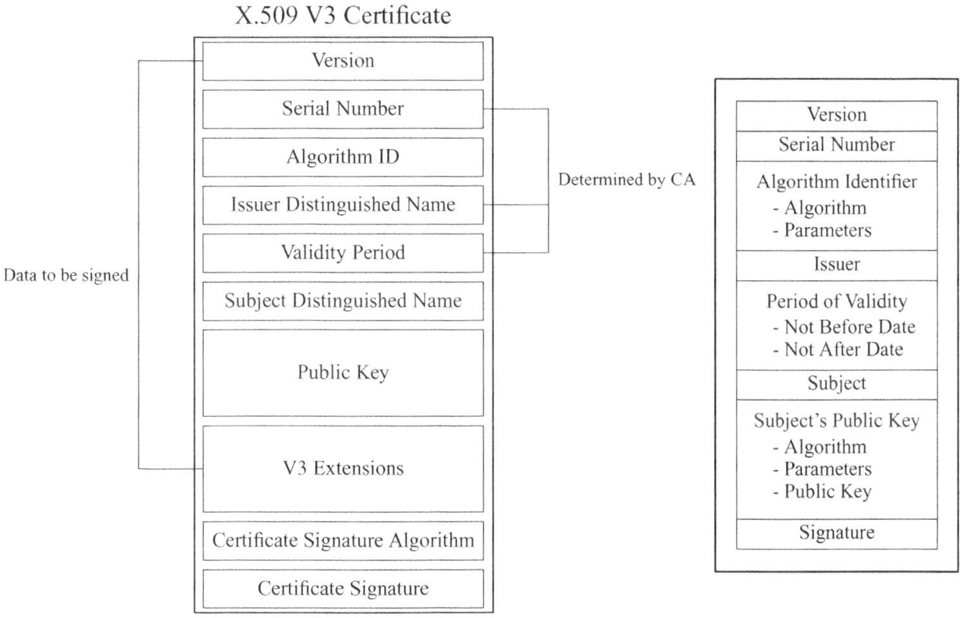

Figure 18.4. X.509 v3 certificate.

Signature algorithm: Identifies the hashing function and encryption algorithm used to derive the signature.

Signature: A signature derived by hashing all fields of the data to be signed and then enciphering using the certificate authority's private key. The hashing functions MD2, MD5, and SHA-1 and the public key cryptosystems RSA (Chapter 6, §6.8) and the digital signature algorithm (DSA) are supported.

The structure of a X.509 v3 certificate is displayed in Figure 18.4.

18.6 THE SECURE SOCKET LAYER (SSL)

SSL was originated by *Netscape* (1995) and was renamed Transport Layer Security (TLS) (1999); it consists of several upper layer protocols by which a pair of users—the client and the server—agree on an key exchange method, an encipherment algorithm for the message, a message digest, and a signature scheme.

In the handshake protocol initiated by the client (Figure 18.5), the two parties agree on the following:

i) A key exchange protocol; RSA, Diffie-Hellman (DH), and the elliptic curve Diffie-Hellman (ECDH) are possible choices.
ii) A data encipherment algorithm; RC4, Data Encryption Standard (DES), Triple DES, and the Advanced Encryption Standard (AES) are possible choices.

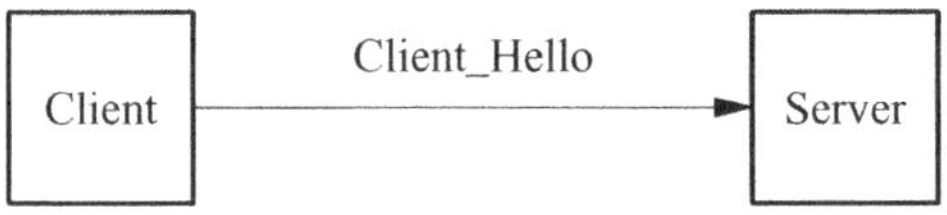

Figure 18.5. SSL phase 1.

iii) A message digest algorithm; Hashed-Based Message Authentication Code (HMAC)-MD5 (or HMAC-SHA for TLS), MD5, and SHA-1 for SSL are possible choices.
iv) A certificate signature scheme; RSA, the DSA, and the elliptic curve digital signature algorithm (ECDSA) are possible choices.

A myriad of details in TLS (Version 1.2) can be found on page 102 [RFC 5246].

18.7 TRUST ON THE WEB · · · TRUST NO ONE OVER 40!

Trust on the Web is a consequence of the validity of certificates. It involves the interaction of the following:

1. Certificate authorities; for example, Verisign, Thawte
2. Browser vendors; for example, Windows Internet Explorer, Mozilla Firefox
3. Secure websites; for example www.amazon.com, www.ual.com

The certificate authority issues a site certificate to vendors (for example, the secure websites) and root certificates to browser vendors.

A. A request by a website to a certificate authority for a certificate is contained in the certificate signing request (CSR), which is a message sent from an applicant to a certificate authority. The CSR is accompanied by other credentials or proofs of identity required by the certificate authority, and the certificate authority may contact the applicant for more information and is signed by the user's private key thus providing proof of the correctness of the website's (private, public) key pair.
B. When a certificate authority such as Verisign registers with a browser vendor such as Mozilla Firefox to supply its root certificate, the browser vendor must verify the correctness of the certificate authority's (private, public) key pair.

Figures 18.6A and 18.6B illustrate the relationship.

Setting aside the issue of what trusted means for a website or certificate authority, there remains the process of checking the certificates. The following steps are required:

- www.trustedwebsite must prove to Trusted Cert Inc. that it is in possession of the correct private key corresponding to the public key contained in its website certificate.

Figure 18.6A. Trusted Cert Inc. issues a certificate to www.trustedwebsite.

Figure 18.6B. Trusted Cert Inc. deposits its root certificate with a browser vendor.

- Trusted Cert Inc. must prove to the browser vendor that it is in possession of the correct private key corresponding to the public key contained in its website root certificate.

In both instances, the proof follows the same paradigm; the certificate data to be signed in Figure 18.4 is hashed (by MD5 or SHA-1) and then enciphered to produce the certificate signature. When a certificate is issued, the correctness relation of the (public, private) key pair is attested to so that

- When a certificate authority is registered with a browser vendor, it must prove the correctness of the (public, private) key pair relation enabling the browser vendor to decipher the certificate authority's root certificate signature and check that it is identical to the hash of the root certificate data.
- When the website www.trustedwebsite requests a certificate, the public key of the certificate authority is revealed enabling the www.trustedwebsite to decipher *the* certificate authority's the certificate signature and check that it is identical to the hash of the www.trustedwebsite's certificate data.

Finally, trusting `User_X` at his/her personal computer (PC) completes the relationship by purchasing some item or service from www.trustedwebsite. The user will reveal personal information, i.e., a credit card number, assuming that it is dealing with a trusted website. The process involves the information exchanges as depicted in Figure 18.6C;

1. `User_X` receives the website certificate of www.trustedwebsite
2. `User_X` either

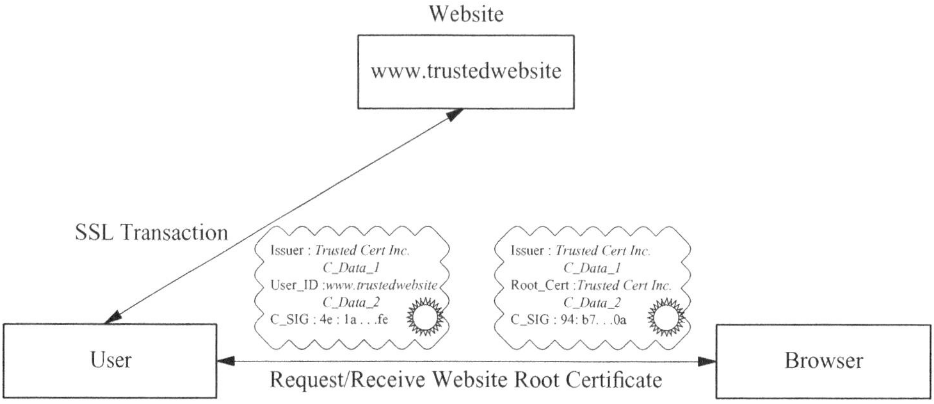

Figure 18.6C. `User_X` Receives {Web Site's Certificate from Web Site / Web Site's Root Certificate from Browser Vendor}

- Has the root certificate of the certificate authority Trusted Cert Inc. resident in its browser
- Receives from the `User_X`'s browser the root certificate of the certificate authority Trusted Cert Inc. from a directory tree; this process is described in [Konheim 2007, Chapter 18, §9].

`User_X` can verify that www.trustedwebsite's certificate was correctly signed by the following steps:

A. Determining the public key of the certificate authority Trusted Cert Inc., which issued www.trustedwebsite's certificate from the root certificate from the user's browser
B. Verifying that www.trustedwebsite's certificate is properly signed by the following steps:
   - Deciphering the signature on www.trustedwebsite's certificate, with
   - Comparing it the hash of the data to be signed on www.trustedwebsite's certificate

Of course, it is is possible to construct a *rogue* certificate for a bona fide website when whatever trust is supplied by certificates disappears.

18.8 MD5

Ronald Rivest is the designer of several message digests; MD2 in 1988 [Kaliski 1992], MD4 in 1990 [Rivest 1992a], and MD5 in 1991 [Rivest 1992b]. To calculate the MD5 message digest MD5[M] of the message M, there is first padding of M (on-the-right) $M \to M, P$ to make its length $|M, P|$ a multiple of 512 bits. Padding begins by appending a 1 to signal the start of the padding, then it appends appending as many

TABLE 18.1. MD5 Initial Chaining Value IHV_0

A	01	23	45	67	B	89	ab	cd	ef	C	fe	dc	ba	98	D	76	54	32	10

as necessary to make its length a multiple of 512 less 64 bits. Finally, the 64-bit encoding of the original message length is appended on the right.

We write $M \to (M_0, M_1, \cdots, M_{15}, \cdots, M_{16r-1})$ to indicate the r 32-bit blocks $\{M_i\}$ of the padded message M.

An MD5 state Ω is the contents of the four 32-bit registers (A, B, C, D); these registers are initialized (in hexadecimal) as bit strings as shown in Table 18.1.

[Rivest 1992b] describes these registers as above with lower order bytes on the left, which indicates the little-endian representation.[3] We follow Stallings [Stallings 1993] and Wang and Yu [Wang and Yu 2005] using the notation 0*x*67452301 for the initial contents shown above of Register A.

The MD5 hash is computed in four MD5-rounds, each consisting of 16 MD5-Steps. If we designate by $(A_{i,j}, B_{i,j}, C_{i,j}, D_{i,j})$ the values in the four 32-bit registers at the start of the ith round and j^{th} step with $0 \leq i < 4$ and $0 \leq j < 16$, then the following occurs:

1. The processing of the 128-bit values in the ith round and j^{th} step result in new values designated by $(\tilde{A}_{i,j}, \tilde{B}_{i,j}, \tilde{B}_{i,j}, \tilde{D}_{i,j})$.
2. After the processing, the new values in the four 32-bit registers $(\tilde{A}_{i,j}, \tilde{B}_{i,j}, \tilde{B}_{i,j}, \tilde{D}_{i,j})$ are circularly right shifted so that at the conclusion of the (i, j) processing

$$\left(\tilde{A}_{i,j}, \tilde{B}_{i,j}, \tilde{B}_{i,j}, \tilde{D}_{i,j}\right) \to \left(\tilde{D}_{i,j}, \tilde{A}_{i,j}, \tilde{B}_{i,j}, \tilde{D}_{i,j}\right) = \left(A_{i,j}, \hat{B}_{i,j}, \hat{C}_{i,j}, \hat{D}_{i,j}\right)$$

3. The iterative structure of MD5 results in the register state

$$\left(\hat{A}_{i,j}, \hat{B}_{i,j}, \hat{B}_{i,j}, \hat{D}_{i,j}\right) = \begin{cases} \left(A_{i,j+1}, B_{i,j+1}, C_{i,j+1}, D_{i,j+1}\right) & \text{if } 0 \leq j < 15 \\ \left(A_{i+1,0}, B_{i+1,0}, C_{i+1,0}, D_{i+1,0}\right) & \text{if } j = 16 \end{cases}.$$

4. We write $\text{IHV}_\ell = (A_{16\ell}, B_{16\ell}, C_{16\ell}, D_{16\ell})$ for $\ell = 0,1, \cdots r$ for the values in the registers after the processing of the ℓth message block M_ℓ of 512-bits and $\text{MD5}[M] = \text{IHV}_r$.

Figure 18.7 is a schematic for the MD5 operations in the ith step, j^{th} round emphasizing its programmatic nature.

The operations in the jth step of the ith MD5 round consist of the following:

1. In (i, j) step #1, round-dependent functions $\{F_j\}$ are nonlinear mappings from 3×32-bits to 32-bits; in the jth round, the function F_j is defined by the following:

[3] Big and little endian come from Jonathan Swift's satiric novel *Gulliver's Travels* [Swift 1726]; **big/little endian** means that the most/least significant byte of a word is stored in memory at the low-/high-address byte position that is, the (big/little end comes first. According to the people of Lilliput, eggs should be broken at the smaller end.

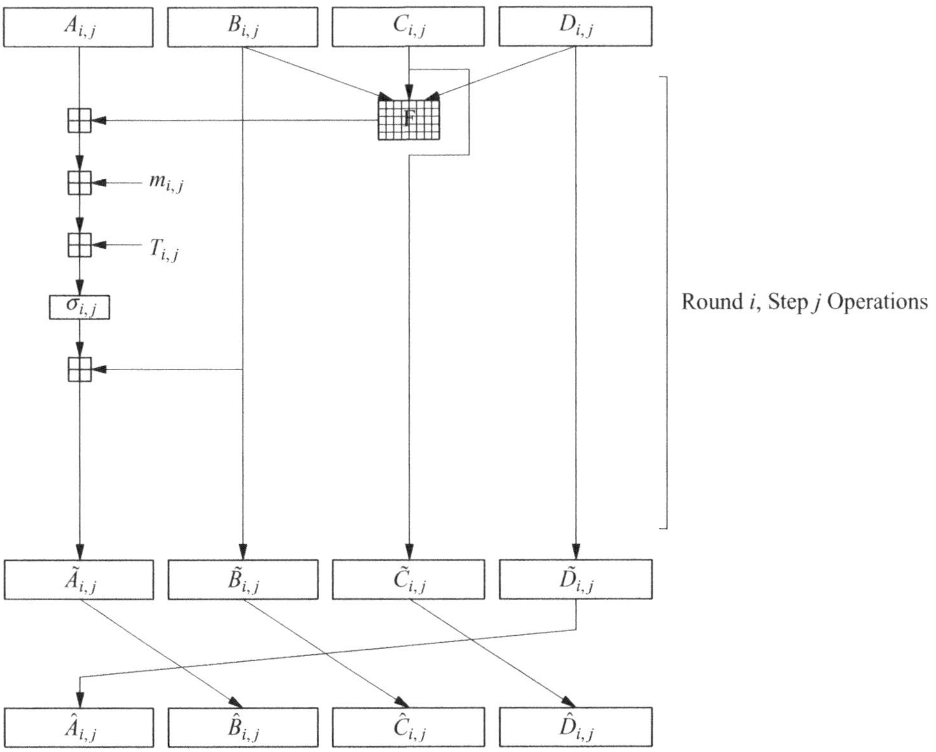

Figure 18.7. MD5 operations in the ith step, j^{th} round.

$\text{Round}_0 : i = 0 \; F_i(B_{i,j}, C_{i,j}, D_{i,j}) = (B_{i,j} \wedge C_{i,j}) \vee (\neg B_{i,j} \wedge D_{i,j})$
$\text{Round}_1 : i = 1 \; F_i(B_{i,j}, C_{i,j}, D_{i,j}) = (B_{i,j} \wedge D_{i,j}) \vee (C_{i,j} \wedge \neg D_{i,j})$
$\text{Round}_2 : i = 2 \; F_i(B_{i,j}, C_{i,j}, D_{i,j}) = (B_{i,j} \oplus C_{i,j} \oplus D_{i,j})$
$\text{Round}_3 : i = 3 \; F_i(B_{i,j}, C_{i,j}, D_{i,j}) = C_{i,j} \oplus (B_{i,j} \vee \neg D_{i,j})$

where the notation $\oplus$ for exclusive OR (XOR), $\wedge$ for AND, and $\vee$ for OR are used.

2. In (i, j) step #2, there is addition modulo 2^{32} $((\boxplus))$ of a message segment, one of the 32-bit words of the ith message block of 512-bits $M_i = (m_{i,0}, m_{i,1}, \cdots, m_{i,15})$ from the message to the output of F_i to the result of the previous step.

   Although all 512 bits of the ℓth message block $M_\ell = (m_{0,0}, m_{0,1}, \cdots, m_{0,15})$ are involved in each MD5 round computation, their order of usage is different, and each message segment is used four times according to the rule

$$m_{(i,j)} = \begin{cases} m_i & \text{if } i = 0, 0 \le j < 16 \\ m_{(1+5j)\,(\text{mod } 16)} & \text{if } i = 1, 0 \le j < 16 \\ m_{(5+3j)\,(\text{mod } 16)} & \text{if } i = 2, 0 \le j < 16 \\ m_{7j\,(\text{mod } 16)} & \text{if } i = 3, 0 \le j < 16 \end{cases} \quad 0 \le i < 4, 0 \le j < 16$$

3. In (i, j) step #3, there is addition modulo 2^{32} ($\boxplus$) of one of 64 different 32-bit constants $\underline{T} = (T[0], T[1], \cdots, T[63])$ to the result of step 2.

   The value of $T[i]$ given in [Rivest 1992b] is the integer part of 42949672976 times the absolute value of $sin(i)$ with $i + 1$ in radians. For example, $T[0] = D76AA478$ (in hexadecimal).

   The constant $T_{(i,j)} = T[16i + j]$ is used in the jth step of the ith round of MD5.
4. In (i, j) step #4, a left cyclic shift is applied to the result of 16 different left step 3, circular shifts $(\sigma_0, \sigma_1, \cdots, \sigma_{15})$ are specified in [Rivest 1992b] by

$$\sigma_{(i,j)} = \begin{cases} (7, 12, 17, 22, 7, 12, 17, 22, \cdots 7, 12, 17, 22) & \text{if } i = 0 \\ (5, 9, 14, 20, 5, 9, 14, 20, \cdots, 5, 9, 14, 20) & \text{if } i = 1 \\ (4, 11, 16, 23, 4, 11, 16, 23, \cdots, 4, 11, 16, 23) & \text{if } i = 2 \\ (6, 10, 15, 21, 6, 10, 15, 21, \cdots, 6, 10, 15, 21) & \text{if } i = 2 \end{cases};$$

5. In (i, j) step #5, there is addition modulo 2^{32} (($\boxplus$)) of the current contents of Register B to the result of step 4.

   (i, j) steps #1–5 complete the processing in the jth step of the ith round of MD5. Finally, the prepare for the next processing step (or round if $j = 16$), the contents of the registers are right circularly shifted.

We differ slightly, but at least consistently, with the notation in Wang and Yu who refer to the contents of an MD5 register A, B, C, D with only an index that corresponds to its step number but without any diacritical marks or subscripts (i, j). Only the contents of the A-register (the leftmost MD5 register) is modified during the ith step processing. Thus $\tilde{B}_i = B_i$, $\tilde{C}_i = C_i$, and $\tilde{D}_i = D_i$. In contrast, Wang and Yu continue to refer to a value, say A_i, even after it moves to the B, C, or D register in steps $i + 1$, $i + 2$, and $i + 3$. Perhaps in the spirit of cryptographic key spaces, we like large subscripts!

Remark 18.1. There are several superficial differences in our MD5 description from that given in [Rivest 1992b], [Schneier 1996], and [Stallings 1993]. All but [Stallings 1993] omit the cyclic shift in (i, j) step #4; [Rivest 1992b, p. 4] instead changes the "names" of the variables involved in the 16 steps in a round. If (R,S,T,U) is a cyclic permutation of (A,B,C,D), then [Rivest 1992b] describes Step #i by writing

> [RSTU i j k] to mean that (register) R's value is replaced by S $\boxplus$ {R $\boxplus$ F[S,T,U] $\boxplus$ m_i $\boxplus$ T_k} $\ll j$.

For example, Rivest writes [ABCD 0 7 1] for the Step 0 in MD5 and [DABC 1 12 2] for Step 2 in MD5. Schneier [Schneier 1996] uses essentially the same notation as Rivest except that he writes R = S $\boxplus$ {R $\boxplus$ F[S,T,U] $\boxplus$ m_i $\boxplus$ T_k} $\ll_j$ again interpreting the equal sign as an assignment of values.

Stalling's description is the closest to ours except that he writes

$$\tilde{A}_i \leftarrow B_i \boxplus \{A_i \boxplus F_j(B_i, C_i, D_i) \boxplus m_i \boxplus T_i\} \lll_{\sigma_i}$$

and Stalling's Figure 12.3 coalesces the intermediate and final registers shown in Figure 18.7.

Exercise 18.1. Prove that if the 96 components of

$$X = (X_0, X_1, \cdots, X_{31}) \quad Y = (Y_0, Y_1, \cdots, Y_{31}) \quad Z = (Z_0, Z_1, \cdots, Z_{31})$$

are independent and identically distributed (0,1)-valued random variables with $Pr\{X_i = 0\} = 1/2$ and

$$U = (U_0, U_1, \cdots, U_{31}) = F_0(X, Y, Z) = (X \wedge Y) \vee (\neg X \wedge Z)$$

then the 32 components of U are are independent and identically distributed (0,1)-valued random variables with

$$Pr\{U_i = 0\} = 1/2$$

Exercise 18.2. Prove that if the 96 components of

$$X = (X_0, X_1, \cdots, X_{31}) \quad Y = (Y_0, Y_1, \cdots, Y_{31}) \quad Z = (Z_0, Z_1, \cdots, Z_{31})$$

are independent and identically distributed (0,1)-valued random variables with $Pr\{X_i = 0\} = 1/2$ and

$$U = (U_0, U_1, \cdots, U_{31}) = F_1(X, Y, Z) = (X \wedge Z) \vee (Y \wedge \neg Z)$$

then the 32 components of U are are independent and identically distributed (0,1)-valued random variables with

$$Pr\{U_i = 0\} = 1/2$$

Exercise 18.3. Prove that if the 96 components of

$$X = (X_0, X_1, \cdots, X_{31}) \quad Y = (Y_0, Y_1, \cdots, Y_{31}) \quad Z = (Z_0, Z_1, \cdots, Z_{31})$$

are independent and identically distributed (0,1)-valued random variables with $Pr\{X_i = 0\} = 1/2$ and

$$U = (U_0, U_1, \cdots, U_{31}) = F_2(X, Y, Z) = X \oplus Y \oplus Z$$

then the 32 components of U are are independent and identically distributed (0,1)-valued random variables with

$$Pr\{U_i = 0\} = 1/2$$

Exercise 18.4. Prove that if the 96 components of

$$X = (X_0, X_1, \cdots, X_{31}) \quad Y = (Y_0, Y_1, \cdots, Y_{31}) \quad Z = (Z_0, Z_1, \cdots, Z_{31})$$

are independent and identically distributed (0,1)-valued random variables with $Pr\{X_i = 0\} = 1/2$ and

$$U = (U_0, U_1, \cdots, U_{31}) = F_3(X, Y, Z) = Y \oplus (X \vee \neg Z)$$

then the 32 components of U are are independent and identically distributed (0,1)-valued random variables with

$$Pr\{U_i = 0\} = 1/2$$

Exercise 18.5. Characterize and enumerate the number of pairs (U, U') with

$$U, U' \in \mathcal{Z}_{32,2} = \{\underline{x} = (x_0, x_1, \cdots, x_{31}) : x_i = 0 \text{ or } 1, 0 \leq i < 32\}$$

for which $U' \boxminus U = 2^{31}$.

Exercise 18.6. Characterize and enumerate the number of pairs (U, U') with

$$U, U' \in \mathcal{Z}_{32,2} = \{\underline{x} = (x_0, x_1, \cdots, x_{31}) : x_i = 0 \text{ or } 1, 0 \leq i < 32\}$$

for which $U' \boxminus U = 2^{15}$.

Exercise 18.7. Characterize and enumerate the number of pairs (U, U') with

$$U, U' \in \mathcal{Z}_{32,2} = \{\underline{x} = (x_0, x_1, \cdots, x_{31}) : x_i = 0 \text{ or } 1, 0 \leq i < 32\}$$

for which $U' \boxminus U = -2^{15}$.

18.9 CRITICISM OF MD5

Why these specific nonlinear functions, constants, shifts, and number of rounds? The answer is not entirely satisfying; the nonlinear functions $\{F_j\}$ enjoy several randomness-like properties (Exercises 18.1–7); if the bits of their inputs X, Y, and Z are independent and unbiased, then the 32-bit output bits are also independent and unbiased. Rivest is careful in his summary [Rivest 1992b, p. 5] writing that "It is *conjectured* [my emphasis] that the difficulty of coming up with two messages having the same message digest is on the order of 2^{64} operations, and the difficulty of coming up with any message having a given message digest is on the order of 2^{128} operations."

Bart Preneel et al. [Preneel, Govaerts, and Vandewalle 1993] discusses design principles for message digests without really arriving at any definitive conclusions or design principles. Although the design of cryptographic hashes and encipherment systems is largely an art, Rivest and others have certainly been influenced by the Horst Feistel's paper [Feistel 1973], which heralded the entry of nongovernmental entities in the design of cryptographic systems. Like DES, MD4 and MD5 iterates a basic transformation to derive its output.

- DES used 16 rounds, but MD5 uses 64 steps, and DES involves half of its input in the same way on each round. MD5 involves all of its input in different ways in each round.

A cryptographic hash function is supposed behave like a random function while still being deterministic and efficiently computable. John von Neumann [von Neumann 1951] once wrote:

> Anyone who considers arithmetical methods of producing random digits is, of course, in a state of sin.

D. H. Lehmer, who is a pioneer in random number generation methodology [Lehmer 1951], wrote

> A random sequence is a vague notion embodying the idea of a sequence in which each term is unpredictable to the uninitiated and whose digits pass a certain number of tests, traditional with statisticians and depending on the uses to which the sequence is to be put.

Because a message digest maps a large message space into a much smaller message space, there are certainly many collisions with a fixed message M. The birthday para*dox* [Chapter 4, §4.5] implies that collisions for an n-bit message digest can be found by random trials using $O(2^{n/2})$ operations. MD5 maps a 512-bit block into 128-bit block; the MD5-hash is a 2^{384}-to-1 mapping so that there are many messages M_0 and M_0' with the same MD5 hash value. A perhaps more modest task is to find near collisions, which are two messages that hash to almost the same value, with a difference of only a few bits. It was believed that this was as difficult, or almost as difficult, as finding a full collision.

It does not seem possible to characterize effectively the set of messages $\mathcal{M}$ having the same the MD5 hash value. Until 2005, the problem of finding a second preimage for MD5[M] seemed also very hard. Will the sun fail to rise tomorrow, if say, `Good_Morning` and `Kssh_Qsvrmrk` have the same MD5 hash-value? Even if MD5[M_0] = MD5[M_0'], this might not even be a serious weakness because message digests are used in applications where the messages (M_0, M_0') are required have special formats, for example, as in an X.509 certificate. Various researches have observed defects in MD5, as follows:

- Bert den Boer and Antoon Bosselaers [1993] and Hans Dobbertin [1996a] and [1996b] discovered some peculiar "near" collisions.
- X. L. Wang, F. D. Guo, X. J. Lai, and H. B. Yu announced collisions for MD4, MD5, HAVAL-128, and Race Integrity Primitives Evaluation Message Digest (RIPEMD) at a rump session of *Crypto'04* [Wang, Lai et al. 2005].

It remained for Wang and Yu's paper [Wang and Yu 2005] to provide details explaining how collisions for MD5 could be found. A. Lenstra and B. de Weger [Lenstra and de Weger 2005] improved on [Wang and Yu 2005] demonstrating how to construct X.509 certificates with the same signatures but with different public keys. Even though this was disconcerting, it did not exactly preclude the use of certificates in SSL transactions.

The ubiquitous phrase "some cryptographers" have even suggested that these "flaws" in MD5 make "further use of the algorithm for security purposes questionable" seems prescient.

18.10 THE WANG-YU COLLISION ATTACK

The Wang-Yu attack produces for any 128-bit-initial hashing value IHV_0, a pair of 512 bit input blocks $(M_1, M_2) \neq (M_1^*, M_2^*)$ such that

$$(IHV_0, M_1) \xrightarrow{MD5} IHV_1 \tag{18.1a}$$

$$(IHV_0, M_2) \xrightarrow{MD5} IHV_1^* \tag{18.1b}$$

$$IHV_{1*} \neq IHV_1^* \tag{18.1c}$$

$$(IHV_1, M_1) \xrightarrow{MD5} MD5[M_1] \tag{18.1d}$$

$$(IHV_1^*, M_2) \xrightarrow{MD5} MD5[M_2] \tag{18.1e}$$

$$MD5[M_1] = MD5[M_2] \tag{18.1f}$$

MD5 and members of the SHA-family (§18.11) are constructed following the Merkle-Damgård construction [Merkle 1979] and [Damgard 1989], which is a paradigm to build cryptographic hash functions. All popular hash functions follow this generic construction.

Their methodology may be used to construct a fixed-length cryptographic hash function for an arbitrarylength message into output. They achieve this by breaking the input up into a series of equal-sized blocks and operating on them in sequence using a one-way (compression) function that transforms a fixed-length input into a shorter, fixed-length output. The compression function might be built from a block cipher or basic functional primitives.

Because the Wang-Yu construction is for any initial chaining value IHV_0, collisions can be combined into larger blocks; that is,

Given: any prefix P and suffix S
There exists: blocks C, C^*
Such that: 1. $MD5[P||C] \neq MD5[P||C^*]$, and
2. $MD5[P||C||S] = MD5[P||C^*||S]$.

where we have taken some notational liberties writing $MD5[P||C]$ and $MD5[P||C^*]$ for the state of the MD5 registers just after the processing of $P||C$ and $P||C^*$. The (common) size of the blocks C, C^* can be more than two (512-bit) input blocks and may even consist of partial blocks.

18.11 STEVEN'S IMPROVEMENT TO THE WANG-YU COLLISION ATTACK

The next step is derived from Stevens' thesis [Stevens 2007] and [Stevens, Lenstra, and de Weger 2007]; the Wang-Yu prefix attack is extended to permit arbitrary initial chaining values.

Given: any pair of prefixes (P, P^*) and a suffix S
There exists: blocks C, C^*
Such that: 1. $MD5[P||C] \neq MD5[P^*||C^*]$, and
2. $MD5[P||C||S] = MD5[P^*||C^*||S]$.

Taking the same notational liberties, $MD5[P||C]$ and $MD5[P||C^*]$ can be written for the state of the MD5 registers just after the processing of $P||C$ and $P^*||C^*$.

Given prefixes P, P^*, whose lengths are not necessarily a multiple of 512 bits, Stevens does the following:

1. First pads P and P^* with bit strings A and A^* so that the lengths $P||A$ and $P^*||A^*$ are the same;
2. Then uses *birthdaying* to pad $P||A$ and $P^*||A^*$ with bit strings B and B^* so that the lengths $P||A||B$ and $P^*||A^*||B^*$ are the same an a multiple of 512-bits;
3. The purpose of the repeated padding in the first two steps is to ensure that $P||A||B$ and $P^*||A^*||B^*$ have a prescribed structure; the padding continues with blocks NC and NC^*, whose lengths are multiple of 512 bits and whose purpose is to result in messages $M = P||A||B||NC$ and $P^*||A^*||B^*||NC^*$ with MD5 near collisions.

The details of the construction are given in [Stevens, Lenstra, and de Weger 2007].

18.12 THE CHOSEN-PREFIX ATTACK ON MD5

In December 2008, Stevens et al. extended the attack in an earlier paper [Stevens 2007] using a cluster of 200 PS3,[4] and announced a method to create a rogue CA certificate that collided with a valid certificate of some commercial CA at the the 25th Chaos Communication Congress (25C3) in Germany on December 30, 2008. Details appear in a preprint on the Web and will be delivered at *Crypto '09* [Stevens et al. 2009].

It is important to understand what Stevens and his colleagues building on the work in [Wang and Yu 2005] have not achieved; their chosen prefix attack does not create a rogue (stand-alone) syntactically correct X.509 certificate issued by Trusted Cert Inc., to the secure website www.trustedwebsite.com, which would be delivered to user, in the course of a transaction from user from www.trustedwebsite.com when requested as in Figure 18.6C from this website.

To create a valid certificate issued by the CA (Trusted Cert Inc.) requires knowledge of this CA's private key.

It is clear that the old-fashioned wire tapping, bribery, and/or theft after breaking and entering might be used to obtain a CA's private key, practice this is not entirely unfamiliar to the intelligence agencies. David Kahn's book [Kahn 2004] relates how Yardley illegally obtained cables reading Japanese diplomatic traffic at the

[4] PS3 is the PlayStation 3, the third home video game console produced by Sony Computer Entertainment. It is the successor to the PlayStation 2 in the PlayStation series. The basic model has a 20-GB hard disk drive (HDD), and a 60-GB hard drive is available in premium model. It is even available in several colors for the color-conscious hacker.

Washington Naval Conference in 1921. The information the Cipher Bureau provided the American delegation was instrumental in getting the Japanese side to agree to a 5:3 ratio instead of the 10:7 ratio the Japanese wanted.

Years later, the U.S. Navy, in collaboration with the Federal Bureau of Investigation and the New York City police, entered the Japanese Consulate in New York City and retrieved the Japanese RED code book.

The analysis of Steven et al. is a significant achievement, blending analytical and programming skills with experimental computer science, which is a subarea of computer science identified as a funding area by a previous Dean of the College of Engineering at UCSB. What Stevens et al. did was to start from a valid certificate issued by Trusted Cert Inc. to www.trustedwebsite.com and create a second valid Trusted Cert Inc. certificate with a different (RSA) distinguished subject name, say www.fakewebsite.com, and a different user private key. In addition to the possible attack scenario to be described, the trust derived from the possession of a valid certificate is brought into doubt.

The success of their attack assumes many things, but it is nevertheless an important achievement. Although not exactly felines, academic cryptographers are a finicky bunch, and any cryptographic blemish causes their fur to stand.[5] The 2007 announcement created a mate to a valid *Cheap_Cert* certificate with a different distinguished subject name; the advance announced in 2009 created a mate to a valid *Cheap_Cert* certificate with a different distinguished subject names and user private keys (RSA). When *Cheap_Cert* issues a certificate, the following occurs:

1. The serial number, issuer distinguished name, and validity period are chosen by *Cheap_Cert*.
2. The distinguished subject name and public key information are at the (partial) discretion of the user.

The complication is that the certificate serial number (Figure 18.4) is specified by the CA issuer.

A Sketch of the Stevens' Chosen Prefix Attack is given in the Appendix 19.

18.13 THE ROGUE CA ATTACK SCENARIO

1. The rogue computer scientist purchases a valid user certificate C_1 from Trusted Cert Inc. with CA = FALSE.
2. The rogue computer scientist fabricates a valid certificate C_2 with the following provisions:
   - Issuer distinguished name = trusted cert Inc.,
   - Distinguished subject name = Fake trusted cert Inc.,
   - Known private (RSA) key

[5]The Electronic Frontier Foundation's $220000 DES cracker proved it possible to find a 56-bit DES-key in 3 days using 1998. It would have taken the DES Cracker 768 days using the 1998 technology, if the key length were $112 = 2 \times 56$ bits as in triple DES (DES3). Perhaps the Web and the world is safer with AES.

- CA = TRUE, path_length = 2
- MD5 hash of the C_1-data_to_besigned = MD5 hash of the C_2-data_to_be_signed

As a consequence, the signature on C_1 and C_2 are the same. C_2 will be the root certificate for the rogue certificate authority Fake_Trusted Cert Inc.

3. After the payment of a suitable gratuity, Fake_Trusted Cert Inc. issues a perfectly valid user certificate C_3 for the distinguished subject name = www.trustedwebsite.com and signed by the issuer distinguished name = Fake_Trusted Cert Inc.
4. When the unsuspecting user attempts to engage in a transaction with the website www.trustedwebsite.com, the rascal rogue engages in a man in the middle attack[6]. This causes his browser to redirect[7] the message from his browser to the fake website www.fakewebsite.com, which graciously supplies the following:
   - The valid user certificate C_3, identifying itself as *www.trustedwebsite.com*
   - The root certificate C_2, allowing User to validate C_3 using the root certificate of *Trusted Cert Inc.* a path length of 2

18.14 THE SECURE HASH ALGORITHMS

The SHA is a family cryptographic hash function following the Merkle-Damgård construction designed by the National Security Agency (NSA) and published as a U.S. government standard [FIPS 180]. The first version published in 1993 is often referred to as SHA-0. SHA-1 is the most commonly used hash function in the family of application protocols including the TLS, SSL, Pretty Good Privacy (PGP), Secure Shell (SSH), and the Internet Protocol Security (IPSec).

SHA-1 first pads the message $M \rightarrow (M_0, M_1, \cdots, M_{15}, \cdots)$ as in MD5 to make its length a multiple of 512 bits and produces a 160-bit hash value SHA-1[M]. The calculation of the SHA-1 hash involves 80 rounds, whose operations modify the contents of five 32-bit registers A,B,C,D,E[8]; the operations in the ith round of SHA-1 are depicted in Figure 18.8.

The SHA-1 operations are as follows:

- A round-dependent nonlinear operation

$$F_i(B,C,D) = \begin{cases} (B \wedge C) \oplus (\neg B \wedge D) & \text{if } 0 \le i < 20 \\ (B \oplus C \oplus D) & \text{if } 20 \le i < 40 \\ (B \wedge C) \oplus (B \wedge D) \oplus (C \wedge D) & \text{if } 40 \le i < 60 \\ B \oplus C \oplus D & \text{if } 60 \le i < 80 \end{cases}$$

[6]A form of active eavesdropping in which an attacker makes independent connections with the victims and relays messages between them, making them believe that they are talking directly to each other over a private connection when in fact the entire conversation is controlled by the attacker. The attacker must be able to intercept all messages going between the two victims and inject new ones, which is possible in many situations in which the communications are not enciphered.

[7]A URL redirection attack connects the browser from one URL to another URL. It is a vulnerability that redirects you to another page freely out of the original website. The redirected malicious page that resembles the original might be used to trick the user into giving their credentials.

[8]SHA uses big-endian byte-ordering for data.

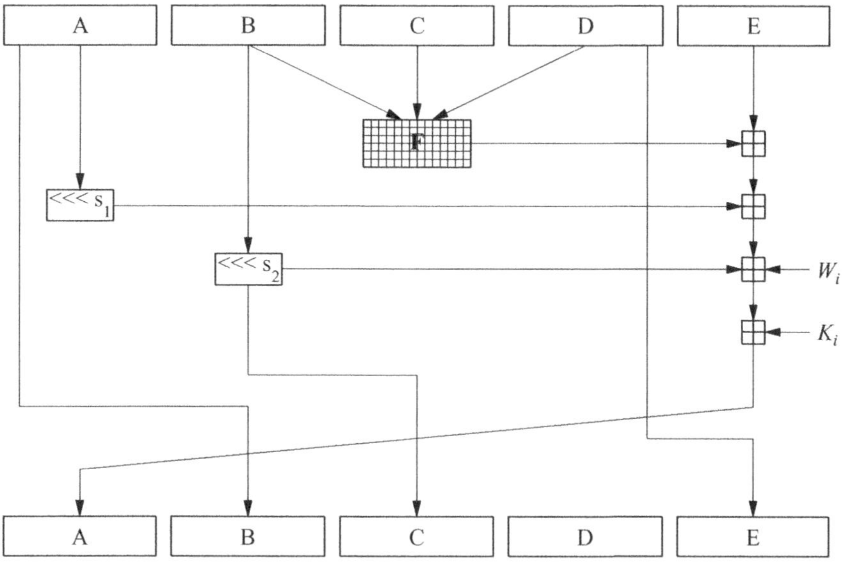

$<<<_s$: Left circular rotation by s bits Addition modulo 2^{32} $\boxplus$

Figure 18.8. The operations in one round of SHA-1.

TABLE 18.2. SHA-1 Initial Hashing Value IHV_0

A 67 45 23 01	B ef cd ab 89	C 98 ba dc fe	D 10 32 54 76	E c3 d2 e1 f0

The registers $IHV_0 \leftarrow (A, B, C, D, E)$ are initialized (in hexadecimal) are shown in Table 18.2.

- Left-circular shifts $<<<_s$ by s bits with $s_i = \begin{cases} 5 & \text{if } i = 1 \\ 30 & \text{if } i = 2 \end{cases}$
- Addition (modulo 2^{32}) ($\boxplus$) of constants $\{K_i\}$ defined by

$$K_i = \begin{cases} \text{5a827999} = \lfloor 2^{30} \times \sqrt{2} & \text{if } 0 \le i < 20 \\ \text{6ed9eba1} = \lfloor 2^{30} \times \sqrt{3} & \text{if } 20 \le i < 40 \\ \text{8f1bbcdc} = \lfloor 2^{30} \times \sqrt{5} & \text{if } 40 \le i < 60 \\ \text{ca62c1d6} = \lfloor 2^{30} \times \sqrt{10} & \text{if } 60 \le i < 80 \end{cases}$$

- Addition (modulo 2^{32}) ($\boxplus$) of a 32-bit word $\{W_i\}$ derived from the message defined by

$$(M_0, M_1, \cdots, M_{15}, \cdots) \rightarrow (W_0, W_1, \cdots, W_{79})$$

$$W_i = \begin{cases} M_i & \text{if } 0 \le i < 16 \\ <<<_1 \langle W_{i-3} \oplus W_{i-8} \oplus W_{i-14} + W_{i-16} \rangle & \text{if } 16 \le i < 80 \end{cases}$$

where $\oplus$ is bitwise XOR.

The first four initialized register values are those of MD5, except that they are in the big-endian byte order in SHA-1.

Thereafter, the values in the register as replaced for each of the 80 rounds as in MD5 for the first 512-bit message block are $IHV_i \to IHV_{i+1}$ for $i = 0,1, \cdots, 79$. If a message M

- Consists of just one block, then SHA-1[M] = IHV_{80};
- Consists of more than one block, then $IHV_{79} \to IHV_{80}$ is used to initialize the register for the second block, and so on.

Exercise 18.8. Show that the (nonlinear) functions $\{F_0(x, y, z), F_1(x, y, z), F_2(x, y, z)\}$ of the bits (x, y, z) are uniformly distributed; that is, $|\{(x, y, z) \in Z_{3,2}:F_i(x, y, z) = 0\}| = 4$ for $i = 0, 1, 2$.

18.15 CRITICISM OF SHA-1

The first attack on SHA-0 was presented as CRYPTO '98 [Chabaud and Joux 1998], and E. Bilham and R. Chen reported *near collisions* in 2004 [Bilham and Chen 1998]. However, in 2005 Wang and his group announced they discovered a way [Wang, Yin and Yu 2005] with a work function of 2^{63} operations, which was a significant improvement over the 2^{80} birthday bound. The Computer Security Division of National Institute of Standards and Technology (NIST)'s Computer Security Resource Center acknowledged this, writing

> "NIST accepts that Prof. Wang has indeed found a practical collision attack on SHA-1" and that "2^{63} operations is plainly within the realm of feasibility for a high resource attacker."
>
> They concluded that "Several steps are now prudent. The first of these is to transition rapidly to the stronger SHA-2 family of hash functions (SHA-224, SHA-256, SHA-384 and SHA-512) for digital signature applications.

In May 2009 the Australians McDonald, Hawkes and Pieprzyk[9] announced [McDonald, Hawkes and Pieprzyk 2009] a significant decrease is the collision-finding work function from $O(2^{63})$ to $O(2^{52})$ operations.

18.16 SHA-2

SHA2 denotes a class of hash functions [FIPS 180-3] as follows:

- SHA-224 and SHA-256 with yield 512-bit hashes
- SHA-384 and SHA-512 with yield 1024-bit hashes

We describe SHA-224 and SHA-256, which are structurally the same, referring to it as SHA-2.

SHA-2 first pads the message M like SHA-1 to make its length a multiple of 512 bits

[9]Faculty in the Centre for Advanced Computing, Algorithms and Cryptography Department of Computing, at *Macquarie University and Qualcomm* (Australia).

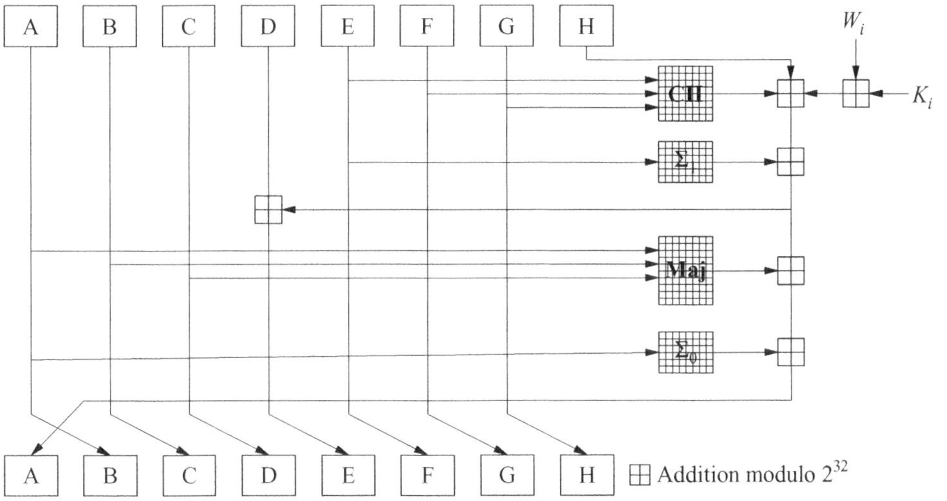

Figure 18.9. The operations in one round of SHA-2.

$$M \rightarrow (M_0, M_1, \cdots, M_{15}, \cdots)$$

and produces a 256-bit hash value SHA-2[M].

One round of SHA-2 is shown in Figure 18.9 where each of A, B, C, D, E, F, G, and H are 32-bit registers that consists of the following operations:

- Nonlinear operations

$$\begin{aligned}
\mathrm{CH}(E, F, G) &= (E \wedge F) \oplus (\neg E \wedge G), \\
\mathrm{Maj}(A, B, C) &= (A \wedge B) \oplus (A \wedge C) \oplus (B \wedge C), \\
\textstyle\sum_0(A) &= \lambda_2(A) \oplus \lambda_{13}(A) \oplus \lambda_{22}(A), \text{ and} \\
\textstyle\sum_1(E) &= \lambda_6(E) \oplus \lambda_{11}(E) \oplus \lambda_{25}(E);
\end{aligned}$$

- Addition (modulo 2^{32}) ($\boxplus$) of constants $\{K_i : 0 \le i < 64\}$ where K_i consists of the first 32 bits of the fractional parts of the cube roots of the first 64 prime numbers $2, 3, 5, 7, \cdots, 311$.
- Addition (modulo 2^{32}) ($\boxplus$) of a 32-bit word $\{W_i : i = 0, 1, \cdots\}$ derived from the message to (A,B,C,D,E,F,G,H) according to

$$\begin{aligned}
&(M_0, M_1, \cdots, M_{15}, \cdots) \rightarrow (W_0, W_1, \cdots, W_{79}) \\
&\qquad s = i \wedge MASK \quad MASK = 0000000\mathrm{F} \quad \text{if } 16 \le i < 80 \\
&\qquad W_s = <<<_1 \langle (W_{s+13} \wedge MASK) \oplus (W_{s+8} \wedge MASK) \oplus (W_{S+2} \wedge MASK) \oplus W_i \rangle
\end{aligned}$$

where λ_s is right-circular rotation by s bits.

The registers A, B, C, D, E, F, G, and H are initialized (in hexadecimal) and are shown in Table 18.3.

TABLE 18.3. SHA-256 Initial Hash Value INV_0

A:	c1	05	9e	d8	B:	36	7c	d5	07	C:	30	70	dd	17	D:	f7	0e	59	39
E:	ff	c0	0b	31	F:	68	58	15	11	G:	64	f9	8f	a7	H:	be	fa	4f	a4

Thereafter, the values in the register as replaced for each of the 80 rounds as in MD5 or SHA-1 for the first 512-bit message block; $IHV_i \to IHV_{i+1}$ for $i = 0, 1, \cdots, 79$. If a message M

- Consists of *just* one block, then SHA-2[M] = IHV_{80};
- Consists of *more* than one block, then $IHV_{79} \to IHV_{80}$ is used to initialize the register for the second block, and so on.

18.17 WHAT NOW?

An open competition for a new SHA-3 function was formally announced in the Federal Register on November 2, 2007. NIST is initiating an effort to develop one or more additional hash algorithms through a public competition, similar to the development process for the AES. Submissions were due October 31, 2008, and the proclamation of a winner and publication of the new standard are scheduled to take place in 2012.

Even though this sounds nifty, what will result? It is important to keep in mind that MD5 was not designed by an amateur or by a committee of insiders; it was the work product of a truly gifted individual. In addition to being a coinventor of one of the only surviving public key cryptosystems, Rivest has a substantial number of refereed publications, not focused in just one area, but spanning several. It is all well and good to pontificate as to the desirable properties in the design of strong cryptographic functions—encipherment or hashing functions; it is quite another to tell in advance which features will lead to compromise. Maybe, the answer is *Bigger is Better*!, and SHA-xyz with $2^{2^{\cdots}}$ rounds and hash size of $2^{2^{\cdots}}$ will suffice forever, and perhaps a stimulus package will be needed for the unemployed cryptographers. However, obesity in design is unlikely to be more a virtue in security than it is in health.

REFERENCES

E. Bilham and R. Chen, Near-Collisions of Sha-0, Proceedings of CRYPTO '04—Advances in Cryptology, Springer-Verlag (Berlin, Germany), 1998, pp. 290–305.

B. den Boer and A. Bosselaers, Collisions for the Compression Function of MD5, Eurocrypt '93, 1993, pp. 293–304.

F. Chabaud and A. Joux, Differentia Collisions in SHa-0, Proceedings of Crypto '98, Springer-Verlag (Berlin, Germany), 1998, pp. 56–71.

I. Damgard, "A Design Principle for Hash Functions", in *Advances in Cryptology CRYPTO 89 Proceedings*, G. Brassard (Editor), Springer-Verlag (Berlin, Germany), 1989, pp. 416–427.

H. Dobbertin, "Cryptanalysis of MD5 Compress", German Information Security Agency (Bonn, Germany), 1996.

H. Dobbertin, "The Status of MD5 After a Recent Attack", *CryptoBytes*, **2**, #2, 1996.

H. Feistel, "Cryptography and Computer Privacy", *Scientific American*, **228**, #5, pp. 15–23, 1973.

FIPS PUB 180-3, Federal Information Processing Standards, Secure Hash Standard (SHS), 2008.

ITU-T Recommendation X.509, version 3, *Information Technology—Open Systems Interconnection—The Directory: Authentication Framework*, 2005.

D. Kahn, *The Reader of Gentlemen's Mail: Herbert O. Yardley and the Birth of American Codebreaking*, Yale University Press (New Haven, CT), 2004.

B. Kaliski, "The MD2 Message Digest", RFC 1319, RSA Laboratories (Bedford, MA), 1992.

A. G. Konheim, *Cryptography and Computer Security*, John Wiley & Sons (New York), 2007.

D. H. Lehmer, "Mathematical Methods on Large-Scale Computing Units", *Annals of the Computation Laboratory of Harvard University*, **26**, pp. 141–146, 1951.

A. K. Lenstra and B. de Weger, *On the Possibility of Constructing Meaningful Hash Collisions for Public Keys*, Australasian Conference on Information Security and Privacy, 2005, pp. 267–279.

C. McDonald, P. Hawkes, and J. Pieprzyk, *Differential Path for SHA-1 with Complexity O(252)*, Eurocrypt 2009 Conference, 2009.

R. C. Merkle, "Secrecy, Authentication, and Public Key Systems", PhD Dissertation, Stanford University, Palo Alto, CA, 1979.

National Institute of Standards and Technology, NIST FIPS PUB 180, Secure Hash Standard, U.S. Department of Commerce (Washington, DC), 1994.

B. Preneel, R. Govaerts, and J. Vandewalle, "Hash Functions Based on Block Ciphers: A Synthetic Approach", Proceedings Crypto '93, 1993, pp. 368–378.

RFC524, The Transport Layer Security (TLS) Protocol (Version 1.2), RTFM, Inc. (San Jose, CA), 2008.

R. L. Rivest, "The MD4 Message Digest", RFC 1320, RSA Data Security Inc. (Bedford, MA), 1992a.

R. L. Rivest, "The MD5 Message Digest", RFC 1321, RSA Data Security Inc. (Bedford, MA), 1992b.

B. Schneier, *Applied Cryptography* (Second Edition), John Wiley & Sons (New York), 1996.

W. Stallings, *Cryptography and Network Security* (Third Edition), Prentice-Hall (Upper Saddle River, NJ), 1993.

M. Stevens, "On Collisions for MD5", MS Thesis, Eindhoven University of Technology, 2007.

M. Stevens, A. Lenstra, and B. de Weger, *Chosen-Prefix Collisions for MD5 and Colliding X.509 Certificates for Different Identities*, Eurocrypt '07, Springer-Verlag (Berlin, Germany), 2007, pp. 1–22.

M. Stevens, A. Sotirov, J. Appelbaum, A. Lenstra, D. Molnar, D. A. Osvik, and B. de Weger, Short Chosen-Prefix Collisions for MD5 and the Creation of a Rogue CA Certificate, Proceedings of Crypto '09, Springer-Verlag (Berlin, Germany), 2009.

J. Swift, *Gulliver's Travels*, Sterling Publishing (New York), 1726.

J. von Neumann, "Various Techniques Used in Connection with Random Digits", *Applied Mathematics Series*, **12**, pp. 36–38, 1951.

Xiaoyun Wang, Xuejia Lai, Dengguo Feng, Hui Chen and Xiuyuan Yu, "Cryptanalysis of the Hash Functions MD4 and RIPEMD", in *Advances in Cryptology EUROCRYPT 2005*, 2005, pp. 1–18.

X. Wang, Y. L. Yin, and X. Yu, *Finding Collisions in the Full Sha-1*, Advances in Cryptology Crypto '05, Springer-Verlag, (Berlin, Germany), 2005, pp. 17–36.

X. Wang and H. Yu, *How to Break MD5 and Other Hash Functions*, Eurocrypt '05, 2005.

APPENDIX 18

Sketch of the Steven's Chosen Prefix Attack

A18.1 BIRTHDAYING[1] [SCHNEIER 1996, PP. 165–166], [OORSCHOT AND WIENER 1999]

The *birthday problem* has already been described in Chapter 4 §4.5. If f is a function on $f: X \to X$ and we perform a (uniformly distributed) *random walk* on X evaluating f at which points visited, then after $M \approx \sqrt{\pi|X|/2}$ steps, we expected to encounter points x_i and x_M for which $f(x_1) = f(x_M)$. Birthdaying is the process of testing randomly chosen values to obtain a collision $f(x) = f(x')$.

A18.2 THE BINARY SIGNED DIGIT REPRESENTATION (BSDR)

Let $x = (x_0, x_1, \cdots, x_{31})$ be a 32-bit word with integer value $x = \sum_{i=0}^{31} x_i 2^i$. The binary signed digit representation [Clark and Liang 1973] of the (decimal value) $x \Leftrightarrow \underline{\bar{x}} = (\bar{x}_0, \bar{x}_1, \bar{x}_2, \cdots, \bar{x}_{31})$ is defined by

$$x = \bar{x}_0 + \bar{x}_1 2 + \bar{x}_2 2^2 + \cdots + \bar{x}_{31} 2^{31} \quad \bar{x}_i \in \{0, \pm 1\} \tag{A18.1}$$

The weight $w(\underline{\bar{x}})$ of the BSDR $\underline{\bar{x}}$ is the number of nonzero $\bar{x}_i$'s.

When an integer x is interpreted modulo 2^{32}, the BSDR of x is not unique.

Example A18.1. If BSDR is used with 4-bit words, then

4-Bit BSDR-Coding of 8
(1,1,1,0) = 7 = 1 + 2 + 4
(1,–1,0,1) = 7 = 1 – 2 + 8
(1,1,–1,1) = 7 = 1 + 2 – 4 + 8
(–1,0,0,1) = 7 = – 1 + 8

Example A18.2. There may be more than one BSDR of n of minimal weight

5-Bit BSDR-Coding of 11
(–1,0,1,1,0) = 11 = – 1 + 4 + 8
(1,0,–1,0,1) = 11 = 1 – 4 + 16

[1] [Schneier 1996, pp. 165–166] and [van Oorschot and Wlener [1999]].

TABLE 18A-1. Trace of the Execution of NAF(x) for x = 8, 9, 13, 14, 15

n_{IN}	e	$\left\{ \overset{?}{3n_{IN} > 2^{e+2}} \right\}$	$2^{e+1} - n_{IN}$	$n_{IN} - 2e$	RV_1	RV_2	n_{OUT}	NAF(n)
15	3	Yes	1		16	–NAF(1)	1	
1	0	No		0	1	NAF(0)	∅	15 = 16 – 1
14	3	Yes	6		8	–NAF(6)	6	
6	2	Yes	2		4	–NAF(2)	2	
2	1	No		0	2	NAF(0)	∅	14 = 16 – 4 + 2
13	3	Yes	3		16	–NAF(3)	3	
3	2	No		1	4	–NAF(1)	1	
1	0	No		0	1	NAF(0)	∅	13 = 16 – 8 + 4 – 1
9	3	No		1	8	NAF(1)	1	
1	0	No		0	1	NAF(0)	∅	9 = 8 + 1
8	3	No		0	8	–NAF(0)	∅	8 = 8

In the non-adjacent form (NAF) of a BSDR, no two nonzero $\{\bar{x}_i\}$ are adjacent. The following (recursive) program computes the NAF of n [Shallit 1992].

Program NAF(n)
if $n \equiv n_{IN} = 0$, then $RV \equiv \text{Return}(\emptyset)$; $\emptyset$ = termination character.
Otherwise, determine (the exponent) e such that $2^e \le n < 2^{e+1}$;

a. If $3n > 2^{e+2}$, then Return = (RV_1, RV_2)

$$RV_1 = 2^{e+1}, \quad RV_2 = -\text{NAF}(n_{OUT}), \quad n_{IN} \to n_{OUT} = 2^{e+1} - n_{IN};$$

b. Otherwise, Return = (RV_1, RV_2)

$$RV_1 = 2^e, \quad RV_2 = \text{NAF}(n_{OUT}), \quad n_{IN} \to n_{OUT} = n_{IN} - 2^e.$$

Sample Traces of NAF(n). Table 18A.1 traces the execution of NAF(n) for n = 8, 9, 13, 14, 15.

18a) Find a formula for the number NAF_n of nonadjacent form sequences $(\bar{x}_0, \bar{x}_1, \bar{x}_2, \cdots, \bar{x}_{n-1})$ of length n.

18b) Prove that
  i) Every positive integer n has an NAF representation.
  ii) The NAF representation is unique.
  iii) The output of NAF(n) has *no* leading[2]

18c) The base-2 representation of the integer $n > 0$ requires $\lfloor \log_2 n \rfloor$ bits. How many bits are required for the NAF-representation of the integer $n > 0$?

18d) Prove that the NAF is of minimal weight.

[2]Zeroes in the least (resp. most) significant positions in a base-2 representation of n are trailing (*resp.* leading) zeros.

A18.3 DIFFERENTIAL ANALYSIS

Differential cryptanalysis was introduced by Bilham and Shamir [Bilham and Shamir 1993] as a tool for the cryptanalysis of a variety of block cipher systems. [Konheim 2007, pp. 304–309] describes its application to the cryptanalysis of Data Encryption Standard (DES).

When differential analysis is used to investigate an encryption system, the term *differential* refers to making use of changes in corresponding plain text and ciphertext to infer information about the unknown key. Wang and Yu's innovation [Wang and Yu 2005] was to use the same idea to construct collisions for a hash function H, that is, to find arguments x_1 and x_2 efficiently, which have the same hash value $H(x_1) = H(x_2)$.

Wang and Yu make use of relationships between the exclusive or XOR-difference (-sum) $\mathrm{MD5}[M] + \mathrm{MD5}[M']$ and the modular difference $\mathrm{MD5}[M] \boxminus \mathrm{MD5}[M']$ of two messages M and M'.

If U and V are (32-bit) words, then

- $U \boxplus V$ denotes the sum mod 2^{32} of U and V (32-bit integers).[3]
- $U \boxminus V$ denotes the difference mod 2^{32} of U and V (32-bit integers).
- $U + V$ denotes the XOR of U and V.

Note that a specific value for $U \boxminus V$ may correspond to several different $U + V$ values. For example, if

$$U \boxminus V = 2^6$$

there are several possibilities for the value of $U + V$; for example,

1. $U = 2^6 = 0_8\ 0_8\ 00000000\ 01000000$
   $V = 0 = 0_8\ 0_8\ 00000000\ 00000000$
   $U + V = 0_8\ 0_8\ 00000000\ 01000000$
2. $U = 2^7 = 0_8\ 0_8\ 00000000\ 10000000$
   $V = 2^6 = 0_8\ 0_8\ 00000000\ 01000000$
   $U + V = 0_8\ 0_8\ 00000000\ 11000000$ Carry = bit effect
3. $U = 2^8 = 0_8\ 0_8\ 00000001\ 00000000$
   $V = 2^6 + 2^7 = 0_8\ 0_8\ 00000000\ 11000000$
   $U + V = 0_8\ 0_8\ 00000001\ 11000000$ Carry = bit effects

with

$$0_8 = \underbrace{00000000}_{8\ 0\text{'s}}$$

Note that if U and V are interchanged, the sign of $U \boxminus V$ changes but $U + V$ remains the same.

[3]The notations $U \pm V$ for modulo 2^{32} addition/subtraction are used in [Stevens, Lenstra and de Weger 2007].

Exercise A18.2. Characterize and enumerate the number of pairs (U, U') with

$$U, U' \in \mathcal{Z}_{32} = \{\underline{x} = (x_0, x_1, \cdots, x_{31}) : x_i = 0 \text{ or } 1, 0 \leq i < 32\}$$

for which $U' \boxminus U = 2^{31}$.

Exercise A18.3. Characterize and enumerate the number of pairs (U, U') with

$$U, U' \in \mathcal{Z}_{32} = \{\underline{x} = (x_0, x_1, \cdots, x_{31}) : x_i = 0 \text{ or } 1, 0 \leq i < 32\}$$

for which $U' \boxminus U = -\ 2^{15}$.

Exercise A18.4. Characterize and enumerate the number of pairs (U, U') with

$$U, U' \in \mathcal{Z}_{32} = \{\underline{x} = (x_0, x_1, \cdots, x_{31}) : x_i = 0 \text{ or } 1, 0 \leq i < 32\}$$

for which $U' \boxminus U = -2^{15}$.

A18.4 OUTLINE OF THE STEVENS CONSTRUCTION

The Stevens chosen prefix attack extended the Wang-Yu attack [Wang and Yu 2005] in §18.10 and previous work by Stevens [2007] and Stevens, Lenstra and de Weger [2007] to permit arbitrary initial chaining values.

Given: any pair of prefixes (P, P')
Construct: suffixes S, S'
Such that: $\text{MD5}[P||S] = \text{MD5}[P'||S']$.

In their construction, the suffixes S, S' are the concatenation (||) of three parts; that is, $S = S_r||S_b||S_c$ and $S' = S'_r\|S'_b\|S'_c$ where

1. S_r (*resp.* S'_r) are random padding bit strings.
2. S_b (*resp.* S'_b) are birthday bit strings where the length $|P\|S_r\|S_p| = |P'\|S'_r\|S'_p| = 512n$ (bits).
3. S_c (*resp.* S'_c) are near collisions blocks (each of the same length, a multiple of 512 bits.
4. $\text{MD5}[P\|S_r\|S_p\|S_c] = \text{MD5}[\text{IHV}_m] = \text{MD5}[\text{IHV}'_m] = \text{MD5}[P'\|S'_r\|S'_p\|S'_c]$ where $m > n$ is the number of 512-bit blocks in $P||S_r||S_p||S_c$ (*resp.* $P'\|S'_r\|S'_p\|S'_c$).

The analysis in [Wang and Yu 2005], [Stevens, Lenstra and de Weger 2007], and [Stevens et al. 2009] produce a collision by adjusting the differences in $\delta\text{IHV}_j \equiv \text{IHV}_j - \text{IHV}'_j$.

Notation. The MD5 transform on 512-bit blocks $M_1, M_2, \cdots, M_m$ can be followed in terms of the initial hashing values (IHVs).

$M = (M_1, M_2, \cdots, M_m)$
IHV0 = `67452302 efcdab89 98badcfe 10325476` IHV1 = $\mathrm{MD5}[\mathrm{IHV}_0, M_1]$ $\mathrm{IHV}_2 = \mathrm{MD5}[\mathrm{IHV}_1, M_2]$. $\ddots$ $\mathrm{IHV}_m = \mathrm{MD5}[\mathrm{IHV}_1, M_m] = \mathrm{MD5}[M]$

1. If M_i (*resp.* M_i') are corresponding 512-bit blocks, then δM_i is the 32-bit word $(M_i - M_i')$ (modulo 2^{32}).
2. If $M = P \| S_r \| S_p \| S_c = (M_1, M_2, \cdots, M_m)$ (*resp.* $M' = P' \| S_r' \| S_p' \| S_c' = (M_1', M_2', \cdots, M_m')$), then $\delta \mathrm{IHV}_m = 0$.

In the [Stevens et al. 2009] chosen prefix attack,

- The prefixes P and P' contain the following:
  - Serial number (specified by the issuing CA)
  - Algorithm identifier, for example, md5WithRSAEncryption
  - Issuer distinguished name, identifier of the issuing CA
  - Validity period, determined by the issuing CA
  - (Different) distinguished subject names
- The suffixes S and S' contain the (different) RSA moduli.

The basic idea in their attack is to use birthdaying to control $\delta\mathrm{IHV}_n$, and to reduce the NAF-weight of the differentials consistently; that is

$$\delta\mathrm{IHV}_{n+j} < \delta\mathrm{IHV}_{n+j-1}$$

by adjoining near-collision blocks so that at the end $\delta\mathrm{IHV}_{n+r}$.

If we birthday with a random walk f on on $X = Z_{32,2}$, when do we encounter the first pair of points $y_i = f(x_i)$ and $y_m = f(x_m)$, for which $\delta(y_i - y_m)$ satisfies some condition, say $\delta(y_i - y_m) = (0_{32}, \delta b, \delta c, \delta d)$ with $\delta b = \delta c = \delta d$?

Birthdaying may be extended, for example, by searching for pairs with low δb NAF weight.

If p is the probability that birthdaying satisfies condition C, the cost $\sqrt{\pi|X|/2}$ is replaced by $\sqrt{\pi|X|/p}$.

The more work done in birthdaying to control $\delta\mathrm{IHV}_n$, the smaller the number of near-collision blocks that need to be appended.

The essential complication with respect to certificate is the size of the RSA moduli, the limit of 2048 bits being imposed. This limits the number of appended near-collision blocks to three.

REFERENCES

E. Bilham and A. Shamir, *Differential Cryptanalysis of the Data Encryption Standard*, Springer-Verlag (Berlin, Germany), 1993.

W. E. Clark and J. J. Liang, "On Arithmetic Weight for a General Radix Representation of Integers", *IEEE Transactions of Information Theory*, **IT-19**, pp. 823–826, 1973.

A. G. Konheim, *Cryptography and Computer Security*, John Wiley & Sons (New York), 2007.

P. C. van Oorschot and M. J. Wiener, "Parallel Collision Search with Cryptanalytic Applications", *Journal of Cryptology*, **12**, #1, pp. 1–28, 1999.

B. Schneier, *Applied Cryptography* (Second Edition), John Wiley & Sons (New York), 1996.

J. Shallit, "A Primer on Balanced Binary Representations", 1992, http://www.cs.uwaterloo.ca/shallit/Papers/bbr.pdf.

M. Stevens, "On Collisions for MD5", MS Thesis, Eindhoven University of Technology, 2007.

M. Stevens, A. Lenstra, and B. de Weger, *Chosen-Prefix Collisions for MD5 and Colliding X.509 Certificates for Different Identities*, Eurocrypt '07, Springer-Verlag (Berlin, Germany), 2007, pp. 1–22.

M. Stevens, A. Sotirov, J. Appelbaum, A. Lenstra, D. Molnar, D. A. Osvik, and B. de Weger, Short Chosen-Prefix Collisions for MD5 and the Creation of a Rogue CA Certificate, Proceedings of Crypto '09, Springer-Verlag (Berlin, Germany), 2009.

X. Wang and H. Yu, *How to Break MD5 and Other Hash Functions*, Eurocrypt '05, 2005.

CHAPTER 19

Hashing and the Secure Distribution of Digital Media

In the end, the only passion is money!, attributed to W. Somerset Maugham.

The theory of Communism may be summed up in the single sentence: Abolition of private property, from The Communist Manifesto (1848) by Karl Marx and Friedrich Engels.

If Karl had made a lot of money instead of writing about capitalism $\cdots$, *we would have all been much better off!*, attributed to Karl Marx's mother.

19.1 OVERVIEW

The use of hash functions to secure the distribution of digital media has direct commercial and financial implications. In the information age, digital media, including music, video, software, books, and data, can offer a high value to its creators, owners, and users. In its native form, the information can be copied, distributed, or altered without the owners or users of the information being aware. However, techniques are now available to limit these activities when desired, and many of these techniques rely heavily on hash functions.

There are at least two different commercial objectives for secure distribution techniques; first, creators and owners of digital media often want to limit the copying, distribution, and use of their assets, and they want to require users to pay for their use. Second and perhaps even more importantly, users of digital media, particularly software, are often concerned about the reliability and authenticity of the data. Among other reasons, unauthorized modifications to software might include viruses, and images might have been deceptively altered. Often, a secure distribution technique addresses both commercial objectives. Although the term digital rights management (DRM) generally refers to the former objective, it can also be used to refer to technologies that perform both functions.

Secure distribution algorithms and techniques are a supplement to legal restraints on the inappropriate use of digital media. Most digital media can theoretically be protected by its rightful owner through the use of copyright laws. Increasingly, courts in the United States are taking a tougher stance on piracy of digital music, videos,

Hashing in Computer Science: Fifty Years of Slicing and Dicing, by Alan G. Konheim

and software.[1] However, outside the United States, owners of digital media often have little legal recourse against inappropriate activity, and even in the United States the ease of copying digital media and the low probability of being prosecuted increases the importance of technical solutions as a supplement to legal ones.

The Business Software Alliance (BSA)[2] defines *software piracy* as the unauthorized copying or distribution of copyrighted software, by copying, downloading, sharing, selling, or installing multiple copies onto personal or work computers. BSA asserts that the purchase of software is just the purchase of a license to use the software and not *ownership* of the actual code. The license specifies the number of times and on which machines the software can be installed. If you make more copies of the software than the license permits, you are pirating.

The impact of the unauthorized distribution of digital media is large; BSA estimates losses from software piracy of over $50 billion per year.[3] Music piracy, according to the Institute of Policy Innovation (IPI) in Lewisville, Texas, causes over $12 billion per year in economic losses and the loss of 71 000 jobs in the United States. When including copyright infringement of motion pictures and electronic games, the IPI believes the loss of American jobs totals more than 370 000 [Siwek 2007a, 2007b].

Given the financial impact of secure software distribution techniques, it's not surprising that many of these techniques are, or at one time were, protected by patents. Consequently, patent filings often serve as appropriate references when studying the operation of the techniques.

This chapter reviews the role of hashing in protecting the intellectual property rights of the developers of digital media. Two excellent references are the survey papers [Abdel-Hamid, Tahar and Aboulhamid 2004] and [Petitcolas, Anderson and Kuhn 1999].

19.2 INTELLECTUAL PROPERTY (COPYRIGHTS AND PATENTS)

A copyright gives the author of an original work and his/her heirs certain basic rights for a certain time; these rights include the following:

- Its reproduction in various forms, such as printed publication or sound recording
- Its public performance, as in a play or musical work
- Recordings of it, for example, in the form of compact discs, cassettes, or videotapes
- Its broadcasting, by radio, cable, or satellite
- Its translation into other languages, or its adaptation, such as a novel into a screenplay

[1] On January 2, 2008, Fox News reported that Jeffrey Howell of Scottsdale was accused of placing 54 music files in a specific shared directory on his personal computer that all users of KaZaA [http://www.kazaa.com] and other peer-to-peer software could access. This was the grounds for a lawsuit by the Recording Industry Association of America (RIAA).

[2] BSA is the voice of the world's software industry and its hardware partners on a wide range of business and policy affairs. BSAs mission is to promote conditions in which the information technology (IT) industry can thrive and contribute to the prosperity, security, and quality of life of all people.

[3] See http://global.bsa.org/globalpiracy2008/index.html.

The concept of copyright, originating with the Statute of Anne (1710) in Britain, established the author of a work as the owner of the exclusive right to copy that work for the term of 14 years. Subsequently the Copyright Clause of the United States Constitution (1787) authorized copyright legislation "To promote the Progress of Science · · ·, by securing for limited Times to Authors · · · the exclusive Right to their · · · Writings." In 1886, the Berne Convention established recognition of copyrights among sovereign nations.

The international economic rights to intellectual property are derived from World Intellectual Property Organization (WIPO) treaties; they have a time limit of 50 years after the creator's death, although national law may establish longer time limits. This limit enables both creators and their heirs to benefit financially for a reasonable period of time. Copyright protection also includes moral rights, which involve the right to claim authorship of a work and the right to oppose changes to it that could harm the creator's reputation.

At the conclusion of the copyright protection period, the work is said to enter the public domain. Copyright applies to any expressible form of an idea or information that is substantive and discrete and fixed in a medium.

The creator—or the owner of the copyright in a work—can enforce rights administratively and in the courts, by inspection of premises for evidence of production or possession of illegally made—"pirated"—goods related to protected works. The owner may obtain court orders to stop such activities, as well as seek damages for loss of financial rewards and recognition.

Copyright protection extends only to expressions and not to ideas, procedures, methods of operation, or mathematical concepts as such. This principle has been confirmed by the Agreement on Trade-Related Aspects of Intellectual Property Rights (TRIPS Agreement) of the World Trade Organization (WTO) as well as the 1996 WIPO Copyright Treaty, which mentions the following two subject matters to be protected by copyright:

i) Computer programs, whatever may be the mode or form of their expression
ii) Compilations of data or other material (databases), in any form, which by reason of the selection or arrangement of their contents constitute intellectual creations

A patent is a grant of an exclusive property right; the full scope of the right to protect the rights of an inventor of a patentable idea. A copyright protects the expression of an idea, whereas a patent protects the idea itself. Different types of patents exist.[4]

The right to patent in the United States is derived from Article I, §8, Clause 8 of the U.S. Constitution, which authorizes Congress

[4] • Utility patents: a new and useful process/machine; perhaps, the food processor/cell-phone combo
• Design patents: a new, original and ornamental design for an article of manufacture
• Plant patents: a new variety of plant; perhaps, the mango-radish

Not all inventions are patentable; ideas, independent of the means to carry them out, are not patentable. A valid patent may not be obtained for an abstract principle, idea, law of nature or scientific truth. You cannot patent gravity, but you could patent a process that uses gravity in a novel way.

> To promote the Progress of Science and useful Arts, by securing for limited Times to Authors and Inventors the exclusive Right to their respective Writings and Discoveries.

The rights of the inventor to exclude others from making, using, offering for sale, or selling the invention throughout the United States or importing their invention into the United States is detailed in 35 U.S.C. §271.

The term of a patent now starts on the date the patent issues and continues to the date for 20 years after the application date per 35 U.S.C. §154. The change from the 17-year term was made in the 1990s to bring U.S. law in line with the patent laws of other countries.

Until 1952, a patentable idea required only novelty and utility. Congress added nonobviousness in that year as another requirement; 35 U. S. C. §103 provides that:

> … a patent may not be obtained, although the invention is not identically disclosed or described as set forth in 35 U. S. C. §102, if the differences between the subject matter sought to be patented and the prior art are such that the subject matter as a whole would have been obvious at the time the invention was to a person having ordinary skill in the art to which the subject matter pertains.

The application of 35 U. S. C. §103 involves the consideration of the following four factors:

1. The scope and content of the prior art[5]
2. Differences between the prior art and the claims at issue
3. The level of ordinary skill in the pertinent art
4. The obviousness or nonobviousness of the subject

The ease and widespread copying of digital media lead, in 1998, to a recognition of the importance of technology to limit the unauthorized copying and use of computer software, music, and video. The 1998 Digital Millennium Copyright Act (DMCA) makes it illegal to circumvent a technological measure that protects copyrighted material [17 U.S.C. §1201]. The DMCA increases the value and impact of DRM techniques, even ones that are technically vulnerable, at least in the United States.

We conclude this section by briefly describing several U.S. patents relating to the use of copyrighted software.

> *"Put Another Nickel In, In The Nickelodeon"* lyric in *Music, Music, Music* by Stephan Weiss & Bernie Baum recorded by Teresa Brewer & The Dixieland All Stars (1950).

Example 19.1. The abstract of Martin Hellman's[6] patent "Software Distribution System" (#4658093, issued on April 14, 1987) reads

> Software (programs, video games, music, movies, etc.) can be authorized for use a given number of times by a base unit after which the base unit (computer, video game base unit, record player, video recorder or video disk player) cannot use that software until

[5] Prior art refers to the disclosure of the contents of the patent's claims prior to the application date of the patent.

[6] Hellman's '093 patent was filled in 1983 well before the appearance of the following hash-functions:
- MD5 (see Chapter 18, §18.7 and [Konheim 2007, pp. 471–473]),
- SHA-1 (see Chapter 18, §18.8 and [Konheim 2007, pp. 473–474]).

the manufacturer sends an authorization for additional uses to the user's base unit. Authorizations may be sent via telephone line, mail, or whatever form of communication is most suited to the application. Authorizations cannot be reused, for example by recording the telephone authorization signal and replaying it to the base unit. Similarly, authorizations can be made base unit specific, so that an authorization for one base unit cannot be transferred to another base unit. This invention also solves the "software piracy problem" and allows telephone sales of software as additional benefits.

Summary of U.S. Patent 468093.

1) A user purchases a software package from the manufacturer; it permits him to set up the mechanism for the software's usage.
2) The user's computer and the software vendor's authorization unit thereupon exchange information to authorize multiple uses of the software, but only on the user's specific computer. The user transmits to the software vendor's authorization unit, over the possibly "insecure" channel, the following:
   2a) The name of the software product; i.e., a "modern" line-editing C++ program
   2b) The serial number of the user's computer, referred to as the base unit
   2c) The number *N* of times the user ways to run the software product
   2d) A random number *R* generated by the base unit
   2e) Billing information, i.e., a credit card

   Oops!, where is Secure Socket Layer (SSL)? Who is being protected?

3) An encrypted authorization code *A* is generated at the software vendor's authorization unit, which is a function of the following:
   3a) A code identifying the software product
   3b) The serial number of the user's computer
   3c) *N* and *R*

   A will enable the user to access the software. Figure 19.1 depicts the processing of the authorization code *A*.

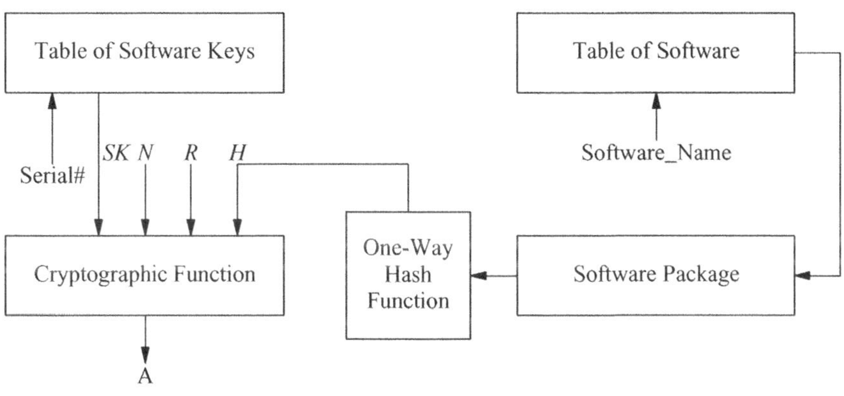

Figure 19.1. Hellman '093 computation of the authorization code.

where

- H is a 64-bit hash of the licensed software package
- A base unit's secret key SK is derived by the authorization unit from its table of software keys is a function of the base unit's serial number
- The inputs SK, N, R, and H to a cryptographic function determine A

4) The term cryptographic function, appearing in Figure 19.1, refers to a function that can be implemented either as a symmetric or public key cryptographic function of the arguments $SK, N, R,$ and H. It is specified in '093 as a modification of DES (see FIPS 46-3) in which SK is "modified" to serve as the key and the input is (N, R, H).
5) The one -way hash function, appearing in Figure 19.1, is specified[7] in Hellman's patent as a modification of the Data Encryption Standard (DES) in which DES is iterated with the original 56-bit key replaced by "larger" key (K_1, K_2, K_3) of at least $168 = 3 \times 56$ independent bits.
6) The authorization code permits use of the specified software product for the specified time at the customer's base unit. To function, the software has some mechanism that must be "unlocked" to function. Hellman '093 refers to a software player, which runs the software and that "will vary from application to application. For example, if the software is recorded music then software player 42 would be a record player; if the software is a computer program, then software player 42 would be a microprocessor or central processing unit (CPU)." Thus, base unit in Hellman '093 would apply to and old-fashioned cell phone that just makes calls.

Several other patents in this area have been issued.

Example 19.2. John H. Ryder, Sr. and Susanna R. Smith are the inventors of U.S. Patent #4953209 titled "Self-Verifying Receipt and Acceptance System for Electronically Delivered Data Objects" issued on August 28, 1990. The abstract of the '209 patent explains the following:

> A system for electronically transmitting data objects such as computer programs with a means for verifying that the computer program was actually received and the terms and conditions of its use accepted by the receiver is presented. In this system, the computer program itself controls the verification for its receipt and acceptance. The sender first modifies the program to be delivered, rendering it non-executable in the form in which it will be received by the user initially. The sender inserts into the program an enabling routine and a verification indicia. The enabling routine is capable of rendering the non-executable program into an executable state if certain prerequisite conditions, contained in the verification and enabling routine, are met. The recipient or receiver inserts or loads the modified, non-executable program into the workstation or computer having a CRT screen display, a printer or the like that allows human

[7] The specification in a patent is a detailed disclosure of the invention. It must provide the best possible mode for implementing the author's ideas, although it need not be the only way to practice the idea.

> observation of certain information that will be presented by the enabling program. The enabling program then displays messages or prompts to the user for entering the user's responses such as acceptance of the terms and conditions of the use of the program. In response to desired indications of acceptance by the user, the enabling program decides whether the prerequisite conditions for enabling the program into an executable form have been met and if they have been met, remodifies the program into a usable, executable form. If the prerequisite conditions are not met or agreed to, the verification and enabling program terminates without rendering the actual program itself into an executable form.

A license agreement must be signed when a customer receives the copyrighted software product in working form. Gary N. Griswold [Griswold 2002] claims that although the usage makes certain the customer understands the licensing rights, it does not prevent misuse.

Example 19.3. Victor Shear is the inventor of U.S. Patent #5825883, titled "Method and Apparatus that Accounts for Usage of Digital Applications," issued on October 20, 1998. The '883 patent abstract reads

> A method and apparatus that accounts for utilization of digital applications on an as used basis is accomplished by compensating a publisher for each use of its digital application instead of a lump sum purchase price. This is done by embedding a tariff file within the digital application where a digital application may be a software application, a video file, a text file, and/or an image file. The embedded tariff file, which includes a digital application identification code and a publisher identification code, is used by a meter module to generate accounting information. The accounting m essage is routed to a collection agency which, in turn, generates debiting information for the user and crediting information for the publishers of the digital application.

Metering of the software usage requires a hardware module that is part of the computer used to access the distributed database. This module records of the intellectual property viewed. Once the module becomes full, it must be removed and delivered to someone who will charge for the usage and set the module back to zero.

19.3 STEGANOGRAPHY

The term steganography refers to methods for concealing messages so that their existence and content is hidden from all but the sender and intended recipients. From the Greek meaning concealed writing, steganographic methods were described in Johannes Trithemius' 1499 treatise on magic *Steganographia*. The first recorded use of steganography occurred earlier in 440 BC when Herodotus of Halicarnassus warned of an attack on Greece in a message written on the wooden backing of a wax tablet before masking it by applying beeswax to its surface. A second example, rivaling even *Fantastic Sams*, is that of Histiaeus, who shaved the head of a slave, wrote a message on it, and sent the slave off after his hair grew back.

Modern-day steganography has the following three applications:

i) The traditional role of secrecy, for spies with or without raincoats
ii) Content authentication, to assure content integrity
iii) To protect intellectual property rights

Modern up-to-date steganography is described in [Petitcolas, Anderson and Kuhn 1999].

19.4 BOIL, BOIL, TOIL AND · · · BUT FIRST, CAREFULLY MIX

Invisible inks were part of high-tech steganography at the start of the 20th century; David Kahn's wonderful book [Kahn 2004] is about the life and achievements of Herbert O. Yardley (1889–1958), the first great American cryptanalyst. Kahn's book tells of the interest by MI-8[8] in constructing and detecting invisible inks used by German spies. Emmett Carver, a student of the American chemistry Nobel laureate Theodore Richards, developed a new ink[9] in 1917 that remained invisible except when exposed to mercury vapors. When the Secretary of State Henry L. Stimson withdrew funding in 1929 for the State Department code-breaking efforts, he claimed, "Gentlemen do not read each others mail." Signal Corps 2nd Lieutenant Herbert O. Yardley and head of the newly created eighth section of military intelligence was permanently furloughed. Without unemployment insurance or another viable employer, Yardley published *The American Black Chamber*, which was a well received and controversial book about his cryptographic triumphs. Publication by people who have held a cryptographic clearance is much more severely restricted today (the Espionage Act, 18 USC 798).[10] Yardley's subsequent literary attempts—spy and mystery novels and movies—were not as successful. Out of work, Yardely returned to his roots in the midwest and established Major Yardley's Secret Ink Incorporated of Worthington, Indiana in 1933.

19.5 SOFTWARE DISTRIBUTION SYSTEMS

A software distribution system provides a mechanism for the sale or licensing of the software by the manufacturers or by a third party. GNU[11] (see Chapter 9, §9.3) Autotools are widely used for distributions that consist of source files written in C++ and the C programming language, but are not limited to these.

Alas, not all software is open, and software piracy is a great threat to software industry, potentially resulting in serious damages to the interests of software devel-

[8]The division of military intelligence responsible for cryptology.
[9]The formula used blood, Rhoadime B extra, alcohol, zinc dust, and hydrogen peroxide.
[10]Whoever knowingly and willfully communicates, furnishes, transmits, or otherwise makes available to an unauthorized person, or publishes, or uses in any manner prejudicial to the safety or interest of the United States or for the benefit of any foreign government to the detriment of the United States any classified information "concerning the nature, preparation, or use of any code, cipher, or cryptographic system of the United States or any foreign government; or · · ·"
[11]GNU is an open system computer operating system, Unix like but free of Unix code. GNU was developed by the GNU Project; the programs released under the auspices of the GNU Project are called GNU packages. The GNU programs are free; all recipients enjoy the right to run, copy, modify, and distribute the programs.

opers or providers and, ultimately, to the users. The problem involves more than software; additionally, digital audio, video, and pictures are available over the Internet and exposed to the same vulnerabilities. The main concern is the copyright of intellectual property; as the availability of digital media increases, the ease with which copies can be made may lead to unauthorized copying. This is of concern to the music, film, book, and software publishing industries.

The Advanced Access Content System (AACS) provide a standard for content distribution and digital rights management. It is intended to restrict access to and copying of the next generation of optical discs and DVDs. The specification released in April 2005 and the standard has been adopted as the access restriction scheme for HD DVD and Blu-ray Disc (BD). It is developed by AACS Licensing Administrator, LLC (AACS LA), a consortium that includes the following:

1. The major producers of DVD equipment, including Matsushita (Panasonic), Toshiba, and Sony
2. The major marketers of DVDs, including Disney and Warner Bros

The Certicom Corporation markets software products based on elliptic curve cryptography (ECC). Elliptic curve cryptosystems seem to offer a considerable efficiency with respect to key size. According to their web site www.certicom.com, the Certicom Intellectual Property portfolio includes over 350 patents and patents pending worldwide covering many key aspects of ECC, including software optimizations, efficient hardware implementations, methods to enhance the security, and various cryptographic,[12] protocols. In their review "The Case for Elliptic Curve Cryptography" of elliptic curve cryptography the National Security Agency pointed out that intellectual property rights are a major roadblocks to the further adoption of elliptic curve cryptography, citing Certicom, which holds over 130 patents in this area. Certicom filed suit in Texas in May 2007 against Sony Corporation claiming that Sony's use of ECC in two of its implemented technologies—AACS and Digital Transmission Content Protection—conceptually violated Certicom's patents for that cryptographic method. May the lawyers with the higher hourly rates prevail!

19.6 WATERMARKS

A watermark is a recognizable pattern in paper that appears as various shades of lightness/darkness when viewed by transmitted light (or when viewed by reflected light, atop a dark background), caused by thickness variations in the paper.

Figure 19.2 is schematic portraying the embedding and extraction of digital watermarking, a process that embeds information into a document; a *visible* watermark is a recognizable pattern, perhaps printed in different shades or colors, used in banknotes, stamps, passports, and motor vehicle titles. It is often used as an anticounterfeiting measure.

For information on the inclusion of watermarks in desktop publishing using digital photocomposition, see http://desktoppub.about.com/od/watermarks/Watermarks.htm.

[12] www.nsa.gov/business/programs/elliptic_curve.shtml.

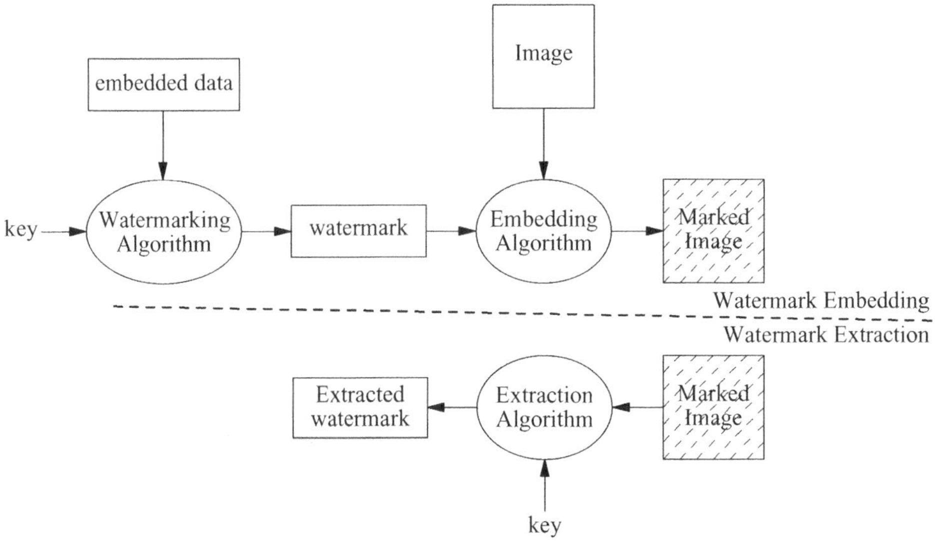

Figure 19.2. Watermarking processes.

Dittman et al. [Dittman, Wohlmacher and Nahrstedt 2001] observe that there are at least three applications of watermarking

1. Confidentiality: prevent release of information
2. Integrity: detection of modifications
3. Availability: denying access to documents

Certainly, a hash function h is a natural candidate to be used in watermarking a document $\mathcal{D}$

$$\mathcal{D} \rightarrow [\mathcal{D}, W] \quad W = h(\mathcal{F})$$

because hashing functions are intended to be sensitive to alterations $[D, W] \rightarrow [D, W]'$.

In contrast, because the watermarking is supposed to be "invisible"—to not materially alter the sound of a sound or picture—perceptual hashing[13] has been suggested.

Digital watermarks are also used to identify positively the source of a digital document and to prove, in their absence or detected alteration, that unauthorized copying has occurred.

[13]*perceptual hash* n. a fingerprint of an audio, video, or image file that is mathematically based on the audio or visual content contained within. Unlike cryptographic hash functions that rely on the avalanche effect of small changes in input leading to drastic changes in the output, perceptual hashes are "close" to one another if the inputs are similar in appearance or sound.

19.7 AN IMAGE-PROCESSING TECHNIQUE FOR WATERMARKING

Kitanovski et al. [Kitanovski, Taskovski and Bogdanova 2007] propose a hashing-based watermark using ideas in Roberts [Roberts 2002]. The pixel luminance[14] at a pixel (or picture element) of a 2-dimensional document $\mathcal{D}$ is an integer in [0, 255]. Roberts suggests the following:

a) $\mathcal{D}$ is divided into contiguous regions R, each of which contains $(2m + 1)$-by-$(2m + 1)$ pixels
b) The luminance $L_{(i,j)}$ is measured at each pixel $(-m \leq i, j \leq m)$ of R, and
c) An average $\bar{L}_R$ is derived for the region R by
   1) Replacing $L_{(i,j)}$ by $\Delta L(i,j) \equiv L_{(0,0)} - L_{(i,j)}$,
   2) Next, computing $\bar{L}_R \equiv \sum_R \Delta L_{(i,j)}$, and finally
d) Defining the bit b_R for R according to $b_R = \begin{cases} 1 & \text{if } \bar{L}_R > 0 \\ 0 & \text{otherwise} \end{cases}$.

The step from item c) number 2 is portrayed in Figure 19.3 with $m = 1$.

Next, suppose $\mathcal{D}$ is partitioned into M-by-M regions By this process, $\mathcal{D}$ produces an M-vector, which [Kitanovshi, Taskovski and Bogdanova 2007] refers to as an image digest Dig($\mathcal{D}$). The image hash is the M-vector resulting from the encipherment Hash($\mathcal{D}$) $\equiv E\{k, \text{Dig}(\mathcal{D})\}$ with a secret key k. Hash($\mathcal{D}$) is a message authentication code (MAC) or digital signature (SIG) of the image $\mathcal{D}$.

In Figure 19.4, note the following:

- The region R_1 in the upper-left top region has slightly less luminance in the center pixel element, which leads to the bit value $b_{R_1} = 1$, whereas
- The region R_2 just below the upper-left top region has slightly more luminance in the center pixel element, which leads to the bit value $b_{R_2} = 0$.

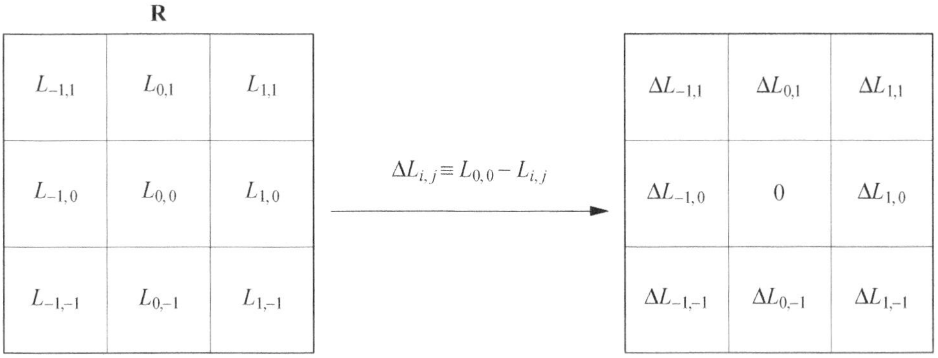

Figure 19.3. The Average Luminance of a 3-by-3 Pixel Region.

[14]The luminance of a pixel is $L = r * 76 + g * 150 + b * 29$, where $r, g, b \in (0, 1)$ are intensities of red, blue, and green. Noting that $255 = 76 + 150 + 29$, the value L is rounded to an integer in [0,255]; i.e., 8 bits.

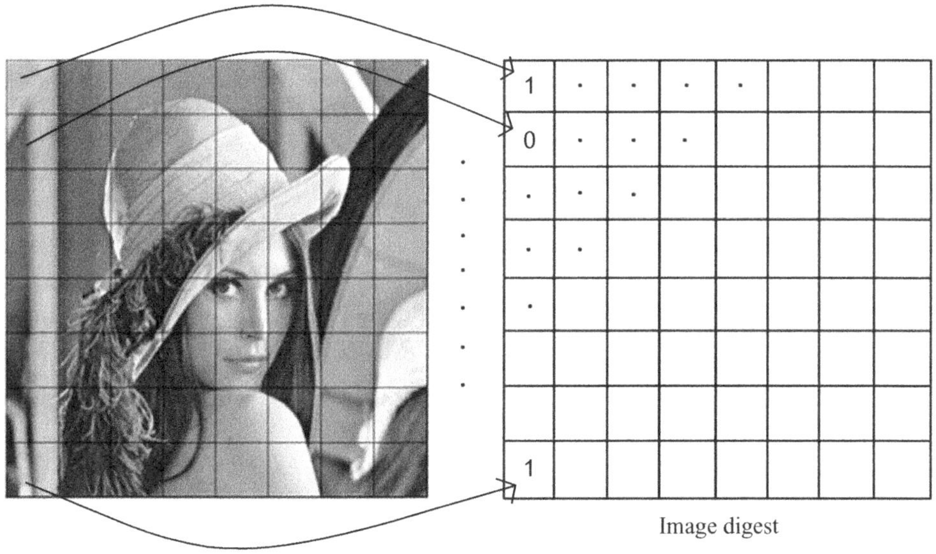

Figure 19.4. An example of the Kitanovski-Roberts' transformation. From [Kitanovski, Taskovski and Bogdanova 2007], reproduced with permission.

The Kitanovski et al. watermarking consists of embedding hash($\mathcal{D}$) into the two-dimensional image $\mathcal{D}$. To explain their embedding, we need to make a short digression.

The Discrete Cosine Transform. The discrete cosine transform (DCT) of the double sequence $\underline{s} = (s_{i,j})$ with $0 \le i \le n_1$, $0 \le j \le n_2$ is defined by

$$\text{DCT}: \underline{s} \rightarrow \hat{\underline{s}} \tag{19.1a}$$

$$\hat{\underline{s}} = (\hat{s}_{\alpha,\beta}) = \frac{2}{\sqrt{n_1 n_2}} f(\alpha) f(\beta) \sum_{i=0}^{n_1-1} \sum_{j=0}^{n_2-1} s_{i,j} \cos\frac{(2i+1)\alpha\pi}{2n_1} \cos\frac{(2j+1)\beta\pi}{2n_2} \quad \begin{cases} 0 \le \alpha < n_1 \\ 0 \le \beta < n_2 \end{cases} \tag{19.1b}$$

$$f(i) = \begin{cases} 2^{-\frac{1}{2}} & \text{if } i = 0 \\ 1 & \text{otherwise} \end{cases} \tag{19.1c}$$

with inverse

$$\text{DCT}^{-1}: \hat{\underline{s}} \rightarrow \underline{s} \tag{19.1d}$$

$$s_{i,j} = \frac{2}{\sqrt{n_1 n_2}} \sum_{\alpha=0}^{n_1-1} \sum_{\beta=0}^{n_2-1} f(\alpha) f(\beta) \hat{s}_{\alpha,\beta} \cos\frac{(2i+1)\alpha\pi}{2n_1} \cos\frac{(2j+1)\beta\pi}{2n_2} \tag{19.1e}$$

A concise and readable review of DCT concepts appears in Watson's paper [Watson 1994]. This paper discusses how the Wolfram's *Mathematica* can be used to effect the calculations. More generally, discrete transforms like $\{\hat{s}_{\alpha,\beta}\}$ can be computed by techniques involving the discrete Fourier transform; for example [Jain 1989].

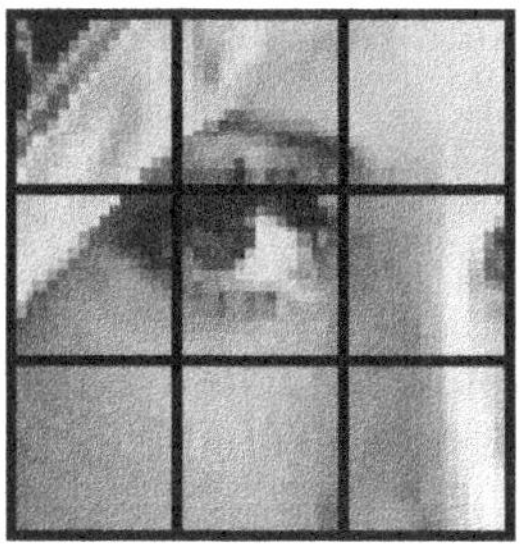

Figure 19.5. A 3-by-3 block of the image in Figure 19.4.

An important observation in [Kitanovski, Taskovski and Bogdanova 2007]'s use of the DCT; whereas the entries in $\underline{s}$ are 8-bit integers, the values of $\hat{\underline{s}}$ are real. If a picture is to be reconstructed from the values in $\hat{\underline{s}}$, some form of quantitization must be used.

The internal base 2 representation[15] of the real number x is $\underline{x}$: sgn(x), n, $x_0x_1 \cdots x_{n-2}x_{n-1}$; it consists of the following:

- The sign of x; $\text{sgn}(x) = \begin{cases} 0 & \text{if } x \geq 0 \\ 1 & \text{if } x \leq 0 \end{cases}$;
- The number n, of bits in the magnitude of x;
- A string of n bits $x_0x_1 \cdots x_{n-2}x_{n-1}$ determining the magnitude of x.

If $x \geq 0$, then $x = x_0 + x_1 2 + \cdots + x_{n-2}2^{n-2} + x_{n-1}2^{n-1}$ with $x_i \in \{0, 1\}$ $(0 \leq i < n)$.

The storage of the DCT of image requires the quantitization of the coefficients in the DCT. The [Kitanovshi, Taskovski and Bogdanova 2007] watermarking involves several ideas, as

a. The image is divided into and array of M-by-M blocks, each of size B-by-B pixels; in the [Kitanovshi, Taskovski and Bogdanova 2007] examples, $B = 48$; Figure 19.5 is an example of a B-by-B block ($B = 4$).
b. The DCT is computed for each block, and
c. The quantized coefficients for each block are modified by a process depending on one bit of the next-to-be-defined hash.

The M-by-M vector of luminance value of a block, say $\mathcal{M}$ of the two-dimensional document $\mathcal{D}$ is $\underline{s} = (s_{i,j})$ with $0 \leq i, j < M$. Kitanovski hide the image hash hash($\mathcal{M}$) by modifying the coefficients of the "low frequencies" in the DCT. [Kitanovski, Taskovski and Bogdanova 2007] used a variant of the quantized index modulation (QIM) as it appears in Chen and Worness [Chen and Worness 2001]. In each block, the N low-frequency cosine transformation coefficients are modified by exclusive or-(XOR)ing to a vector $K_i(i = 0, 1)$ depending on the corresponding bit in hash($\mathcal{M}$).

Certainly, the inclusion of the watermark introduces distortion in the image, but the human eye may not be able to detect the changes with a careful embedding.

[15]The GNU Multiple Precision Arithmetic Library, which is described as www.swox.comb/gmp, refers to the digits as limbs. A number x is referred to as `mpz_to`, it has a sign, a number of limbs `_mp_size`, and, if this last number is positive, a pointer to dynamically allocated array for `_mp_d` data.

If we ignore the effect of quantitization of the coefficients of the DCT and write

$$\mathcal{M} \rightarrow \mathcal{M}' = [\mathcal{M}, W_{\mathcal{M}}] \quad (19.2a)$$

$$W_{\mathcal{M}} = F_{W_{\mathcal{M}}}(k, \text{Dig}(\mathcal{D})) \quad (19.2b)$$

where $W_{\mathcal{M}}$ is the watermark of block $\mathcal{M}$ determined by the $\mathcal{M}$-bit of hash($\mathcal{D}$), then the low-order frequency DCT coefficients $\mathcal{M}$ and $\mathcal{M}'$ should differ only in those determined by the $\mathcal{M}$-bit of hash($\mathcal{D}$).

This suggests a test to verify whether $\mathcal{D}'$ is the true watermarked version of $\mathcal{D}$; namely, to test if $\mathcal{D}^*$ is a valid watermarked copy of $\mathcal{D}$, then

1. Repeat the computation in equations (19.2*a* and 19.2*b*) replacing $\mathcal{M}$ and $\mathcal{D}$ by $\mathcal{M}^*$ and $\mathcal{D}^*$
2. Compare (the Hamming distances between the DCT's of) $W_{\mathcal{M}}$ and $W_{\mathcal{M}^*}$ until the distance "converges".

I have simplified the argument that is offered in more detail in [Kitanovshi, Taskovski and Bogdanova 2007]. As in real life, one cannot prove that this form of watermarking will always detect forgery or other evil deals, but examples are offered in [Kitanovshi, Taskovski and Bogdanova 2007] to support their approach.

19.8 USING GEOMETRIC HASHING TO WATERMARK IMAGES

Geometric hashing [Wolfson and Rigoutsos 1997] was developed as a tool to recognize two-dimensional objects. The basic problem is to recognize whether some object (or scene) O^* is present in a database of objects $\mathcal{O} = \{O_i :\in \mathcal{D}\}$. The object O^* might be the transformation (scaled, rotation, or partial occlusion) of an object in $\mathcal{O}$. The question to be answered is:

> Is there a transformed (rotated, translated and scaled) subset of some model point-set $O_i \in \mathcal{O}$ which matches a subset of the scene point-set O^*?

A methodology requires preprocessing of the objects in $\mathcal{O}$, which are referred to as models. For each $O_i \in \mathcal{O}$, the process steps are as follows:

1. Introduce a mesh in some Cartesian coordinate system, to permit the quantization of the coordinates of O_i
2. Identify a(finite!) set of feature points
3. For each pair of feature points (p_1, p_2), sometimes referred to as base points:
   a. Define $p_x = (p_2 - p_1)$, $p_y = \text{ROT}_{90}(p_x)$, $p_0 = \frac{1}{2}(p_1 + p_2)$;
   b. Write every other feature point p of $\mathcal{O}_i$ in terms of p_x, p_y as $p = p_0 + up_x + vp_y$.

Example 19.4. Figure 19.6 shows the letter "L" (the object O^*) with six feature points.

As base points, we choose $p_1 = (-0.5, -0.8)$ and $p_2 = (-0.3, -0.6)$ so that $p_x = (0.2, 0.2)$ and $p_y = (-0.2, 0.2)$.

If p is a feature point of O^*, then we write

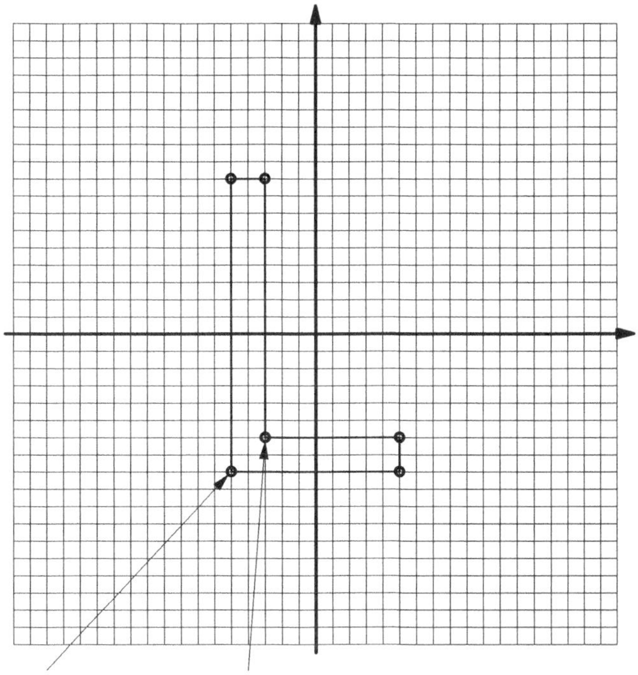

Figure 19.6. The letter "L," a set of feature points · and p_1, p_2 and p_0 = (–0.4, –0.7).

TABLE 19.1. (*u*, *v*) Coding of the Feature Points of O* for the Base Points p_1, p_2

	p	u	v		p	u	v
C_1	(–0.5, 0.9)	$\frac{15}{4}$	$\frac{17}{4}$	C_4	(0.5, –0.8)	$\frac{5}{2}$	–2
C_2	(–0.5, –0.8)	$-\frac{1}{2}$	0	C_5	(–0.3, –0.6)	$\frac{1}{2}$	0
C_3	(0.5, –0.8)	2	$-\frac{5}{2}$	C_6	(–0.3, 0.9)	$\frac{17}{4}$	$\frac{15}{4}$

$$p = p_0 + up_x + vp_y \tag{19.3a}$$

then

$$u = \|p_x\|^{-2}(p_x, p - p_0) \tag{19.3b}$$

$$v = \|p_y\|^{-2}(p_y, p - p_0) \tag{19.3c}$$

where ||· · ·|| is the norm of the vector · · · and (a, b) is the inner product of the vectors a and b. Table 19.1 lists the values of (u, v) for each of the feature points.

Exercise 19.1. Prove the geometric hashing coding $\mathcal{G}_{p_1,p_2} : p \to (u, v)$ satisfies the following:

a) If c is any fixed vector and $p_i \to p_i + c$ $(i = 1, 2)$, then (u, v) remains unchanged; i.e., $\mathcal{G}_{p_1+c,p_2+c} = \mathcal{G}_{p_1,p_2}$;
b) if c is any nonzero constant and $p_i \to cp_i$ $(i = 1, 2)$, then (u, v) remains unchanged; i.e., $\mathcal{G}_{cp_1,cp_2} = \mathcal{G}_{p_1,p_2}$;
c) More generally, if A is any 2-by-2 similarity transformation[16] and $p_i \to p_iA$ $(i = 1, 2)$, then (u, v) remains unchanged; i.e., $\mathcal{G}_{p_1A,p_2A} = \mathcal{G}_{p_1,p_2}$.

The use of geometric hashing to recognize two-dimensional objects hash two phases, preprocessing and recognition.

Preprocessing Phase. For each object $O_i \in \mathcal{O}$, Wolfson and Rigoutos refer to them as models; a data structure $\mathcal{D}$ is constructed according to the steps

1. Identify n (say) feature points
2. For each pair of base points (p_1, p_2) and each feature point p of O calculate (u, v) according to equations (19.3) and quantities $(u, v) \to (\bar{u}, \bar{v})$
3. Use a hash function H to locate the "bin" of $(\bar{u}, \bar{v})$ and insert $(\mathrm{ID}_{\mathcal{O}}, p_1, p_2)$ (in a linked list), where $\mathrm{ID}_{\mathcal{O}}$ is some identifier of the model

Recognition Phase. For each object O^*, to be identified, do the following:

1. Detect s (say) feature points S.
2. Select a pair of base points (p_1^*, p_2^*); for each feature point p^* of O^* calculate to (u^*, v^*) according to equations (19.3) and then quantize $(u^*, v^*) \to (\bar{u}^*, \bar{v}^*)$.
3. Search $\mathcal{D}$ by hashing to the "bin" of $(\bar{u}^*, \bar{v}^*)$.
4. Vote[17]! Count the number of entries $(\mathrm{ID}_{\mathcal{O}_i}, p_{i,1}, p_{i,2})$ in the linked list pointed to by this hash-table bin.
5. Make a histogram of the number of entries and decide on the winner! For example,
   - Examine the pair(s) $(\mathrm{ID}_{\mathcal{O}_i}, p_{i,1}, p_{i,2})$ with the highest score
   - Examine the pairs $(\mathrm{ID}_{\mathcal{O}_i}, p_{i,1}, p_{i,2})$ with a score above some threshold
   - Find the pair $(\mathrm{ID}_{\mathcal{O}_i}, p_{i,1}, p_{i,2})$ that is best least squares fit.

Errors in recognition are caused by a variety of reasons including the following:

- Quantitization error
- Partial occlusion of feature points
- The effect of various compression techniques

[16]A 2-by-2 similarity transformation is a matrix $A = \begin{pmatrix} a_{1,1} & a_{1,2} \\ a_{2,1} & a_{2,2} \end{pmatrix}$ of determinant $\neq 0$.

[17]The nongeometric hashing instruction, "Vote early and vote often" is often attributed to the late Richard J. Daley, who was a former mayor of Chicago. However, some claim that the late famous Chicago gangster Al Capone offered the advice, "Vote early, vote often"; they also note that he opined, "You can get much further with a kind word and a gun than you can with a kind word alone." Of course, Capone frequently ignored this last advice.

A substantial part of [Wolfson and Rigoutsos 1997] develops a model to evaluate the performance of geometric hashing. A previous paper [Lamdan and Wolfson 1991] examines the error analysis of geometric hashing and contains experimental results.

We conclude our discussion of watermarking by noting that geometric hashing may be applied by combining [Hel-Or, Yitzhaki and Hel-Or 2001] the modification of the low-frequency DCT coefficients as in §19.7 ([Kitanovshi, Taskovski and Bogdanova 2007]) and a watermarking technique of [Hel-Or and Butman 2001] in which the image is divided into 8-by-8 blocks. Then the following is true:

a. A random number generator is used to select certain blocks.
b. Certain predetermined low-frequency components of the DCT in a favored block are nullified, i.e., "a value statistically close to zero."
c. The inverse DCT is applied to embed the watermark in the image.

Detection of the authentic watermarked proceeds as in §19.7,

[Hel-Or, Yitzhaki and Hel-Or 2001] replaced the "random" selection of blocks by another mechanism; they started with a fixed pattern of blocks and used a random transformation to choose the favored blocks whose DCT coefficients will be modified.

19.9 BIOMETRICS AND HASHING

Yagoz Sutcu et al. [Sutcu, Sencar and Memon 2005] provides a nice comparison of the security of various schema that I have modified and present in Table 19.2.

The danger of basing security only on cryptographic methods—knowledge of a key—has led to the study of biometrics, which are methods to provide access control

TABLE 19.2. The Cost for Security in the 21st Century

Method	Ingredients	Advantages and/or Drawbacks
What you know?	User_ID	Might be shared.
	Password	May be easy to guess?
	User_PIN	Might be forgotten.
What you possess	Card	Might be shared.
	Badge	Might be duplicated.
	Key	Might be stolen.
Knowledge + Possession	ATM card + ID_PIN	The old reliable stick-up. "Your card and your PIN or your life! Pause *Wait, I'm thinking!*
Something unique to the user	Fingerprint	Difficult to forge.
	Iris scan,	Not possible to share.
	Voice, smell	Cannot be lost or stolen ··· *Will I be able to withdraw funds, if I (do not) shower very often?*

based on measurements of one or more intrinsic physical or behavioral human traits. The measured traits include are of two types:

Physiological. fingerprint, face recognition, DNA, hand and palm geometry, iris recognition, voice pitch, and odor/scent.

Behavioral. typing rhythm, gait,[18] and voice.

Biometric data might be combined with cryptography in the following ways:

- Access to the files and processing capabilities on a laptop might be enabled by a fingerprint scan.[19]
- Smart-card technology with biometric measurements stored thereon to provide a user access to a resource, i.e., software, a building.
- Opening the locks at a facility might be enalded by locks activated by biometric measurements.

The essential issues include the following:

1. *Uniqueness*: How accurate is a human being identified by a characteristic?
2. *Acceptability*: Will people approve of some characteristic?
3. *Acquisition*: Is a characteristic easy to measure?
4. *Permanence*: Like a fine California wine, does the characteristic age well?

Additionally, there are implementation issues including the following:

1. The noisiness of measurement data
2. The concern about databases of biometric measurements, if the implementation requires such
3. The social stigma attached to fingerprinting

If biometrics are to replace the (User_ID, User_PIN), there must be some mechanism to compare the measurements when entry is attempted and the valid measurements when a user enrolls. Either these latter values are retrieved from some database, or they are stored on a token, such as an biometric ATM card.

In both cases, the questions include the protection of the reference data and how these data are retrieved.

Several researchers ([Sutcu, Sencar and Memon 2005] and [Ratha, Connell and Bolle 2001]) suggest using a one-way function to map a biometric measurement vector that is stored on a smart card. A User_PIN is required to access the data on the card.

[18]Who can forget the World War II movies in which the Gestapo officer i) kills the true beautiful secretagent radio-operator, ii) sends a fake message, iii) and is recognized by the Allies as fraudulent by noting the differences in his "hand." *Morse characteristics analysis* (MOCA) refers to the study and cataloging of recorded manual morse transmissions to identify individual Morse operators by their sending operator characteristics.

[19]Toshiba and Identix unveiled the Toshiba PC Card Fingerprint Reader, which will allow users to log on to their Toshiba laptops by placing a finger in the device. This fingerprint reader will provide a higher level of security as well as convenience, because users will not have to remember passwords nor key in their user names.

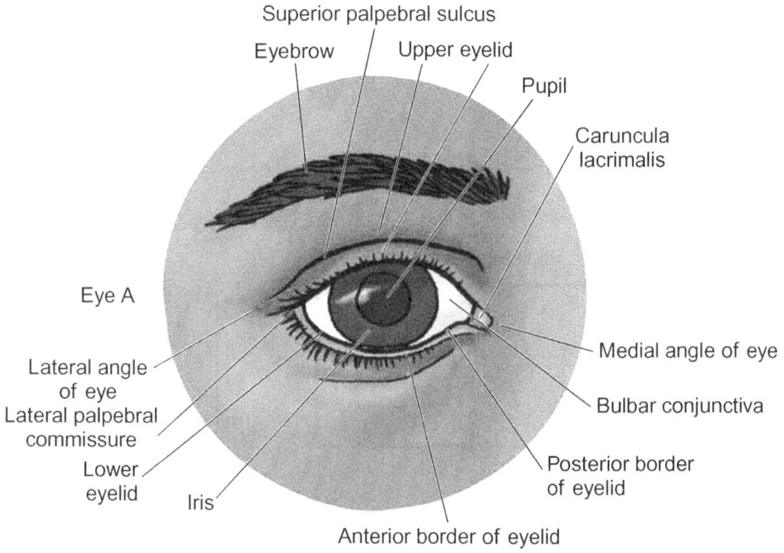

Figure 19.7. Certainly Kein Ayin Hora![21] Courtesy of *Anatomy Atlases* (www.anatomyatlases.org).

Hao, Anderson and Daugman [2005] present a well-written description of the technical problems and propose "the first practical and secure way to integrate the iris biometric into cryptographic applications" with an extensive bibliography.

They describe the use of biometric measurements of the eye. John Daugman [Daugman 2004] of the Computer Laboratory (University of Cambridge) notes that the eye begins to form in the third month of gestation and is largely completed by the eighth month of gestation. Figure 19.7 is a schematic of the eye; the distinctive features of the iris include arching ligaments, furrows, ridges, crypts, rings, corona, freckles, and zigzag collarette.[20] Its color is determined by mainly by the density of melanin pigment.

The papers ([Daugman 2004] and [Williams and Ezekiel 2005]) describe the iris measurement and feature-extraction processes. A charge coupled device (CCD) camera converts optical brightness into electrical amplitude signals using a plurality of CCDS and, analog shift registers. They reproduce the image of a subject using the electric signals. [Daugman 2004] uses only *phase information* because the amplitude of the signal depends on other external factors, i.e., imaging contrast, illumination, and camera gain. The phase is coded into 2 bits as shown in Table 19.3.

[20]An irregular jagged line dividing the anterior surface of the iris into two regions.

[21]No Evil Eye! (*Yiddish*) May the evil eye stay away! Spit, at least three times. You are probably now safe, but remain diligent; watch out for ladders and black cats and make a fig with your fingers behind your back.

The evil eye is truly multi-cultural; in India, it is the *drishti* or *nazar*, however, spitting is replaced by Aarti, which is a Hindu ritual. Not to be outdone, the website www.phoenixorion.com/phoenixorion/oriental2a.htm offers evil eye pendants from \$3.95 + shipping. According to Chinese mythology and Feng Shui, such charms are beneficial to prevent the ill effects arising from the evil thoughts of a jealous person.

TABLE 19.3. Daugman 2-Bit Phase Encoding

Phase	Code	Phase	Code
$[0, \frac{\pi}{4})$	(1, 1)	$[\frac{\pi}{4}, \frac{\pi}{2})$	(0, 1)
$[\frac{\pi}{2}, \frac{3\pi}{4})$	(0, 0)	$[\frac{3\pi}{4}, \pi)$	(1, 0)

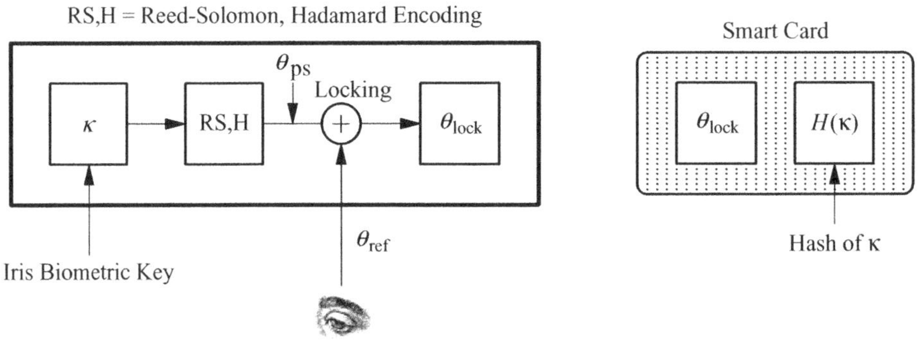

Figure 19.8. Creation of the iris biometric smart card.

For Iris_A, [Daugman 2004] computes ($\text{Code}_{\text{Iris\_A}}$, $\text{Mask}_{\text{Iris\_A}}$) where

1. $\text{Code}_{\text{Iris\_A}}$ is the 2048 phase code bits (256 bytes) for each iris.
2. $\text{Mask}_{\text{Iris\_A}}$ is a 2048-bit (0,1)-vector; 0 at non-iris regions and 1 at iris regions supplying information regarding whether any iris region is obscured by eyelids, contains any eyelash occlusions, has specular refections, contains boundary artifacts of hard contact lenses, or demonstrates a poor signal-to-noise ratio.

The basic tool for comparing $\text{Code}_{\text{Iris\_A}}$ and $\text{Code}_{\text{Iris\_B}}$ is the fractional Hamming distance, which is defined by

$$\rho(\text{Code}_{\text{Iris\_A}}, \text{Code}_{\text{Iris\_B}}) = \frac{\|(\text{Code}_{\text{Iris\_A}} \oplus \text{Code}_{\text{Iris\_B}}) \cap \text{Mask}_{\text{Iris\_A}} \cap \text{Mask}_{\text{Iris\_B}}\|}{\|\text{Mask}_{\text{Iris\_A}} \cap \text{Mask}_{\text{Iris\_B}}\|} \tag{19.4}$$

Because it is reasonable to assume that bits in $\text{Code}_{\text{Iris\_A}}$ and $\text{Code}_{\text{Iris\_B}}$ are equally to agree, the expected fraction of agreeing bits is likely to be 0.500. [Daugman 2004] reports remarkable experimental results; their Figure 19.4 shows the distribution of $\rho(\text{Code}_{\text{Iris\_A}}, \text{Code}_{\text{Iris\_B}})$ for 9 060 003 pairs of irises. The measured mean and standard deviation was 0.499 and 0, 0317, respectively, which is consistent with a binomial distribution with 249 = 256 − 7 degrees of freedom.

We conclude by describing how Hao, Anderson and Daugman [2005] combine cryptography and iris biometrics to produce "the first practical and secure way to integrate the iris biometric into cryptographic applications."

Figure 19.8 depicts the process creating the iris biometric/smarty card; the following steps are required:

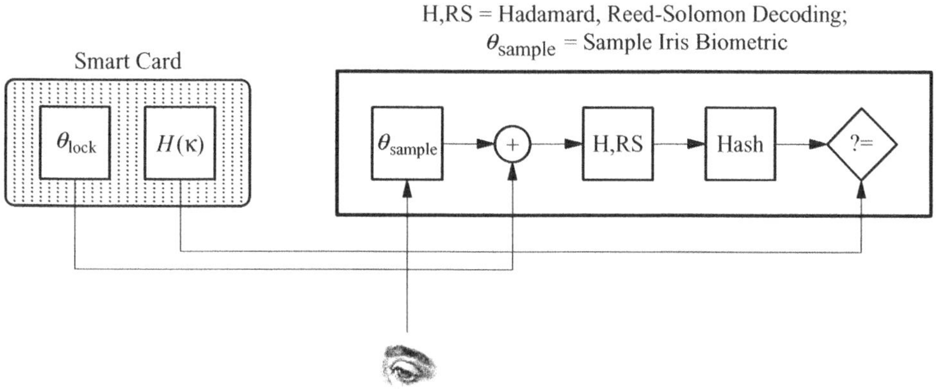

Figure 19.9. Recognition with the Iris biometric/smart card.

1. A random 140-bit iris biometric key κ is generated.
2. κ is encoded using a Reed-Solomon and then a Hadamard error-correcting code producing a pseudo-iris code, a 2048-bit vector θ_{ps}.
3. θ_{ps} is XORed with a real 2048-bit iris code (user reference code) θ_{ref} created during enrollment to produce a lock vector $\theta_{lock} = \theta_{ps} \oplus \theta_{ref}$.
4. θ_{lock} and a hash function $H(\kappa)$ of the iris biometric key are stored on the smart card.

The steps in the recognition process are depicted in Figure 19.9 and are listed as follows:

1. A sample iris biometric reference θ_{sample} is measured.
2. The lock vector θ_{lock} is read from the smart card.
3. θ_{sample} is XORed $\theta_{lock} \oplus \theta_{sample}$ yielding a presumptive pseudo iris code θ_{ps}^*.
4. θ_{ps}^* is decoded using a Reed-Solomon and then a Hadamard error-correcting code produces a presumptive iris key κ^*.
5. κ^* is hashed.
6. $H(\kappa^*)$ is compared with the true value $H(\kappa)$ read from the smart card.

The person's identity is confirmed if there is agreement in the Step 6 comparison.

Remarks. The use of hashing here is clever and different from its other applications; κ is a key but not a cryptographic key used to encipher data, and it does not need to be memorized by the user. Its randomness serves to make $H(\kappa)$ "hard" to guess.

Protecting against the inherent fuzziness in biometric measurements is addressed by a two-stage coding of biometric iris key κ

Stage 1 (Reed-Solomon Coding): to correct burst errors

Stage 2 (Hadamard Coding): to correct random bit errors

Why isn't one layer of coding sufficient? Professor Hao has pointed out to me (in a private communication) that the Reed-Solomon code is used to correct errors at the block level. Sometimes, the errors are clustered in certain regions, if the subject "blinks," it can cause specular reflections[22] and other abnormalities. The Hao, Anderson and Daugman algorithm [2005] took this into consideration by permuting the iris-code bits before doing error encoding and decoding. Permutation helps to disperse the errors but still may not be enough. Statistically, there are also certain error bits clustered in some blocks. Because recognition cannot afford to have even a single block with errors, two layers of coding are used. Note that although the increase in coding reduces the bit rate, this is irrelevant because the pseudo-iris code is not transmitted but stored on the smart card.

Appendix 19 summarizes the principal features of the Reed-Solomon and Hadamard Codes.

Hao, Anderson and Daugman [2005] encode the 140-bit iris biometric key κ into a 2048-bit pseudo-iris code vector θ_{ps} as follows:

- First, the 140-bit iris biometric key κ consisting of 20 symbols (each of $m = 7$ bits) is expanded by the Reed-Solomon code RS(n, k) with $(n, k, t) = (32, 20, 6)$ $[n - k = 2t]$ into 32 symbols (each of m bits).

$$\kappa \xrightarrow{\text{RS}} \kappa^*$$

- This RS coding can correct up to six symbol errors.
- Next, κ^* consisting of 32 symbols (each of m bits) is expanded by the Hadamard code $\mathcal{G}_{2^k}$ with $k = 6$ into 32 blocks each of 64 bits

$$\kappa^* \xrightarrow{\text{Hadamard}} \theta_{ps}$$

- This Hadamard coding will correct up to 15 bit-errors in each block of 64 bits.

Why isn't Reed-Solomon or Hadamard coding alone sufficient? Dr. Hao[23] was kind enough to point out that, if the subject "blinks" or if their are specular reflections, there are error-bursts. Because encipherment will generally cause a small bit difference to result in a large difference (in ciphertext), additional error-correcting codes are needed.

Coda. Is iris recognition secure? In analyzing the security of their system, Hao, Anderson and Daugman [2005] point out that it combines two factors, a biometric measurement and the smart card, which is a token. Even though it may be possible in the fashion of James Bond to capture and duplicate a copy of someone's iris, the token is also needed. If only the token is available, a complicated analysis is required to reverse engineer and to discover the iris value. Of course, if the smart card is not as smart as the phrase token resistance suggests, there might be an exposure. To make things really hard, Hao et al. suggest requiring a password to activate the card. They also observe the certification by the Australian Ministry of Defense

[22]A *specular reflection* is the mirror-like surface reflection of light; the light from a single incoming direction is reflected into a single outgoing direction.

[23]In a private communication.

of a camera to measure the iris, which can distinguish between real and fake eyes. Perhaps spitting may not be necessary.

19.10 THE DONGLE[24]

We conclude this chapter, this section, and this book by describing a very early technology to protect software that, perhaps has a less hashing flavor.

The ease of piracy of digital medium relates principally to the fact that the information in soft electronic form is easy to copy. Consequently, many techniques to prevent unauthorized use of software introduce additional hardware components that complement the software; if the hardware is not present, then the software does not run.

Early techniques to protect computer software relied on the use of a *dongle*, which is a small piece of hardware that connects to a computer. Dongles are still used for higher priced software, where the cost of manufacturing the dongle in production quantities is small relative to the price paid for the software. The dongle shown in Figure 19.10 is generally a portable device usually connected via a USB port.

To keep costs low, the earliest dongles had a simple identification number either hard wired or contained in a read-only memory (ROM). The software program would periodically poll the dongle for the presence of the ROM. If the ROM was not connected, or the identification number did not match the number in the software, the software would either stop running or enter an altered mode of operation, perhaps limiting functionality.

In the mid-1990s, the Intel Corporation added an operations code (opcode) to its Pentium III Central Processing Unit™ (CPU) to query a unique processor identification number. This processor number could be used by a software application in much the same as a ROM-based dongle. However, because of privacy concerns,

Figure 19.10. A dongle.

[24]Written with the assistance of Mr. Andrew Pickholtz, Esq.

particularly in the European Union, Intel removed the opcode from subsequent processors. Advanced Micro Devices (AMD) never included a comparable opcode in their CPUs.

Unfortunately, it is relatively easy to replicate a dongle that contains a simple identification ROM. A more secure system is described in U.S. Patent #4593353 [Pickholtz 1986]. In the method of the '353, the software contains not one but two authorization codes. The dongle includes a processor or other circuitry configured with a class of functions that take the first authorization code as input and output a result. The software periodically queries the dongle with the first authorization code and confirms that the response from the dongle matches the second authorization code. If it does match, the program continues uninterrupted.

The '353 dongle, as well as other dongles, are often encased in an epoxy resin, so that the physical hardness makes the dongle difficult to reverse engineer. Similarly, the function used inside the dongle is hard to determine just based on the first and second authorization codes. Some dongles merely rely on security through obscurity, using a single or handful of complicated algorithms that the dongle manufacturer hopes will not become known. However, security through obscurity approaches seldom last for long; there are just too many enterprizing pirates.

In the '353 approach, the function used in each dongle is from a parameterized class of functions, each dongle containing one function in the class. Ideally, the functions are chosen so that it is difficult to determine what the function is from any given pair of first and second authorization codes. Hence, the dongle hardware is not compromised even if one instance of software protected by the dongle is. In this sense, it is desirable to use a class of function with properties similar to one-way hash functions. For a one-way hash function, it is difficult to determine a message from the hash value. Here, it is difficult to determine the specific function in the class solely from the input and output of the dongle.

Given the hardness objectives for this type of dongle, it is not surprising that good candidates for the choice of functions are cryptographically secure one-way hash functions with each dongle having a unique key.

The '353 protection has the additional virtue that copies of the software can be made for backup or for use by other computers. But because the dongle is necessary to run it, only one copy can be used per dongle. Other techniques allow a single dongle on a server to allow as many copies to run over a network as are licensed.

REFERENCES

A. T. Abdel-Hamid, S. Tahar, and E. M. Aboulhamid, "A Survey on IP Watermarking Techniques", *Design Automation for Embedded Systems*, **9**, #3, pp. 211–227, 2004.

B. Chen and G. W. Worness, "Quantization Index Modulation: A Class of Provably Good Methods for Digital Watermarking and Information Embedding", *IEEE Transactions on Information Theory*, **47**, #4, pp. 1423–1443, 2001.

J. Daugman, "How IRIS Recognition Works", *IEEE Transactions on Circuits and Systems for Video Technology*, **14**, #1, pp. 21–30, 2004.

J. Dittman, P. Wohlmacher and K. Nahrstedt, "Using Cryptographic and Water-Marking Algorithms", *IEEE Multimedia*, **8**, #3, pp. 54–65, 2001.

Federal Information Processing Standards Publication 46-1, Data Encryption Standard (DES), National Bureau of Standards, January 22, 1988; superceded by Federal Information Processing Standards Publication 46-2, December 30, 1993; reaffirmed as FIPS PUB 46-3, October 25, 1999.

G. N. Griswold, "A Method for Protecting Copyright on Networks", 2002, http://www.cni.org/dox/ima.ip-workshop/Griswold.html.

F. Hao, R. Anderson, and J. Daugman, *Combining Cryptography with Biometrics Effectively*, Technical Report 640, University of Cambridge, 2005.

H. Hel-Or and M. Butman, "Multi-Level Watermarking with Independent Decoding", IEEE International Conference on Image Processing, pp. 514–517, 2001.

H. Z. Hel-Or, I. Yitzhaki, and Y. Hel-Or, "Geometric Hashing Techniques for Watermarking", Proceedings of the 2001 International Conference on Image Processing, pp. 498–501, 2001.

A. K. Jain, *Fundamentals of Digital Image Processing*, Prentice Hall (Englewood Cliffs, NJ), 1989.

D. Kahn, *The Reader of Gentlemen's Mail*, Yale University Press (New Haven, CT), 2004.

Vlado Kitanovski, Dimitar Taskovski and Söfija Bogdanova, "Combined Hashing/Watermarking Method for Image Authentication", *International Journal of Signal Processing*, **3**, #3, pp. 223–229, 2007.

A. G. Konheim, *Computer Security and Cryptography*, John Wiley & Sons (New York), pp. 288–302, 2007.

Y. Lamdan and H. J. Wolfson, "On the Error Analysis of Geometric Hashing", Proceedings of the IEEE Conference on Computer Vision and Pattern Recognition, pp. 22–27, 1991.

F. A. P. Petitcolas, R. J. Anderson, and M. G. Kuhn, "Information Hiding—A Survey", *Proceedings of the IEEE*, **87**, #7, pp. 1062–1078, 1999.

A. Pickholtz, U.S. Patent 4593353, "Software Protection Method and Apparatus", issued June 3, 1986.

N. Ratha, J. Connell, and R. Bolle, "Enhancing Security and Privacy in Biometrics-Based Authentication Systems", *IBM Systems Journal*, **40**, #3, pp. 614–634, 2001.

D. K. Roberts, "Security Camera Video Authentication", 10th IEEE Digital Signal Processing Workshop, pp. 125–130, 2002.

S. E. Siwek, "The True Cost of Sound Recording Piracy to the U.S. Economy", IPI Policy report, 2007. Available at http://www.ipi.org/ipiupfulltext/5C2EE3D2107A4C228625733E0053A1F4.

Y. Sutcu, H. T. Sencar, and N. D. Memon, "A Secure Biometric Authentication Scheme Based on Robust Hashing", Proceedings of the 7th Workshop on Multi-Media Security, pp. 111–116, 2005.

A. B. Watson, "Image Compression Using the Discrete Cosine Transformation", *Mathematica Journal*, **4**, #1, pp. 81–88, 1994.

J. Williams and S. Ezekiel, "Wavelet Based Iris Recognition", *The Pennsylvania Association of Computer and Information Science Educators* (PACISE), 2005.

H. J. Wolfson and I. Rigoutsos, "Geometric Hashing: An Overview", *IEEE Computational Science and Engineering*, **4**, #4, pp. 10–21, 1997.

APPENDIX 19

Reed-Solomon and Hadamard Coding

A.19.0 CODES

There are as many books and papers on coding as there are fish in the sea or stars in the sky. My favorite is Peterson's book [Peterson 1961], but there are more recent books, the two such as volumes of MacWilliams and Sloane [MacWilliams and Sloane 1977] and the more up-to-date books by Reed and Chen [Reed and Chen 1999], McEliece [McEliece 2002], and Lin and Costello [Lin and Costello 2004].

A(0,1)-code is a set of codewords $\underline{x} = (x_0, x_1, \cdots, x_{n-1})$ $(x_i \in \{0\ 1\}$. The class of Hamming codes was introduced in 1950 by Richard W. Hamming (1915–1998), then at Bell Telephone Laboratories [Hamming 1950]. The simplest Hamming code Ham[7, 4] consists of sixteen 7-bit codewords defined by

$$\begin{pmatrix} x_0 \\ x_1 \\ x_3 \end{pmatrix} = \begin{pmatrix} 1 & 1 & 0 & 1 \\ 1 & 0 & 1 & 1 \\ 1 & 0 & 1 & 1 \\ 0 & 1 & 1 & 1 \end{pmatrix} \begin{pmatrix} x_2 \\ x_4 \\ x_5 \\ x_6 \end{pmatrix} \tag{A19.1}$$

The Hamming distance ρ_H between code words $\underline{x} \neq \underline{x}^*$ is at least three so that:

- The Hamming code Ham[7, 4] can correct any single-bit error.
- The Hamming code Ham[7, 4] can detect an single- or double-bit error.

Ham[7, 4] transmits 4 information bits (x_2, x_4, x_5, x_6) after including 3 parity bits (x_0, x_1, x_3).

Of course, there is no reason to restrict the notion of codes to binary sequences, and the Reed-Solomon codes are nonbinary codes in which the codewords are composed of s-valued symbols. These codes are best understood within the framework of fields and extension fields as discussed in Chapter 5, §5.5. The cases $s = q^m$, where q is a prime and $m \geq 1$, provide natural examples for the size of the symbol space. Even more natural is the case $q = 2$, in which case the symbols are m-bit (0,1)-sequences and elements of $\mathcal{Z}_{m,2}$.

A.19.1 REED-SOLOMON CODES

Published in 1960 ([Reed and Solomon 1960]) by the MIT Lincoln Laboratory researchers Irving Reed (born in 1923) and Gustave Solomon (1930–1996), the codewords of the Reed-Solomon code RS(n, k) are each of length n(symbols = mn bits with $m > 2$) with k of information symbols and $n - k$ parity-check symbols. The values of n, k, m are constrained by

$$0 < k < n < 2^m + 2 \qquad \text{(A19.2)}$$

If our example $n - k = 2t$, where t is the symbol-error correcting capability. These Reed-Solomon codes achieve the largest possible minimum distance between (distinct) codewords for given n, m, k;

$$\min_{\underline{x} \neq \underline{x}^*} \rho_H(\underline{x}, x^*) = n - k + 1$$

where ρ_H is the symbol Hamming distance, the number of symbols in which $\underline{x}$ and $\underline{x}^*$ disagree.

The Reed-Solomon codes are cyclic codes determined by a generator polynomial

$$g(x) = g_0 + g_1 x + \cdots + g_{n-k-1} x^{n-k-1} + x^{n-k} \quad \{g_i \in \mathcal{Z}_{m,2} : 0 \le j < n-k\} \qquad \text{(A19.3}a\text{)}$$

of degree $n - k$ whose m-bit symbol coefficients $\{g_j \in \mathcal{Z}_{m,2}: 0 \le j < n - k\}$ are the $n - k$ parity check coefficients.

To encipher the k-information symbols

$$\underline{d} = (d_0, d_1, \cdots, d_{k-1}) \quad \{d_j \in \mathcal{Z}_{m,2} : 0 \le j < k\} \qquad \text{(A19.3}b\text{)}$$

with corresponding polynomial

$$d(x) = d_0 + d_1 x + \cdots + d_{k-1} x^{k-1} \qquad \text{(A19.3}c\text{)}$$

the polynomial $d(x)$ is multiplied by x^{n-k} (right-shifted $n - k$ positions), we divide $x^{n-k}d(x)$ by $g(x)$ yielding

$$x^{n-k} d(x) = q(x) g(x) + r(x) \quad \deg(q) = k, \quad 0 \le \deg(r) < n-k \qquad \text{(A19.3}d\text{)}$$

The n-symbol codeword

$$\underline{c} = (c_0, c_1, \cdots, c_{n-1}) \quad \{c_j \in \mathcal{Z}_{m,2} : 0 \le j < n\} \qquad \text{(A19.3}e\text{)}$$

with corresponding polynomial

$$c(x) = c_0 + c_1 x + \cdots + c_{n-1} x^{n-1} \qquad \text{(A19.3}f\text{)}$$

is defined by

$$c(x) = x^{n-k} d(x) - r(x) \qquad \text{(A19.3}g\text{)}$$

Remarks. Equation (A19.3g) writes the codeword polynomial $c(x)$ as the difference of two polynomials. Many definition of the Reed-Solomon code differ from mine in a trivial notational way; they replace the minus sign by addition where the addition of symbols, say s and s^*, means to add modulo 2 the binary sequences, which are represented by s and s^*. In any event, $r(x)$ is chosen so that $c(x)$ is a multiple of the generator $g(x)$; that is,

$$d(x) \to c(x) \tag{A19.3h}$$

$$r(x) = x^{n-k} d(x) \pmod{g(x)} \tag{A19.3i}$$

$$c(x) = 0 \pmod{g(x)} \tag{A19.3j}$$

Thus, the codewords of RS(n, k) are polynomial multiples of $g(x)$.

Exercise 19.2. Prove the converse; if $c(x)$ of degree $\leq n - 1$ satisfies $0 = c(x) \pmod{g(x)}$, then $d(x)$ exists such that they are related as in equations (A19.3)

How do we decode? Suppose the following is true:

- $\underline{c} = (c_0, c_1, \cdots, c_{n-1})$ is the transmitted codeword and
- $\underline{c}^* = (c_0^*, c_1^*, \cdots, c_{n-1}^*)$ is the received codeword, then
- $\varepsilon = (\varepsilon_0, \varepsilon_1, \cdots, \varepsilon_{n-1})$ with $\varepsilon_j \in \mathcal{Z}_{m,2}$ is the error vector with $c_j^* = c_j + \varepsilon_j$ and the corresponding polynomial $\varepsilon(x) = \varepsilon_0 + \varepsilon_1, + \cdots, +\varepsilon_{n-1}x^{n-1}$.

It follows that:

1. If $0 = \varepsilon(x) \pmod{g(x)}$, then $c^*(x)$ is also a codeword of RS(n, k) and therefore ε is an instance of a nondetectable error.
2. If $0 \neq \varepsilon(x) \pmod{g(x)}$, then $\underline{\varepsilon}$ is an instance of a detectable error.

Example A19.1. Let $(n, k, m) = (7, 3, 3)$ and

$$g(x) = g_0 + g_1 x + g_2 x^2 + g_3 x^3 + x^4 \quad g_k = \begin{cases} (1, 1, 0) & \text{if } k = 0 \\ (0, 1, 0) & \text{if } k = 1 \\ (0, 0, 1) & \text{if } k = 2 \\ (1, 1, 0) & \text{if } k = 3 \end{cases} \tag{A19.4a}$$

Query. How did we get this polynomial and these coefficients?

Answer. A polynomial $p(x) \in \mathcal{Z}_{m,2}[x]$ of degree n is primitive if has a root, say β such that the powers $1, \beta, \beta^2, \cdots, \beta^{2^m-2}$ are the nonzero elements of $\mathcal{Z}_{m,2}[x]$. For reasons explained in Chapter 5, §5.5, RS(n, k) gives more bang for the parity-check symbol if $g(x)$ is a primitive polynomial in the field whose elements are in $\mathcal{Z}_{m,2}$. When $g(x) \in \mathcal{Z}_{m,2}[x]$ of degree n is primitive, the smallest integer e for which $0 = (x^e - 1) \pmod{g(x)}$ is $e = 2^m - 1$.

The extension field $\mathcal{Z}_{3,2}$ may be generated by adjoining to $\mathcal{Z}_2$ the root, say α of a primitive polynomial f; one such example is $f(x) = 1 + x + x^3$.

Table A19.1 is a coding for the normal basis $1, \alpha, \alpha^2, \cdots, \alpha^7$ of the field $\mathcal{Z}_{3,2}$. which gives the formula

TABLE A19.1. Coding for the Normal Basis of $\mathcal{Z}_{3,2}$

α^0	(1, 0, 0)	α^4	(0, 1, 1)
α^1	(0, 1, 0)	α^5	(1, 1, 1)
α^2	(0, 0, 1)	α^6	(1, 0,1)
α^3	(1, 1, 0)	α^7	(1, 0, 0)

TABLE A19.2. Relationships for the Normal Basis of $\mathcal{Z}_{3,2}$

$\alpha^3 = 1 + \alpha$	$\alpha^4 = \alpha + \alpha^2$
$\alpha^5 = 1 + \alpha + \alpha^2$	$\alpha^6 = 1 + \alpha^2$
$\alpha^7 = 1$	

TABLE A19.3. Checking Primitivity for the Normal Basis of $\mathcal{Z}_{3,2}$

$0 = f(\alpha)$	$0 = f(\alpha^2) = 1 + \alpha^2 + \alpha^6$
$0 \neq f(\alpha^3) = 1 + \alpha^3 + \alpha^9 = 1 + \alpha^2 + \alpha^3$	$0 = f(\alpha^4) = 1 + \alpha^4 + \alpha^{12} = 1 + \alpha^4 + \alpha^5$
$0 \neq f(\alpha^5) = 1 + \alpha^5 + \alpha^{15} = 1 + \alpha + \alpha^5$	$0 \neq f(\alpha^6) = 1 + \alpha^6 + \alpha^{18} = 1 + \alpha^2 + \alpha^4$
$0 \neq f(\alpha^7) = 1 + \alpha^7 + \alpha^{21} = 1$	

$$g(x) = \alpha^3 + \alpha x + \alpha^0 x^2 + \alpha^3 x^3 + x^4 \tag{A19.4b}$$

It is easy to verify the entries in Table A19.2,

In fact, because

$$0 = 1 + \alpha + \alpha^3 \rightarrow \alpha^3 = 1 + \alpha \tag{A19.5a}$$

Multiplying both sides of equation (A19.9*a*) by α gives

$$\alpha^3 = 1 + \alpha \rightarrow \alpha^4 = \alpha + \alpha^2 \tag{A19.5b}$$

and repeating the same process

$$\alpha^4 = \alpha + \alpha^2 \rightarrow \alpha^5 = \alpha^2 + \alpha^3 = \alpha^2 + 1 + \alpha \tag{A19.5c}$$

$$\alpha^5 = 1 + \alpha + \alpha^2 \rightarrow \alpha^6 = \alpha + \alpha^2 + \alpha^3 \tag{A19.5d}$$

$$\alpha^6 = 1 + \alpha^2 \rightarrow \alpha^7 = \alpha + \alpha^3 = 1 \tag{A19.5e}$$

From equations (A19.5) and $\alpha^k = \alpha^{k(\text{mod } 7)}$, Table A19.3 shows

The column x^i (*modulo* $g(x)$) in Table A19.4 lists the values of x^i (mod $g(x)$) for $g(x)$ given by equation (A19.4*a*). The entry for the polynomial $E_{i+1}(x) \equiv x^{i+1}$ (mod $g(x)$) of degree $3 = \deg(g) - 1$ is computed from the entry for $E_i(x) \equiv x^i$ (mod $g(x)$) using the following steps:

TABLE A19.4. x^i (modulo $g(x)$) $1 \le i \le 7$

i	x_i(modulo g(x))	Notes
1	x	
2	x^2	
3	x^3	
4	$\alpha^3 + \alpha x + x^2 + \alpha^3 x^3$	
5	$\alpha^6 + \alpha^6 x + x^2 + \alpha^2 x^3$	$\alpha^3(1+\alpha) = \alpha^6$; $\alpha(1+\alpha^2) = \alpha^7 = 1$; $(1+\alpha^6) = \alpha^{-1}(1+\alpha) = \alpha^2$
6	$\alpha^5 + \alpha^4 x + x^2 + \alpha^4 x^3$	$\alpha^3 + \alpha^6 = (1+\alpha+\alpha^6) = (\alpha+\alpha^2) = \alpha^4$; $1+\alpha^5 = \alpha^{-2}(1+\alpha^2) = \alpha^4$
7	1	

1. Multiply E_i by x
2. Replace the coefficient c_4 of x^4 in the result by $c_4(\alpha^3 + \alpha x + x^2 + \alpha^3 x^3)$
3. Simplify using the Notes in Table A19.4, express each coefficient of $E_{i+1}(x)$ as a power of α

The column *Notes* in Table A19.4 indicates the nature of the simplification in step 3 in the previous list.

It is easy to prove in general [Peterson 1961, p. 103] that $g(x) \in \mathcal{Z}_{3,2}[x]$ is primitive if and only if at least one of its roots is primitive. Using Table A19.4, we can show the coefficients of the polynomial

$$g(x) = (x-\alpha)(x-\alpha^2)(x-\alpha^3)(x-\alpha^4) \tag{A19.6}$$

are given in equation (A19.4*a*) in *Example A19.1*; Table A19.4 shows that $g(x)$ is primitive.

To derive the RS(7, 3)-codeword for the data symbols, take the following steps:

$$\underline{d} = (d_0, d_1, d_2) \quad d_k = \begin{cases} \alpha = (0,1,0) & \text{if } k = 0 \\ \alpha^3 = (1,1,0) & \text{if } k = 1 \\ \alpha^5 = (1,1,1) & \text{if } k = 2 \end{cases} \tag{A19.7}$$

1. First, compute $x^4 d(x) = d_0 x^4 + d_1 x^5 + d^2 x^6 = \alpha x^4 + \alpha^3 x^5 + \alpha^5 x^6$
2. Next, compute the remainder $r(z) = x^4 d(x) \pmod{g(z)} = \alpha^0 + \alpha^2 x + \alpha^4 x^2 + \alpha^6 x^3$
3. Finally, add $d(x) + r(x) = \alpha^0 + \alpha^2 x + \alpha^4 x^2 + \alpha^6 x^3 + \alpha x^4 + \alpha^3 x^5 + \alpha^5 x^6$

The computation in step 2 in the previous list can be carried out in hardware using a linear feedback shift register (LFSR) as depicted in Figure A19.1.

The operation of the LFSR is clocked; in each clock-cycle, the following occurs:

1. A computation in the field $\mathcal{Z}_{3,2}$ consisting of a multiplication of the value on the feedback line, and an addition of the value in two adjacent cells
2. A single $\mathcal{Z}_{3,2}$-symbol transfer takes place, either from the input message line or to the output message line.

The destination of the symbol transfer is determined by the position of switch #2, the nature of the feedback symbol by the position of switch #1 as follows;

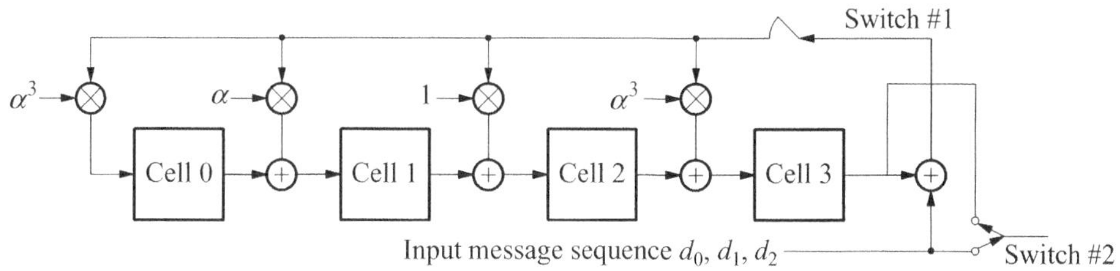

Figure A19.1. The linear feedback shift register for example A19.1.

1. Switch #1 is closed for the first $n-k$ clock cycles to load the $n-k$ data symbols $d_{n-k-1}, \cdots, d_1, d_0$ (in this order) into $\text{Cell}_0, \text{Cell}_1, \cdots, \text{Cell}_{n-k-1} = \text{Cell}_3$
2. Switch #2 is in the down position during this time so that these data symbols are also transmitted as output
3. After the first k clock cycles, switch #1 is open, switch #2 is in the up position, and the following two events take place:
   a. The contents of each cell multiplied by the feedback symbol (the taps) and then added to the contents of the cell to the right
   b. The contents of the rightmost cell is the output symbol

If we write the following:

- $\text{OUT}[\tau]$ for the symbol output in the τth clock cycle,
- $\text{Cell}_j[\tau]$ for the symbol stored in the jth cell just before the τth clock cycle,
- $F[\tau]$ for the feedback symbol just before the τth clock cycle,
- $\underline{g} = (g_0, g_1, g_2, g_3) = (\alpha^3, \alpha, 1, \alpha^3)$ for the coefficients of the generator polynomial $g(x)$, then the recursion describing the Figure A19.1 LFSR is

$$\text{OUT}[\tau] = \begin{cases} d_{k-\tau-1} = d_{2-\tau} & \text{if } 0 \le \tau < k = 3 \\ \text{Cell}_{n-k-1}[\tau] = \text{Cell}_3[\tau] & \text{if } 3 = k \le \tau \end{cases} \tag{A19.8a}$$

$$\text{Cell}_j[\tau] = \begin{cases} F|\tau|g_0 & \text{if } j = 0 \\ \text{Cell}_{j-1}[\tau-1] + F[\tau-1]g_j & \text{if } 1 \le j \le n-k-1 = 3 \end{cases} \tag{A19.8b}$$

Because

$$F[\tau] = \text{Cell}_3[\tau] \quad 2 < \tau \tag{A19.8c}$$

equation (A19.8b) yields

$$F[\tau] = F[\tau-3]g_0 + F[\tau-2]g_1 + F[\tau-1]g_2 \quad 2 < \tau \tag{A19.8d}$$

or equivalently

$$0 = -F[\tau] + F[\tau-3]g_0 + F[\tau-2]g_1 + F[\tau-1]g_2 \quad 2 < \tau \tag{A19.8e}$$

Suppose the (polynomial) codeword $c(x) = x^{n-k}d(x) + p(x)$ is transmitted and the (polynomial) codeword $r(x) = c(x) + \varepsilon(x)$ is received. If $0 = \varepsilon(x)$ (mod $g(x)$) where $g(x)$ is the generator of RS(n, k), then $\varepsilon(x)$ is an undetectable (polynomial) error. If $0 \neq \varepsilon(x)$ (mod $g(x)$), then the error is detectable[1] and may be correctable.

The syndrome of the (polynomial) error $\varepsilon(x)$ is $S_i \equiv S(\alpha^i) = \varepsilon(\alpha^i)$ where $S(x) = \varepsilon(x)$ (mod $g(x)$). Because $g(\alpha^i) =$ for $1 \leq i \leq 2t$, the error is undetectable if and only if $S_i = 0$ $(1 \leq i \leq 2t)$.

Suppose that the following is true:

- t or fewer symbols are in error.
- The syndromes $(S_1, S_2, \cdots, S_{2t})$ are not identically 0.
- $\varepsilon(x) = \varepsilon_{j0}x^{j1} + \varepsilon_{j2}x^{j2} + \cdots + \varepsilon_{jt}x^{jt}$ where $0 \leq j_1 < j_2 < \cdots < j_t < n$, and

$$\begin{Bmatrix} S_1 = \varepsilon(\alpha) = \chi_1\beta_1 + \chi_2\beta_2 + \cdots + \chi_t\beta_t \\ S_2 = \varepsilon(\alpha^2) = \chi_1\beta_1^2 + \chi_2\beta_2^2 + \cdots + \chi_t\beta_t^2 \\ \ddots \\ S_i = \varepsilon(\alpha^i) = \chi_1\beta_1^i + \chi_2\beta_2^i + \cdots + \chi_t\beta_t^i \\ \ddots \\ S_{2t} = \varepsilon(\alpha^{2t}) = \chi_1\beta_1^{2t} + \chi_2\beta_2^{2t} + \cdots + \chi_r\beta_t^{2t} \end{Bmatrix} \text{with} \begin{cases} \chi_s \equiv \varepsilon_{j_s} \\ \beta_s \equiv \alpha^{j_s} \end{cases} \quad 1 \leq s \leq t \qquad \text{(A19.9)}$$

The error-locator polynomial is $\sigma(x) = \sigma_t + \sigma_1 x + \sigma_0 x^t = (\gamma_1 + x)(\gamma_2 + x) \cdots (\gamma_t + x)$ where $\gamma_s = \alpha_s^{-1} \equiv \beta_s$ for $1 \leq s \leq t$. Note that $\sigma(\alpha^{j_s}) = 0$ for $1 \leq s \leq t$.

Claim. If no more than t symbols are in error, then the syndromes $(S_1, S_2, \cdots, S_{2t})$ determine the locations of the symbols in error.

Proof. Write

$$\sigma(x) = \sigma_t + \sigma_{t-1}x + \sigma_{t-2}x^2 + \cdots + \sigma_0 x^t \qquad \text{(A19.10}a\text{)}$$

For each s with $1 \leq s \leq \deg(g)$ we have $\sigma(\alpha^{j_s}) = 0$ and so that

$$0 = \sigma_t + \sigma_{t-1}\beta_s + \sigma_{t-2}\beta_s^2 + \cdots + \sigma_0\beta_s^t \qquad \text{(A19.10}b\text{)}$$

Multiply by χ_s and sum over s with $1 \leq s \leq \deg(g)$ ad used equation (A19.9) to obtain

$$0 = S_j\sigma_t + S_{j+1}\sigma_{t-1} + S_{j+2}\sigma_{t-2} + \cdots + S_{j+t} \quad 1‘ \leq j \leq t \qquad \text{(A19.10}c\text{)}$$

[1] Error correction is not free. When a detectable error is encountered, the (packetized) message is retransmitted when the underlying network protocol being used is X.25, ITU-T standard protocol suite for packet switched wide area network (WAN) communications. For security reasons, the Hao et al. [Hao, Anderson and Daugman 2005] protocol wants to correct the errors, if possible, rather than requiring another biometric measurement of the iris key.

which we may write in the form

$$(0)_{t+1} = \begin{pmatrix} S_1 & S_2 & S_3 & \cdots & S_{t-1} & S_t & S_{t+1} \\ S_2 & S_3 & S_4 & \cdots & S_t & S_{t+1} & S_{t+2} \\ & & & \ddots & & & \\ S_{t-1} & S_t & S_{t+1} & \cdots & S_{2t-3} & S_{2t-2} & \\ S_t & S_{t+1} & S_{t+2} & \cdots & S_{2t-2} & S_{2t-1} & \end{pmatrix} \begin{pmatrix} \sigma_t \\ \sigma_{t-1} \\ \sigma_{t-2} \\ \vdots \\ \sigma_2 \\ \sigma_1 \\ \sigma_0 \end{pmatrix} \tag{A19.10d}$$

Any solution $\underline{\sigma}$ of equation (A19.10*d*) determines the error locator. The first solution was essentially given by W. W. Peterson [1960] but is complexity limited its usefulness. The real breakthrough appeared in Massey's paper [1969] who noticed the similarity of equations (A19.8*e*) and (A19.10*d*). Therefore, to find the *taps*

- $\{g_j\}$ in equation (A19.8*e*), and
- $\{\sigma_j\}$ in equation (A19.10*d*)

to produce either the feedback $\{F[j]\}$ satisfying equations (A19.8) or the syndromes $\{S_j\}$ satisfying equation (A19.10*d*), the equations may be solved using the Berlekamp-Massey algorithm [1969]. We end our lengthy digression here, referring the reader to the excellent presentations in [Massey 1969], [Reed and Chen 1999], and [Lin and Costello 2004].

A.19.3 HADAMARD CODES

Hadamard Matrices and Codes. These codes were named in honor of the illustrious French mathematician Jacques Salomon Hadamard (1865–1963). Excellent references are Agaian [Agaian 1986] and the more recent book by Horadam [Horadam 2007]. The codes are defined in terms of Hadamard matrices $\{H_k: 0 \le k < \infty\}$ where

$$H_1 = (1) \tag{A19.11a}$$

$$H_2 = \begin{pmatrix} 1 & 1 \\ 1 & -1 \end{pmatrix} \tag{A19.110b}$$

$$H_{2^k} = \begin{pmatrix} H_{2^{k-1}} & H_{2^{k-1}} \\ H_{2^{k-1}} & -H_{2^{k-1}} \end{pmatrix} \quad 1 < k < \infty \tag{A19.11c}$$

Note the following:

1. H_{2^k} is of dimension 2^k-by-2^k.
2. The rows of H_{2^k}, say $\underline{r}_{2^k,0}, \underline{r}_{2^k,1}, \cdots, \underline{r}_{2^k,2^k-1}$ are orthogonal, i.e., $(\underline{r}_{2^k,i}, \underline{r}_{2^k,j}) = 0$ if $i \neq j$,
3. $H_{2^k} H_{2^k}^T = 2^k I$

where ^{T} denotes the transpose and I is the identity matrix.

Start with a Hadamard matrix H_{2^k}; the codewords of $\mathcal{H}_{2^k}$ are the rows of the array $\mathcal{H}_{2^k} = \begin{pmatrix} H_{2^k} \\ -H_{2^k} \end{pmatrix}$ in which -1 entries are replaced by 0.

Example A19.2. The Hadamard matrices and codes H_{2^k}, $\mathcal{H}_{2^k}$ for $k = 1(1)3$.

$$H_2 = \begin{pmatrix} 1 & 1 \\ 1 & -1 \end{pmatrix} \quad \mathcal{H}_2 = \begin{pmatrix} 1 & 1 \\ 1 & 0 \\ 0 & 0 \\ 0 & 1 \end{pmatrix}$$

$$H_4 = \begin{pmatrix} 1 & 1 & 1 & 1 \\ 1 & -1 & 1 & -1 \\ 1 & 1 & -1 & -1 \\ -1 & 1 & -1 & 1 \end{pmatrix} \quad \mathcal{H}_4 = \begin{pmatrix} 1 & 1 & 1 & 1 \\ 1 & 0 & 1 & 0 \\ 1 & 1 & 0 & 0 \\ 0 & 1 & 0 & 1 \\ 0 & 0 & 0 & 0 \\ 0 & 1 & 0 & 1 \\ 0 & 0 & 1 & 1 \\ 1 & 0 & 1 & 0 \end{pmatrix}$$

$$H_8 = \begin{pmatrix} 1 & 1 & 1 & 1 & 1 & 1 & 1 & 1 \\ 1 & -1 & 1 & -1 & 1 & -1 & 1 & -1 \\ 1 & 1 & -1 & -1 & 1 & 1 & -1 & -1 \\ -1 & 1 & -1 & 1 & -1 & 1 & -1 & 1 \\ 1 & 1 & 1 & 1 & -1 & -1 & -1 & -1 \\ 1 & -1 & 1 & -1 & -1 & 1 & -1 & 1 \\ 1 & 1 & -1 & -1 & -1 & -1 & 1 & 1 \\ -1 & 1 & -1 & 1 & 1 & -1 & 1 & -1 \end{pmatrix} \quad \mathcal{H}_8 = \begin{pmatrix} 1 & 1 & 1 & 1 & 1 & 1 & 1 & 1 \\ 1 & 0 & 1 & 0 & 1 & 0 & 1 & 0 \\ 1 & 1 & 0 & 0 & 1 & 1 & 0 & 0 \\ 0 & 1 & 0 & 1 & 0 & 1 & 0 & 1 \\ 1 & 1 & 1 & 1 & 0 & 0 & 0 & 0 \\ 1 & 0 & 1 & 0 & 0 & 1 & 0 & 1 \\ 1 & 1 & 0 & 0 & 0 & 0 & 1 & 1 \\ 0 & 1 & 0 & 1 & 1 & 0 & 1 & 0 \\ 0 & 0 & 0 & 0 & 0 & 0 & 0 & 0 \\ 0 & 1 & 0 & 1 & 0 & 1 & 0 & 1 \\ 0 & 0 & 1 & 1 & 0 & 0 & 1 & 1 \\ 1 & 0 & 1 & 0 & 1 & 0 & 1 & 0 \\ 0 & 0 & 0 & 0 & 1 & 1 & 1 & 1 \\ 0 & 1 & 0 & 1 & 1 & 0 & 1 & 0 \\ 0 & 0 & 1 & 1 & 1 & 1 & 0 & 0 \\ 1 & 0 & 1 & 0 & 0 & 1 & 0 & 1 \end{pmatrix}$$

Stated without proof, the following theorem summarizes the principal properties of Hadamard codes.

Theorem A19.1.

a) The Hadamard code $\mathcal{H}_{2^k}$ contains 2^{k+1} codewords of length 2^k.
b) The minimum distance between any two codewords of the Hadamard code $\mathcal{H}_{2^k}$ is 2^{k-1}.
c) The Hadamard code $\mathcal{H}_{2^k}$ will correct any error pattern of length $\leq 2^{k-2} - 1$.
d) If the 2^{k+1} rows of $\mathcal{H}_{2^k} = \left(\underline{c}_{2^k,0}, \underline{c}_{2^k,1}, \cdots, \underline{c}_{2^k,2^{k+1}-1}\right)$ are numbered by i with $0 \leq i < 2^{k+1}$, then if $0 \leq i \leq j < 2^{k+1}$.
   - $\underline{c}_{2^k,i}$ and $\underline{c}_{2^k,i}$ agree in k coordinates,
   - 0 positions if $j = i + 1$, and
   - $\frac{k}{2}$ positions, otherwise.
e) If the Hamming distance ρ_H of a(0,1)-vector $\underline{v}$ of length 2^k to $\underline{c}_{2^k,i}$ is $\leq 2^{k-2}$, then $\rho_H\left(\underline{v}, \underline{c}_{2^k,i}\right) < \min_{j \neq i} \rho_H\left(\underline{v}, \underline{c}_{2^k,i}\right)$, that is, $\underline{v}$ is closer to $\underline{c}_{2^k,i}$ than to any other vector in $\mathcal{H}_{2^k}$.
e) The Hadamard code $\mathcal{H}_{2^k}$ enciphers $k + 1$ bits into a codeword of 2^k bits and will correct any error pattern of length $<2^k$.

Hadamard Encoding. Any 1-to-1 mapping of the set $\mathcal{Z}_{k+1,2}$ of the 2^{k+1} message vectors $\underline{d} = (d_0, d_1, \cdots, d_k)$ into the rows $\{\underline{c}_i : 0 \leq i < 2k+1\}$ of $\mathcal{H}_{2^k}$ is a Hadamard encoding of $\underline{d}$; for example, the mapping

$$\underline{d} = (d_0, d_1, \cdots, d_k) \to \underline{c}_i \quad \text{with } i = d_0 + 2d_1 + \cdots + 2^k d_k$$

is a Hadamard encoding of $\underline{d}$.

The rate of a code (see [Lin and Costello 2005, p. 4]) is the ratio of message bits to code word length; the rate for $\mathcal{H}_{2^k}$ is

$$\frac{k+1}{2^k}$$

which decreases rapidly with k. The Hadamard code $\mathcal{H}_{32}$ with rate $\frac{6}{32} \approx 0.19$ was used in 1969 to encode data transmitted from the *Mariner* spacecraft in 1969 ··· *When you care to receive only the most accurate pictures!*

Hadamard Decoding. Suppose the following:

- The message vector $\underline{d} \in \mathcal{Z}_{2^k,2}$ is encoded into the codeword $\hat{\underline{d}} \in \mathcal{Z}_{k+1,2}$ of $\mathcal{H}_{2^k}$ and received as $\underline{v} \in \mathcal{H}_{k+1,2}$.

The rows of $\mathcal{H}_{2^k}$ are searched for a row $\underline{c}_i$ that differs from $\underline{v}$ in at most $2^{k-2} - 1$ positions.

Success. there is such a row; it unique by Theorem A19.1b or

Failure. an error is detected.

The labels *Success* and *Failure* are somewhat misleading; if $d_{\text{Hamming}}(\hat{\underline{d}}, \underline{c}_i) \leq 2^{k-2} - 1$, then $\underline{c}_i$ is the most likely transmitted codeword, meaning that if all 2^{k+1} message vectors in $\mathcal{Z}_{k+1,2}$ are equally likely to have been encoded, then

$$Pr\left\{\hat{\underline{d}} \text{ transmitted}/\underline{c}_i \text{ received}\right\} > \max_{\substack{d^* \in \mathcal{Z}_{k+1,2} \\ d^* \neq \underline{d}}} Pr\left\{\widehat{\underline{d}^*} \text{ transmitted}/\underline{c}_i \text{ received}\right\}$$

REFERENCES

S. S. Agaian, *Hadamard Matrices and Their Applications*, Springer-Verlag (Berlin, Germany), 1986.

R. W. Hamming, "Error Correcting and Detecting Codes", *Bell System Technical Journal*, **26**, #2, pp. 147–160, 1950.

Feng Hao, Ross Anderson, John Daugman, "Combining Cryptography with Biometrics Effectively", Technical Report UCAM-CL-TR-640, University of Cambridge Computer Laboratory, July 2005.

K. J. Horadam, *Hadamard Matrices and Their Applications*, Princeton University Press (Princeton, NJ), 2007.

S. Lin and D. J. Costello, *Error Control Coding* (Second Edition), Prentice-Hall (Upper Saddle River, NJ), 2005.

J. L. Massey, "Shift-Register Synthesis and BCH Decoding", *IEEE Transactions on Information Theory*, **IT-15**, #1, pp. 122–127, 1969.

F. J. MacWilliams and N. J. A. Sloane, *The Theory of Error Correcting Codes (Parts I & II)*, North Holland Publishing (Amsterdam, The Netherlands), 1977.

R. J. McEliece, *The Theory of Information and Coding*, Cambridge University Press (Cambridge, UK), 2002.

W. W. Peterson, "Encoding and Error-Correction Procedures for Bose-Chaudhuri Codes", *IEEE Transactions on Information Theory*, **IT-6**, pp. 459–470, 1960.

W. W. Peterson, *Error Correcting Codes*, MIT Press (Cambridge, MA), 1961.

I. S. Reed and X. Chen, *Error Control Coding for Data Networks*, Kluwer Academic Publishers (New York), 1999.

I. S. Reed and G. Solomon, "Polynomial Codes Over Certain Finite Fields", *Journal Society of Industrial and Applied Mathematics*, **8**, #2, pp. 300–304, 1960.

Exercises and Solutions

PART II, CHAPTER 7

Exercise 7.1. (Model for performance evaluation of simple storage). Assume the following:

- m records $R_0, R_1, \cdots, R_{m-1}$, each requiring for storage a single cell (1 unit), have been stored in a block of n contiguous cells with starting address A.
- If an existing record is selected for access, then each of the m records is chosen with equal probability.

The procedure SEARCH(k) to locate the record with key k = KEY requires a C comparison.

7.1a) Calculate the probability distribution and average number of C needed to execute SEARCH(k) if such the record with key k = KEY is currently in storage.

7.1b) Assuming that no record with key k = KEY has been stored, calculate the average number of comparisons needed to execute SEARCH(k).

Solution. Let C he number of comparisons used by SEARCH(k) to locate the record with key k = KEY *assuming* it has already been stored. Since this record may be any of the m records in the single chain with equal probability

$$Pr\{C = r\} = \frac{1}{m} \quad 0 \le r < m \qquad \text{(Ex7.1}a\text{)}$$

and

$$E\{C\} = \frac{m+1}{2} \qquad \text{(Ex7.1}b\text{)}$$

In **7.1b**, $C = m + 1$ comparisons are needed by SEARCH(k) since all of the keys in the records in the *chain* need to be tested. ■

PART II, CHAPTER 8

Exercise 8.1. Assume the following:

H1. The information of m records has been entered into a hash table of size n-entries with $n \gg m$

H2. All possible hash values $h(K) \in \{0, 1, \cdots, n-1\}$ are equally likely

H3. The CRP causes all m-element subsets of the n-address hash table to be equally likely to occur

Calculate the probability that the insertion into the hash table of the information of an $(m+1)^{st}$ record is collision free.

Solution. Hypothesis H3 means that all $\binom{n}{m}$ subsets of the hash-table are equally likely to be occupied. Let $Pr_{n,m}\{CF\}$ denote the probability that the insertion into the hash-table of the information of an $(m+1)^{st}$-record be collision-free. This event requires that the $(m+1)^{st}$-record *not* be hashed to to one of the m occupied addresses in the hash-table and by H2 is therefore given by

$$Pr_{n,m}\{CF\} = \frac{1}{\binom{n}{m}}\left(1 - \frac{m}{n}\right) \tag{Ex8.1}$$

■

PART II, CHAPTER 10

Exercise 10.1. (Deletions with separate chaining): Write pseudocode describing the deletion of the hash-table entries of KEY = k if separate chaining as in Figure 10.1 is used. Assume the following:

10.1a) $h(k) = h(k_i)(0 \le i < L)$ for the L keys $k_0, k_1, \cdots, k_{L-1}$ in the linked list pointed to by HEAD[$h(k)$];

10.1b) LINK[$h(k_i)$] is in the link-field in the node of the key k_i $(0 \le i < L)$.

10.1c) LINK–ADD[i] is in the link-field of record R_i $(0 \le i < L)$.

Solution

For $i = 0$ to $L -$ do
{
Compare k and k_i $k = k_i$?
if YES, then
if $0 < i < L - 1$, set LK–ADD[$i - 1$] ← LK–ADD[$i + 1$], else
if $0 < i = L - 1$, set LK–ADD[$i - 1$] ← $\emptyset$, else
if $0 = i < L - 1$, set $k_0 \leftarrow k_1$ and ADD[k_1] ← ADD[k_2], else
if $0 = i = L - 1$, clear the flag and delete the entries in the hash-table at hash-table address $h(k)$.

if NO, then $i \leftarrow i + 1$
}

Exercise 10.2 (Technical details for separate chaining)
Ex10.2a) Prove

$$v_{m,r}^{(0)} = \sum_{\underline{\ell} \in \Upsilon_{m,r}^{(0)}(\underline{i})} \binom{m}{\ell_0 \ell_1 \cdots \ell_{r-1}} \quad 1 \le r \le m \qquad \text{(Ex10.1.}a\text{)}$$

with boundary values $v_{m,0}^{(0)} = \begin{cases} 1 & \text{if } m = 0 \\ 0 & \text{if } m > 0 \end{cases}$. In addition, probe

$$v_{m,r} = \sum_{s=0}^{r-1} v_{m,r}^{(s)} \quad v_{m,r}^{(r)} \equiv 0 \qquad \text{(Ex10.1.}b\text{)}$$

$$v_{m,r}^{(s)} = \binom{r}{s} v_{m,r-s}^{(0)} \quad 0 \le s < r \qquad \text{(Ex10.1.}c\text{)}$$

and conclude

$$n^m = \sum_{r=0}^{m} \binom{n}{r} v_{m,r}^{(0)} \quad v_{n,m,0}^{(0)} = 0 \qquad \text{(Ex10.1.}d\text{)}$$

Ex10.2b) Prove

$$v_{m,r}^{(0)} = \sum_{s=0}^{r} \binom{r}{s} (-1)^s (r-s)^m = r!S_{m,r} \qquad \text{(Ex10.2)}$$

where $S_{m,r}$ is the Stirling number of the second kind (see Chapter I/1, §1.9 and Theorem 2.3 in Chapter I/2, §2.7).

Ex10.2c) Write a program to evaluate $v_{m,r}^{(0)}$ for $m = 1(1)10, r = 1(1)m$ and tabulate[1] the results.

Solutions (**Ex10.2a**): The multinomial coefficient $\binom{m}{\ell_0 \ell_1 \cdots \ell_{r-1}}$ is the number of partitions of m into r *non-negative* parts of sizes $(\ell_0\ \ell_1 \cdots \ell_{r-1})$(see Chapter I/1, §1.6 and [Feller 1957, pp. 36–37]). The formula in equation (Ex10.1.a) follows from the definition of $v_{m,r}^{(0}$ and $\Upsilon_{m,r}^{(0}$ in (Ex10.1.a, b).

Since every $\underline{\ell} \in \Upsilon_{m,r}$ contains s components ℓ_j equal to 0, for *some* s with $0 \le s \le r$, the formula in equation (Ex10.1.b) results.

If $\underline{\ell} \in \Upsilon_{m,r}$ contains s ℓ_j equal to 0, these components may be any of $\binom{r}{s}$ indices of ℓ_j proving the formula in equation (Ex10.1.c).

[1] $a = b(1)c$ is the tablemaker's notation indicating that the parameters of the value being tabulated range from $a = b$ to $a = c$ in steps of 1.

Finally, every (n,m) h-sequence $\underline{h}$ has r distinct names $\underline{i} = (i_0, i_1, \cdots, i_{r-1})$ for *some* r with $1 \le r \le m$. Since there are $\binom{n}{r}$ ways of choosing the names $\underline{i}$. The formula in equation (Ex10.1.*d*) completing the proof.

(**Ex10.2b**): In Chapter I/I, §1.10 (also see [Riordan 1968; Chapter 2, p. 43]), we gave an example of an **inverse relation**, an invertible transformation between sequences $\underline{x} = (x_0, x_1 \cdots)$ and $\underline{y} = (y_0, y_1 \cdots)$ involving binomial coefficients.

$$x_t = \sum_{s=0} (-1)^s \binom{t}{s} y_s \leftrightarrow y_t = \sum_{s=0}^{t} (-1)^s \binom{t}{s} x_s \qquad \text{(Ex10.3.}a\text{)}$$

If we set

$$x_s = (t-s)^m \quad y_s = (-1)^s v_{m,s}^{(0)} \qquad \text{(Ex10.3.}b\text{)}$$

this yields the relations

$$v_{m,r} = \sum_{s=0}^{r} v_{m,r}^{(s)} = \sum_{s=0}^{r} \binom{r}{s} v_{m,r-s}^{(0)} \leftrightarrow v_{m,r}^{(0)} = \sum_{s=0}^{r} \binom{r}{s} (-1)^s (r-s)^m \qquad \text{(Ex10.3.}c\text{)}$$

The proof of (Ex10.2.*b*) is completed by using Theorem 2.3 (in Chapter I/2) which asserts

$$S_{m,r} = \frac{1}{r!} \sum_{s=0}^{r} \binom{r}{s} (-1)^s (r-s)^m = \frac{1}{r!} v_{m,r}^{(0)} \qquad \text{(Ex10.3.}d\text{)}$$

Ex10.2c) Table Ex10.1 contains the values of $v_{m,r}^{(0)}$ for $m = 1(1)10$, $r = 1(1)m$.

Exercise 10.3. Prove

$$rS_{m,r} = \sum_{k=1}^{m-(r-1)} \binom{m}{k} (n)_r S_{m-k,r-1} \quad 1 \le k \le m-(r-1) \qquad \text{(Ex10.3)}$$

Exercise 10.4 (Deletions with coalesced chaining): Write a pseudocode describing a deletion of the hash-table entries of KEY = k if coalesced chaining as in Figures 10.4 and 10.5 is used. Assume the current CP value points to $j \ge 0$ cells bottom of cellar. Distinguish the cases in which CP is above and below the top of cellar.

PART II, CHAPTER 11

Exercise 11.1. Construct a perfect hash for `SUNDAY`, `MONDAY`, $\cdots$, `SATURDAY` dropping the suffix `DAY` and using the letters k_0 and k_1 of the key $\underline{k} = (k_0, k_1, \cdots, k_{L-1})$.

Solution. The letter counts for k_0 and k_1 are given in Table Ex11.1.

TABLE EX10.1. $\nu_{m,r}^{(0)}$ **for m = 1(1)10, r = 1(1)m**

		r →								
m↓	1	2	3	4	5	6	7	8	9	10
1	1									
2	1	2								
3	1	6	6							
4	1	14	36	24						
5	1	30	150	240	120					
6	1	62	540	1560	1800	720				
7	1	126	1806	8400	16800	15120	5040			
8	1	254	5796	40824	126000	191520	141120	40320		
9	1	510	18150	186480	834120	1905120	2328480	1451520	362880	
10	1	1022	55980	818520	5103000	16435440	29635200	30240000	16329600	3628800

TABLE EX11.1. Letter Counts

S	U	M	O	T	H	W	E	F	R	A
2	2	1	1	2	1	1	1	1	1	1

yield the initial modified Chicelli hash-value assignments with $f_1 = f_2$ = letter-counts shown in Table Ex11.2.

TABLE EX11.2. Variant of Cichelli's Initial Hashing-Value Assignment

k	L	$f_1(k_0)$	$f_1(k_1)$	$h^{(0)}(k)$
SUN	3	2	2	7
MON	3	1	1	5
TUES	4	2	2	8
WEDNES	6	1	1	8
THURS	5	2	1	8
FRI	3	1	1	5
SATUR	5	2	1	8

While there are several possible solutions, one choice for f_1 is shown as in Table Ex11.3 yielding the final hash-value assignments in Table Ex11.4.

TABLE EX11.3. f_1 Letter Values

$f_1(k)$										
S	U	M	O	T	H	W	E	F	R	A
1	1	1	1	1	1	3	0	1	0	2

TABLE EX11.4. Variant of Cichelli's Final Hashing-Value Assignment

k	L	$f_1(k_0)$	$f_1(k_1)$	$h(k)$
SUN	3	1	1	5
MON	3	0	0	3
TUES	4	1	1	6
WEDNES	6	3	0	9
THURS	5	1	1	7
FRI	3	1	0	4
SATUR	5	1	2	8

PART II, CHAPTER 12

Exercise 12.1. Prove that insertion of a the key k in a hash table of capacity n by uniform hashing can be described by the following *nonconstructive* algorithm:

procedure U-INSERT

1. Success := 0.
2. Repeat while {Success = 0};
   choose a random cell A with the uniform distribution $Pr\{A = a\} = \frac{1}{n}$.
   If empty, then set Success := 1 and insert k into hash-table address a.

end

Solution. The proof is by induction on m; the case $m = 1$ is obvious. Let Γ be the set of hash-table addresses at which the m keys $k_0, k_1, \cdots, k_{m-1}$ are inserted

$$\Gamma = \{x_0, x_1, \cdots, x_{m-1}\}$$

and suppose x_j is the address at which U-INSERT places key k_{m-1}.

$$\Gamma_j = \{x_0, x_1, \cdots, x_{m-1}\} - \{x_j\} \quad 0 \le j < m$$

By the induction hypothesis

$$Pr\{\Gamma_j\} = \frac{1}{\binom{n}{m-1}} \tag{Ex12.1.a}$$

so that

$$\begin{aligned} Pr\{\Gamma\} &= \sum_{j=0}^{m-1} Pr\{\Gamma_j\} \sum_{i=0}^{\infty} \left(\frac{m-1}{n}\right)^i \left(\frac{1}{n}\right) \\ &= \sum_{j=0}^{m-1} \frac{1}{\binom{n}{m-1}} \frac{1}{n-m+1} = \frac{1}{\binom{n}{m}} \quad \blacksquare \end{aligned} \tag{Ex12.1.b}$$

Exercise 12.2. Suppose keys $k_0, k_1, \cdots, k_{m-1}$ are inserted into a hash table of capacity n (keys) with uniform hashing.

E12.2a) Calculate the probability of the event $E(m, n)$: The insertion of the key k_{m-1} is collision free.

E12.2b) Let CF(m, n) be the total number of collision-free insertions when = m keys are inserted. Find a recursion for $p_r(m, n) \equiv Pr\{\text{CF}(m, n) = r\}$ for $1 \le r \le m \le n$.

E12.2c) Calculate the expectation $\bar{R}(m, n)$ of CF(m, n).

E12.2d) Calculate the generating function $\text{CF}(m, n, z) \equiv \sum_{r=1}^{m} p_r(m, n) z^r$ of the sequence $\{p_r(m, n)\}$.

E12.2e) Tabulate the values of $p_r(m, n)$ for $n = 1(1)10$, $m = 0(1)n$, $r = 0(1)m$.

E12.2f) Browse through the entries in the table just constructed and make some reasonable conjectures; for example, monotonicity in the parameters m, n, r. Even better, prove some interesting properties of $\{\rho_r(m, n)\}$.

Solution. Equation (Ex12.1.b) gives

$$Pr\{E(m,n)\} = \left(1-\frac{m}{n}\right) \quad \text{(Ex12.2.}a\text{)}$$

If $\chi_{\{E\}}$ is the **indicator function** of the event E; 1, if E is true and 0, otherwise, then

$$E\{\text{CF}(m,n)\} = \sum_{k=0}^{m-1} \chi\{E(k,n)\} \quad \text{(Ex12.2.}b\text{)}$$

$$E\{\text{CF}(m,n)\} = \sum_{k=0}^{m-1}\left(1-\frac{k}{n}\right) = m - \frac{m(m-1)}{n} \quad \text{(Ex12.2.}c\text{)}$$

For **Ex12.2b** we begin by computing a few values.

$$\rho_1(m,n) = \begin{cases} 1 & \text{if } m=1 \\ \frac{1}{n} & \text{if } m=2 \\ \frac{1}{n}\frac{2}{n} & \text{if } m=3 \\ \frac{1}{n}\frac{2}{n}\frac{3}{n} & \text{if } m=4 \end{cases} \quad \text{(Ex12.3.}a\text{)}$$

$$\rho_2(m,n) = \begin{cases} 0 & \text{if } m=1 \\ 1-\frac{1}{n} & \text{if } m=2 \\ \frac{1}{n}\left(1-\frac{2}{n}\right)+\frac{2}{n}\left(1-\frac{1}{n}\right) & \text{if } m=3 \\ \frac{1}{n}\frac{2}{n}\left(1-\frac{3}{n}\right)+\frac{1}{n}\frac{3}{n}\left(1-\frac{2}{n}\right)+\frac{2}{n}\frac{3}{n}\left(1-\frac{1}{n}\right) & \text{if } m=4 \end{cases} \quad \text{(Ex12.3.}b\text{)}$$

$$\rho_3(m,n) = \begin{cases} 0 & \text{if } m=1,2 \\ \left(1-\frac{1}{n}\right)\left(1-\frac{2}{n}\right) & \text{if } m=3 \\ \frac{1}{n}\left(1-\frac{2}{n}\right)\left(1-\frac{3}{n}\right)+\frac{2}{n}\left(1-\frac{1}{n}\right)\left(1-\frac{3}{n}\right)+\frac{3}{n}\left(1-\frac{1}{n}\right)\left(1-\frac{2}{n}\right) & \text{if } m=4 \end{cases}$$

(Ex12.3.c)

$$\rho_4(m,n) = \begin{cases} 0 & \text{if } m=1,2,3 \\ \left(1-\frac{1}{n}\right)\left(1-\frac{2}{n}\right)\left(1-\frac{3}{n}\right) & \text{if } m=4 \end{cases} \quad \text{(Ex12.3.}d\text{)}$$

Equations (Ex12.3.*a*–*d*) suggest that $\rho_r(m, n)$ is the product of $m-1$ terms, either $\frac{j}{n}$ or $1-\frac{j}{n}$ with $1 \le j < m$ yielding the explicit formula

$$\rho_r(m,n) = \frac{(m-1)!}{n^{m-1}} \sum_{1 \le j_1 < j_2 \cdots < j_{r-1} < m} \prod_{\ell=1}^{r-1} \frac{n-j_\ell}{j_\ell} \quad 1 \le r \le m \le n \qquad \text{(Ex12.4)}$$

To derive the recurrence, we note that there are r collision-free entries when the first m keys are inserted into the hash-table of size n, if either

- there are $r-1$ collision-free entries when the first $m-1$ hash-table entries are made *and* the key k_m is hashed to *none* of the already occupied $m-1$ cells resulting in an additional collision-free insertion, or
- there are r collision-free entries when the first $m-1$ hash-table entries are made *and* the key k_m is hashed to *one* of the already occupied $m-1$ cells.

This gives the recursion

$$\rho_r(m,n) = \begin{cases} 0 & \text{if } r = 0 \text{ or } r > m \\ 1 & \text{if } r = 0 \text{ and } 1 \le m \le n \\ \rho_{r-1}(m-1,n)\left(1-\frac{m-1}{n}\right) + \rho_r(m-1,n)\frac{m-1}{n} & \text{if } 0 \le r \le m \le n \end{cases} \qquad \text{(Ex12.5)}$$

To obtain the expectation $\bar{R}(m, n)$, we write

$$\begin{aligned} \bar{R}(m,n) &\equiv E\{\mathrm{CF}(m,n)\} \equiv \sum_{r=0}^{m} r\rho_r(m,n) \\ &= \sum_{r=0}^{m}\left[\left(1-\frac{m-1}{n}\right)(r-1+1)\rho_{r-1}(m-1,n) + \left(\frac{m-1}{n}\right) r\rho_r(m-1,n)\right] \\ &= \left\{\left(1-\frac{m-1}{n}\right)\sum_{r=1}^{m-1}(r-1+1)\rho_{r-1}(m-1,n)\right\} + \left\{\left(\frac{m-1}{n}\right)\sum_{r=0}^{m} r\rho_r(m-1,n)\right\} \\ &= \bar{R}(m-1,n)\left(1-\frac{m-1}{n}\right) \end{aligned}$$

leading to

$$\bar{R}(m,n) = \bar{R}(m-k,n) + \sum_{\ell=1}^{k}\left(1-\frac{m-\ell}{n}\right) \quad 1 \le k < n \qquad \text{(Ex12.6.}a\text{)}$$

Since $\bar{R}(1, n) = 1$, we finally obtain

$$\bar{R}(m,n) = 1 + \sum_{\ell=1}^{m-1}\left(1-\frac{m-\ell}{n}\right) = m + \frac{m-1}{2n} \qquad \text{(Ex12.6.}b\text{)}$$

Equations (Ex12.3.*a*–*d*) and the remarks following it yield the generating function of the sequence $\rho_1(m, n), \rho_2(m, n), \cdots, \rho_m(m, n)$

$$\mathrm{CF}(m,n,z) \equiv \sum_{r=0}^{m} \rho_r(m,n) z^r = z \frac{1}{n^{m-1}} \prod_{j=1}^{m-1} (j+(n-j)z) \qquad \text{(Ex12.7)}$$

The values of $\rho_r(m, n)$ for $n = 1(1)10$, $m = 0(1)n$, $r = 0(1)m$ are gives in Tables Ex12.1–2. The r-local maxima (maximum) in each row (m-value) is (are) underlined.

For **Ex12.2e**, the entries in Tables Ex12.1–2 suggest a number of properties of the sequences $\{\rho_r(m, n)\}$; for example, monotonicity and unimodality[2].

Property #1 (Symmetry): The relationship

$$\rho_r(n,n) = \rho_{n+1-r}(n,n) \quad 1 \le r \le n, 1 \le n < \infty \qquad \text{(Ex12.8)}$$

TABLE EX12.1. $\rho_r(m, n)$ $n = 1(1)6$, $m = 1(1)n$, $r = 1(1)m$

	r →					
	1	2	3	4	5	6
m↓	**n = 1**					
1	1.000000					
m↓	**n = 2**					
1	1.000000					
2	0.500000	0.50000				
m↓	**n = 3**					
1	1.000000					
2	0.333333	0.666667				
3	0.222222	0.555556	0.22222			
m↓	**n = 4**					
1	1.000000					
2	0.250000	0.750000				
3	0.125000	0.500000	0.375000			
4	0.093750	0.406250	0.406250	0.093750		
m↓	**n = 5**					
1	1.000000					
2	0.200000	0.800000				
3	0.080000	0.440000	0.480000			
4	0.048000	0.296000	0.464000	0.192000		
5	0.038400	0.246400	0.430400	0.246400	0.038400	
m↓	**n = 6**					
1	1.000000					
2	0.166667	0.833333				
3	0.055556	0.388889	0.555556			
4	0.027778	0.222222	0.472222	0.277778		
5	0.018519	0.157407	0.388889	0.342593	0.092593	
6	0.015432	0.134259	0.350309	0.350309	0.134259	0.015432

[2] A sequence $a_1, a_2, \cdots, a_m$ is **unimodal** if for some value r^*, it is monotonically *increasing* for $1 \le i < r^*$ and monotonically *decreasing* for $r^* \le i \le m$.

TABLE EX12.2. $\rho_r(m, n)$ $n = 6(1)10$, $m = 1(1)n$, $r = 1(1)m$

	r →									
	1	2	3	4	5	6	7	8	9	10
m↓	**n = 7**									
1	<u>1.000000</u>									
2	0.142857	<u>0.857143</u>								
3	0.040816	0.346939	<u>0.612245</u>							
4	0.017493	0.172012	<u>0.460641</u>	0.349854						
5	0.009996	0.105789	0.336943	<u>0.397334</u>	0.149938					
6	0.007140	0.078420	0.270899	<u>0.380080</u>	0.220622	0.042839				
7	0.006120	0.068237	0.243402	<u>0.364482</u>	0.243402	0.068237	0.006120			
m↓	**n = 8**									
1	<u>1.000000</u>									
2	0.125000	<u>0.875000</u>								
3	0.031250	0.312500	<u>0.656250</u>							
4	0.011719	0.136719	<u>0.441406</u>	0.410156						
5	0.005859	0.074219	0.289062	<u>0.425781</u>	0.205078					
6	0.003662	0.048584	0.208496	<u>0.374512</u>	0.287842	0.076904				
7	0.002747	0.037354	0.168518	<u>0.333008</u>	0.309509	0.129639	0.019226			
8	0.002403	0.033028	0.152122	<u>0.312447</u>	<u>0.312447</u>	0.152122	0.033028	0.002403		

TABLE EX12.2. *(Continued)*

	r →									
	1	2	3	4	5	6	7	8	9	10
m↓	**n = 9**									
1	1.000000									
2	0.111111	0.888889								
3	0.024691	0.283951	0.691358							
4	0.008230	0.111111	0.419753	0.460905						
5	0.003658	0.053955	0.248285	0.438043	0.256059					
6	0.002032	0.031601	0.161916	0.353706	0.336941	0.113804				
7	0.001355	0.021745	0.118478	0.289776	0.342529	0.188183	0.037935			
8	0.001054	0.017214	0.096982	0.251710	0.330806	0.222482	0.071323	0.008430		
9	0.000937	0.015418	0.088118	0.234518	0.322018	0.234518	0.088118	0.015418	0.000937	
m↓	**n = 10**									
1	1.000000									
2	0.100000	0.900000								
3	0.020000	0.260000	0.720000							
4	0.006000	0.092000	0.398000	0.50400						
5	0.002400	0.040400	0.214400	0.440400	0.302400					
6	0.001200	0.021400	0.127400	0.327400	0.371400	0.151200				
7	0.000720	0.013320	0.085000	0.247400	0.353800	0.239280	0.060480			
8	0.000504	0.009540	0.063496	0.198680	0.321880	0.273636	0.114120	0.018144		
9	0.000403	0.007733	0.052705	0.171643	0.297240	0.283285	0.146023	0.037339	0.003629	
10	0.000363	0.007000	0.048208	0.159749	0.284680	0.284680	0.159749	0.048208	0.007000	0.000363

is proved by starting with the generating function CF(n, n, z) (equation (Ex12.7)), replacing z by $1/z$ and then multiplying the result by z^n

$$\begin{aligned} \mathrm{CF}(n,n,z) &= z\frac{1}{n^{n-1}}\prod_{j=1}^{n-1}(j+(n-j)z) \\ &\overset{[z\to z^{-1}]}{\to} z^{-1}\frac{1}{n^n}\prod_{j=1}^{n-1}(j+(n-j)z^{-1}) \\ &\overset{[\times z^n]}{\to} z\frac{1}{n^n}\prod_{j=1}^{n-1}(jz+(n-j)) \end{aligned}$$

Property #2 (Monotonicity Relations):

MR1.

$$\rho_{r-1}(m-1,n) < \rho_r(m-1,n) \Leftrightarrow \rho_r(m,n) < \rho_r(m-1,n) \tag{Ex12.9.a}$$

If the inequality on the left-hand-side of (Ex12.7) holds, then equation (Ex12.2.4) yields

$$\begin{aligned} \rho_r(m,n) &= \rho_{r-1}(m-1,n)\left(1-\frac{m-1}{n}\right)+\rho_r(m-1,n)\frac{m-1}{n} \\ &< \rho_r(m-1,n)\left(1-\frac{m-1}{n}\right)+\rho_r(m-1,n)\frac{m-1}{n} \\ &= \rho_r(m-1,n) \end{aligned} \tag{Ex12.9.b}$$

shows the inequality on the right-hand-side of (Ex12.2.9a) holds. The relationship in (Ex12.2.9a) holds if the comparison relation < is replaced by any of the comparison relations ≤ = ≥ or >.

MR2. Subtracting the recursion of equation (Ex12.2.5) for r from $r + 1$ yields

$$\rho_{r+1}(m,n) = \rho_r(m-1,n)\left(1-\frac{m-1}{n}\right)+\rho_{r+1}(m-1,n)\frac{m-1}{n}$$

$$\rho_r(m,n) = \rho_{r-1}(m-1,n)\left(1-\frac{m-1}{n}\right)+\rho_r(m-1,n)\frac{m-1}{n}$$

and we may conclude

$$\begin{aligned} \rho_{r+1}(m,n)-\rho_r(m,n) &= \left(1-\frac{m-1}{n}\right)[\rho_r(m-1,n)-\rho_{r-1}(m-1,n)] \\ &\quad +\frac{m-1}{n}[\rho_{r+1}(m-1,n)-\rho_r(m-1,n)] \end{aligned} \tag{Ex12.10.a}$$

from which we may infer

$$\left\{\begin{array}{l} \rho_r(m-1,n)-\rho_{r-1}(m-1,n) \gtreqless 0 \\ \rho_{r+1}(m-1,n)-\rho_r(m-1,n) \gtreqless 0 \end{array}\right. \Rightarrow \rho_{r+1}(m,n)-\rho_r(m,n) \gtreqless 0 \tag{Ex12.10.b}$$

As in the case of MR1, the pair of comparisons relations $\gtrless$ can be replaced by several complementary pairs, for example $\gtrless$.

Property #3 (Monotonicity and Unimodality): The entries in Tables Ex12.1–2 appear to suggest that $\rho_1(m, n), \rho_2(m, n), \cdots, \rho_m(m, n)$

1. is *non*decreasing for $1 \le r \le m$ and $1 \le m \le n^*$;
2. is unimodal; strictly *in*creasing for $1 \le r \le r^*(m, n)$ and strictly *de*creasing for $r^*(m, n) \le r \le m$ for $1 \le m < n$;
3. unimodality continues to hold for $m = n$ except that for odd values of n, while for even values of n, the symmetry (Property #1) dictates two local maxima;
4. the position $r^*(m, n)$ of the (first) r-local maxima is *non*decreasing in m;

Outline of the Proof: Since $\rho_0(m, n) = 0$, equation (Ex12.2.4) yields

$$\rho_1(m,n) = \rho_1(m-1,n)\frac{m-1}{n} \qquad \text{(Ex12.11.}a\text{)}$$

$$\rho_2(m,n) = \rho_1(m-1,n)\left(1-\frac{m-1}{n}\right) + \rho_2(m-1,n)\frac{m-1}{n} \qquad \text{(Ex12.11.}b\text{)}$$

we can derive the formulae

$$\rho_2(m,n) = \gamma(m,n)\rho_1(m,n) \quad \gamma(m,n) > 1 \quad \text{for} \quad n \ge 2 \qquad \text{(Ex12.11.}c\text{)}$$

For example

$$\rho_2(2,n) = (n-1)\rho_1(2,n) > \rho_1(2,n)$$

$$\rho_2(3,n) = \left(\frac{3}{2}n-2\right)\rho_1(3,n) > \rho_1(3,n)$$

$$\rho_2(4,n) = \left(\frac{11}{6}n-3\right)\rho_1(4,n) > \rho_1(4,n)$$

$$\rho_2(5,n) = \left(\frac{25}{12}n-4\right)\rho_1(5,n) > \rho_1(5,n)$$

Equation (Ex12.11c) is proved by mathematical induction, using $0 = \rho_0(m,n) < \rho_1(m, n)$ and equation (Ex12.6) with $r = 1$. Thus, the entries in *all* rows of the $\rho_r(m,n)$-array increase, at least initially. We claim that they ultimately decrease; define

$$T(m,n) \equiv \frac{\rho_{m-1}(m,n)}{\rho_m(m,n)} \qquad \text{(Ex12.12.}a\text{)}$$

Using equation (Ex12.4)

$$T(m,n) = \sum_{j=1}^{m-1} \frac{j}{n-j} \qquad \text{(Ex12.12.}b\text{)}$$

TABLE EX12.3. T(m, n) m = 2(1)(n − 1) , n = 3(1)9

m →	2	3	4	5	6	7	8	9
n = 3	0.5000							
n = 4	0.3333	1.3333						
n = 5	0.2500	0.9167	2.4167					
n = 6	0.2000	0.7000	1.7000	3.7000				
n = 7	0.1667	0.5667	1.3167	2.6500	5.1500			
n = 8	0.1429	0.4762	1.0762	2.0762	3.7429	6.7429		
n = 9	0.1250	0.4107	0.9107	1.7107	2.9607	4.9607	8.4607	
n = 10	0.1111	0.3611	0.7897	1.4563	2.4563	3.9563	6.2897	10.2897

Table Ex12.3 lists the values of $T(m, n)$.

For example, assume for $n = 6$ that we have shown

$$\rho_1(2,6) < \rho_2(2,6) \qquad \text{(Ex12.13.}a\text{)}$$

$$\rho_1(3,6) < \rho_2(3,6) < \rho_2(3,6) \qquad \text{(Ex12.13.}b\text{)}$$

To determine the ordering of $\rho_1(4,6), \rho_2(4,6) \cdots, \rho_4(4,6)$, we begin by using equation (Ex12.12.c)

$$\rho_1(4,6) < \rho_2(4,6)$$

Applying MR2 to equation (Ex12.12.b) yields

$$\rho_1(4,6) < \rho_2(4,6) < \rho_3(4,6)$$

Finally, the $(m, n) = (4, 6)$-entry of 1.7000 in Table Ex12.3 is >1 which implies

$$\rho_1(4,6) < \rho_2(4,6) < \rho_3(4,6) > \rho_4(4,6) \qquad \text{(Ex12.13.}c\text{)}$$

To determine the ordering of $\rho_1(5,6), \rho_2(5,6) \cdots, \rho_5(5,6)$, we begin by using equation (Ex12.11.c)

$$\rho_1(5,6) < \rho_2(5,6)$$

Applying MR2 to equation (Ex12.12.c) yields

$$\rho_1(5,6) < \rho_2(5,6) < \rho_3(5,6)$$

The $(m, n) = (5, 6)$-entry of 3.7000 in Table Ex12.3 is >1 which implies

$$\rho_1(5,6) < \rho_2(5,6) < \rho_3(5,6)\ ?\ \rho_4(5,6) > \rho_5(5,6)$$

where the proper comparison relation ? is not yet known.

PART II, CHAPTER 13

Exercise 13.1. (Collision). Let $N_{m,n}$ be the number r of comparisons needed to locate the mth record when keys $\{k_0, k_1, \cdots, k_{m-1}\}$ are inserted into the hash table.

13.1a) Calculate the probability $Pr\{N_{m,n} = 0\}$ that the insertion of k_{m-1} is collision free.

13.1b) Calculate the expected number $\text{CF}_{m,n}$ of collision-free insertions experienced by the insertions of the m keys $\{k_0, k_1 \cdots, k_{m-1}\}$.

13.1c) Certainly, as the hash-table size n and the number of keys inserted m grows, $E\{\text{CF}_{m,n}\}$ will grow. Calculate the asymptotic growth as $n \to \infty$ of $E\{\text{CF}_{m,n}\}$ as $n, m \to \infty$ with $m \approx \mu n$.

13.1d) Write a recursion for $Pr\{\text{CF}_{m,n} = r\}$.

Solution. If the key k_{m-1} is assigned actual address $a_{m-1} = j$, the event $\{N_{m,n} = 0\}$ requires that keys $\{k_0, k_1, \cdots, k_{m-2}\}$ where inserted into the hash-table in such a way as to leave address j unoccupied. There are $T_{m-2,m-1} = \begin{cases} n-1 & \text{if } m=1 \\ (n-1)^{m-2} & \text{if } m \geq 2 \end{cases}$ hashing sequences which meet this condition. Thus, the probability of the event $\{N_{m,n} = 0\}$ conditioned by (the event) $\{a_{m-1} = j\}$ is

$$Pr\{N_{m,n} = 0/a_{m-1} = j\} = \left(1 - \frac{1}{n}\right)^{m-1} \qquad \text{(Ex13.1.}a\text{)}$$

Since there are n (equiprobable) choices for j

$$Pr\{N_{m,n} = 0\} = \left(1 - \frac{1}{n}\right)^{m-1} \qquad \text{(Ex13.1.}b\text{)}$$

for (Ex13.1.*a*), the expected number of collision-free insertions is

$$E\{\text{CF}_{m,n}\} = \sum_{j=1}^{m} \left(1 - \frac{1}{n}\right)^{j-1} = n\left\{1 - \left(1 - \frac{1}{n}\right)^{m-1}\right\} \qquad \text{(Ex13.1.}c\text{)}$$

If $n, m \to \infty$ with $m \approx \mu n$, then

$$\left\{1 - \left(1 - \frac{1}{n}\right)^{m-1}\right\} = m + \frac{m(m-1)}{2n}$$

so that

$$E\{\text{CF}_{m,n}\} \approx \mu n \qquad \text{(Ex13.1.}d\text{)}$$

The event $\{\text{CF}_{m,n} = r\}$ can occur in two (mutually exclusive) ways:

E_1: $\{\text{CF}_{m-1,n} = r\}$ and the insertion of k_{m-1} is *not* collision-free;

$$Pr\{E_1\} = Pr\{\mathrm{CF}_{m-1,n} = r\}\frac{m-1}{n}.$$

E_2: $\{\mathrm{CF}_{m-1,n} = r - 1\}$ and the insertion of k_{m-1} is collision-free;

$$Pr\{E_2\} = Pr\{\mathrm{CF}_{m-1,n} = r-1\}\left(1-\frac{m-1}{n}\right).$$

Therefore

$$Pr\{\mathrm{CF}_{m,n} = r\} = Pr\{\mathrm{CF}_{m-1,n} = r\}\frac{m-1}{n} + Pr\{\mathrm{CF}_{m-1,n} = r-1\}\left(1-\frac{m-1}{n}\right) \qquad \text{(Ex13.2)}$$

■

Exercise 13.2. If m keys $K = \{k_0, k_1, \cdots, k_{m-1}\}$ have been inserted into the hash table of n cells, calculate the probability of the event

> $\mathcal{O}_{[a,a+b)}$: the insertion has caused the contiguous block of $a + b$ cells $\mathrm{HT}_a, \cdots, \mathrm{HT}_{a+b-1}$ to be occupied where $0 \le a \prec b$ and $1 \le b < m$.

Solution. $\mathcal{O}_{[a,a+b)}$ requires that for some integers $r, s \ge 0$ with $\hat{b} \le m$ with $\hat{b} \equiv b + r + s$.

C1a. a set $\Gamma \subseteq K$ of $1 \le \hat{b}$ keys are inserted into the $\hat{b} + 1$ cells $\mathrm{HT}_{a-r}, \cdots, \mathrm{HT}_{a+b+s-1}$ leaving HT_{a+b+s} unoccupied, and

C1b. the remaining set $K - \Gamma$ of $m - \hat{b} \ge 0$ keys are inserted into the remaining $n - \hat{b} - 1$ cells $\mathrm{HT}_0, \cdots, \mathrm{HT}_{a-r-1}$ and $\mathrm{HT}_{a+b+s}, \cdots, \mathrm{HT}_{n-1}$ leaving cells HT_{a-r-1} and HT_{n-1} unoccupied.

as depicted in Figure Ex13.1, assuming the parameters (a, b, r, s) satisfy $r \le a < a + b + s < n$.

The number of ways such an occupancy configuration can occur is the product of three terms:

1a. the number $\binom{m}{\hat{b}}$ of possible subsets $\Gamma \subset K$;

1b. the number $(\hat{b} + 1)^{\hat{b}-1}$ of hashing sequences $\underline{h}(\Gamma)$ for the $\hat{b}$ keys in Γ into $\hat{b} + 1$ cells leaving the rightmost cell HT_{a-r-1} unoccupied;

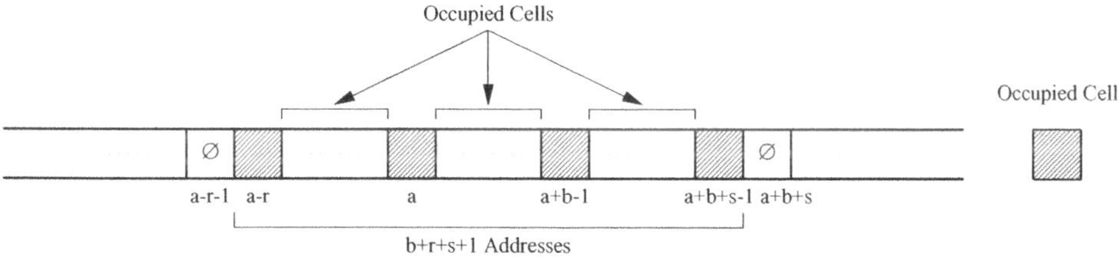

Figure Ex13.1. The Contiguous Block of b Cells $\mathrm{HT}_a, \cdots, \mathrm{HT}_{b-1}$ Occupied.

2. the number $(n - \hat{b} - 1)^{m-\hat{b}-1}(n - m)$ of hashing sequences $\underline{h}(K - \Gamma)$ for the remaining $m - \hat{b}$ keys in $K - \Gamma$ into the remaining $n - \hat{b} - 1$ cells leaving the rightmost cell unoccupied.

so that

$$Pr\{\mathcal{O}_{[a,a+b]}\} = \frac{1}{n^m} \sum_{\substack{r,s\geq 0 \\ r+s\leq m-b}} \binom{m}{\hat{b}} (\hat{b}+1)^{\hat{b}-1} (n-\hat{b}-1)^{m-\hat{b}-1} (n-m-1)$$
$$= \frac{m-b+1}{n^m} \sum_{t=0}^{m-b} \binom{m}{b+t} (b+t+1)^{b+t-1} (n-b-t-1)^{m-b-t-1} (n-m-1) \quad \text{(Ex13.3)}$$

■

Exercise 13.3. If m keys $K = \{k_0, k_1, \cdots, k_{m-1}\}$ have been inserted into the hash table of n cells, calculate the probability of the event

$\mathcal{U}_{[a,a+b)}$: the contiguous block $HT_a, \cdots, HT_{a+b-1}$ of exactly b cells are unoccupied with $0 < a \prec b$ and $1 \leq b < n - m$.

Solution. The event $\mathcal{U}_{[a,a+b)}$ requires the m keys to be inserted into the remaining $n - b$ cells $HT_0, \cdots, HT_{a-1}$ and $HT_{a+b}, \cdots, HT_{n-1}$ leaving cell HT_{a-1} unoccupied.

Applyiong *Hash-Table Decomposition Principle* and Equation (7.4) yields

$$Pr\{\mathcal{U}_{[a,a+b)}\} = \frac{(n-b)^{m-1}(n-m-b)}{n^m} = \left(1 - \frac{b}{n}\right)^{m-1}\left(1 - \frac{m+-b}{n}\right) \quad \text{(Ex13.4)}$$

■

Exercise 13.4. How does $Pr\{\mathcal{U}_{[a,a+b)}$ change if $\mathcal{U}_{[a,a+b)}$ is replaced by

$\overline{\mathcal{U}}_{[a,a+b)}$: the contiguous block of at least b cells including $HT_a, \cdots, HT_{b-1}$ are unoccupied with a $\prec b$ and $1 \leq b \leq n - m$?

Exercise 13.5. Prove that the distribution of the number of probes for SEARCH(k) is the same for each $c \geq 1$, which is relatively prime to the table length n.

Solution. Let h_c be the hashing function $h_c(k) = ch_1(k)$ (*modulo*); then, the sequence of comparisons for any sequence of keys $K_0, K_1, \cdots, K_j$ are related as follows

$$h_1(k_0), h_1(k_1), \cdots, h_1(k_j) \xrightarrow{h_1} a_0, a_1, \cdots, a_j$$

$$a_j = (h_1(k_j) + \delta_j)(modulo\ n) \Leftrightarrow (h_1(k_j) + \ell)(modulo\ n) \begin{cases} \in \{a_0, a_1, \cdots, a_{j-1}\} & \text{if } 0 \leq \ell < \delta_j \\ \notin \{a_0, a_1, \cdots, a_{j-1}\} & \text{if } \ell = \delta_j \end{cases}$$

$$h_c(k_0), h_c(k_1), \cdots, h_c(k_j) \xrightarrow{h_c} \tilde{a}_0, \tilde{a}_1, \cdots, \tilde{a}_j$$

$$\tilde{a}_j = (h_c(k_j) + \tilde{\delta}_j c)(modulo\ n) \Leftrightarrow (h_c(k_j) + \tilde{\ell} c)(modulo\ n) \begin{cases} \in \{\tilde{a}_0, \tilde{a}_1, \cdots, \tilde{a}_{j-1}\} & \text{if } 0 \leq \ell < \tilde{\delta}_j \\ \notin \{\tilde{a}_0, \tilde{a}_1, \cdots, \tilde{a}_{j-1}\} & \text{if } \tilde{\ell} = \tilde{\delta}_j \end{cases}$$

Mathematical induction shows that

$$\tilde{a}_i = ca_i \quad 0 \le i \le j$$

■

Exercise 13.6. (Technical preparation). Suppose $K = \{k_0, k_1, \cdots, k_{N-1}\}$ is a set of keys with with equal hashing value; that is, $Pr\{h(K_i) = j\}$ is independent of i. The repeated use of SEARCH(k) with $k \in K$ may determine a chain of cells (of length ≥ 4) where we have suppressed the subscript 0 on L_0.

$$\mathcal{C}_{h(k)} : \mathrm{HT}_{h(k)} \xrightarrow{\mathrm{L}} \mathrm{HT}_{\mathrm{LINK}[h(k)]} \xrightarrow{\mathrm{L}} \mathrm{HT}_{\mathrm{LINK}^2[h(k)]} \xrightarrow{\mathrm{L}} \mathrm{HT}_{\mathrm{LINK}^3[h(k)]}$$

For example with $N = 4$, if the keys (k_0, k_1, k_2, k_3) have the same hashing value $h(k_0) = h(k_1) = h(k_2) = h(k_3) = h(k)$ (say) and are inserted in the order $\underline{x} = (3, 1, 0, 2)$, then they produce a single chain with

k_3 is inserted in $\mathrm{HT}_{h(k)}$	k_1 is inserted in $\mathrm{HT}_{\mathrm{LINK}[h(k)]}$
k_0 is inserted in $\mathrm{HT}_{\mathrm{LINK}^2[h(k)]}$	k_2 is inserted in $\mathrm{HT}_{\mathrm{LINK}^3[h(k)]}$

The position of a particular k^* in $\mathcal{C}_{h(k)}$ may change as a result of relinking. We model these changes by a discrete-time stationary Markov chain $\mathcal{M}_N = \{\underline{X}_n : n = 0, 1, \cdots\}$ where $\underline{X}_n = (x_0, x_1, \cdots, x_{N-1})$, which is a permutation of the integers $0, 1, \cdots, N-1$.

Let $\mathcal{M}_N = \{X_n : n = 0, 1, \cdots\}$. We imagine $\mathcal{M}_N$ to be a synchronous time process whose states are observed at times $0, \tau, 2\tau, \cdots$, at which point SEARCH* is carried out; let q_j be the probability that SEARCH*(k_j) is requested at time $n\tau$.

The rules governing changes in the table state $\underline{x}$ are as follows:

- If at time n, (SEARCH*(k_{x_0}) is requested, $\underline{x}$ remains unchanged, with probability q_{x0}

$$Pr\{\underline{X}_n = \underline{x}/\underline{X}_{n-1} = \underline{x}\} = q_{x0}$$

- If at time n, (SEARCH*(k_{x_j}) for $0 < j < N$ is requested, the state makes the transition

$$Pr\{\underline{X}_n = \underline{x}_{[j-1\leftrightarrow j]}/\underline{X}_{n-1} = \underline{x}\} = q_{x_j}$$

$\underline{x} = (x_0, x_1, \cdots, x_{j-1}, x_j, \cdots, x_{N-1}) \rightarrow \underline{x}_{[j-1\leftrightarrow j]} = (x_0, x_1, \cdots, x_j, x_{j-1}, \cdots, x_{N-1})$ where $\underline{x}_{[j-1\leftrightarrow j]} : (\cdots, x_j, x_{j-1}, \cdots) \rightarrow (\cdots, x_{j-1}, x_j, \cdots)$ indicates that the $(j-1)^{\mathrm{st}}$ and j^{th} components of $\underline{x}$ are interchanged for $1 \le j < N$.

The probabilities $\{q_j\}$ determine the simplex

$$\mathcal{S}_N : q_j \ge 0 \quad 1 = q_0 + q_1 + \cdots + q_{N-1}$$

Ex13.4a) Construct the state transition diagrams of $\mathcal{M}_N$ for $N = 2, 3$ and conclude that $\mathcal{M}_N$ is irreducible for $N = 2, 3$.

Ex13.4b) Prove that $\mathcal{M}_N$ is irreducible for general $N \geq 2$.

Ex13.4c) Write the backward equations for $\mathcal{M}_N$ for $N = 2, 3$.

Ex13.4d) Write the backward equations for $\mathcal{M}_N$ with general N.

Ex13.4e) Show that

$$\pi_{x_0,x_1,\cdots,x_{N-1}}(N) = \frac{Q_{\underline{x}}}{\sum_{\underline{y}\in \mathcal{P}_N} Q_{\underline{y}}} \quad Q_{\underline{x}} = \prod_{j=0}^{n-1} q_{x_j}^{n-j-1} \tag{A}$$

$$\frac{\pi_{(x_0,x_1,\cdots,x_{j-2},x_{j-1},x_j,x_{j+1},\cdots,x_{N-1})}}{\pi_{(x_0,x_1,\cdots,x_{j-2},x_j,x_{j-1},x_{j+1},\cdots,x_{N-1})}} = \frac{q_{x_{j-1}}}{q_{x_j}} \quad 1 \leq j < N \tag{B}$$

Ex13.4f) Use the results in **13.4e** to show that $\pi_{x_0,x_1,\cdots,x_{N-1}}(N)$ given by equation (A) is the solution of the backward equation – the equilibrium probabilities for $\mathcal{M}_N$.

Solution. Figure Ex13.2 shows the state transitions for $N = 2, 3$. next.
from which it is easily seen that $\mathcal{M}_N$ is irreducible for $N = 2, 3$. proving **Ex13.4a**. Since the transpositions

$$(x_0, x_1, \cdots, x_{j-2}, x_{j-1}, x_j, x_{j+1}, \cdots, x_{N-1}) \to (x_0, x_1, \cdots, x_{j-2}, x_j, x_{j-1}, x_{j+1}, \cdots, x_{N-1})$$

generate $\mathcal{M}_N$ (see [Ledermann 1953]), the transitions in $\mathcal{M}_N$ implies $Pr\{\underline{x}_{n+k}/\underline{x}_n\} > 0$ for all states $\underline{x}$ for k large enough proving **Ex13.4b**. It is not difficult to show that $\mathcal{M}_N$ is aperiodic.

For **Ex13.4c** we write

$\underline{N = 2}$

$$\begin{pmatrix} \pi_{0,1} \\ \pi_{1,0} \end{pmatrix} = \begin{cases} q_0\pi_{0,1} + q_0\pi_{1,0} \\ q_1\pi_{1,0} + q_1\pi_{0,1} \end{cases} \tag{Ex13.5.a}$$

with solution

$$\pi_{0,1} = q_0 \quad \pi_{1,0} = q_1 \tag{Ex13.5.b}$$

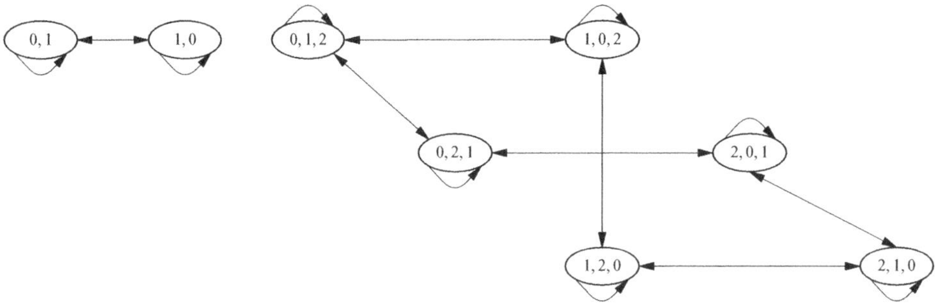

Figure Ex13.2. State Transitions for N = 2, 3.

$\underline{N=3}$

$$\begin{pmatrix} \pi_{0,1,2} \\ \pi_{0,2,1} \\ \pi_{1,0,2} \\ \pi_{1,2,0} \\ \pi_{2,0,1} \\ \pi_{2,1,0} \end{pmatrix} = \begin{cases} q_0\pi_{0,1,2} + q_0\pi_{1,0,2} + q_1\pi_{0,2,1} \\ q_0\pi_{0,2,1} + q_0\pi_{2,0,1} + q_2\pi_{0,1,2} \\ q_1\pi_{1,0,2} + q_1\pi_{0,1,2} + q_0\pi_{1,2,0} \\ q_1\pi_{1,2,0} + q_1\pi_{2,1,0} + q_2\pi_{1,0,2} \\ q_2\pi_{2,0,1} + q_2\pi_{0,2,1} + q_0\pi_{2,1,0} \\ q_2\pi_{2,1,0} + q_2\pi_{1,2,0} + q_1\pi_{2,0,1} \end{cases} \tag{Ex13.6.a}$$

Equation (Ex13.6.*a*) correspond to the matrix equation

$$\underline{\pi}_{\underline{x}} = \mathbf{P}\,\underline{\pi}_{\underline{x}} \qquad \mathbf{P} \equiv \begin{array}{c} \\ 0,1,2 \\ 0,2,1 \\ 1,0,2 \\ 1,2,0 \\ 2,0,1 \\ 2,1,0 \end{array} \begin{array}{c} \begin{array}{cccccc} 0,1,2 & 0,2,1 & 1,0,2 & 1,2,0 & 2,0,1 & 2,1,0 \end{array} \\ \begin{pmatrix} q_0 & q_1 & q_0 & 0 & 0 & 0 \\ q_2 & q_0 & 0 & 0 & q_0 & 0 \\ q_1 & 0 & q_1 & q_0 & 0 & 0 \\ 0 & 0 & q_2 & q_1 & 0 & q_1 \\ 0 & q_2 & 0 & 0 & q_2 & q_0 \\ 0 & 0 & 0 & q_2 & q_1 & q_2 \end{pmatrix} \end{array} \tag{Ex13.6.b}$$

with

$$\underline{\pi}_{\underline{x}} = (\pi_{0,1,2}, \pi_{0,2,1}, \pi_{1,0,2}, \pi_{1,2,0}, \pi_{2,0,1}, \pi_{2,1,0}) \tag{Ex13.6.c}$$

For **Ex13.4d** we note that the states which make a transition to $\pi_{x_0,x_1,\cdots,x_{j-2},x_{j-1},x_j,x_{j+1},\cdots,x_{N-1}}$ are

$$\begin{array}{c} \pi_{(x_0,x_1,\cdots,x_{j-2},x_j,x_{j-1},x_{j+1},\cdots,x_{N-1})} \\ \downarrow \qquad\qquad 0 \le j < N \\ \pi_{(x_0,x_1,\cdots,x_{j-2},x_{j-1},x_j,x_{j+1},\cdots,x_{N-1})} \end{array}$$

Summing over these states yields the backward equation

$$\begin{aligned} \pi_{(x_0,x_1,x_2,\cdots,x_{N-1})} = {} & q_{x_0}\left[\pi_{(x_0,x_1,x_2,\cdots,x_{N-1})} + \pi_{(x_1,x_0,x_2,\cdots,x_{N-1})}\right] \\ & + \sum_{j=2}^{N-2} q_{x_j}\pi_{(x_0,x_1,\cdots,x_j,x_{j-1},x_{j+1},\cdots,x_{N-1})} \end{aligned} \tag{Ex13.7}$$

If

$$\pi_{x_0,x_1,\cdots,x_{N-1}}(N) = \frac{Q_{\underline{x}}}{\sum_{\underline{y}\in\mathcal{P}_N} Q_{\underline{y}}} \tag{Ex13.8.a}$$

with

$$Q_{\underline{x}} = \prod_{j=0}^{n-1} q_{x_j}^{n-j-1} \tag{Ex13.8.b}$$

then

$$\frac{\pi_{(x_0,x_1,\cdots,x_{j-2},x_{j-1},x_j,x_{j+1},\cdots,x_{N-1})}}{\pi_{(x_0,x_1,\cdots,x_{j-2},x_j,x_{j-1},x_{j+1},\cdots,x_{N-1})}} = \frac{q_{x_{j-1}}}{q_{x_j}} \quad 1 \le j < N \tag{Ex13.8.c}$$

If

$$\underline{x} = (x_0, x_1, \cdots, x_{j-2}, x_{j-1}, x_j, x_{j+1}, \cdots, x_{N-1}) \tag{Ex13.8.d}$$

$$\underline{x}_{[j-1 \leftrightarrow j]} = (x_0, x_1, \cdots, x_{j-2}, x_j, x_{j-1}, x_{j+1}, \cdots, x_{N-1}) \tag{Ex13.8.e}$$

then

$$q_{x_j Q(\underline{x}_{[j-1 \leftrightarrow j]})} = q_{x_{j-1}} Q(\underline{x}) \tag{Ex13.8.f}$$

which proves **Ex13.4e,f.**

To prove **Ex13.2g**, note that if

$$\text{Num}(\underline{x}) = Q_{\underline{x}} \quad \text{Denom} = \sum_{\underline{y} \in \mathcal{M}_N} Q_{\underline{y}} \tag{Ex13.9.a}$$

$$\pi_{\underline{x}} = \frac{\text{Num}(\underline{x})}{\text{Denom}} \tag{Ex13.9.b}$$

$$X = \sum_{\underline{x} \in \mathcal{M}_N} \left\{ \sum_{j=0}^{N-1} (j+1) q_{x_j} \right\} \pi_{\underline{x}} \tag{Ex13.10.a}$$

$$Y = \sum_{\underline{x} \in \mathcal{M}_N} \left\{ \sum_{j=0}^{N-1} (N-j) q_{x_j} \right\} \pi_{\underline{x}} \tag{Ex13.10.b}$$

then

$$X + Y = (N+1) \tag{Ex13.10.c}$$

so that

$$\sum_{\underline{x} \in \mathcal{M}_N} \pi_{\underline{x}} \left\{ \sum_{j=0}^{N-1} (j+1) q_{x_j} \right\} \le \frac{N+1}{2} \Leftrightarrow \sum_{\underline{x} \in \mathcal{M}_N} \pi_{\underline{x}} \sum_{j=0}^{N-1} (N-j) q_{x_j} \ge \frac{N+1}{2} \tag{Ex13.11}$$

which completes the proof. ■

PART II, CHAPTER 14

Exercise 14.1. Suppose m keys are inserted into a hash table of n cells by uniform hashing, n is a prime, and (i, c) satisfy $0 \le i < n, 1 \le c < n$.

14.1a) Prove if $r < m$, then the probability $p_{m,n}(r)$ of the event

- Uniform hashing inserts keys into all the cells HT_i, HT_{i-c}, $\cdots$, $HT_{i-(r-2)c}$.
- Uniform hashing does not insert a key into the cell $HT_{i-(r-1)c}$ is

$$p_{m,n}(r) = \frac{m}{n}\frac{m-1}{n-1}\cdots\frac{m-(r-2)}{n-(r-2)}\left(1-\frac{m-(r-1)}{n-(r-1)}\right)$$

14.1b) Prove $p_{m,n}(r) \to \mu^{r-1}(1-\mu)$ if $n, m \to \infty$ with $m = \mu n$ and $0 < \mu < 1$.

14.1c) Conclude that the number of probes using double or uniform hashing to insert a key into the hash table created by uniform hashing have the same probability distribution as $n \to \infty$.

Solution. The set of progressions

$$B(1) = \{C_i(\ell, c): HT_{i+\ell c} \to \cdots \to HT_{i+2c} \to HT_{i+c} : i + jc \notin \{1, 2, \cdots, \ell\}, 1 \le j \le \ell\}$$

are those independent of $C_i(l, 1)$, while for $1 \le d < n$

$$B(d) = \{C_i(\ell, c): HT_{i+\ell c} \to \cdots \to HT_{i+2c} \to HT_{i+c} : i + jc \notin \{d, 2d, \cdots, \ell d\}, 1 \le j \le \ell\}$$

are those independent of $C_i(l, d)$. When n is a prime

$$c \in B(1) \Leftrightarrow cd \in B(d)$$

proving that for each destination

- each of the r sets of independent strides contain the same number of progressions (vertices), and
- this number is the same for each destination. ■

Exercise 14.2. Prove that ~ defined on the progressions $C_i(\ell, c)$ in $\mathcal{G}(i, n, \ell)$ by $c_1 \sim c_2$ if and only if $C_i(\ell, c_1) \cap C_i(\ell, c_2) = \emptyset$. Is an equivalence relation?

Solution. Since ~ determines a partition of the set of progressions of length ℓ to HT_i, it follows that ~ is an equivalence relation. ■

Exercise 14.3 (Monotonicity of $\succ$). Let $N_i(\underline{x})$ be the number of chains of any length or stride leading to cell HT_i if the hash table state is $\underline{x}$. Prove that $\underline{x}_1 \succ \underline{x}_2$ implies $N(\underline{x}_1) \ge N(\underline{x}_2)$.

Solution. If $\underline{x}_1 = (x_{1,0}, x_{1,1}, \cdots, x_{1,n-1}) \succ \underline{x}_2 = (x_{2,0}, x_{2,1}, \cdots, x_{2,n-1})$and

$$C_i(\ell, c): HT_{i+\ell c} \to HT_{i+(\ell-1)c} \to \cdots \to HT_{i+c}$$

is a progression leading to cell HT_i for the table state $\underline{x}_2$, then $x_{2,j} = 1$ for $j \in \{i + lc, i + (l-1)c, \cdots, i + c\}$.

Since $\underline{x}_1 \succ \underline{x}_2$, we have $x_{1,j} = 1$ for $j \in \{i + \ell c, i + (\ell - 1)c, \cdots, i + c\}$so that $\mathcal{C}_i(\ell, c)$ is a progression leading to cell HT_i for the table state $\underline{x}_1$, proving $N(\underline{x}_1) \geq N(\underline{x}_2)$. ■

PART II, CHAPTER 15

Exercise 15.1. Prove that for a fixed cell HT_i, an $(m - d, n)$ scenario $\underline{h} = (h_0, h_1, \cdots, h_{m-d-1})$ is a skeleton if and only if $h_j \in \mathcal{P}'[\underline{h}_{[0,j)}]$ for $0 \leq j < m - d$.

Exercise 15.2. Prove that if X_0, X_1 are independent random variables with the Poisson distributions

$$Pr\{X_i \leq s\} = e^{-\lambda_i} \sum_{j=0}^{s} \frac{\lambda_i^j}{j!} \quad i = 0, 1, 0 \leq s < \infty \tag{Ex15.1.a}$$

then

$$Pr\{X_0 + X_1 \leq s\} = e^{-(\lambda_0+\lambda_1)} \sum_{j=0}^{s} \frac{(\lambda_0 + \lambda_1)^j}{j!} \quad 0 \leq s < \infty \tag{Ex15.1.b}$$

Solution. Equation (Ex15.1.*a*) and independence gives

$$Pr\{X_0 + X_1 \leq s\} = e^{-(\lambda_0+\lambda_1)} \sum_{j=0}^{s} \frac{\lambda_0^s}{j!} \frac{\lambda_1^{j-s}}{j-s!} \tag{Ex15.2}$$

Equation (Ex15.1.*b*) is an immediate consequence of equation (Ex15.1.*a*) and the Binomial Theorem (Chapter I/1, Theorem 1.2). ■

Exercise 15.3. If $|z| \leq 1$, $z \neq -1$, then

$$\sum_{j=0}^{r} z^j \leq \sum_{j=0}^{r} \frac{\left(\log \frac{1}{1-z}\right)^j}{j!} \quad 0 \leq r < \infty \tag{Ex15.3.a}$$

Solution. Since $\sum_{j=0}^{\infty} z^j = \frac{1}{1-z}$ and $\exp(\log 1/(1-z)) = \frac{1}{1-z}$, we have

$$\sum_{j=0}^{\infty} z^j = \sum_{j=0}^{\infty} \frac{\left(\log \frac{1}{1-z}\right)^j}{j!}$$

Writing the McLauren series expansion [Abramowitz and Stegun 1972, p.68] of $\log \frac{1}{1-z}$

$$\log\frac{1}{1-z}=z+\frac{1}{2}z^2+\frac{1}{3}z^3+\cdots=z\left(1+\frac{1}{2}z+\frac{1}{3}z^2+\cdots\right)\quad |z|\le 1,\ z\ne -1$$

$$\frac{\left(\log\frac{1}{1-z}\right)^j}{j!}=\frac{z^j}{j!}\left(1+\frac{1}{2}z+\frac{1}{3}z^2+\cdots\right)^j \qquad \text{(Ex15.3.}b\text{)}$$

so that

$$\sum_{j=0}^{\infty}z^j=\sum_{j=0}^{\infty}\frac{\left(\log\frac{1}{1-z}\right)^j}{j!}=\sum_{j=0}^{\infty}\frac{z^j}{j!}\left(1+\frac{1}{2}z+\frac{1}{3}z^2+\cdots\right)^j$$

Since the left-hand side of equation (Ex15.3.*a*) is a polynomial in z of degree r and the coefficients of

$$\sum_{j=r+1}^{\infty}\frac{z^j}{j!}\left(1+\frac{1}{2}z+\frac{1}{3}z^2+\cdots\right)^j$$

are positive, the bound in equation (Ex15.3*a*) is proved. ■

Exercise 15.4. Prove the sequences $\{E_d(x)\}$ and $\{f_d(x)\}$ given by

$$E_d(x)=\sum_{t=d}^{\infty}\frac{\lambda^t}{t!}\quad 0\le d,x<\infty,\quad \lambda=\log(1-x) \qquad \text{(Ex15.4.}a\text{)}$$

$$f_d(x)=\lambda-e^{-\lambda}\sum_{t>d}(t-d)\frac{\lambda^t}{t!}\quad 0\le d,x<\infty,\quad \lambda=\log(1-x) \qquad \text{(Ex15.4.}b\text{)}$$

are related by

$$f_d(x)=\lambda-e^{-\lambda}[\lambda E_d(x)-dE_{d+1}(x)]\quad d=0,1,\cdots\ \lambda=-\log(1-x) \qquad \text{(Ex15.4.}c\text{)}$$

Solution

$$\begin{aligned} f_d(x)&=\lambda-e^{-\lambda}\sum_{t>d}(t-d)\frac{\lambda^t}{t!}\\ &=\lambda-e^{-\lambda}\left(\sum_{t>d}\frac{\lambda^t}{(t-1)!}-d\sum_{t\le q}\frac{\lambda^t}{t!}\right)\\ &=\lambda-e^{-\lambda}(\lambda E_d(x)-dE_{d+1}(x)) \end{aligned}$$

Exercise 15.5. Prove that $f_d(x)=\lambda-\frac{\lambda^{d+1}}{(d+1)!}+o(\lambda^{d+2})$ for $d\ge 1$.

Solution. Start with equation (Ex15.4*c*)

$$f_d(x)-\lambda = e^{-\lambda}[dE_{d+1}(x)-\lambda E_d(x)]$$
$$= e^{-\lambda}\left\{d\left(\frac{\lambda^{d+1}}{(d+1)!}+\frac{\lambda^{d+2}}{(d+2)!}+\cdots\right)-\left(\frac{\lambda^{d+1}}{d!}+\frac{\lambda^{d+2}}{(d+1)!}+\cdots\right)\right\}$$

Since $e^{-\lambda} = 1-\lambda+\frac{1}{2!}\lambda^2-\frac{1}{3!}\lambda^3+\cdots+(-1)^t\frac{1}{t!}\lambda^t-(-1)^t\cdots$, we have

$$f_d(x)-\lambda = d\frac{\lambda^{d+1}}{(d+1)!}-\frac{\lambda^{d+1}}{d!}+o(\lambda^{d+2}) = -\frac{\lambda^{d+1}}{(d+1)!}$$

■

PART III, CHAPTER 16

Exercise 16.1. Use Lemma 5.2a (Chapter 11) to prove $\prod_{p\in\mathcal{P}_n} p > 2^n$ if $n \geq 49$.

Proof. Since

$$e^{n-2.05282\sqrt{n}} \geq 2^n \Leftrightarrow \sqrt{n} \geq \frac{2.05282}{1-\log 2} \approx \frac{2.05282}{0.306852} \approx 6.6899352$$

Lemma 5.2a implies that

$$\prod_{p\in\mathcal{P}_n} p > 2^n \quad n \geq 49 \quad ■$$

Exercise 16.2. Tabulate $n, \pi(n), 2^n$ and $Q_n \equiv \prod_{p\in\mathcal{P}_n} p$ for $n = 29(1)48$ to prove $\prod_{p\in\mathcal{P}_n} p > 2^n$ if $29 \leq n \leq 48$.

Proof

TABLE Ex16.1. (n, π(n), 2ⁿ, Qₙ) for n = 29(1)48

n	Q_n	2^n
29	6,915,878,970	268,435,456
30	6,915,878,970	536,870,912
31	21,4392,248,070	1,073,741,824
32	21,4392,248,070	2,147,483,648
33	21,4392,248,070	4,294,967,296
34	21,4392,248,070	8,589,934,592
35	21,4392,248,070	17,179,869,184
36	21,4392,248,070	34,359,738,368
37	7,932,513,178,590	68,719,476,736
38	7,932,513,178,590	137,438,953,472
39	7,932,513,178,590	274,877,906,944
40	7,932,513,178,590	549,755,813,888
41	325,233,040,322,190	1,099,511,627,776
42	325,233,040,322,190	2,199,023,255,552

TABLE Ex16.1 (*Continued*)

n	Q_n	2^n
43	13,985,020,733,854,170	4,398,046,511,104
44	13,985,020,733,854,170	8,796,093,022,208
45	13,985,020,733,854,170	17,592,186,044,416
46	13,985,020,733,854,170	35,184,372,088,832
47	657,295,974,491,145,990	70,368,744,177,664
48	657,295,974,491,145,990	140,737,488,355,328

■

Exercise 16.3. Prove that if $n \geq 29$ and $\eta \leq 2^n$, then η has fewer that $\pi(n)$ different prime divisors.

Proof. Let $p_1, p_2, \cdots, p_r$ be the distinct prime divisors of η and assume that $r \geq \pi(\eta)$. If $\eta \leq 2^n$, then $2^n \geq p_1 p_2 \cdots p_r$. If, on the contrary, we assume $\pi(\eta) > \pi(n)$, then

$$2^n \geq p_1 p_2 \cdots p_r = p_1 p_2 \cdots p_{\pi(\eta)} \geq p_1 p_2 \cdots p_{\pi(n)} = \prod_{p \in \mathcal{P}_n} p$$

Finally, Lemma 1.1a gives $\prod_{p \in \mathcal{P}_n} p > 2^n$ which provides the contradiction $2^n > 2^n$ completing the proof. ■

Exercise 16.4. Prove that if $\underline{y}$ and $\underline{z}$ are (0,1)-strings of the same length, then $\mathbf{K}_{\underline{y}} = \mathbf{K}_{\underline{z}}$ if and only if $\underline{y} = \underline{z}$.

Query: Did we need to add the quantifier "· · · of the same length"?

Proof. If $n \geq 1$, $\underline{y} = (y_0, y_1, \cdots, y_n)$ and $\underline{z} = (y_0, y_1, \cdots, y_{n-1})$

$$\mathbf{K}_{\underline{y}} = \begin{pmatrix} a & b \\ c & d \end{pmatrix} \quad \text{then}$$

$$\mathbf{K}_{\underline{z}} \times \mathbf{K}_0 = \begin{pmatrix} a+b & b \\ c & c+d \end{pmatrix}$$

$$\mathbf{K}_{\underline{z}} \times \mathbf{K}_1 = \begin{pmatrix} a & a+b \\ c+d & d \end{pmatrix}$$

from which it follows that

$$\mathbf{K}_{\underline{y}} \times \mathbf{K}_{y_n} = \begin{pmatrix} A & B \\ C & D \end{pmatrix} \quad B > A \Leftrightarrow y_n = 1$$

$$\mathbf{K}_{\underline{y}} \times \mathbf{K}_{y_n} = \begin{pmatrix} A & B \\ C & D \end{pmatrix} \quad A > B \Leftrightarrow y_n = 0$$

which proves Exercise 16.4 at least for strings $\underline{y}$ and $\underline{z}$ of the same length. The general case may be proved by mathematical induction; if

- $\mathbf{K}_{\underline{y}} = \mathbf{K}_{\underline{z}}$ and
- $\underline{y} = u, \underline{y}'$ and $\underline{z} = u, \underline{z}'$ or $\underline{y} = \underline{y}', u$ and $\underline{z} = \underline{z}', u$ with $u \in \{0, 1\}$

then the induction argument yields $\underline{y}' = \underline{z}'$. The only remaining case to be eliminated is

$$\underline{y} = u, \underline{y}', v \quad \underline{z} = u, \underline{z}', v \quad u \neq v \quad u, v \in \{0, 1\} \tag{Ex16.1.a}$$

Assuming, as we may(???) that $u = 0$, $v = 1$ and

$$\mathbf{K}_{\underline{y}'} = \begin{pmatrix} a & b \\ c & d \end{pmatrix} \quad \mathbf{K}_{\underline{z}'} = \begin{pmatrix} A & B \\ C & D \end{pmatrix}$$

equation (ExIII/1–1.a) gives

$$\begin{pmatrix} a+b & b \\ a+b+c+d & b+d \end{pmatrix} = \begin{pmatrix} A+C & A+B+C+D \\ C & C+D \end{pmatrix} \tag{Ex16.1.b}$$

Since the entries in $\mathbf{K}_{\underline{y}'}$ and $\mathbf{K}_{\underline{z}'}$ are clearly non-negative integers, equation (Ex1.4b) yields a contradiction. ■

Exercise 16.5. Prove that if $|\underline{y}| = n$, then each element in $\mathbf{K}_{\underline{y}}$ is bounded above by F_n.

Hint: Use mathematical induction.

Proof. Write

$$\mathbf{K}_{\underline{y}}\underline{u}_0 = A_{\underline{y}}\underline{u}_0 + B_{\underline{y}}\underline{u}_1$$

and assume by mathematical induction that when $|underline y| = n - 1$ and $F_{-1}0$, the following bolds hold.

$$\left|(A_{\underline{y}}, B_{\underline{y}})\right| \leq \begin{cases} (F_{n-2}, F_{n-1}) & \text{or} \\ (F_{n-1}, F_{n-2}) \end{cases} \tag{Ex16.2}$$

then by equations (1.8a–b)

$$\begin{aligned} \mathbf{K}_{0,\underline{y}}\underline{u}_0 &= A_{\underline{y}}\underline{u}_0 + \left(A_{\underline{y}} + B_{\underline{y}}\right)\underline{u}_1 \\ &\leq \begin{cases} (F_{n-2}, F_{n-2} + F_{n-1}) = (F_{n-2}, F_n) \leq (F_{n-1}, F_n) & \text{or} \\ (F_{n-1}, F_{n-2} + F_{n-1}) = (F_{n-1}, F_n) \end{cases} \end{aligned}$$

Similar bounds on the multipliers of $\underline{u}_0$ and $\underline{u}_1$ hold for the three vectors $\mathbf{K}_{1,\underline{y}}\underline{u}_0$, $\mathbf{K}_{0,\underline{y}}\underline{u}_1$ and $\mathbf{K}_{1,\underline{y}}\underline{u}_1$. But the bound in (Ex16.5) shows that entries in $\mathbf{K}_{\underline{y}}$ with $|\underline{y}| = n$ are bounded by F_n. ■

PART III, CHAPTER 17

Exercise 17.1. Prove that the Haar wavelet transformation $\underline{f} \to \mathcal{H}[\underline{f}]$ is invertible.

Exercise 17.2. Calculate $\mathcal{H}[\underline{f}]$ if $\underline{f} = (c)_8 \equiv \underbrace{(c, c, \cdots, c)}_{8 \text{ copies}}$..

Solution. $\mathcal{H}[\underline{f}] = 4, (0)_7$. ■

Exercise 17.3. Calculate $\mathcal{H}[\underline{f}]$ if

17.3a) $\underline{f} = (c)_4, (d)_4$;
17.3b) $\underline{f} = (c)_4, (d)_3$.

Solution.

$$\underline{f} = (c)_4, (d)_4 \Rightarrow \mathcal{H}[\underline{f}] = \frac{1}{2}(c+d), \frac{1}{2}(-c+d), (0)_6.$$

$$\underline{f} = (c)_5, (d)_3 \Rightarrow \mathcal{H}[\underline{f}] = \frac{1}{8}(5c+3d), \frac{1}{8}(3c-3d), 0, \frac{1}{4}(c-d), 0, 0, \frac{1}{2}(c-d), 0. \quad ■$$

Exercise 17.4. Prove that if $\underline{f} = (f_0, f_1, \cdots, f_{2^n-1})$, then $\mathcal{H}[\underline{f}] = \frac{1}{2^n}[f_0 + f_1 + \cdots + f_{2^n-1}], \cdots$.

Solution. By mathematical induction. ■

Exercise 17.5. Prove

Lemma 17.1. Prove the formula

$$\text{sim}(C_i, C_j) = \frac{{}_{(1,1)}n_{i,j}}{{}_{(1,1)}n_{i,j} + {}_{(1,0)}n_{i,j} + {}_{(0,1)}n_{i,j}}$$

and show that

$$\text{sim}(C_i, C_j) = \frac{|C_i| + |C_j| - d_H(C_i, C_j)}{|C_i| + |C_j| + d_H(C_i, C_j)}$$

where $d_H(C_i, C_j)$ is the Hamming distance between columns C_i and C_j).

Proof

17.1a $|\mathcal{R}_i \cap \mathcal{R}_j| = {}_{(1,1)}n_{i,j} = d_H(C_i, C_j)$;
17.1b $C_i \cup C_j = C_i \cap \bar{C}_j + \bar{C}_i \cap C_j + C_i \cap C_j = |C_i| + |C_j| - d_H(C_i, C_j)$;
17.1c $|C_i \cap \bar{C}_j| = {}_{(1,0)}n_{i,j}$ and $|\bar{C}_i \cap C_j| = {}_{(0,1)}n_{i,j}$. ■

Exercise 17.6. Prove

Lemma 17.2. If the rows of X are permuted randomly[3], the random variables $H = (H_0, H_1, \cdots, H_{N-1})$ satisfy

$$Pr\{H_i = H_j\} = \mathrm{sim}(\mathcal{C}_i, \mathcal{C}_j)$$

Proof. Let $\pi = (\pi_0, \pi_1, \cdots, \pi_{n-1})$ be the row permutation; then

17.2a H_i and H_j agree if the first row in either column $\mathcal{C}_i$ or column $\mathcal{C}_j$ has a 1 is of type a, and

17.2b H_i and H_j disagree if the first row in either column $\mathcal{C}_i$ or column $\mathcal{C}_j$ has a 1 is of type b or c. ■

Exercise 17.7. Use the counting methods in Chapter I/1 (Counting) to give a proof of

Theorem 17.3.

$$Pr\{H(\mathcal{C}_i) = r\} = \frac{{}_{(1)}n_i}{n} \frac{\binom{n - {}_{(1)}n_i}{r}}{\binom{n-1}{r}} \qquad 0 \le r \le n - {}_{(1)}n_i$$

Proof. The event $\{H(\mathcal{C}_i) = r\}$ requires $x_{\pi_k, i} = \begin{cases} 0 & \text{if } 0 \le k < r \\ 1 & \text{if } k = r \end{cases}$ so that

$$Pr\{H(\mathcal{C}_i) = r\} = \frac{({}_{(0)}n_i)_r \times ({}_{(1)}n_i)_1 \times ([{}_{(0)}n_i - r] + [{}_{(1)}n_i - 1])!}{n!} \qquad 0 \le r \le {}_{(0)}n_i \qquad \text{(Ex17.1.}a\text{)}$$

where $n = {}_{(0)}n_i + {}_{(1)}n_i$ and the falling factorials (Chapter I/1, equation (1.3a)) are

$$({}_{(0)}n_i)_r = \begin{cases} 1 & \text{if } r = 0 \\ {}_{(0)}n_i \times ({}_{(0)}n_i - 1) \times \cdots \times ({}_{(0)}n_i - r + 1) & \text{if } 0 < r \le {}_{(0)}n_i \end{cases} \qquad \text{(Ex17.1.}b\text{)}$$

$$({}_{(1)}n_i)_1 = {}_{(1)}n_i \qquad \text{(Ex17.1.}c\text{)}$$

Simplifying equation (Ex17.1.a) using equations (ExIII/2–1.b,c) yields

$$Pr\{H(\mathcal{C}_i) = r\} = \frac{(n - {}_{(1)}n_i)!}{(n - {}_{(1)}n_i - r)!} \, {}_{(1)}n_i \, \frac{(n - r - 1)!}{n!} \qquad 0 \le r < {}_{(0)}n_i \qquad \text{(Ex17.1.}d\text{)}$$

which may be rearranged to complete the proof. ■

Exercise 17.8. Prove equation (17.9) and indicate the range of integer parameters (j, m, x, z) for which it is valid.

[3]using the uniform distribution on the $n!$ permutation of rows.

Proof. By mathematical induction; if $m = j$, then

$$\frac{\binom{z}{j}}{\binom{x}{j}} = \frac{z!(x-j)!}{x!(z-j)!} \qquad \text{(Ex17.2.}a\text{)}$$

while the right hand side of equation (Ex17.2.a) is

$$\frac{x+1}{x-z+1}\left\{\frac{\binom{z}{j}}{\binom{x+1}{j}} - \frac{\binom{z}{j+1}}{\binom{x+1}{j+1}}\right\} = \frac{x+1}{x-z+1}$$

$$\frac{z!(x-j)!}{(x+1)!(z-j)!}\{(x+1-j)-(z-j)\} \qquad \text{(Ex17.2.}b\text{)}$$

from which the equality for $m = j$ is obvious. In this example of mathematical induction, proving the *base case* is all that is required, since the induction step is clearly demonstrated in exactly the same way. ■

Exercise 17.9. Prove

Theorem 17.4.

$$E\{H(C_i)\} = \frac{n - {}_{(1)}n_i}{{}_{(1)}n_i + 1}$$

Proof. Starting with equation (17.8)

$$E\{H(C_i)\} = \sum_{r=0}^{n_{(1)}n_i} r \frac{{}_{(1)}n_i}{n} \frac{\binom{n - {}_{(1)}n_i}{r}}{\binom{n-1}{r}} \qquad \text{(Ex17.3.}a\text{)}$$

we write $r = (r - n) + n$ and obtain

$$r\frac{\binom{n - {}_{(1)}n_i}{r}}{\binom{n-1}{r}} = n\left\{\frac{\binom{n - {}_{(1)}n_i}{r}}{\binom{n-1}{r}} - \frac{\binom{n - {}_{(1)}n_i}{r}}{\binom{n}{r}}\right\} \qquad \text{(Ex17.3.}b\text{)}$$

Complete the proof using the identity in equation (17.10). ■

Exercise 17.10. Use the methods in Chapter I/3 (Asymptotic Analysis) to show

Theorem 17.5.

$$\lim_{\substack{n\to\infty \\ \frac{(1)n_i}{n}=\alpha}} Pr\{H(C_i)=r\}=\alpha(1-\alpha)^r$$

$$\lim_{\substack{n\to\infty \\ \frac{(1)n_i}{n}=\alpha}} E\{H(C_i)=\frac{1-\alpha}{\alpha}$$

Proof. From equation (17.8)

$$\lim_{\substack{n\to\infty \\ \frac{(1)n_i}{n}=\alpha}} Pr\{H(C_i)=r\}=\alpha \lim_{\substack{n\to\infty \\ \frac{(1)n_i}{n}=\alpha}} \frac{n(1-\alpha)!(n-1-r)!}{(n(1-\alpha)-r)!n!} \tag{Ex17.4}$$

Using Stirling's formula $n!\approx\sqrt{2\pi}n^{n+\frac{1}{2}}e^{-n}$ yields equation (Ex17.10) which leads immediately to the formula for $\lim_{\substack{n\to\infty \\ \frac{(1)n_i}{n}=\alpha}} E\{H(C_i)\}$. ■

PART III, CHAPTER 18

Exercise 18.1. Prove that if the 96 components of

$$X=(X_0,X_1,\cdots,X_{31}) \quad Y=(Y_0,Y_1,\cdots,Y_{31}) \quad Z=(Z_0,Z_1,\cdots,Z_{31})$$

are independent and identically distributsed (0,1)-valued random variables with $Pr\{X_i=0\}=1/2$ and

$$U=(U_0,U_1,\cdots,U_{31})=FF(X,Y,Z)=(X\wedge Y)\vee(\neg X\wedge Z)$$

then the 32 components of U are are independent and identically distributed (0,1)-valued random variables with

$$Pr\{U_i=0\}=1/2$$

Exercise 18.2. Prove that if the 96 components of

$$X=(X_0,X_1,\cdots,X_{31}) \quad Y=(Y_0,Y_1,\cdots,Y_{31}) \quad Z=(Z_0,Z_1,\cdots,Z_{31})$$

are independent and identically distributed (0,1)-valued random variables with $Pr\{X_i=0\}=1/2$ and

$$U=(U_0,U_1,\cdots,U_{31})=GG(X,Y,Z)=(X\wedge Z)\vee(Y\wedge\neg Z)$$

then the 32 components of U are are independent and identically distributed (0,1)-valued random variables with

$$Pr\{U_i=0\}=1/2$$

Solution

$$U_i = (X_i \wedge Z_i) \vee (Y_i \wedge \neg Z_i) = \begin{cases} 0 & \text{if } (Y_i, Z_i) = (0,0) \text{ wp } 1/4 \\ X_i & \text{if } (Y_i, Z_i) = (0,1) \text{ wp } 1/4 \\ \neg X_i & \text{if } (Y_i, Z_i) = (1,0) \text{ wp } 1/4 \\ 1 & \text{if } (Y_i, Z_i) = (1,1) \text{ wp } 1/4 \end{cases}$$

which implies that $Pr\{U_i = 0\} = 1/2$. Since the components of X, Y and Z are independent, the components of U are independent. ■

Exercise 18.3. Prove that if the 96 components of

$$X = (X_0, X_1, \cdots, X_{31}) \quad Y = (Y_0, Y_1, \cdots, Y_{31}) \quad Z = (Z_0, Z_1, \cdots, Z_{31})$$

are independent and identically distributed (0,1)-valued random variables with $Pr\{X_i = 0\} = 1/2$ and

$$U = (U_0, U_1, \cdots, U_{31}) = HH(X, Y, Z) = X \oplus Y \oplus Z$$

then the 32 components of U are are independent and identically distributed (0,1)-valued random variables with

$$Pr\{U_i = 0\} = 1/2$$

Solution

$$U_i = X_i \oplus Y_i \oplus Z_i = \begin{cases} X_i & \text{if } (Y_i, Z_i) = (0,0) \text{ wp } 1/4 \\ X_i \oplus 1 & \text{if } (Y_i, Z_i) = (0,1) \text{ wp } 1/4 \\ X_i \oplus 1 & \text{if } (Y_i, Z_i) = (1,0) \text{ wp } 1/4 \\ X_i & \text{if } (Y_i, Z_i) = (1,1) \text{ wp } 1/4 \end{cases}$$

which implies that $Pr\{U_i = 0\} = 1/2$. Since the components of X, Y and Z are independent, the components of U are independent. ■

Exercise 18.4. Prove that if the 96 components of

$$X = (X_0, X_1, \cdots, X_{31}) \quad Y = (Y_0, Y_1, \cdots, Y_{31}) \quad Z = (Z_0, Z_1, \cdots, Z_{31})$$

are independent and identically distributed (0,1)-valued random variables with $Pr\{X_i = 0\} = 1/2$ and

$$U = (U_0, U_1, \cdots, U_{31}) = II(X, Y, Z) = Y \oplus (X \vee \neg Z)$$

then the 32 components of U are are independent and identically distributed (0,1)-valued random variables with

$$Pr\{U_i = 0\} = 1/2$$

Solution

$$U_i = Y_i \oplus (X_i \vee \neg Z_i) = \begin{cases} 1 & \text{if } (Y_i, Z_i) = (0,0) \text{ wp } 1/4 \\ X_i & \text{if } (Y_i, Z_i) = (0,1) \text{ wp } 1/4 \\ 0 & \text{if } (Y_i, Z_i) = (1,0) \text{ wp } 1/4 \\ 1 \oplus X_i & \text{if } (Y_i, Z_i) = (1,1) \text{ wp } 1/4 \end{cases}$$

which implies that $Pr\{U_i = 0\} = 1/2$. Since the components of X, Y and Z are independent, the components of U are independent. ■

Exercise 18.5. Characterize and enumerate the number of pairs (U, U') with

$$U, U' \in \mathcal{Z}_{32} = \{\underline{x} = (x_0, x_1, \cdots, x_{31}) : x_i = 0 \text{ or } 1, 0 \le i < 32\}$$

for which $U' \boxminus U = 2^{31}$.

Solution. If $U = (u_0, u_1, \cdots, u_{31})$, $U' = (u'_0, u'_1, \cdots, u'_{31})$ with $U' \boxminus U = 2^{31}$ then

$$U' = (1, u_1, u_2, \cdots, u_{31}) \quad U = (0, u_1, u_2, \cdots, u_{31}) \quad U + U' = (1, (0)_{31})$$

so that there are 2^{31} pairs (U, U') which satisfy $U' \boxminus U = 2^{31}$. ■

Exercise 18.6. Characterize and enumerate the number of pairs (U, U') with

$$U, U' \in \mathcal{Z}_{32} = \{\underline{x} = (x_0, x_1, \cdots, x_{31}) : x_i = 0 \text{ or } 1, 0 \le i < 32\}$$

for which $U' \boxminus U = 2^{15}$.

Solution. If $U = (u_0, u_1, \cdots, u_{31})$, $U' = (u'_0, u'_1, \cdots, u'_{31})$ with $U' \boxminus U = 2^{15}$ then for $1 \le r_0 \le 16$

$$\begin{aligned} U' &= \left(u_0, u_1, \cdots, u_{16-r_0}, 1, (0)_{r_0}, u_{17}, u_{18}, \cdots, u_{31}\right) \\ U &= \left(u_0, u_1, \cdots, u_{16-r_0}, 0, (1)_{r_0}, u_{17}, u_{18}, \cdots, u_{31}\right) \quad U + U' = \left((0)_{16-r_0}, (1)_{r_0+1}, (0)_{15}\right) \end{aligned}$$

so that

- for *each* r_0 with $1 \le r_0 \le 16$ there are 2^{32-r_0} pairs (U, U') which satisfy $U' \boxminus U = 2^{15}$, and
- a total of $2^{32} - 1 = \sum_{r_0=1}^{32} 2^{32-r_0}$ pairs (U, U') which satisfy $U' \boxminus U = 2^{15}$. ■

Exercise 18.7. Characterize and enumerate the number of pairs (U, U') with

$$U, U' \in \mathcal{Z}_{32} = \{\underline{x} = (x_0, x_1, \cdots, x_{31}) : x_i = 0 \text{ or } 1, 0 \le i < 32\}$$

for which $U' \boxminus U = -2^{15}$.

Solution. If $U = (u_0, u_1, \cdots, u_{31})$, $U' = (u'_0, u'_1, \cdots, u'_{31})$ with $U' \boxminus U = -2^{15}$ then for $1 \le r_0 \le 16$

$$U' = \left(u_0, u_1, \cdots, u_{16-r_0}, 0, (1)_{r_0}, u_{17}, u_{18}, \cdots, u_{31}\right)$$
$$U = \left(u_0, u_1, \cdots, u_{16-r_0}, 1, (0)_{r_0}, u_{17}, u_{18}, \cdots, u_{31}\right) \quad U + U' = \left((0)_{16-r_0}, (1)_{r_0+1}, (0)_{15}\right)$$

so that

- for *each* r_0 with $1 \le r_0 \le 16$ there are 2^{32-r_0} pairs (U, U') which satisfy $U' \boxminus U = -2^{15}$, and
- a total of $2^{32} - 1 = \sum_{r_0=1}^{32} 2^{32-r_0}$ pairs (U, U') which satisfy $U' \boxminus U = -2^{15}$. ■

Exercise 18.8. Show that the (nonlinear) functions $\{F_0(x, y, z), F_1(x, y, z), F_2(x, y, z)\}$ of the bits (x, y, z) are uniformly distributed; that is, $|\{(x, y, z) \in \mathcal{Z}_{3,2} : F_i(x, y, z) = 0\}| = 4$ for $i = 0, 1, 2$.

Solution

F₀(x, y, z)	F₁(x, y, z)	F₁(x, y, z)
(0,0,0) 0	(0,0,0) 0	(0,0,0) 0
(0,0,1) 1	(0,0,1) 1	(0,0,1) 0
(0,1,0) 0	(0,1,0) 1	(0,1,0) 0
(0,1,1) 1	(0,1,1) 0	(0,1,1) 1
(1,0,0) 0	(1,0,0) 1	(1,0,0) 0
(1,0,1) 1	(1,0,1) 0	(1,0,1) 1
(1,1,0) 1	(1,1,0) 0	(1,1,0) 1
(1,1,1) 0	(1,1,1) 1	(1,1,1) 1

■

Exercise A18.1.

Ex18.1a) Find a formula for the number NAF_n of nonadjacent form sequences $(\bar{x}_0, \bar{x}_1, \bar{x}_2, \cdots, \bar{x}_{n-1})$ of length n.

Ex18.1b) Prove that

i) Every positive integer n has an NAF representation.

ii) The NAF representation is unique.

iii) The output of NAF(n) has *no* leading[4]

Ex18.1c) The base-2 representation of the integer $n > 0$ requires $\lfloor \log_2 n \rfloor$ bits. How many bits are required for the NAF-representation of the integer $n > 0$?

Ex18.1d) Prove that the NAF is of minimal weight.

[4]Zeroes in the least (*resp.* most) significant positions in a base-2 representation of n are trailing (*resp.* leading) zeros 0's.

TABLE Ex.18.1. NAF$_n$ n = 1(1)7

n	1	2	3	4	5	6	17	8
NAF$_n$	1	1	3	5	11	21	43	85

Solution Ex18.1a. There are three possibilities when $(\bar{x}_0, \bar{x}_1, \bar{x}_2, \cdots, \bar{x}_{n-1})$ is an NAF sequence

1. $\bar{x}_0 = 0$,
2. $(\bar{x}_0, \bar{x}_1) = (0, 1)$, or
3. $(\bar{x}_0, \bar{x}_1) = (0, 11)$.

This leads to the recurrence

$$\text{NAF}_{n,1} = \begin{cases} 0 & \text{if } n = 0 \\ 1 & \text{if } n = 1, 2 \\ \text{NAF}_{n-1} + 2\text{NAF}_{n-2,1} & \text{if } 3 \le n < \infty \end{cases} \tag{Ex18.1}$$

Equation (Ex18.1) gives Table Ex.18.1 of values for $\{\text{NAF}_n : 1 \le n \le 8\}$.

If you carefully look at these values, it is not too difficult to guess the formula for NAF_n but it may easily be derived by the method of generating functions.

Defining NAF(z) by

$$\text{NAF}(z) = \sum_{n=0}^{\infty} \text{NAF}_n z^n \tag{Ex18.2}$$

Equation (Ex18.1) gives

$$\text{NAF}(z) - z - z^2 = z(\text{NAF}(z) - z) + z^2\text{NAF}(z) \tag{Ex18.3.a}$$

which when rearranged yields

$$\text{NAF}(z) = \frac{z}{1 - z - 2z^2} \tag{Ex18.3.b}$$

The denominator of equations (Ex18.3.b) factors

$$(1 - z - 2z^2) = (1 - 2z)(1 + z)$$

leading to the partial fraction expansion

$$\text{NAF}(z) = -\frac{1}{3}\frac{1}{1+z} + \frac{1}{3}\frac{1}{1-2z} \tag{Ex18.3.c}$$

and the formula

$$\text{NAF}_n = \frac{1}{3}\left[(-1)^{n+1} + 2^n\right] \tag{Ex18.4}$$

Solution Ex18.1b. The proof of existence is by induction on n; there are two cases to be examined;

C1. $0 = n$ (*modulo* 2)
Concatenate the unique NAF-representation of $\frac{n}{2}$ on the right by 0, 1);
C2. $1 = n$ (*modulo* 4)
Concatenate the unique NAF-representation of $\frac{n-1}{4}$ on the right by 0, 1);
C3. $-1 = n$ (*modulo* 4)
Concatenate the unique NAF-representation of $\frac{n+1}{4}$ on the right by 0, −1).

To prove uniqueness, we also use induction on n. Assume, on the contrary, that two NAF-representations of n exist.

$$\underline{x} = (\bar{x}_0, \bar{x}_1, \bar{x}_2, \cdots, \bar{x}_{m-1}) \quad \underline{y} = (\bar{y}_0, \bar{y}_1, \bar{y}_2, \cdots, \bar{y}_{m-1})$$

1. If $0 = n$ (*modulo* 2), then $\bar{x}_0 = \bar{y}_0$ and the induction hypothesis can be applied.
2. If $0 \neq n$ (*modulo* 2), there are only two possibilities
   - $(\bar{x}_0, \bar{x}_1) = (\bar{y}_0, \bar{y}_1) = (1, 0)$, or
   - $(\bar{x}_0, \bar{x}_1) = (\bar{y}_0, \bar{y}_1) = (-1, 0)$

   to which we may again apply the induction hypothesis.

The final assertion of Ex18.1b *iii*) concerning leading, orinfacttrailing, 0's is obvious; the output of NAF(n) are the powers of 2 with sign appearing in the NAF-representations of n.

Solution Ex18.1c. If $2^e \leq n < 2^{e+1}$, them $e = \lfloor \log_2 n \rfloor$. Therefore, $3n > 2^{e+2} \Leftrightarrow n \leq \lceil \log_2 n \rceil$.

Solution Ex18.1d. Use induction on n as in the proof of Ex18.1b *ii*). ■

Exercise A18.2. Characterize and enumerate the number of pairs (U, U') with

$$U, U' \in \mathcal{Z}_{32} = \{\underline{x} = (x_0, x_1, \cdots, x_{31}) : x_i = 0 \text{ or } 1, 0 \leq i < 32\}$$

for which $U' \boxminus U = 2^{31}$.

Solution. If $U = (u_0, u_1, \cdots, u_{31})$, $U' = (u'_0, u'_1, \cdots, u'_{31})$ with $U' \boxminus U = 2^{31}$ then

$$U' = (1, u_1, u_2, \cdots, u_{31}) \quad U = (0, u_1, u_2, \cdots, u_{31}) \quad U + U' = (1, (0)_{31})$$

so that there are 2^{31} pairs (U, U') which satisfy $U' \boxminus U = 2^{31}$. ■

Exercise A18.3. Characterize and enumerate the number of pairs (U, U') with

$$U, U' \in \mathcal{Z}_{32} = \{\underline{x} = (x_0, x_1, \cdots, x_{31}) : x_i = 0 \text{ or } 1, 0 \leq i < 32\}$$

for which $U' \boxminus U = 2^{15}$.

Solution. If $U = (u_0, u_1, \cdots, u_{31})$, $U' = (u_0', u_1', \cdots, u_{31}')$ with $U' \boxminus U = 2^{15}$ then for $1 \le r_0 \le 16$

$$U' = (u_0, u_1, \cdots, u_{16-r_0}, 1, (0)_{r_0}, u_{17}, u_{18}, \cdots, u_{31})$$
$$U = (u_0, u_1, \cdots, u_{16-r_0}, 0, (1)_{r_0}, u_{17}, u_{18}, \cdots, u_{31}) \quad U + U' = ((0)_{16-r_0}, (1)_{r_0+1}, (0)_{15})$$

so that

- for *each* r_0 with $1 \le r_0 \le 16$ there are 2^{32-r_0} pairs (U, U') which satisfy $U' \boxminus U = 2^{15}$, and
- a total of $2^{32} - 1 = \sum_{r_0=1}^{32} 2^{32-r_0}$ pairs (U, U') which satisfy $U' \boxminus U = 2^{15}$. ■

Exercise A18.4. Characterize and enumerate the number of pairs (U, U') with

$$U, U' \in \mathcal{Z}_{32} = \{\underline{x} = (x_0, x_1, \cdots, x_{31}) : x_i = 0 \text{ or } 1, 0 \le i < 32\}$$

for which $U' \boxminus U = -2^{15}$.

Solution. If $U = (u_0, u_1, \cdots, u_{31})$, $U' = (u_0', u_1', \cdots, u_{31}')$ with $U' \boxminus U = -2^{15}$ then for $1 \le r_0 \le 16$

$$U' = (u_0, u_1, \cdots, u_{16-r_0}, 0, (1)_{r_0}, u_{17}, u_{18}, \cdots, u_{31})$$
$$U = (u_0, u_1, \cdots, u_{16-r_0}, 1, (0)_{r_0}, u_{17}, u_{18}, \cdots, u_{31}) \quad U + U' = ((0)_{16-r_0}, (1)_{r_0+1}, (0)_{15})$$

so that

- for *each* r_0 with $1 \le r_0 \le 16$ there are 2^{32-r_0} pairs (U, U') which satisfy $U' \boxminus U = -2^{15}$, and
- a total of $2^{32} - 1 = \sum_{r_0=1}^{32} 2^{32-r_0}$ pairs (U, U') which satisfy $U' \boxminus U = -2^{15}$. ■

PART III, CHAPTER 19

Exercise 19.1. Prove the geometric hashing coding $\mathcal{G}_{p1,p2} : p \to (u, v)$ satisfies the following:

19.1a) If c is any fixed vector and $p_i \to p_i + c$ $(i = 1, 2)$, then (u, v) remains unchanged; i.e., $\mathcal{G}_{p1+c,p2+c} = \mathcal{G}_{p1,p2}$;

19.1b) If c is any nonzero constant and $p_i \to cp_i$ $(i = 1, 2)$, then (u, v) remains unchanged; i.e., $\mathcal{G}_{cp1,cp2} = \mathcal{G}_{p1,p2}$;

19.1c) My generally, if A is any 2-by-2 similarity transformation[5] and $p_i \rightarrow p_i A$ ($i = 1, 2$), then (u, v) remains unchanged; i.e., $\mathcal{G}_{p_1 A, p_2 A} = \mathcal{G}_{p_1, p_2}$.

Solution. If $p_i \rightarrow p_i + c$ ($i = 1, 2$) and $p = p_0 + up_x + vp_y$, then

1. $p - p_0$ is unchanged since
   $$p_0 = \frac{1}{2}(p_1 + p_2) \rightarrow p_0 + c$$
   $$p \rightarrow p + c$$
2. p_x and p_y are both unchanged since
   $$p_x = (p_2 - p_1) \rightarrow p_x,$$
   $$p_y = \text{ROT}_{90} p_x \rightarrow p_y$$

proving Exercise 19.1a. The argument with obvious modifications proves 19.1b–c. ■

REFERENCES

Milton Abramowitz and Irene A. Stegun, (Editors), *Handbook of Mathematical Functions with Formulas, Graphs, and Mathematical Tables*, Dover Publications (New York), 1972.

William Feller, *An Introduction to Probability Theory and Its Applications*, Volume 1, (Second Edition), John Wiley & Sons, 1957; (Third Edition), John Wiley & Sons (New York), 1967.

Walter Ledermann, *Introduction to the Theory of Finite Groups*, Oliver and Boyd, 1953.

John Riordan, *Combinatorial Identities*, John Wiley & Sons (New York), 1968.

[5]A 2-by-2 *similarity* transformation is a matrix $A = \begin{pmatrix} a_{1,1} & a_{1,2} \\ a_{2,1} & a_{2,2} \end{pmatrix}$ of determinant $\neq 0$.

INDEX

Printed and bound by CPI Group (UK) Ltd, Croydon, CR0 4YY
27/10/2024
14580259-0003